S0-BZL-443

INTRODUCTION TO

ANALYSIS

TEXTBOOKS in MATHEMATICS

Series Editors: Al Boggess and Ken Rosen

PUBLISHED TITLES

ABSTRACT ALGEBRA: A GENTLE INTRODUCTION
Gary L. Mullen and James A. Sellers

ABSTRACT ALGEBRA: AN INTERACTIVE APPROACH, SECOND EDITION
William Paulsen

ABSTRACT ALGEBRA: AN INQUIRY-BASED APPROACH
Jonathan K. Hodge, Steven Schlicker, and Ted Sundstrom

ADVANCED LINEAR ALGEBRA
Hugo Woerdeman

ADVANCED LINEAR ALGEBRA
Nicholas Loehr

ADVANCED LINEAR ALGEBRA, SECOND EDITION
Bruce Cooperstein

APPLIED ABSTRACT ALGEBRA WITH MAPLE™ AND MATLAB®, THIRD EDITION
Richard Klima, Neil Sigmon, and Ernest Stitzinger

APPLIED DIFFERENTIAL EQUATIONS: THE PRIMARY COURSE
Vladimir Dobrushkin

A BRIDGE TO HIGHER MATHEMATICS
Valentin Deaconu and Donald C. Pfaff

COMPUTATIONAL MATHEMATICS: MODELS, METHODS, AND ANALYSIS WITH MATLAB® AND MPI, SECOND EDITION
Robert E. White

A COURSE IN DIFFERENTIAL EQUATIONS WITH BOUNDARY VALUE PROBLEMS, SECOND EDITION
Stephen A. Wirkus, Randall J. Swift, and Ryan Szypowski

A COURSE IN ORDINARY DIFFERENTIAL EQUATIONS, SECOND EDITION
Stephen A. Wirkus and Randall J. Swift

DIFFERENTIAL EQUATIONS: THEORY, TECHNIQUE, AND PRACTICE, SECOND EDITION
Steven G. Krantz

PUBLISHED TITLES CONTINUED

PUBLISHED TITLES CONTINUED

THE MATHEMATICS OF GAMES: AN INTRODUCTION TO PROBABILITY
David G. Taylor

A MATLAB® COMPANION TO COMPLEX VARIABLES
A. David Wunsch

MEASURE AND INTEGRAL: AN INTRODUCTION TO REAL ANALYSIS, SECOND EDITION
Richard L. Wheeden

MEASURE THEORY AND FINE PROPERTIES OF FUNCTIONS, REVISED EDITION
Lawrence C. Evans and Ronald F. Gariepy

NUMERICAL ANALYSIS FOR ENGINEERS: METHODS AND APPLICATIONS, SECOND EDITION
Bilal Ayyub and Richard H. McCuen

ORDINARY DIFFERENTIAL EQUATIONS: AN INTRODUCTION TO THE FUNDAMENTALS
Kenneth B. Howell

PRINCIPLES OF FOURIER ANALYSIS, SECOND EDITION
Kenneth B. Howell

REAL ANALYSIS AND FOUNDATIONS, FOURTH EDITION
Steven G. Krantz

RISK ANALYSIS IN ENGINEERING AND ECONOMICS, SECOND EDITION
Bilal M. Ayyub

SPORTS MATH: AN INTRODUCTORY COURSE IN THE MATHEMATICS OF SPORTS SCIENCE AND SPORTS ANALYTICS
Roland B. Minton

TRANSFORMATIONAL PLANE GEOMETRY
Ronald N. Umble and Zhigang Han

TEXTBOOKS in MATHEMATICS

INTRODUCTION TO ANALYSIS

Corey M. Dunn

California State University

San Bernardino, California, USA

CRC Press is an imprint of the
Taylor & Francis Group an **informa** business
A CHAPMAN & HALL BOOK

CRC Press
Taylor & Francis Group
6000 Broken Sound Parkway NW, Suite 300
Boca Raton, FL 33487-2742

Printed on acid-free paper
Version Date: 20170429

International Standard Book Number-13: 978-1-4987-3201-7 (Hardback)

Library of Congress Cataloging-in-Publication Data

Names: Dunn, Corey M., 1978-
Title: Introduction to analysis / Corey M. Dunn.
Description: Boca Raton : CRC Press, 2017.
Identifiers: LCCN 2016028980 | ISBN 9781498732017
Subjects: LCSH: Mathematical analysis--Textbooks. | Numbers, Real--Textbooks.
Classification: LCC QA300 .D848 2017 | DDC 515--dc23
LC record available at https://lccn.loc.gov/2016028980

Visit the Taylor & Francis Web site at
http://www.taylorandfrancis.com

and the CRC Press Web site at
http://www.crcpress.com

This book is for my family: Mom, Dad, Ken, Sharon, Alex, Steve, Tracy, Jeffrey, Evie, Elizabeth, Matthew, Hannah, Holden, the Mellors, Shastin, Tobey, my loving and supportive wife, Sharisa, and our wonderful daughters, Sequoia and Tenaya. Thank you all for your continued support through the years.

Contents

Preface

The goal of this book is to provide a one-semester course in introductory Analysis, including an introduction to proof. There are many analysis and proof texts with this aim, although I have endeavored to write this book entirely from the student's perspective: there is enough rigor to challenge even the best students in the class, but also enough explanation and detail to meet the needs of a struggling student. To this end, there are a variety of features that I hope makes this book a good resource:

- **Informal summaries of the idea of proof provided before each result, and before a solution to a practice problem.** One of the most common questions I am asked is to make clearer a certain concept, example, or proof. I also find that at this level, it is sometimes the case that a certain level of rigor needs to be instilled in even the best students: what must a good proof include? In making proofs more clear, I generally find that presenting a short and nontechnical description of the goal of a proof helps the student know a little more about what to expect in the argument, and can therefore understand and retain more of it. I have therefore included a brief and nontechnical summary of the goals of a proof or solution for each of the results and practice problems in this book. It is possible that the student might mistake these informal summaries for proofs, and so they are clearly marked as "Idea of Proof," or as "Methodology," followed by a clearly marked formal proof or solution.

- **More context is included to help the student place the importance and use of each result.** This subject is generally the place where we expect the student to depart from the rote formulaic calculation done in lower division courses, and to move on to explaining their findings with solid reasoning. This is a jarring experience for some students, since simple memorization is no longer adequate to excel. Moreover, the solutions to many problems require the student to synthesize a variety of notions and to produce their own argument going forward. To help the student in this transition, I have endeavored to include more contextual reasoning as to why a certain subject is studied, and even why a subject is studied at that time. The hope is that the student will be able to better retain and access the *meaning* behind the material, bolstering their proof-writing skills in the process.

- **There are very many references to previous results.** Some of the easiest questions for me to answer are some of the hardest for my students to answer: at some point in an important proof, we recall a result that was previously proven, and the student may either not remember the result, or perhaps he or she is enticed into wondering why such a fact is true now that he or she has had some reason to use it. This book contains many references to previous definitions and results so that the interested student (or professor!) can easily see where exactly they can learn more about a certain concept. In electronic versions, these references should be hyperlinked, so that a simple click can take you immediately there. These references begin to phase out in the final parts of the book so as to not distract from the presentation.

- **A list of the important results appears at the end of each section.** This list of results is to help the student and professor see a quick summary–with references–of the material covered in each section.

- **A "Troubleshooting Guide" appears at the end of each chapter that answers common questions.** No matter how good a text is, there are still questions that a student might have. Instead of slowing the presentation of the material to answer potential questions that a student might have about a topic, I have begun a list of commonly asked questions by students, along with an answer to their question and a suggestion where they could look within the book for more help. I hope that both students and professors will share with me the common questions they encounter so that I may add to these troubleshooting guides.

Generally, I expect that the well-prepared student will have completed the calculus series and therefore has had some exposure to various notions such as continuity, differentiation, and infinite series. With that being said, only a nodding acquaintance of polynomials, logarithms, and trigonometric functions is required to fully grasp everything the book has to offer.

Before presenting a short message to instructors and to students, I would like to acknowledge the many wonderful analysis texts that already exist that I have benefitted from as I researched this text. In this text, some of these reference a specific contribution, while others are simply excellent resources. These resources not specifically cited in the text include [1, 4, 7, 9, 10, 13, 15–17, 25, 28], As I mention in the Acknowledgments, there are a core of texts that I have routinely consulted in the preparation of this book. Some proofs and ideas I have attempted to improve upon, although many of these are what I would consider commonplace. Appropriate acknowledgment is given in the text to certain other proofs or ideas. With this being said, I am solely responsible for any errors found within these pages, and I would appreciate being informed of any as they are found.

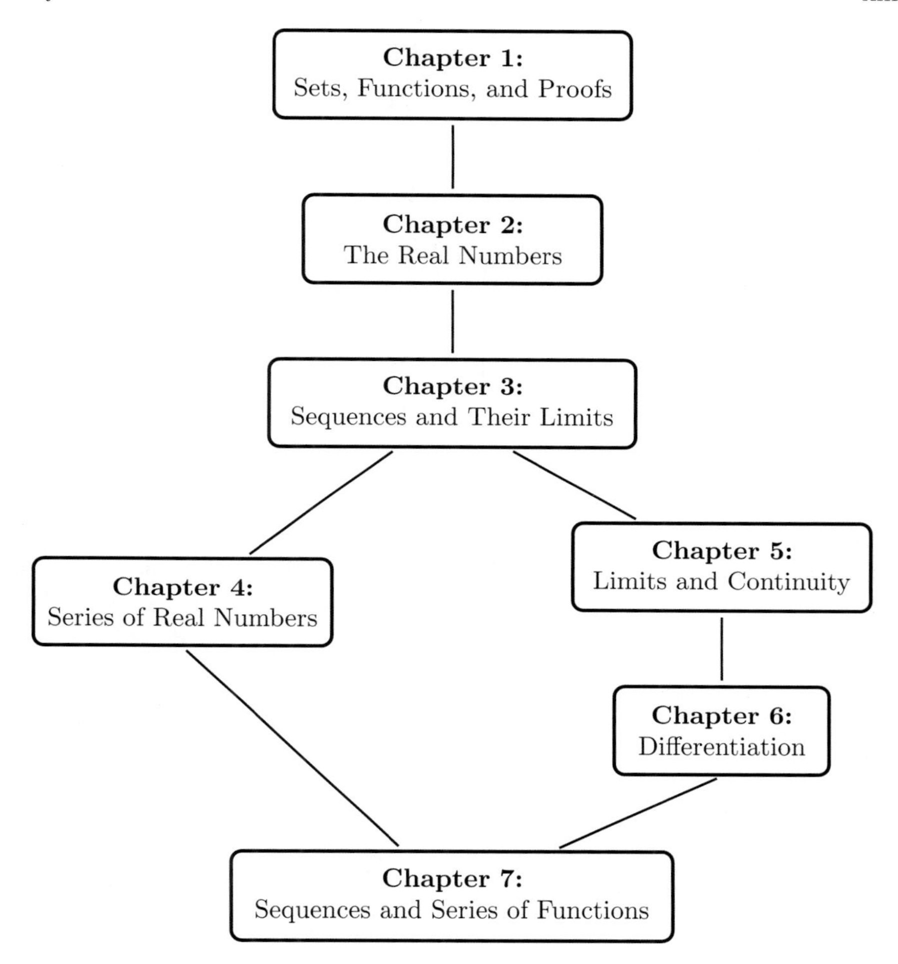

Outline

Chapter 1 is a basic introduction to logic and proofs. We begin with a brief introduction to logic and some suggestions for writing proofs in the first section. In the following sections, we fill out the basic requirements for the subsequent study of analysis with a primer on set theory, proof by Mathematical Induction, and functions. We close the chapter with an excellent application of basic set theory and functions: a discussion of cardinality, where we prove that the real numbers have a cardinality strictly larger than that of the rationals.

Chapter 2 is a rigorous treatment of the real number system and the Completeness Axiom. Since we do not give a construction of the real numbers, we assume as axioms that they satisfy the field and order axioms. Later, we assume the real numbers satisfy the Completeness Axiom, and go on to establish the Archimedean Property. Basic facts about the supremum and infimum are discussed in detail.

Chapter 3 is the largest chapter in the book and forms the foundation of our further study. The main motivation here is to rigorously develop a number of aspects concerning convergence and divergence of sequences for later use, especially divergence to $\pm\infty$. To this end, a basic introduction to sequences begins the chapter to establish notation, but also to ensure that this material is present for those students who somehow have never had an introduction to sequences. Section 3.2 defines convergence of a sequence, and describes in detail elementary aspects of both convergence and divergence. In keeping with one of the goals of this book, there are many examples presented, each with a complete solution. Section 3.3 presents basic results that make computing limits easier. These include basic results concerning *equality* (for instance, that the sum of two convergent sequences converges to the sum of their limits), and *inequality* (for example, the Squeeze Theorem). Information about subsequences and monotone sequences are presented next in preparation for the Bolzano-Weierstrass Theorems in Section 3.6. After introducing Cauchy sequences in Section 3.7, we discuss the important subject of infinite limits in Section 3.8. This section simply adapts each of the important previous convergence results to its infinite analogue. An optional section covering the lim sup and lim inf ends the chapter. We remark that this material is not needed until Section 7.4, since we assume a certain limit exists in our proof of the Root Test in Chapter 4.

After our careful preparation in Chapter 3, the remainder of the text can be done quite rapidly. Chapter 4 defines convergence of infinite series of real numbers. This short chapter presents the common tests for convergence and absolute convergence, and concludes with an optional section on rearrangements.

Chapter 5 introduces the concept of a limit and its basic properties. These are generally proven using the characterization of a limit in terms of sequences, although alternate independent proofs are sometimes given as well. The same treatment is given in our discussion of infinite limits in Section 5.3. The important property of continuity is then presented, followed by some of its major consequences. The chapter ends with a discussion of uniform continuity.

Chapter 6 introduces the derivative and its immediate uses. These include the Mean Value Theorem, Taylor's Theorem, and L'Hôpital's Rule.

The final chapter, Chapter 7, uses what we have developed to study sequences and series of functions. Uniform and pointwise convergence is defined, and important preservation theorems are presented in Section 7.2. These preservation theorems are applied to series of functions in the next section, followed by a short introduction to Power Series in the final Section 7.4.

To the instructor

Although the goal of this book is to assist the student in learning and retaining as much material as possible, I have tried to arrange the presentation of this material in an attempt to make the book easier to teach out of.

First, I have included a summary of information wherever it seemed appropriate, so you can tell at a glance what important concepts are to be covered. At the beginning of every chapter is a short chapter summary, followed by a brief abstract of each section. Hopefully, you can use these to easily pick and choose what material you would like to cover as you plan your course. Also, at the end of each section is a list of all important concepts found within the section, to further help you plan each class. So, although the book is long, I hope this will easily direct you to the important issues in each section and chapter.

There are several ways this book could be used in a one-semester course, depending on how much background in proof is to be covered. If there is no proof background or set theory assumed, one can cover Chapters 1 to 3, and then continuing to Chapters 5 and 6, skipping the material on infinite series of numbers in Chapter 4. If there is no need to go over the material in Chapters 1 or 2, a one-semester course can cover Chapters 3 to 7. It is up to you when you cover Chapter 4, either directly after Chapter 3 or directly before Chapter 7. Of course, depending on the pace you choose, it is possible to cover all of Chapters 1 through 7.

One final note: the theory of integration has been purposely left out, since I cannot seem to find a one-semester course in introductory Analysis that includes it. With that being said, I still wanted to provide some flexibility for you to cover some very interesting topics. Thus, at certain times I choose to use the word "antidifferentiation" instead of "integration." A brief reference to definite integration is made in Chapter 7, but otherwise this book proceeds independently from that theory. In an aim to improve this text over time, I would be very happy to hear your suggestions for content to be included in future editions.

To the student

I have a confession to make: I have not always found the subject of analysis easy, and there are parts of it that still frighten me. I also vividly recall what it is like to sit in an Analysis class and be either disinterested or lost. In these times I found myself asking, "What is all of this for?" or "I don't have any idea what's going on." This book is designed to help students who find themselves asking the same sorts of questions, but is also designed to challenge the brightest of students.

In this message to you, I want to tell you how I think you can get the most out of this book. Here are some suggestions:

- **Do not just skim between the main results. Read the rest of**

the material, too. I understand that many math texts seem terse in their presentation or lack some context to help see the use in the results found inside. It may surprise you to know that this terse presentation is actually a goal of good mathematical writing; however, I would use the term "efficient" instead of "terse." Efficient mathematical writing is modeled in this book in each of our proofs and solutions. A nontechnical summary of these is given directly before to help you to get a better sense of what the structure of the argument is, and is not to be confused with an actual proof. In addition, each major concept is described in a way that is designed to help you to see not just what it is, but what it is for. All of this has been done to make it easier for you to actually read what each concept is about.

- **When you come upon a result or a practice problem, cover up the proof or solution and try to do it yourself first. If that does not work, then uncover just the "Idea of proof" or "Methodology" and try again.** It is a lot to ask to expect a student in Analysis to be able to immediately prove all of the results presented. However, with the right hints, the attempt can be extremely illuminating. This may take just a little more time as you read through each section. But it will be well worth it, especially as you attempt each practice problem, since these are specifically geared toward increasing your foundational skills.

- **Read through the troubleshooting section at the end of each chapter.** You may find answers to questions you did not even know you had! In my experience teaching these courses, I find that certain questions get asked again and again. I have logged these over several years and included their answers for you in these troubleshooting guides. Even if you are not struggling, you might still enjoy reading through these sections. If you have a suggestion for a question you think I should include, please feel free to send it to me via email to `cmdunn@csusb.edu`.

- **The book contains references to other results so that you can easily look these up if you cannot remember why a certain referenced fact is true. Follow these references if you need to.** Sometimes it is the case that we cannot recall why a certain statement in a proof holds. This is not limited to students in Analysis: the majority of time I spend reading proofs is asking myself why a certain statement holds! To alleviate some of this wonder, I have routinely included references to exactly where you can find more information about a certain statement. In electronic versions these references are hyperlinked, you can simply click on the reference to be taken there immediately.

- **Every aspect of every proof in this book is done for a reason. If you cannot find that reason, then you might be missing something.** I should say that this comment is not limited to this book, and is really just a study suggestion in all of your proof-based mathematics courses. Most of the main aspects in each proof are at least mentioned in the "Idea of proof" paragraphs, but in the event you come across a statement that does not seem to fit, sit down and really ask yourself why such a statement is important: the answer can sometimes be very instructive.

I genuinely hope you get a lot out of this book. I hope that it helps you to learn what you need to know in your Analysis course, but I also hope that you gain a number of important skills that will help you to learn future mathematics as well.

Dr. Corey M. Dunn
California State University, San Bernardino
March 2016

Acknowledgments

This project would not have been possible without the help of many people. Perhaps the largest debt of thanks that I owe is to my family, to whom this book is dedicated. I would especially like to thank my parents for providing for me the opportunities that have helped me to get to where I am, and for their constant support: thank you for everything you have done. I would also like to thank my wife, Sharisa: without her patience and understanding this project would have never finished. My beautiful children, Sequoia and Tenaya, are a constant inspiration to me.

I would like to give credit to the many well-written analysis books that are already in print. I consulted a number of these in researching this book. In particular, I found the presentation in the following texts [3, 8, 20, 24, 27] to be quite helpful, and often times I could not improve upon their efficient organization of the material. I also owe a debt of thanks to the Taylor & Francis Group and CRC Press for publishing this work, Amber Donley, Michele Dimont, Ken Rosen, the copyeditors, and Bob Ross for their thoughtful suggestions in helping to make this book a better one. I would also like to thank California State University, San Bernardino, for approving my sabbatical in Fall of 2012 to allow me the time to start this project. I am now at a loss for what I shall do for my next sabbatical.

I am also grateful to two people who are largely responsible for fostering my mathematical progress. My advisor, Dr. Peter Gilkey, patiently supported and nurtured my mathematical growth through graduate school and beyond. He is surely the one I must thank for teaching me how to be a mathematician and how to write. Dr. John Gorman is another mentor that instilled my mathematical enthusiasm, and helped me to see that mathematics was my calling. This book would not be possible without either of them.

Finally, there are a number of individuals that have directly or indirectly helped in the writing of this book in their own way that I would like to acknowledge: my analysis students who have repeatedly asked that I write a textbook in analysis, Matt Miller, Max Wakefield, Jim McCann, Madeleine Jetter, Giovanna Llosent, Chuck Stanton, Zahid Hasan, John Sarli, Bob Stein, Shawn McMurran, Belisario Ventura, Joe "there are four lights" Sutliff-Sanders, the Moffetts, Eric Merchant, Hayden Harker, Bonnie Amende, Ted Coleman, and Rollie Trapp. I would also like to thank Jeremy Aikin for a number of things, but most notably, his LaTeX assistance; without it, this book would surely be handwritten. Thank you all for your encouragement.

List of Figures

1

Sets, Functions, and Proofs

Chapter Summary:

This chapter lays out the basic information required for a study of mathematical analysis: a primer on logic, set theory, Mathematical Induction, and cardinality.

Section 1.1: Logic and an Introduction to Proof. This section introduces basic logical concepts, including the negation, conjunction ("and"), and disjunction ("or") of statements. We go on to describe implications, logical equivalence, and the relationship to the converse and contrapositive of an implication. We negate the conjunction and disjunction in DeMorgan's Laws (Theorem 1.2), and close the section with a collection of proof writing suggestions.

Section 1.2: Sets and their Operations. This section introduces the notion of a set, and goes on to define various set relations and operations, such as the union, intersection, difference (complement), and cross-product of sets. DeMorgan's Laws (Theorem 1.12) are given, which describe how the union and intersection of sets are affected by set complement. The section ends with some suggestions about the level or rigor one should include in proofs.

Section 1.3: Mathematical Induction. We introduce a method of proof called Mathematical Induction. We assume the Well-Ordering Principle (Axiom 1.14) as an axiom that falls outside of the standard axioms of set theory. We then demonstrate that Mathematical Induction is a valid method of proof in Theorem 1.15 and give several examples.

Section 1.4: Functions. The definition of a function is given here, and the image and preimage of sets are introduced. Various facts about unions and intersections of images and preimages are presented in Propositions 1.26 and 1.28. Combinations of functions are also introduced–including the composition of functions–and how the property of being injective, surjective, or bijective is preserved through composition.

Section 1.5: Cardinality. The cardinality of sets is introduced in terms of set equivalence. Basic facts and supporting material about finite, countable, countably infinite, and uncountable sets are presented. In particular, $\mathbb{N}$ is infinite (Theorem 1.42), the countable union of countable sets is countable (Theorem 1.56), and $[0, 1]$ is uncountable (Theorem 1.58).

Section 1.6: Troubleshooting Guide for Sets, Functions, and Proofs. A list of frequently asked questions and their answers, arranged by section. Come here if you are struggling with any material from this chapter.

1.1 Logic and an Introduction to Proof

A mathematical proof is something that should convince someone that a particular statement is indeed true. This is inherently tricky, since what is convincing to one person may not be convincing to another. Or, perhaps the inclusion of more information or detail might obscure one's overall message.

In fact, proofs should be more than just a persuasive discussion: they should teach readers something that perhaps they did not know. This could be because they are now convinced that a certain statement is true, or, because they have learned something about the line of reasoning that is used.

This section serves as an introduction to proof. We introduce statements as those objects that we aim to prove, and discuss different ways to combine statements. The Logical DeMorgan's Laws will indicate how to negate some of these statements. We will also discuss the logical relationships between an implication, its converse, and its contrapositive. We end the section with some helpful proof writing suggestions.

1.1.1 Logical statements

A fundamental goal in the field of mathematics is simply to better understand the objects one encounters and why they work the way they do. Part of understanding why these things work involves proving that certain outcomes must always hold. Thus, proof (and proving things) is a fundamental aspect of the study of mathematics, and we begin our introduction to proof by first figuring out exactly what we should be trying to prove. A *statement* is a declarative sentence that is either true or false, and not both true and false. In mathematics, we only seek to prove statements, since they are either true or false.

Example 1.1. The following are some sentences. Which ones are statements?

(a) "This apple weighs less than 8 pounds."

(b) "What is the square root of two?"

(c) "There are four lights."

(d) "I find politics repellant."

The first sentence is a statement, since it can be determined to be true or false, but not both. Thus, we could try to prove the first statement. The second sentence is not a declarative sentence, it is a question. Thus, we cannot attempt to prove it, even if we could answer it. The third sentence is a statement: there are either four lights or there are not. The final sentence is not a statement, even though it is declarative. Although there is a general understanding of what the word "repellant" means, different people may rate different things as repellant, and so for some people the sentence may be true and for some others the sentence may be false. ■

For every statement, there is an opposite statement called its *negation*. If P is a statement, then its negation $\neg P$ is sometimes read as "not P." If P is true, then $\neg P$ is false, and if P is false, then $\neg P$ is true. For simplicity, we can summarize this relationship in what is called a *truth table*. Figure 1.1.1 is a truth table summarizing the relationship between P and its negation $\neg P$, where T and F mean "True" and "False," respectively. As we develop more complicated sorts of statements in the next subsection, we will often summarize the veracity of them in truth tables. It should be noted that the symbol $\neg$ is called a *logic symbol*, and one generally does not include these symbols in formal mathematical writing since they are often difficult to read. Other logical shorthands that are to be avoided are the symbols $\forall$ (meaning "for all" or "for every") and $\exists$ (meaning "there exists"). There are some exceptions to this, but generally within most proofs, they are to be avoided.

P	$\neg P$
T	F
F	T

FIGURE 1.1.1
A truth table for P and $\neg P$.

1.1.2 Combining statements

It is often the case that certain statements are more complicated than those given in Example 1.1. For example, you could combine the two statements "This apple weighs less than 8 pounds" and "There are four lights" by making the compound statement "This apple weighs less than 8 pounds AND there are four lights." In requiring that both statements hold, we are creating a logical *conjunction* of statements. If P and Q are statements, then we might write $P \wedge Q$ to be this conjunction, and this is sometimes read "P and Q."

Similarly, we could have used the word "or" instead of "and" to combine these statements, and the logical *disjunction* of statements P and Q, written $P \vee Q$, is sometimes read as "P or Q." Again, the logic symbols $\wedge$ and $\vee$ are rarely written in the course of proofs. The new statement "P and Q" is true exactly when both P and Q are, and the new statement "P or Q" is true exactly when one of P or Q (or both) are true. We summarize these findings in truth tables in Figure 1.1.2.

Another way of combining two statements is for one statement to imply another. For instance, if statement P is "I will win the lottery tomorrow," and statement Q is "I will pay off my student loans," then one can form the *implication*, or *conditional statement*, "If I win the lottery tomorrow, then I

P	Q	$P \wedge Q$ (P and Q)	$P \vee Q$ (P or Q)
T	T	T	T
T	F	F	T
F	T	F	T
F	F	F	F

FIGURE 1.1.2
A truth table for P, Q, P and Q, and P or Q.

will pay off my student loans." Such an implication is written $P \Rightarrow Q$, read "P implies Q," and it is somewhat more common to use the symbol $\Rightarrow$ in writing proofs but is still generally discouraged. In the conditional statement $P \Rightarrow Q$, P is sometimes called the *hypothesis*, and Q is sometimes called the *conclusion*.

Let us study the veracity of the conditional statement concerning winning the lottery and paying off student loans. There are four possibilities: by tomorrow, I will win the lottery or I will not (P is true or false, respectively), and that I will or will not pay off my student loans (Q is true or false, respectively). Let us suppose that I actually win the lottery tomorrow (P is true). If I go on to pay off my student loans (Q is true), then I will have told the truth by declaring "If I win the lottery tomorrow, then I will pay off my student loans" ($P \Rightarrow Q$ is true). If, however, I win the lottery tomorrow (P is true), but I do not pay off my student loans (Q is false), then I will have lied in making the statement, "If I win the lottery tomorrow, then I will pay off my student loans" ($P \Rightarrow Q$ is false). Now suppose that I do not win the lottery tomorrow (P is false). Since I only promised that I would pay off my student loans in the event that I won the lottery, and I did not win the lottery, I will not be a liar whether I end up paying my student loans or I do not. Therefore, the conditional statement $P \Rightarrow Q$ is true whenever P is false.

In building implications, we could have just as easily considered $Q \Rightarrow P$, which is also written as $P \Leftarrow Q$. The statement $Q \Rightarrow P$ is called the *converse* of $P \implies Q$. Two statements are *equivalent* if $P \Rightarrow Q$ and its converse $Q \Rightarrow P$ are both true or both false, and in this event, we sometimes write $P \Leftrightarrow Q$. In addition, the equivalence $P \Leftrightarrow Q$ is sometimes stated as "P if and only if Q", or "P is a necessary and sufficient condition for Q." Figure 1.1.3 illustrates the truth values of implications and equivalences. Note that when P and Q have the same truth values (that is, either both true or both false), the equivalence $P \Leftrightarrow Q$ is true and vice versa–this is the reason we describe $P \Leftrightarrow Q$ as an equivalence of logical statements.

It is important to remember, as can be readily seen in Figure 1.1.3, that $P \Rightarrow Q$ and $Q \Rightarrow P$ generally do not depend on one another. It is a common mistake to think that they are logically equivalent, so that proving one is true amounts to proving the other is true, or vice versa. To see this, consider

P	Q	$P \Rightarrow Q$	$Q \Rightarrow P$	$P \Leftrightarrow Q$
T	T	T	T	T
T	F	F	T	F
F	T	T	F	F
F	F	T	T	T

FIGURE 1.1.3
A truth table for implications.

the statement P : "I am at the South Pole," and the statement Q : "I am in Antarctica." Certainly $P \Rightarrow Q$ is true: the South Pole is located in Antarctica. The converse of $P \Rightarrow Q$ reads, "If I am in Antarctica, then I am at the South Pole," and may not be true depending on where exactly you are in Antarctica (Figure 1.1.4 summarizes this information, the "contrapositive" of a statement will be discussed shortly). Thus, one statement and its converse need not have the same truth values.

P :	"I am at the South Pole."
Q :	"I am in Antarctica."
$P \Rightarrow Q$:	"If I am at the South Pole, then I am in Antarctica."
$Q \Rightarrow P$:	"If I am in Antarctica, then I am at the South Pole."

FIGURE 1.1.4
If $P \Rightarrow Q$ is true, its converse $Q \Rightarrow P$ may or may not be true.

Now that we have the background to formally discuss it, we can consider what happens when we negate the statement "P and Q": if it is not the case that P and Q are true, then one of P or Q must be false. Thus, the negation of "P and Q" is logically equivalent to the negation of P or the negation of Q. Similarly, negating the statement "P or Q" would lead us to the requirement that both of P and Q are false, so that the negation of "P or Q is logically equivalent to the negation of P and the negation of Q. These two formulations are sometimes called *Logical DeMorgan's Laws* (there is an interesting comparison between the DeMorgan's Laws listed below for logic, and those in Theorem 1.12 concerning sets). We prove the first of these equivalencies with a truth table, and leave the second one as an exercise (see Exercise 4).

Theorem 1.2 (Logical DeMorgan's Laws). *Suppose P and Q are statements.*

1. $\neg(P \wedge Q) \Leftrightarrow (\neg P) \vee (\neg Q)$.
2. $\neg(P \vee Q) \Leftrightarrow (\neg P) \wedge (\neg Q)$.

Idea of proof. *One statement is logically equivalent to the other exactly when they take on the same truth values. So we will make a truth table exhibiting their truth values, and observe that this is indeed the case.*

Proof of Theorem 1.2. Figure 1.1.5 illustrates that $\neg(P \wedge Q)$ and $(\neg P) \vee (\neg Q)$ have the same truth values, and are therefore logically equivalent. The proof of the second assertion is left as Exercise 4.

P	Q	$\neg P$	$\neg Q$	$\neg(P \wedge Q)$	$(\neg P) \vee (\neg Q)$	$\neg(P \wedge Q) \Leftrightarrow (\neg P) \vee (\neg Q)$
T	T	F	F	F	F	T
T	F	F	T	T	T	T
F	T	T	F	T	T	T
F	F	T	T	T	T	T

FIGURE 1.1.5
A truth table for $\neg(P \wedge Q)$ and $(\neg P) \vee (\neg Q)$.

□

One final type of implication is called the *contrapositive*, and it is sometimes confused with the converse. The contrapositive of the statement $P \Rightarrow Q$ is the statement $\neg Q \Rightarrow \neg P$ (it is easy to see why there is sometimes a confusion between this and the converse). Consider again the statements P and Q above concerning Antarctica and the South Pole. We agreed that the statement $P \Rightarrow Q$ ("If I am at the South Pole, then I am in Antarctica") is true. Imagine if the conclusion "I am in Antarctica" were false ($\neg Q$): would it be possible to be at the South Pole? Since the South Pole is located in Antarctica, it would not be possible ($\neg P$). Thus, if $P \Rightarrow Q$ is true, then its contrapositive $\neg Q \Rightarrow \neg P$ is also true. Figure 1.1.6 summarizes this.

P :	"I am at the South Pole."
Q :	"I am in Antarctica."
$P \Rightarrow Q$:	"If I am at the South Pole, then I am in Antarctica."
$(\neg Q) \Rightarrow (\neg P)$:	"If I am not in Antarctica, then I am not at the South Pole."

FIGURE 1.1.6
The statement $P \Rightarrow Q$ is true precisely when its contrapositive $(\neg Q) \Rightarrow (\neg P)$ is true.

The following truth table in Figure 1.1.7 illustrates that $P \Rightarrow Q$ is equivalent to its contrapositive $\neg Q \Rightarrow \neg P$.

There is a common form of proof called *proof by contradiction* whose validity rests in the fact that an implication and its contrapositive are equivalent. We give an example of this in Example 1.3.

Example 1.3. Consider the following statements concerning a whole number n:

P	Q	$\neg P$	$\neg Q$	$P \Rightarrow Q$	$\neg Q \Rightarrow \neg P$	$(P \Rightarrow Q) \Leftrightarrow (\neg Q \Rightarrow \neg P)$
T	T	F	F	T	T	T
T	F	F	T	F	F	T
F	T	T	F	T	T	T
F	F	T	T	T	T	T

FIGURE 1.1.7
A statement and its contrapositive are equivalent.

$P:$	$3n + 7$ is even.
$Q:$	n is odd.

We prove $P \Rightarrow Q$ by contradiction. That is, we prove $\neg Q \Rightarrow \neg P$: If n is even, then $3n + 7$ is odd (generally, writing down the contrapositive statement is a good place to start for proofs of this sort). If n is even, then it is a multiple of 2, so that there must be another whole number k where $n = 2k$. Then

$$3n + 7 = 3(2k) + 7 = 6k + 7.$$

The number $6k$ is even since it is a multiple of 2: $6k = 2(3k)$. The number 7 is odd, and the sum of an even number and an odd number is odd. Therefore $3n + 7$ is odd. Exercise 6 asks you to prove the converse $Q \Rightarrow P$. ∎

There are other forms of proof by contradiction that are more subtle, in the sense that sometimes the conclusion $\neg P$ of the contrapositive is something not necessarily mentioned in the implication we are asked to prove, and that we negate something that is otherwise so obvious that we do not mention it within the statement. As an example of this, we prove that $\sqrt{2}$ cannot be written as the quotient of two whole numbers in Practice 1.4. We shall discuss in Section 1.2 that the set of *rational numbers* is the set of all quotients of whole numbers (and their negatives), so that Practice 1.4 demonstrates that $\sqrt{2}$ is not rational (called *irrational*).

Practice 1.4. Prove that $\sqrt{2}$ is not the quotient of two whole numbers.

Methodology. *We prove this by contradiction: we assume that $\sqrt{2}$ can be expressed as such a quotient, say $\sqrt{2} = \frac{a}{b}$ for some whole numbers a and b. We will manipulate this expression and deduce that $\frac{a}{b}$ cannot be reduced into lowest terms. Since every fraction can be reduced into lowest terms, it must not be possible to express $\sqrt{2}$ in this way. Notice that this same proof can show that $-\sqrt{2}$ is also not rational.*

Solution. Suppose that $\sqrt{2} = \frac{a}{b}$ for whole numbers a and b. Suppose also that this fraction is written in lowest terms, so that there are no common factors of a or b that could be canceled out. By squaring both sides and multiplying by b^2, we must have $2b^2 = a^2$. Since a^2 is a multiple of 2, a^2

must be even. Therefore by Exercise 2, a is even. Since a is even, it must be a multiple of 2, say $a = 2k$ for some whole number k. Then since $2b^2 = a^2 = (2k)^2 = 4k^2$, we must have $b^2 = 2k^2$. Since b^2 is a multiple of 2, by the same reasoning, b must be a multiple of 2, and so $b = 2\ell$ for some whole number ℓ. But then $a = 2k$ and $b = 2\ell$, and a and b share the common factor 2, which is a contradiction since we assumed that $\frac{a}{b}$ was written in lowest terms, so that they had no factors in common. ■

1.1.3 Suggestions for writing proofs

To close this section, we give some helpful hints for writing proofs. Keep in mind that there is no recipe or golden method to prove any statement, rather, different statements require different methods, and the following are just some things to keep in mind as you prove things throughout this subject and beyond. In general, **the goal of a proof is to clearly express why a certain statement is true.**

1. **In every proof, you must use complete sentences, and use correct grammar, punctuation, and spelling.** There is absolutely no reason not to strive to accomplish this, since using these writing tools only makes more precise the reasoning you are trying to transmit. In addition, you should generally not use logical symbols as part of your writing[1] (unless, of course, the symbols are part of what you are writing about).

2. **Write your proofs for the correct audience.** You may often find yourself wondering, "what does my professor want me to write?" Or, "Do I really need to say this? I'm sure my professor knows why this is true." Do not write your proofs for your professors: they already presumably know everything about what you are writing. Instead, a suggestion is to write your proof at the level of someone who is in your class, but about a week behind you. This ensures that your proof emphasizes all of the correct points, but does not laboriously dwell on certain other more obvious facts.

3. **Generally, you should justify each of the claims you make as part of your proof.** Taking too big of a logical leap or sparsely justifying your logic is a common problem in proof writing. In looking at the previous suggestion, if you are not sure if you take leaps which are too big, you might show your proof to a friend and get his or her opinion. The more proofs you write, the better you will become at navigating through this issue.

4. **If you are relying on a fact whose proof can be found elsewhere, you almost always have to cite that statement by**

---

[1]There are exceptions to this, but generally a proof with logical symbols is a little harder to read.

name and even sometimes where its proof can be located. Observe that we did this in the solution to Practice 1.4, where we referenced the fact that if a^2 is even, then a is even as being found somewhere else, and referencing exactly where it was. Throughout this book you will see a multitude of further examples of this, and it is a good practice in proof writing.

5. **If you are about to embark on a complicated calculation or obscure reasoning, it may be helpful to alert the reader about what it is you will be doing in advance.** Sometimes people can get lost while reading a proof because they ask themselves "why are they doing *that*?" To help make the main message of your ideas more readable and clear–especially the complicated ones–at times it can be helpful to foreshadow your ideas.

6. **If you do not know where to start, try to determine if you really understand what is being asked.** Perhaps the most common difficulty in proof writing stems from not having a firm idea as to what you want to say. This can be because your thoughts are not organized well enough to begin an actual proof, but it can also be because perhaps you do not truly understand why a certain mathematical fact must be true. If your problem is the former, then try outlining the major reasoning points on a separate sheet of paper. If your problem is the latter, try doing a couple of examples to better understand what is happening.

7. **Proof by contradiction is great, but it should not be overused.** Generally, the first type of proof you should try is a direct one. If you find it difficult to complete such a proof, *then* try something different such as a proof by contradiction.

This book has been arranged to help you not only understand what goes into proofs, but also how to write them for yourself. As a result, each of the results presented in the book will include a short "*Idea of proof*" or "*Methodology*" paragraph that informally describes the main ideas that go into our reasoning. These are not meant to be formal proofs, but rather, designed to help you see what we will be trying to accomplish in a formal proof or solution. It is these formal proofs or solutions that will follow that you might find helpful as you attempt the exercises at the end of each section, both in content and as an example of the appropriate level of rigor. At the end of each chapter is a place you can go if you are having trouble. These final "troubleshooting" sections anticipate some common problems that students have and are designed to point you in the right direction.

1.1.4 Review of important concepts from Section 1.1

1. A statement is a declarative sentence that is either true or false (not both). The negation of a statement takes on the opposite truth value.
2. There are various ways to combine statements to make other statements, for example, "P and Q" ($P \wedge Q$), "P or Q" ($P \vee Q$), "P implies Q" ($P \Rightarrow Q$) and other variants.
3. The Logical DeMorgan's Laws (Theorem 1.2) tell us how to negate $P \wedge Q$ and $P \vee Q$.
4. A statement and its converse do not always have the same truth values. A statement and its contrapositive are logically equivalent (see Figure 1.1.7).
5. There is no one method to proving any statement. There are, however, some suggestions listed at the end of this section.

Exercises for Section 1.1

1. Prove that if a is even, then a^2 is even.
2. Prove that if a^2 is even, then a is even.
3. Prove that a is odd if and only if a^2 is odd.
4. Use a truth table to prove the second part of Theorem 1.2: If P and Q are statements, prove that $\neg(P \vee Q) \Leftrightarrow (\neg P) \wedge (\neg Q)$.
5. Using the logical symbols presented in this section, find a different way to express $\neg(P \Rightarrow Q)$. Use a truth table to express the logical equivalence of $\neg(P \Rightarrow Q)$ and your new expression.
6. Prove the converse $Q \Rightarrow P$ from Example 1.3: If n is odd, then $3n + 7$ is even.
7. Suppose that $x = \frac{a}{b} \neq 0$ can be expressed as the quotient of two whole numbers, and suppose α cannot be expressed as such. Prove that $x+\alpha$ and $x\alpha$ both cannot be expressed as the fraction of whole numbers.
8. (a) Prove that $\sqrt{3}$ cannot be expressed as the quotient of two whole numbers. *[Hint: See Practice 1.4 for a blueprint as to how you might proceed. Be careful, though, at some point you will likely have to prove that if a^2 is a multiple of 3, then a is a multiple of 3.]*

   (b) At what specific point does the argument you give in the previous part break down in proving that $\sqrt{4}$ cannot be expressed as a quotient of two whole numbers?

1.2 Sets and Their Operations

The notion of a set will be of central importance to us throughout this book, and a firm understanding of set theory is crucial to understanding much of the later material. In this section we introduce sets and various aspects of set theory. We define the common sets that will be used throughout this book, and introduce set notation and set relations. We also introduce the union, intersection, difference (complement), and cross-product of sets. DeMorgan's Laws are proven in Theorem 1.12, which describes the effect that union and intersection have in the presence of set complement. We conclude the section with more suggestions about the level of rigor when it comes to proofs about sets.

1.2.1 Sets

We begin by considering what must be a familiar concept: the notion of a set. For our purposes, we will define a *set* to be any well-defined collection of objects. By *well-defined*, we simply mean that an object is either in or not in the set (not both).

It should be pointed out that there is a large amount of careful mathematics that has been done to accurately define what we mean by a set so as to avoid certain unwanted paradoxes. For example, going strictly by our definition above, we could encounter a set that has itself as a member (for example, the set containing "everything," including itself), and this leads to what is known as "Russell's Paradox", named after Bertrand Russell (see Exercise 7). This concerted effort has been collected into what are now called the *Zermelo-Fraenkel axioms of set theory*, which meticulously lay out those basic properties of sets that we need, and most all of mathematics is based on this important list. We do not discuss these axioms here since they are generally self-evident, and one usually does not run into any paradoxes which we would specifically need to avoid by formally introducing them. For a nice description of the Zermelo-Fraenkel axioms, see page 91 of [19].

One can define a set by simply describing its elements. One does this in two common ways: as a list of objects (an object is not allowed to appear more than once on this list), or through some sort of description of what qualifies an element to be in your set. One usually puts this list or description between the symbols $\{$ and $\}$. For example, if we wish to form a set containing the numbers $2, 4$, and 6, where S is the name of this set, we could write $S = \{2, 4, 6\}$.

There are several sets that you are already familiar with that we place in Definition 1.5 for easy reference. We shall give a better description of the real numbers in Chapter 2, as what is given below is somewhat vague.

Definition 1.5. The following symbols will be used throughout this book:

1. The *natural numbers*, denoted $\mathbb{N}$, the set $\{1, 2, 3, \ldots\}$.

2. The *integers*, denoted $\mathbb{Z}$, the set $\{\ldots, -2, -1, 0, 1, 2, \ldots\}$.

3. The *rational numbers* $\mathbb{Q}$, the set of all fractions of integers with nonzero denominator.

4. The *real numbers* $\mathbb{R}$, the set of all quantities between $-\infty$ and ∞.

There is a mathematical shorthand that is useful and that is commonly used indicating that a certain element is a member of a set: one uses the symbol $\in$ and writes $x \in S$ to mean "x is an element of S." We write $x \notin S$ to mean "x is not an element of S." One could use this sort of shorthand to build a set in terms of a description as we mentioned above. For example, suppose you wanted to build a set that only had certain rational numbers in it, but that you did not know exactly what those numbers were. Say that you wanted to build a set containing all rational numbers x that satisfied the equation $(x^2 - 2)(2x - 1) = 0$. You could describe this set by first indicating that you are only considering rational numbers for inclusion in the set, and then by giving a condition that qualifies a rational number to be in the set. The set of all rational numbers that are solutions to the equation $(x^2 - 2)(2x - 1) = 0$ can be expressed in Figure 1.2.1.

$$\{ \underset{\substack{\uparrow \\ \text{type of element}}}{x \in \mathbb{Q}} \quad | \quad \underset{\substack{\uparrow \\ \text{condition on that element}}}{(x^2 - 2)(2x - 1) = 0} \}$$

FIGURE 1.2.1
Interpreting set theoretic notation.

The description $x \in \mathbb{Q}$ before the "|" indicates that the only type of elements you are considering are rational numbers (whose name is x), and the description $(x^2 - 2)(2x - 1) = 0$ after the "|" is the condition that qualifies the rational number x for inclusion in the set. Example 1.6 gives some examples of sets written in this way.

Example 1.6. We give several examples of sets written in the style of Figure 1.2.1 and write out the elements in those sets.

(a) The set $\{x \in \mathbb{N} | 2 < x < 6\}$ contains only natural numbers, since it is given that $x \in \mathbb{N}$. In addition, these elements must be greater than 2, and less than 6. Thus, this set is $\{3, 4, 5\}$.

(b) The set $\{x \in \mathbb{Q} | x^3 - 2x = 0\}$ contains only rational numbers, since it is given that $x \in \mathbb{Q}$. In addition, these rational numbers must satisfy $x^3 - 2x = 0$. By factoring out an x, we see that $x^3 - 2x = x(x^2 - 2)$, and this is zero only when $x = 0$ or when $x = \pm\sqrt{2}$. But both of $\sqrt{2}$ and $-\sqrt{2}$ are not rational (see Practice 1.4). So this set contains only the number 0.

(c) The set $\{x \in \mathbb{R} | 0 \leq x < 6\}$ is made up of all real numbers between 0 and 6, where we include 0 in this set and we do not include 6. Below we will describe a notation to better express this set.

Before moving on, there are two items that need to be discussed. First, there is a special set known as the *empty set* that contains no elements at all. We will denote this set as $\emptyset$. Second, we will want a convenient way to describe sets of real numbers that contain all numbers between two others, such as in the third item in Example 1.6. The interval notation (a, b) for real numbers a and b will be defined as the set of all real numbers strictly between a and b:

$$(a, b) \text{ means } \{x \in \mathbb{R} | a < x < b\}. \tag{1.1}$$

The (and) symbols indicate that the endpoints a and b are not to be included in the set. We use [to include the left endpoint, and] to include the right endpoint, so that

$$\begin{array}{lll} [a, b) & \text{means} & \{x \in \mathbb{R} | a \leq x < b\}, \\ (a, b] & \text{means} & \{x \in \mathbb{R} | a < x \leq b\}, \text{and} \\ [a, b] & \text{means} & \{x \in \mathbb{R} | a \leq x \leq b\}. \end{array}$$

For instance, one could use this notation to write $[0, 6)$ as a different way of expressing $\{x \in \mathbb{R} | 0 \leq x < 6\}$ as in Example 1.6.

1.2.2 Set relations

Given two sets, there are a variety of ways to relate them. The most basic relation between two sets is whether or not they are *the same set*. Two sets S and T are *equal* if S and T contain exactly the same elements. We would write $S = T$ if this is the case, and $S \neq T$ if not. We could therefore rewrite the declaration in Equation (1.1) as

$$(a, b) = \{x \in \mathbb{R} | a < x < b\}.$$

Another way to relate sets would be to observe that every element of one set is also a member of another set. For instance, if every element of S is also an element of T, then we would say that S is a *subset* of T, and write $S \subseteq T$. In the event that $S \subseteq T$ but that $S \neq T$ (so that there are elements of T that are not elements of S), we say that S is a *proper subset* of T, and write either $S \subset T$ or $S \subsetneq T$, where the latter notation is sometimes reserved for situations where there is some emphasis on the fact that $S \neq T$. One writes $\subseteq$ to allow for the possibility that the set containment is proper, since often that distinction is unnecessary. Generally one uses the symbols $\supseteq$ and $\supset$ in the same way: for example, $S \supsetneq T$ means that T is a proper subset of S. Using this notation, one can say that $S = T$ **if and only if** $S \subseteq T$ **and** $T \subseteq S$**.** Proving that $S = T$ by showing $S \subseteq T$ and $T \subseteq S$ is referred to as *double inclusion*.

Practice 1.7. In the following, find an appropriate set relation symbol to relate the given sets S and T.

(a) $S = (-1, 1], T = \{x \in \mathbb{R} | x^2 < 1\}$.

(b) $S = \{n \in \mathbb{Z} | n > 0\}, T = \mathbb{N}$.

(c) $S = \{\sqrt{n} | n \in \mathbb{N}\}, T = \mathbb{Q}$.

Methodology. *To determine the needed symbol, we consider whether or not one set is a subset of the other, and describe why this is the case. It should be noted that the solutions below are very detailed to better illustrate the idea, and one does not always need such a level of detail for descriptions of this sort. The discussion at the end of this section on page 20 discusses detail and rigor in proofs.*

Solution.

(a) Every element of T is also an element of S: if $x \in T$, then $x^2 < 1$. Thus, $-1 < x < 1$, and so $x \in (-1, 1]$. So at least $T \subseteq S$. Not every element of S is an element of T: $1 \in S$; however, $1 \notin T$, since $1^2 \not< 1$. So, S is not a subset of T, so these sets are not equal. Therefore the appropriate relation is $T \subseteq S$. Some other ways to express this would be $S \supseteq T$, $T \subset S$, or $T \subsetneq S$, although the latter two expressions are somewhat uncommon.

(b) Every element of S is an element of T: if $n \in \mathbb{Z}$ (that is, n is an integer), and $n > 0$, then n must be one of $1, 2, 3, \ldots$. Therefore $n \in \mathbb{N}$, so $S \subseteq T$. But every element of T is also an element of S: if $n \in T = \mathbb{N}$, then $n \in \mathbb{Z}$ and $n > 0$, so $n \in S$. Therefore, since $S \subseteq T$ and $T \subseteq S$, we conclude $S = T$.

(c) There are some elements of S that are also elements of T. For instance, $3 \in S$, since $n = 9 \in \mathbb{N}$ and $\sqrt{9} = 3$. In addition, $3 \in T = \mathbb{Q}$ since $3 = \frac{3}{1}$, the quotient of two integers (with nonzero denominator). But there are other elements of S that are not elements of T: $\sqrt{2} \in S$ since $2 \in \mathbb{N}$, and elements in S are comprised of the square root of any natural number. It was shown in Practice 1.4 that $\sqrt{2} \notin \mathbb{Q} = T$. So, we initially conclude that S is not a subset of T.

We have the same situation as we compare elements of T to those in S. There are some elements of T that are also elements of S, for example, the number 3 is in both sets as described above. However, there are other elements of T that are not elements of S: For example, the number $\frac{1}{2} \in \mathbb{Q} = T$, however, $\frac{1}{2} \neq \sqrt{n}$ for any natural number n. We prove this claim by contradiction. If there were such a number n with $\frac{1}{2} = \sqrt{n}$, then by squaring both sides we would have $\frac{1}{4} = n$, and this is not a natural number (our original assumption was that $n \in \mathbb{N}$). So, T is not a subset of S. Therefore, the only appropriate relation between S and T is $S \neq T$. ■

1.2.3 Operations on sets

Given two sets S and T, there are a variety of ways to create new sets from them. Perhaps the most basic of ways to do this is to create a new set that combines all of the elements of both sets into one "larger" set. This is called the *union* of S and T, written $S \cup T$. Another basic way to create a new set is to choose only those elements that are in both sets S and T. This is called the *intersection* of S and T, written $S \cap T$. Definition 1.8 summarizes this information for easy reference, and the picture below Definition 1.8 shades what $S \cup T$ and $S \cap T$ and is called a *Venn diagram*.

Definition 1.8. Let S and T be sets.

1. The *union* of S and T, written $S \cup T$, is the set of all elements from either S or T. In set notation,

$$S \cup T = \{x | x \in S \text{ or } x \in T\}.$$

2. The *intersection* of S and T, written $S \cap T$, is the set of all elements that are in both S and T. In set notation,

$$S \cap T = \{x | x \in S \text{ and } x \in T\}.$$

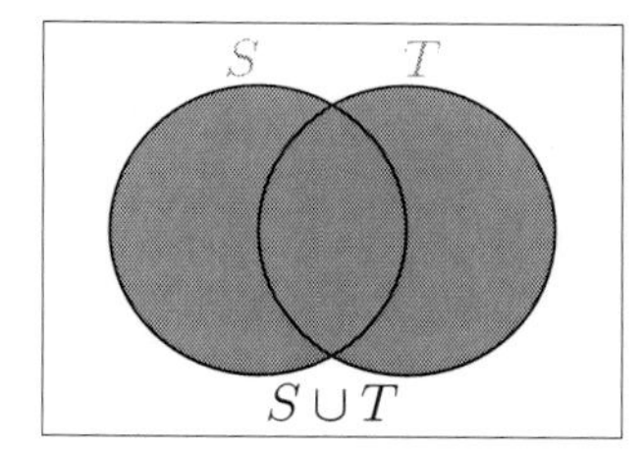

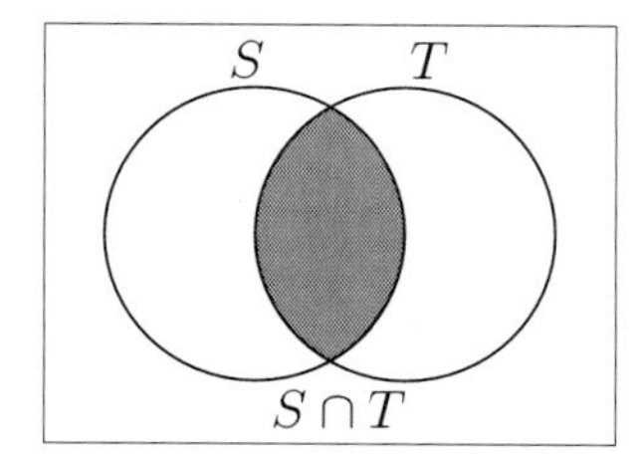

The following is good practice with unions and intersections of sets.

Practice 1.9. Suppose that

$$S = (-4, 4), \quad T = [0, 9].$$

(a) Compute $S \cup T$.

(b) Compute $S \cap T$.

Methodology. *We simply follow Definition 1.8: the union of those sets will combine the elements from both sets, while the intersection will only pick out the elements common to both sets. We deduce the correct answers and show that the sets are equal by double inclusion.*

Solution.

(a) If $x \in S \cup T$, then $x \in S$ or $x \in T$. If $x \in S$, then $-4 < x < 4$. If $x \in T$, then $0 \leq x \leq 9$. If x could satisfy either of these, then $-4 < x \leq 9$, or differently put, $x \in (-4, 9]$. We have shown that $x \in S \cup T$ implies $x \in (-4, 9]$, so that $S \cup T \subseteq (-4, 9]$. Now suppose $x \in (-4, 9]$. Then, among other descriptions, $-4 < x < 4$, or $0 \leq x \leq 9]$, so that $x \in S \cup T$. We have now shown $x \in (-4, 9]$ implies $x \in S \cup T$. Therefore $(-4, 9] \subseteq S \cup T$, and so by double inclusion, $S \cup T = (-4, 9]$.

(b) If $x \in S \cap T$, then $x \in S$ and $x \in T$. Therefore $-4 < x < 4$ and $0 \leq x \leq 9$. Since both of these inequalities must hold, $0 \leq x < 4$. We have shown that $S \cap T \subseteq [0, 4)$. Now suppose $x \in [0, 4)$. It follows that $-4 < x < 4$ and $0 \leq x \leq 9$, so that $x \in S$ and $x \in T$. We have now shown that $x \in [0, 4)$ implies $x \in S \cap T$, or that $[0, 4) \subseteq S \cap T$. Therefore, $[0, 4) = S \cap T$ by double inclusion. ■

We will sometimes need to union or intersection more than just two sets, and we shall need a notation to illustrate this. If there are only several sets to union or intersect, say 3 or 4 sets, then we could repeat the notation we have used earlier: if S, T, and W are sets, then we would write $S \cup T \cup W$ as the union of all three sets, and $S \cap T \cap W$ as the intersection of all three sets. Notationally,

$$\begin{aligned} S \cup T \cup W &= \{x | x \in S \text{ or } x \in T \text{ or } x \in W\}, \text{ and} \\ S \cap T \cap W &= \{x | x \in S \text{ and } x \in T \text{ and } x \in W\}. \end{aligned}$$

If, however, there are a great number more to union or intersect–perhaps infinitely many–we express such a combination by using the appropriate symbol ($\cup$ or $\cap$), with some indication of what will index the union or intersection in the subscript and/or superscript, followed by the type of set to be unioned or intersected. For example, if $S_1, S_2, S_3, \ldots$ are sets, we denote

$$\begin{aligned} S_1 \cup S_2 \cup S_3 \cup \cdots &= \cup_{n=1}^{\infty} S_n, \text{ and} \\ S_1 \cap S_2 \cap S_3 \cap \cdots &= \cap_{n=1}^{\infty} S_n. \end{aligned} \tag{1.2}$$

The above example could be looked at in the following way: for every element $k \in \mathbb{N}$, we have a set S_k. The notation in Equation (1.2) demonstrates how to union or intersect over all S_k for every $k \in \mathbb{N}$. In this case, the set $\mathbb{N}$ is called the *indexing set*. We could encounter a need to union or intersect a collection of sets via some other indexing set. For instance, if for every element α in some indexing set A, we have the set S_α, we could express the union and intersection of the S_α over all α in A as

$$\begin{aligned} \cup_{\alpha \in A} S_\alpha &= \{x | x \in S_\alpha \text{ for some } \alpha \in A\}, \text{ and} \\ \cap_{\alpha \in A} S_\alpha &= \{x | x \in S_\alpha \text{ for all } \alpha \in A\}. \end{aligned}$$

In Example 1.10, we present two examples of how this notation might be used.

Example 1.10. Here are two examples to illustrate the use of indexing sets when forming unions and intersections.

(a) For every $n \in \mathbb{N}$, define the set $S_n = (n-1, n+1)$. We determine $\cup_{n=1}^{\infty} S_n$ by first noting that, by definition, $x \in \cup_{n=1}^{\infty} S_n$ if and only if $x \in S_n$ for at least one n in the indexing set $\mathbb{N}$. To get an idea as to what such an x might be, we consider listing out a few of the sets that are in this infinite union:

$$\begin{array}{ccccccccc} \cup_{n=1}^{\infty} S_n & = & (0,2) & \cup & (1,3) & \cup & (2,4) & \cup & \cdots \\ & & \uparrow & & \uparrow & & \uparrow & & \\ & & n=1 & & n=2 & & n=3 & & \cdots \end{array}$$

We observe that $(0,2) \cup (1,3) = (0,3)$, and that

$$(0,2) \cup (1,3) \cup (2,4) = (0,3) \cup (2,4) = (0,4).$$

So it seems that for any $M \in \mathbb{N}$, $\cup_{n=1}^{M} S_n = (0, M+1)$, and so we conjecture that $\cup_{n=1}^{\infty} S_n = (0, \infty)$. To prove this conjecture, we use the method of double inclusion.

First, let $x \in \cup_{n=1}^{\infty} S_n$. Then $x \in S_n = (n-1, n+1)$ for some n. Since $n \geq 1$, the number $n-1 \geq 0$, and so $x \in (0, \infty)$. We have shown that $\cup_{n=1}^{\infty} S_n \subseteq (0, \infty)$.

Next, let $x \in (0, \infty)$. Find the first natural number n that is greater than or equal to x; that is, find $n \in \mathbb{N}$ so that $n - 1 < x \leq n$ (that such an n exists is rigorously proved in Theorem 2.35). Then $x \in (n-1, n+1) = S_n$, so that $x \in \cup_{n=1}^{\infty}$. We have now shown that $(0, \infty) \subseteq \cup_{n=1}^{\infty} S_n$. Therefore, by double inclusion, $\cup_{n=1}^{\infty} S_n = (0, \infty)$.

(b) For every $\alpha \in (0, \infty)$, define the set $S_\alpha = (-\alpha, \alpha)$. We determine $\cap_{\alpha \in (0,\infty)} S_\alpha$ by first noting that, by definition, $x \in \cap_{\alpha \in (0,\infty)} S_\alpha$ if and only if $x \in S_\alpha = (-\alpha, \alpha)$ for every α in the indexing set $(0, \infty)$. Thus,

$$-\alpha < x < \alpha \text{ for every } \alpha \in (0, \infty).$$

There is only one value for x that seems to work: $x = 0$. We prove that $\cap_{\alpha \in (0,\infty)} S_\alpha = \{0\}$ by double inclusion. First let $x \in \cap_{\alpha \in (0,\infty)} S_\alpha$. Then $x \in S_\alpha$ for every $\alpha \in (0, \infty)$, and so $-\alpha < x < \alpha$ for every $\alpha \in (0, \infty)$. We prove that $x = 0$ by contradiction: if any number other than $x = 0$ could satisfy this, we shall run into a contradiction. So suppose $x \neq 0$, but that $-\alpha < x < \alpha$ for every $\alpha \in (0, \infty)$. If $x > 0$, then choose $\alpha = \frac{x}{2} \in (0, \infty)$, so that $\alpha < x$, a contradiction. If $x < 0$, then choose $\alpha = -\frac{x}{2} \in (0, \infty)$, so that $x < -\alpha$. Since we find a contradiction with every other possibility, it must be the case that $x = 0$. We have shown that $\cap_{\alpha \in (0,\infty)} S_\alpha \subseteq \{0\}$.

The opposite containment of $\{0\} \subseteq \cap_{\alpha \in (0,\infty)} S_\alpha$ is easy to verify, since there is only one element $0 \in \{0\}$, and $0 \in (-\alpha, \alpha) = S_\alpha$ for every $\alpha \in (0, \infty)$. We have now shown that $\{0\} \subseteq \cap_{\alpha \in (0,\infty)} S_\alpha$. Therefore, by double inclusion, $\cap_{\alpha \in (0,\infty)} S_\alpha = \{0\}$. ■

Besides the union and intersection, one can create new sets from old ones using the *set difference* or *set complement* operations.

If S and T are sets, then the *set difference* (also called the *complement of* S *in* T) $T - S$ is defined to be the set of all things in T that are not in S. In other words,

$$T - S = \{x \in T | x \notin S\}.$$

This construction allows you to begin with a desired set T, and remove those elements that you do not wish to include. The term "set difference" is usually used in this context. For example, the set of *irrational numbers* (those real numbers that are not rational) could be denoted by $\mathbb{R} - \mathbb{Q}$. Another way to use this construction is to begin with a set S and to consider only those elements which lie outside of S. The term "complement" is more commonly used in this context, and *the complement of* S *in* T is written as S^c. The only issue with the latter is that one needs to be certain that the set T is defined properly, for if it is not, confusion can ensue.

For example, suppose that $S = \{0\}$. If you want to form the set of all objects not in S, one may think that you are talking about the set $(-\infty, 0) \cup (0, \infty)$: this is the set of all real numbers that are not 0. If this is indeed what is meant, then you are describing the set difference $\mathbb{R} - \{0\}$, and implicitly assuming that $T = \mathbb{R}$. Strictly speaking, however, the color green is not in the set S, so does it not belong in "the set of all objects that are not in S"? In the latter perspective, it is not entirely clear what the set T is, and this sort of confusion needs to be avoided. Since this book deals with the real numbers, *unless it is specified otherwise, the set* S^c *is the complement of* S *in* $T = \mathbb{R}$. In other situations beyond this book, it will either be implicitly clear what the set T is, or it will be specified explicitly.

Practice 1.11. Let $T = \mathbb{R}$. Let $S_1 = (-\infty, 4)$ and let $S_2 = (2, 8)$.

(a) Determine S_1^c, S_2^c, and $S_1 - S_2$.

(b) Determine $(S_1 \cup S_2)^c$.

(c) Determine $S_1^c \cap S_2^c$.

Methodology. *We will determine these sets according to the definition of set difference and complement. Since the sets involved are not complicated, we do not give a solution as detailed in terms of double inclusion arguments as some of the others we have seen up to this point.*

Solution.

(a) The set $S_1^c = \mathbb{R} - S_1 = [4, \infty)$. The set $S_2^c = (-\infty, 2] \cup [8, \infty)$. Finally, the set $S_1 - S_2$ is the set of all objects in S_1 that are not in S_2. Thus, $S_1 - S_2 = (-\infty, 2]$.

(b) The set $S_1 \cup S_2 = (-\infty, 4) \cup (2, 8) = (-\infty, 8)$. So, $(S_1 \cup S_2)^c = (-\infty, 8)^c = [8, \infty)$.

(c) We determined in Part (a) that $S_1^c = [4, \infty)$ and $S_2 = (-\infty, 2] \cup [8, \infty)$. The intersection $S_1^c \cap S_2^c$ is the set of all elements that are in both S_1^c and S_2^c, which is the set $[8, \infty)$.

■

Notice that in Practice 1.11, the sets $(S_1 \cup S_2)^c = S_1^c \cap S_2^c$. It turns out that this (and more) is always true: DeMorgan's Laws for sets describes the relationship between complements of unions and intersections of sets and the complements of the sets themselves. It has a visible similarity to the Logical DeMorgan's Laws (Theorem 1.2).

Theorem 1.12 (De Morgan's Laws). *Let S_1 and S_2 be sets.*

1. $(S_1 \cup S_2)^c = S_1^c \cap S_2^c$.
2. $(S_1 \cap S_2)^c = S_1^c \cup S_2^c$.

Idea of proof. *We prove the first of these, and leave the second for Exercise 3. We will prove that these sets are equal by double inclusion. To prove each inclusion, we simply unravel what each of the set operations means, and reinterpret this description.*

Proof of Theorem 1.12. We prove the first assertion by double inclusion, and leave the second for Exercise 3. Let $x \in (S_1 \cup S_2)^c$. Then $x \notin S_1 \cup S_2$, and so $x \notin S_1$ and $x \notin S_2$. Therefore $x \in S_1^c$ and $x \in S_2^c$, and so $x \in S_1^c \cap S_2^c$. We have shown that $(S_1 \cup S_2)^c \subseteq S_1^c \cap S_2^c$.

Now let $x \in S_1^c \cap S_2^c$. Then $x \in S_1^c$ and $x \in S_2^c$, so $x \notin S_1$ and $x \notin S_2$. Therefore $x \notin S_1 \cup S_2$, and so $x \in (S_1 \cup S_2)^c$. We have now shown that $S_1^c \cap S_2^c \subseteq (S_1 \cup S_2)^c$, and so by double inclusion, $(S_1 \cup S_2)^c = S_1^c \cap S_2^c$. □

The final set operation we consider here is called the *cross-product* of sets (sometimes also called the *Cartesian product* of sets). By definition, if S and T are sets, then the cross-product $S \times T$ is defined to be the set of all ordered pairs, where the first in the pair comes from S, and the second in the pair comes from T:

$$S \times T = \{(s, t) | s \in S, t \in T\}.$$

Example 1.13. In this example, we give two illustrations of cross-products.

(a) Let $S = \{2, 3, 4, 5, 6, 7, 8, 9, 10, J, Q, K, A\}$, and let $T = \{\clubsuit, \blacklozenge, \blacktriangledown, \spadesuit\}$. It is easy to create some sample elements in $S \times T$: simply choose an element of S and an element of T and make an ordered pair from them. For example, $(A, \blacktriangledown), (Q, \spadesuit) \in S \times T$. In this example, there is a correspondence between $S \times T$ and a standard deck of cards (without the jokers): $(A, \blacktriangledown)$ could correspond to the Ace of hearts, $(Q, \spadesuit)$ could correspond to the Queen of spades, and so on.

(b) Let $S = T = \mathbb{N}$. We can begin to list out the elements of $\mathbb{N} \times \mathbb{N}$:

$$\begin{aligned} \mathbb{N} \times \mathbb{N} &= \{(n,m) | n, m \in \mathbb{N}\} \\ &= \{(1,1),(1,2),(1,3),(1,4),\ldots, \\ &\quad (2,1),(2,2),(2,3),(2,4),\ldots, \\ &\quad (3,1),(3,2),(3,3),(3,4),\ldots, \\ &\quad \ldots\}. \end{aligned}$$

(c) Let $S = T = \mathbb{R}$. The reader is probably already familiar with $\mathbb{R} \times \mathbb{R}$:

$$\mathbb{R} \times \mathbb{R} = \{(x,y) | x, y \in \mathbb{R}\}.$$

This is sometimes written as $\mathbb{R}^2$, although we caution that the notation does not always work like this. For example, if S^1 is the set of unit vectors in $\mathbb{R}^2$, and S^2 is the set of unit vectors in $\mathbb{R}^3$, then $S^1 \times S^1 \neq S^2$! A fun mental exercise is to imagine what $S^1 \times S^1$ "looks like", and how it differs from S^2; this is one topic among others that one discusses in the subject of topology. ■

We close this section with a brief description of what sort of detail is required in writing proofs involving sets. Generally speaking, when one is asked to prove that two sets are equal, it is typical that a double-inclusion sort of proof is done, as in Practice 1.9, Example 1.10, and in the proof of DeMorgan's Laws (Theorem 1.12). However in Practice 1.11 leading up to DeMorgan's laws, we did not supply a double-inclusion argument, which may lead the careful reader to wonder: *"What level of detail is expected of me when I prove things?"*

This is an important issue that does not usually have a clear answer. Generally, one should at least follow the guidelines given on page 8, in particular, suggestion number 2 that states that your proofs should be written for the correct audience. To prove two sets are equal, it is generally expected that one should use a double-inclusion proof unless it seems that the sets involved are so uncomplicated that such a set equality is more or less obvious. If you have any doubt as to whether or not such rigor is required, it is always safe to err on the side of rigor, and to include such an argument.

More generally, the rigor you supply in proofs should follow the same idea: any statement that you feel is not completely obvious needs to be thoroughly explained, and if you have any question as to whether or not you are giving a sufficient level of detail it is always a good idea to err on the side of rigor.

1.2.4 Review of important concepts from Section 1.2

1. A set is, roughly speaking, a well-defined collection of objects. We introduced $\mathbb{N}$ (the natural numbers), $\mathbb{Z}$ (the integers), $\mathbb{Q}$ (the rational numbers), $\mathbb{R}$ (the real numbers), and $\emptyset$ (the empty set).
2. Expressing a set in set notation is a convenient way to define a set, and is very commonly used.

3. Set relations describe a relationship between two sets. These symbols include $\subseteq, =, \neq$, etc.
4. Double-inclusion is a very common method of proving that two sets are equal.
5. There are several common operations on sets: the union, intersection, set difference (complement), and cross-product. In this book, when $S \subseteq \mathbb{R}$, the complement of S is assumed to be $\mathbb{R} - S$ unless otherwise specified.

Exercises for Section 1.2

1. What is the difference between $\emptyset$ and $\{\emptyset\}$?
2. Let S be a set. Prove that $S = \cup_{x \in S}\{x\}$.
3. Prove the second DeMorgan's Law: If S_1 and S_2 are sets, then $(S_1 \cap S_2)^c = S_1^c \cup S_2^c$. *[Hint: The proof proceeds very similarly to the proof of the first DeMorgan's Law in Theorem 1.12.]*
4. Prove that DeMorgan's Laws hold for arbitrary unions and intersections of sets. Let S_α be an arbitrary collection of sets ($\alpha \in A$).
   (a) Prove that $(\cup_{\alpha \in A} S_\alpha)^c = \cap_{\alpha \in A} S_\alpha^c$.
   (b) Prove that $(\cap_{\alpha \in A} S_\alpha)^c = \cup_{\alpha \in A} S_\alpha^c$.
5. Let S and T be sets. For every $x \in S$, define the set $T_x = \{(x,t) | t \in T\}$. Prove that
$$S \times T = \cup_{x \in S} T_x.$$
6. If S is a set, then the *power set on* S is the set of all subsets of S, denoted $\mathcal{P}(S)$. If $S = \{1, 2, 3\}$, list the elements of $\mathcal{P}(S)$.
7. (Russell's Paradox) Let S be the set of all sets that are not members of themselves. Prove the paradoxical statement $S \in S \Leftrightarrow S \notin S$. *[Luckily, we do not encounter such situations in this book.]*
8. Suppose there is an island inhabited only by people who need to shave (or be shaved). There is one barber on the island, and by definition, the barber shaves all those who do not shave themselves. Who shaves the barber?

1.3 Mathematical Induction

It is fairly often that we will need to demonstrate that a certain collection of facts are true, in particular, in Chapter 3. In this section, we introduce a powerful method of proving such statements called *Mathematical Induction.* We begin the section with an example illustrating just how this method of proof works, followed by a formal demonstration that it indeed is a valid form of proof. In order to do this, we shall have to assume an axiom of set theory called the *Well-ordering Principle.* We close the section with some remarks about the flexibility one has in using this method, along with some examples of its use.

1.3.1 An example

We begin this section with an example illustrating the reasoning as to why a certain type of proof (proof by *Mathematical Induction*) is valid. Suppose we consider a natural number n, and the equation

$$1+2+3+\cdots+n=\frac{n(n+1)}{2}. \tag{1.3}$$

The task is to give a proof that this statement holds for every natural number n. Clearly, if one were to test only certain values of n, we would not encounter any false statements. For example, if $n=1$, our statement would read $1=\frac{1(1+1)}{2}=1$, which is true. If $n=2$, our statement would read $1+2=\frac{2(2+1)}{2}=3$, which is again true. With a little bit of time, we could even verify the $n=100$ statement:

$$1+2+3+\cdots+100=\frac{100(100+1)}{2}=5050.$$

How then can we prove that Equation (1.3) holds for *any* n? Simply checking that it holds for a few values of n is certainly not a proof that it holds for *every* n.

We suggest the following reasoning: prove that Equation (1.3) holds for every $n \in \mathbb{N}$ by contradiction: suppose that there were some value of n for which Equation (1.3) fails, and find a contradiction.

So suppose that Equation (1.3) failed for some n. In fact, it could be the case that Equation (1.3) fails for a multitude of different n, so let us find the *first* place at which it fails (that we are finding the first place it fails is an important assumption). We have already checked that this place of first failure is not $n=1$, since we have explicitly verified that the first statement (that is, the $n=1$ statement) holds. So at the very least, the first n for which this statement is false must be greater than or equal to 2 (we have also verified the $n=2$ statement above, although this turns out not to be necessary).

Since Equation (1.3) fails for the first time at n, it must be the case that the statement previous to that is true. Since $n \geq 2$, there exists a statement that is previous to it. We will call this the $(n-1)^{\text{st}}$ statement[2]. Let us actually determine what the $(n-1)^{\text{st}}$ statement is. If the nth statement is Equation (1.3), then the $(n-1)^{\text{st}}$ statement can be found by replacing $n-1$ with n in Equation (1.3):

$$1+2+3+\cdots+(n-1) = \frac{(n-1)n}{2},$$

and we recall that this statement is assumed to be true. But if this is the case, then

$$\begin{aligned}
1+2+3+\cdots+(n-1) &= \tfrac{(n-1)n}{2} \\
1+2+3+\cdots+(n-1)+n &= \tfrac{(n-1)n}{2}+n \\
&= n\left(\tfrac{n-1}{2}+1\right) \\
&= n\left(\tfrac{n+1}{2}\right) \\
&= \tfrac{n(n+1)}{2}.
\end{aligned}$$

In other words, we have proven that if the $(n-1)^{\text{st}}$ statement is true, then the nth statement is true. But this is a contradiction to our assumption that Equation (1.3) failed at this particular n. Thus, the assumption that there existed an n for which the statement fails must not be true. In other words, for every $n \in \mathbb{N}$, the statement is true and we have accomplished our goal.

Let us briefly review the main parts of our argument. First, we checked that the first of these statements held. Second, we proved that if any one statement holds, then the next statement holds as well (generally, one assumes that the nth statement is true and proves that the $(n+1)^{\text{st}}$ statement is true). This method of proof is what we call a *proof by Mathematical Induction.*

Before we proceed to demonstrate that this method is a valid technique of proof, we pause to make an important distinction between what we call "Mathematical Induction" and the term "induction". The term "induction" can refer to a number of topics across the sciences, in particular, logicians refer to "inductive reasoning" as the method of drawing probable conclusions from a given set of premises. "Mathematical Induction" is a method of proof which deduces the certain truth of a statement (or collection of statements) from a given collection of hypotheses, which is a method of reasoning better described as "deductive". As a result, when using the method of Mathematical Induction it is generally appropriate to list "Mathematical Induction" as your method of proof, as opposed to the form of reasoning simply referred to as "induction".

---

[2]This is not the number $n-1$ raised to the *st* power, it is how we describe statement number $n-1$. It is read "the n minus first statement".

1.3.2 The Well-Ordering Principle and Mathematical Induction

Upon closer inspection of the reasoning used in the example above, there was one very subtle assumption we made: out of all of the possible places at which Equation (1.3) failed, we were able to select the *first* place at which it failed. More generally, given a nonempty subset of $\mathbb{N}$, we selected a *least* element in that set. It turns out that the ability to do this is not something that can be shown to always be possible, as such an activity cannot be shown to be true by the standard axioms of set theory. A set is called *well-ordered* if a complete ordering of its elements can be done, and the *Well-Ordering Principle* asserts that every set can be well-ordered. We shall assume that $\mathbb{N}$ and all of its nonempty subsets are well-ordered:

Axiom 1.14 (Well-Ordering Principle)**.** *Every nonempty set can be well-ordered. In particular, every nonempty subset of* $\mathbb{N}$ *has a smallest element.*

With the well-ordering of $\mathbb{N}$ in hand, we can now demonstrate that the method of Mathematical Induction is a valid form of proof. For simplicity, if we are given a statement that depends on a natural number n, we may denote that statement by $P(n)$. Assumption 1 in Theorem 1.15 is sometimes called the *base case* or *trivial case*, and Assumption 2 is sometimes called the *induction hypothesis*, while proving that Assumption 2 is satisfied is sometimes called the *inductive step.*

Theorem 1.15 (Mathematical Induction is a valid form of proof)**.** *For any* $n \in \mathbb{N}$*, suppose that* $P(n)$ *is a statement that depends on* n*. Suppose also that*

1. $P(1)$ *is true.*
2. *If* $P(n)$ *is true for some* $n \geq 1$*, then* $P(n+1)$ *is also true.*

If these two conditions are met, then $P(n)$ *is true for every* $n \in \mathbb{N}$*.*

Idea of proof. *The idea of this proof follows along the same lines as our example that began the section. If there were some natural number at which one of these statements fails, then by the Well-Ordering Principle, we could find the first place this statement fails to be true. Since* $P(1)$ *is assumed to be true, this "first failure" is not when* $n = 1$*, and so there is a statement previous to this that is therefore assumed to be true. The second assumption now concludes that the assumption that our original statement fails is false.*

Proof of Theorem 1.15. By assumption, $P(1)$ is true. Now assume to the contrary that $P(n)$ is false for some $n \in \mathbb{N}$. Let S be the set of all natural numbers n for which $P(n)$ is false. By assumption, S is a nonempty subset of $\mathbb{N}$, and therefore has a least element $n+1$ for some $n \in \mathbb{N}$ (this least element cannot be 1 since $P(1)$ is true, and for every $n \in \mathbb{N}$, $n+1 \geq 2$). Since $n+1$ is the least element of S, and $n < n+1$ we have $n \notin S$. Therefore $P(n)$ is true. The second assumption implies that $P(n+1)$ is true, a contradiction. □

Before giving some examples of how to use Mathematical Induction, there are several remarks that are in order. First, there is nothing special about the number 1 when one uses Mathematical Induction. To be more specific, one could replace the first assumption in Theorem 1.15 with "$P(N)$ is true," and the second assumption with "If $P(n)$ is true for some $n \geq N$, then $P(n+1)$ is also true" for some integer N. In this sense, it does not matter at what point your statements begin to be true. Practice 1.16 is an example where our statements are true for $n \geq 0$, and in Practice 1.17 our statements are true for $n \geq 4$.

Second, if one looks closely at the proof of Theorem 1.15, one sees that it is not just the case that the second assumption guarantees just $P(n)$ to be true. Since $P(n+1)$ is the first place these statements fail, it is actually the case that *all* previous statements can be assumed to be true, not just $P(n)$. This type of reasoning is sometimes called *strong Mathematical Induction* and is equivalent to Theorem 1.15.

Finally, very many statements to be proven using Mathematical Induction are phrased in a way to make the meaning of the statement more apparent at the expense of creating a statement which does not necessarily list all assumptions explicitly. For example, Exercise 1 asks you to "Use Mathematical Induction to prove $2+4+6+\cdots+2n = n(n+1)$." This could be rephrased as the less readable but more mathematically sound sentence "Use Mathematical Induction to prove that if $n \in \mathbb{N}$, then $2+4+6+\cdots+2n = n(n+1)$." The latter statement clearly lists the hypothesis ("if $n \in \mathbb{N}$") and the conclusion ("then $2+4+6+\cdots+2n = n(n+1)$"). Generally there is no confusion that results from this effort to make our statements more readable, although it is always a good idea to try to rephrase the statement for yourself if you are confused.

1.3.3 Examples of Mathematical Induction

Here are several practice problems that illustrate the use of Mathematical Induction. The first fact is important in the study of geometric sequences and series.

Practice 1.16. Let $r \neq 1$ be any real number. Prove that for $n \geq 0$ we have $1 + r + r^2 + \cdots + r^n = \frac{1-r^{n+1}}{1-r}$.

Methodology. *We use the method of Mathematical Induction to prove this. The base case is easy to check, and the inductive step follows very similarly in reasoning to the example that began this section. In this example, as is sometimes a good idea to do, we specifically state what we are assuming and what we are proving in the inductive step.*

Solution. We prove this using Mathematical Induction. When $n = 0$, we can check that this statement is true by noting that $\frac{1-r^{1-1}}{1-r} = 1$. Now suppose that for some $n \geq 0$, we have

$$1 + r + r^2 + \cdots + r^n = \frac{1 - r^{n+1}}{1 - r}.$$

We wish to prove that

$$1 + r + r^2 + \cdots + r^n + r^{n+1} = \frac{1 - r^{n+2}}{1 - r}.$$

But

$$\begin{aligned} 1 + r + r^2 + \cdots + r^n + r^{n+1} &= \tfrac{1-r^{n+1}}{1-r} + r^{n+1} \\ &= \tfrac{1-r^{n+1}}{1-r} + \tfrac{r^{n+1}(1-r)}{1-r} \\ &= \tfrac{1-r^{n+1}+r^{n+1}-r^{n+2}}{1-r} \\ &= \tfrac{1-r^{n+2}}{1-r}. \end{aligned}$$

■

This example of Mathematical Induction involves an inequality, rather than an equality.

Practice 1.17. Prove that for all $n \geq 4$, $2^n \leq n!$.

Methodology. *We will use Mathematical Induction. In proving the inductive step, in contrast to the last practice problem, we will have to overestimate our expression, and we do so with the overestimate $2 \leq n + 1$; after all, this statement only holds for those $n \geq 4$, and so this is a valid inequality.*

Solution. We prove this using Mathematical Induction. When $n = 4$, we can check that $2^4 = 16 \leq 24 = 4!$. Now suppose that $2^n \leq n!$ for some $n \geq 4$. We aim to prove that $2^{n+1} \leq (n+1)!$. But using the induction hypothesis and the fact that $2 \leq n + 1$, we have

$$\begin{aligned} 2^{n+1} &= 2^n \cdot 2 \\ &\leq n! \cdot 2 \\ &\leq n! \cdot (n + 1) \\ &= (n + 1)!. \end{aligned}$$

■

This final example of Mathematical Induction extends one of the DeMorgan Laws.

Practice 1.18. Let $n \geq 1$. For sets $S_1, \ldots, S_n$, prove that $(\cup_{k=1}^n S_k)^c = \cap_{k=1}^n S_k^c$.

Methodology. *We use Mathematical Induction, and the fact that we have already proven that this statement holds when there are only two sets involved in Assertion 1 of Theorem 1.12.*

Solution. We prove this by Mathematical Induction. If $n = 1$, this statement is a tautology, and is therefore true. Now suppose that for some $n \geq 1$, we have $(\cup_{k=1}^{n} S_k)^c = \cap_{k=1}^{n} S_k^c$. We wish to prove that $(\cup_{k=1}^{n+1} S_k)^c = \cap_{k=1}^{n+1} S_k^c$. But

$$\cup_{k=1}^{n+1} S_k = \cup_{k=1}^{n} S_k \cup S_{n+1},$$

and so by DeMorgan's Law (Assertion 1 of Theorem 1.12), we have

$$(\cup_{k=1}^{n} S_k \cup S_{n+1})^c = (\cup_{k=1}^{n} S_k)^c \cap S_{n+1}^c.$$

By the induction hypothesis, $(\cup_{k=1}^{n} S_k)^c = \cap_{k=1}^{n} S_k^c$. So,

$$(\cup_{k=1}^{n+1} S_k)^c = (\cup_{k=1}^{n} S_k)^c \cap S_{n+1}^c = \cap_{k=1}^{n} S_k^c \cap S_{n+1}^c = \cap_{k=1}^{n+1} S_k^c.$$

■

We end this section by highlighting something subtle but important that we have done in this section. Generally speaking, the theory of Analysis and mathematics in general proceeds from several axioms that are supposed to be "self-evident," and that we accept as true without proof. These are the Zermelo-Fraenkel axioms of set theory, discussed briefly on page 11. In this section, we have assumed that an additional axiom holds: The Well-Ordering Principle (Axiom 1.14). This has its advantages and its disadvantages. Without using the Well-Ordering Principle, we cannot prove that Mathematical Induction is a valid form of proof (Theorem 1.15). This is an advantage in accepting the Well-Ordering Principle. On the other hand, it can be shown [23] that the Well-Ordering Principle is equivalent to the Axiom of Choice:

Axiom 1.19 (Axiom of Choice)**.** *If S_α ($\alpha \in I$) is any collection of nonempty sets, then it is possible to choose $s_\alpha \in S_\alpha$ for each $\alpha \in I$.*

Thus, an explicit assumption that the Well-Ordering Principle is true means that we must also accept the Axiom of Choice, as they are logically equivalent. The Banach-Tarski Paradox uses the Axiom of Choice to deconstruct a solid ball into a finite number of pieces, and reassemble them into two solid balls of the same size! This goes against our intuition, but if we accept the Well-Ordering Principle, we must also accept the Axiom of Choice and subsequently accept the Banach-Tarski Paradox. Luckily, we shall not encounter any paradoxes of this sort in this book.

1.3.4 Review of important concepts from Section 1.3

1. Mathematical Induction is a powerful method of proof when asked to prove a collection of statements.
2. Generally, it does not matter where these statements start to be true: you must check that the initial statement is true, and that the inductive step holds.
3. The Well-Ordering Principle is an additional axiom we have assumed, which is a necessity in demonstrating that proof by Mathematical Induction is a valid form of proof.

Exercises for Section 1.3

1. Use Mathematical Induction to prove $2+4+6+\cdots+2n = n(n+1)$.
2. Prove that $1^2 + 2^2 + 3^2 + \cdots + n^2 = \frac{n(n+1)(2n+1)}{6}$.
3. Let S be a set containing exactly n elements. Prove that the set of all subsets of S (the power set $\mathcal{P}(S)$ of S) has exactly 2^n elements.
4. Let $n \geq 1$. For sets $S_1, \ldots, S_n$, prove that $(\cap_{k=1}^n S_k)^c = \cup_{k=1}^n S_k^c$. *[Hint: This proof is very similar to the one in Practice 1.18.]*
5. Prove that if proof by Mathematical Induction is valid, then every subset of $\mathbb{N}$ has a smallest element.
6. Find the error in the following Mathematical Induction "proof" that all dogs have the same color. We induct on the number of dogs. If there is only one dog, then clearly every dog in this group has the same color. Now suppose that for some $n \geq 1$, every group of n dogs has the same color. Consider a set of $n+1$ dogs. Remove one dog, leaving a group of n dogs behind. As a group of n dogs, they all have the same color. Now replace the dog you took and remove a different dog. Again, the remaining group of n dogs must all have the same color. Thus, all $n+1$ dogs have the same color.

1.4 Functions

In this section we formally define the notion of a *function* in Definition 1.20. After going through some examples and some helpful new notation, we discuss injective, surjective, and bijective functions in Definition 1.33, and demonstrate how these properties persist in the context of composition of functions.

1.4.1 Definitions and examples

A fundamental object to mathematicians is a *function.* We formally define this object below.

Definition 1.20. Let X and Y be sets. A *function* f from X to Y is an association of every element in X each to a unique element in Y. In such a situation, we write $f : X \to Y$. The set X is called the *domain* of f and the elements of X are sometimes called *inputs.* The set Y is called the *range* of f, and the elements of Y are sometimes called *outputs.* If there is a formula given for expressing this association, it is called a *rule* for f.

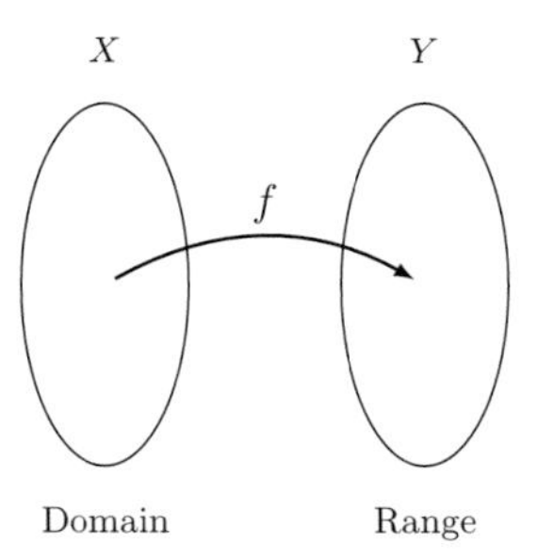

It should be pointed out that the phrase "an association of every element in X each to a unique element in Y" means two things. First, it means that every element in X must be sent to something in Y by f. Second, it means that every $x \in X$ must only be sent to one element in Y by f, as opposed to there being some $x \in X$ that gets sent to several values in Y. It should also be pointed out that it is sometimes cumbersome to always have to list the domain and range of every function we encounter. For that reason, unless the domain of a given function is specifically listed, we assume that the domain of the function is the largest set of real numbers that makes sense to plug into the given rule. This implicit domain is sometimes called the *natural domain* of the function. In this event, the range of the function is understood to be all outputs one gets by using inputs from the natural domain.

Although we expect the reader is well familiar with the notion of a function, we give two brief examples below. The first example is a basic example of a function, and the second is more of a nonexample of a function, illustrating how Definition 1.20 could be violated.

Example 1.21. Let $X = \mathbb{R}$ and $Y = \mathbb{R}$. For each $x \in X = \mathbb{R}$, define $f(x) = x^3$. Thus, Definition 1.20 is satisfied, and this defines a function $f : X \to Y$. Notice here that the natural domain of the function $f(x) = x^3$ is all of $\mathbb{R}$, and so we could have omitted the first sentence of this example. ■

Example 1.22. (We would like to thank Dr. Eric Merchant for suggesting this example.) Let $X = \mathbb{Q}$, and let $Y = \mathbb{Z}$. For $\frac{a}{b} \in \mathbb{Q}$ (where $a, b \in \mathbb{Z}$ and $b \neq 0$), define the association $f(\frac{a}{b}) = a + b$. At first glance, this appears to be a function. But notice that $f(\frac{1}{2}) = 1 + 2 = 3$, whereas $f(\frac{2}{4}) = 2 + 4 = 6$. Since $\frac{1}{2} = \frac{2}{4}$, we have found one input $x = \frac{1}{2} = \frac{2}{4}$ that is assigned more than one output, violating Definition 1.20. ■

1.4.2 Expressing the image and preimage of sets

It will be often convenient to have a notation for a set that is derived by considerations involving a function. The two main examples of this are in the next definition. A visual interpretation of these definitions can be seen in Figure 1.4.1.

Definition 1.23. Let $f : X \to Y$.

1. If $S \subseteq X$, then the *image* of S by f is denoted $f(S)$ and defined as
$$f(S) = \{y \in Y | \text{there is an } x \in X \text{ with } f(x) = y\}.$$

2. If $T \subseteq Y$, then the *preimage* (or *inverse image*) of T under f is denoted $f^{-1}(T)$ and defined as
$$f^{-1}(T) = \{x \in X | f(x) \in T\}.$$

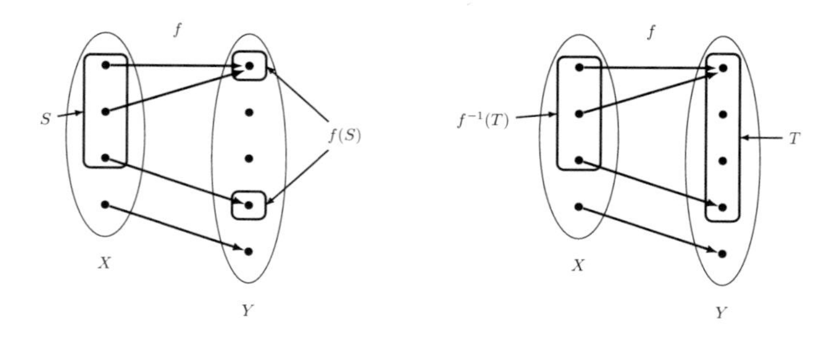

FIGURE 1.4.1
The sets S and $f(S)$ are indicated on the left, and the sets T and $f^{-1}(T)$ are indicated on the right.

Remark 1.24. It should be strongly emphasized that **the inverse image notation $f^{-1}(T)$ has absolutely nothing to do with the notion of the function inverse to f, should it exist.** As we shall see in Exercise 11, there are certain instances where an inverse to a function exists, although this notation does not refer to such a function. An easy way to see that these concepts and notations are different is to notice that $f^{-1}(T)$ refers to the inverse image of the *set* T, while $f^{-1}(y)$ would be what f^{-1} sends the *number* y to.

Nontechnically summarizing Definition 1.23, if S is a collection of inputs, then the image $f(S)$ is the set of all things they get sent to. If T is a collection of possible outputs, then the preimage $f^{-1}(T)$ is the set of all inputs that get mapped into T.

Practice 1.25. Let $f : \mathbb{N} \to \mathbb{N}$ be defined as $f(n) = n^2$.

(a) Let $S = \{1, 2, 6\}$. Determine $f(S)$.

(b) Let $T = \{1, 2, 3, 4\}$. Determine $f^{-1}(T)$.

Methodology. *We use Definition 1.23 to determine these sets. In particular, for the first part, to find $f(S)$ we simply plug each element into f and form a set containing the results. For the second part, for each element in T, we attempt to find an element of S that maps to it. Since it is given that the domain of this function is $\mathbb{N}$, we will only look for natural numbers that map to each element in T. This is an instance for which the natural domain of a function differs from the given domain.*

Solution. (a) Since $f(1) = 1, f(2) = 4$, and $f(6) = 36$, the set $f(S) = \{1, 4, 36\}$.

(b) Although there are two real numbers that map to each of the elements of T, it is an important part of the problem that not all of these numbers are in the given domain of $\mathbb{N}$. Only the natural number 1 maps to 1 ($f(1) = 1$). There are no natural numbers that map to 2 or to 3. Only the natural number 2 maps to 4 ($f(2) = 4$). Therefore, $f^{-1}(T) = \{1, 2\}$. ■

There are a variety of facts concerning how images and preimages behave in the presence of unions, intersections, and complements. The next two propositions illustrate this.

Proposition 1.26. *Let $f : X \to Y$, and let $S_1, S_2, S \subseteq X$. We have*

1. $f(S_1 \cup S_2) = f(S_1) \cup f(S_2)$,
2. $f(S_1 \cap S_2) \subseteq f(S_1) \cap f(S_2)$, *and*
3. $f(X - S) \supseteq f(X) - f(S)$.

Idea of proof. *These facts will follow by correctly interpreting Definition 1.23, and by using a standard inclusion (Assertions 2 and 3) or double-inclusion (Assertion 1) proof. Notice that each of the image sets we consider is a subset of the range Y. It is for this reason that it seems more appropriate to consider an element y in our inclusion proofs, although this is not an essential part of the reasoning.*

Proof of Proposition 1.26. We begin with proving Assertion 1 by a double-inclusion argument. Let $y \in f(S_1 \cup S_2)$. There exists an $x \in S_1 \cup S_2$ with

$f(x) = y$. Thus, $x \in S_1$ or $x \in S_2$. If $x \in S_1$, then $f(x) \in f(S_1)$. Similarly, if $x \in S_2$, then $f(x) \in f(S_2)$. Since $f(x)$ is in either $f(S_1)$ or $f(S_2)$, and $f(x) = y$, it follows that $y \in f(S_1) \cup f(S_2)$.

To prove the other inclusion, we consider an element $y \in f(S_1) \cup f(S_2)$, so that $y \in f(S_1)$ or $y \in f(S_2)$. If $y \in f(S_1)$, then there exists an $x_1 \in S_1$ for which $f(x_1) = y$. If $y \in f(S_2)$, there exists an $x_2 \in S_2$ for which $f(x_2) = y$. In either case, there exists an element $x \in S_1 \cup S_2$ for which $f(x) = y$. Thus, $y \in f(S_1 \cup S_2)$, and this completes the proof of Assertion 1.

To prove Assertion 2, let $y \in f(S_1 \cap S_2)$. There exists an $x \in S_1 \cap S_2$ so that $f(x) = y$. Since x is in both S_1 and S_2, $f(x) = y$ is in both $f(S_1)$ and $f(S_2)$. So, $y \in f(S_1) \cap f(S_2)$.

To prove Assertion 3, let $y \in f(X) - f(S)$. So, $y \in f(X)$ but $y \notin f(S)$. Since $y \in f(X)$, there exists an $x \in X$ for which $f(x) = y$. But if it were the case that $x \in S$ as well, then $f(x) = y \in f(S)$, and we have assumed that this is not the case. So, $x \in X$ but $x \notin S$, so $x \in X - S$. Therefore, $f(x) = y \in f(X - S)$. □

Remark 1.27. There are examples of functions for which the containments in Assertions 2 and 3 of Proposition 1.26 are proper. See Exercises 2 and 3.

Generally speaking, preimages of this sort behave better than images do, compare Assertions 2 and 3 of Proposition 1.26 above with Assertions 2 and 3 of Proposition 1.28 below.

Proposition 1.28. *Let $f : X \to Y$, and let $T_1, T_2, T \subseteq Y$. We have*

1. $f^{-1}(T_1 \cup T_2) = f^{-1}(T_1) \cup f^{-1}(T_2)$,
2. $f^{-1}(T_1 \cap T_2) = f^{-1}(T_1) \cap f^{-1}(T_2)$, *and*
3. $f^{-1}(Y - T) = X - f^{-1}(T)$.

Idea of proof. *Much like the proof to Proposition 1.26, we correctly interpret Definition 1.23 and prove each of these assertions with a double-inclusion argument. Notice that each of the sets below is a subset of the domain X.*

Proof of Proposition 1.28. We prove each of these with a double-inclusion argument. For Assertion 1, let $x \in f^{-1}(T_1 \cup T_2)$. Then $f(x) \in T_1 \cup T_2$, and so $f(x) \in T_1$ or $f(x) \in T_2$. If $f(x) \in T_1$ then $x \in f^{-1}(T_1)$ If $f(x) \in T_2$, then $x \in f^{-1}(T_2)$. Since x is in $f^{-1}(T_1)$ or $f^{-1}(T_2)$, we have $x \in f^{-1}(T_1) \cup f^{-1}(T_2)$.

To prove the other inclusion, let $x \in f^{-1}(T_1) \cup f^{-1}(T_2)$. Then $x \in f^{-1}(T_1)$ or $x \in f^{-1}(T_2)$. Thus, $f(x) \in T_1$ or $f(x) \in T_2$, so that $f(x) \in T_1 \cup T_2$, and so $x \in f^{-1}(T_1 \cup T_2)$.

The proof of Assertion 2 is left as Exercise 5.

To prove Assertion 3, let $x \in f^{-1}(Y - T)$. Then $f(x) \in Y - T$, so that $f(x) \in Y$ but $f(x) \notin T$. Since $f(x) \notin T, x \notin f^{-1}(T)$ (if it were, then $f(x)$ *would* be in T). Our choice of x assumes that $x \in X$, and so since $x \notin f^{-1}(T)$, $x \in X - f^{-1}(T)$.

To prove the reverse inclusion, let $x \in X - f^{-1}(T)$. Then, $x \notin f^{-1}(T)$, and so $f(x) \notin T$ (if $f(x)$ were in T, then x *would* be in $f^{-1}(T)$). Since $f(x) \notin T$ and by definition of f, $f(x) \in Y$, we have $f(x) \in Y - T$, so that $x \in f^{-1}(Y - T)$. □

1.4.3 Operations on functions

There are five different familiar ways that one may combine old functions to make new ones. Four of them relate to the four arithmetic operations of addition, subtraction, multiplication, and division. The final method, composition, is unique to functions.

Definition 1.29. Let $f, g : X \to \mathbb{R}$. We define the following functions:

$$\begin{aligned} (f+g)(x) &= f(x) + g(x) \\ (f-g)(x) &= f(x) - g(x) \\ (fg)(x) &= f(x)g(x) \\ \left(\frac{f}{g}\right)(x) &= \frac{f(x)}{g(x)}, \text{ for } g(x) \neq 0. \end{aligned}$$

If $f : X \to Y$, and $g : Y \to Z$, we define the composition $g \circ f$ as

$$(g \circ f)(x) = g(f(x)).$$

These five constructions construct a new rule for the function, but largely leave the domain and range of the functions involved alone. There are two other common constructions of functions that leave the rule of the function alone, and only alter the domain (and sometimes, the range).

Definition 1.30. Let $f : X \to Y$, and let $S \subset X \subset \tilde{X}$. The *restriction of f to* S, denoted $f|_S$, is defined as the function $f|_S : S \to Y$, where $f|_S(s) = f(s)$. An *extension of f to* $\tilde{X}$, denoted $\tilde{f}$, is a function $\tilde{f} : \tilde{X} \to \tilde{Y}$ for which $\tilde{f}|_X = f$. In the case of extensions, the range $\tilde{Y} \subseteq Y$ may need to be extended as well.

Remark 1.31. The functions $f|_S$ and $\tilde{f}$ are sometimes simply written as f when confusion would not result. We do this in Practice 1.34 below. One may also consider a function with a restricted range to be another sort of restriction.

Extensions and restrictions of functions are important in many circumstances. We give a brief example below.

Example 1.32. Define the function $f : (0, \infty) \to \mathbb{R}$ by $f(x) = \frac{1}{1+x^2}$. A restriction or extension of this function would simply alter the domain to be smaller (for restrictions), or larger (for extensions). For example, if $S = [1, \infty)$, the function $f|_S$ is the same as the function f, although we are forcing the domain of $f|_S$ to be the smaller set S. So, if $s \in [1, \infty)$, then $f|_S(s) = \frac{1}{1+s^2}$. Remember this is technically a different function because we have altered the domain, even if the rule of the function is the same.

Similarly, we can extend this function by making the domain larger.

Whether or not we use the same rule for the function would have to be made explicit. So, let us define an extension of f as follows. Let $\tilde{X} = \mathbb{R}$, and define $\tilde{f}(x) = \frac{1}{1+x^2}$. This function has the same rule as f does, and with a larger domain $\tilde{X}$. The range of f is $\mathbb{R}$, and so there was no need to adjust it to account for what might be more outputs. There was nothing forcing us to use the same rule, we could also have defined $\tilde{f}$ by the piecewise defined function below:

$$\tilde{f}(x) = \begin{cases} \frac{1}{1+x^2} & \text{for } x \in (0, \infty) \\ 1 & \text{for } x \in (-\infty, 0] \end{cases}.$$

Both are considered to be extensions of the same function f. ■

1.4.4 Types of functions

Although there are many different types of functions within all of mathematics, we focus on three very important types: injective (or one to one), surjective (or onto), and bijective functions.

Definition 1.33. Let $f : X \to Y$.

1. The function f is *injective* (or *one-to-one*) if $f(x_1) = f(x_2)$, then $x_1 = x_2$.

2. The function f is *surjective* (or *onto*) if for every $y \in Y$ there exists an $x \in X$ with $f(x) = y$.

3. The function f is *bijective* if it is both one to one and onto.

Here are a couple practice questions illustrating these properties.

Practice 1.34. Let $f : \mathbb{R} \to \mathbb{R}$ be defined as $f(x) = x^2$.

(a) Is $f(x)$ injective?

(b) Is $f(x)$ surjective?

(c) Find a restriction of the domain of f so that the resulting function is injective.

(d) Find a restriction of the range of f so that the resulting function is surjective.

Methodology. *For Part (a), since there are two different inputs that map to the same output, the function is not injective according to Definition 1.33–we will need to exhibit such inputs. Since there are outputs that are not targeted by any input, this function is not surjective according to Definition 1.33–we will have to exhibit such an output and demonstrate that no input is mapped to it by f. There are lots of answers to Part (c) since there are many different restrictions we could choose, although for this particular function, negative numbers are mapped to the same place as their positive counterpart, and so removing the negative numbers from the domain should work. For Part (d),* *we observe that any non-negative number has a square root in $\mathbb{R}$, whereas no negative numbers do. So, no negative numbers are targeted by this function and removing them from the range should produce the desired effect.*

Solution.

(a) This function is not injective, since $f(-2) = f(2) = 4$, but $2 \neq -2$.

(b) This function is not surjective. If it were, then there would exist a real number x for which $f(x) = x^2 = -1 \in \mathbb{R}$ (-1 is in the given range of f). Since the square of any real number is always nonnegative, and -1 is negative, there is no such x (this fact is proven in Part 1 of Proposition 2.6).

(c) Restrict the domain of f to $S = [0, \infty)$. We claim the restricted function $f|_S : [0, \infty) \to \mathbb{R}$ is injective. For convenience, write this function $f|_S$ as f, as in Remark 1.31 following Definition 1.30. To prove that f (with this restricted domain) is injective, suppose that $x_1, x_2 \in [0, \infty)$ satisfy $f(x_1) = f(x_2)$, so that $x_1^2 = x_2^2$. Then

$$\begin{aligned} x_1^2 &= x_2^2, \\ x_1^2 - x_2^2 &= 0, \\ (x_1 - x_2)(x_1 + x_2) &= 0. \end{aligned}$$

So either $x_1 = x_2$, or $x_1 = -x_2$. If $x_1 = x_2$, then we are done as this is what we are aiming to show. Now suppose instead that $x_1 = -x_2$. If one of these numbers is nonzero, say $x_2 \neq 0$, then since $x_2 \in [0, \infty)$, it must be the case that $x_2 > 0$. Since $x_1 = -x_2$ it follows that $x_1 < 0$. This cannot be the case because we assumed that $x_1 \in [0, \infty)$.

(d) Clearly it is not the case that $f(x) = x^2 < 0$ for any real x (this was discussed above). Moreover, for every $y \geq 0$, $\sqrt{y} = x$ is a number for which $f(x) = f(\sqrt{y}) = (\sqrt{y})^2 = y$. So if we restrict the domain of f to $[0, \infty)$, every element in this new range is targeted by a real number by f, so the resulting function is surjective. ■

The properties of being injective, surjective, and bijective are preserved by compositions (Theorem 1.35 immediately below). Conversely, these properties may only be partially preserved (Theorem 1.36 further below).

Theorem 1.35. *Let $f : X \to Y$, and let $g : Y \to Z$.*

1. *If f and g are injective, then $g \circ f$ is injective.*
2. *If f and g are surjective, then $g \circ f$ is surjective.*
3. *If f and g are bijective, then $g \circ f$ is bijective.*

Idea of proof. *Assertions 1 and 2 can be proven directly using Definition 1.33. For Assertion 1, we suppose that $(g \circ f)(x_1) = (g \circ f)(x_2)$ and try to* *show that $x_1 = x_2$. First we use the assumption that g is injective, and then we use the assumption that f is injective to conclude this. The same idea is employed in Assertion 2, although in the context of surjectivity, and is left to the reader as Exercise 8. Assertion 3 follows from Assertions 1 and 2.*

Proof of Theorem 1.35. To prove Assertion 1, as assume that $(g \circ f)(x_1) = (g \circ f)(x_2)$ and try to prove that $x_1 = x_2$. Since g is injective and $(g \circ f)(x_1) = g(f(x_1)) = g(f(x_2)) = (g \circ f)(x_2)$, it follows that $f(x_1) = f(x_2)$. Since f is injective, it follows that $x_1 = x_2$.

The proof of Assertion 2 is left as Exercise 8, and Assertion 3 follows directly from Assertions 1 and 2. □

Theorem 1.36. *Let $f : X \to Y$, and let $g : Y \to Z$.*

1. *If $g \circ f$ is injective, then f is injective.*
2. *If $g \circ f$ is surjective, then g is surjective.*

Idea of proof. *These are best proven by contradiction. We leave Assertion 1 as Exercise 9. For Assertion 2, we suppose that g is not surjective, and try to show that $g \circ f$ is not surjective. If g is not surjective, then there is some element in Z that is not targeted by g from anything in Y. Namely, this element is not targeted by $g \circ f$ from anything in X.*

Proof of Theorem 1.36. Assertion 1 is left as Exercise 9. For Assertion 2, we suppose that g is not surjective. Then there is some element $z \in Z$ for which $g(y) \neq z$ for any $y \in Y$. But by assumption, $g \circ f$ is surjective, so there exists an $x \in X$ for which $(g \circ f)(x) = g(f(x)) = z$. But this is a contradiction, since now $y = f(x)$ is an element of Y that g maps to z. □

1.4.5 Review of important concepts from Section 1.4

1. A function is a fundamental object to mathematicians and is defined formally in Definition 1.20.
2. The image $f(S)$ and the preimage $f^{-1}(T)$ are convenient ways of expressing where elements are sent (image) or where they come from (preimage) by a function f.
3. There are a variety of ways in which unions, intersections, and complements of sets interact with images and preimages of functions. See Propositions 1.26 and 1.28.
4. In addition to the standard arithmetic combinations of functions $f \pm g$, fg, and $\frac{f}{g}$, there are several other useful constructions of functions: the composition of functions, and a restriction or an extension of a function.

5. Composition of functions preserves the property of being injective, surjective, or bijective (these properties are defined in Definition 1.33). The reverse is not necessarily the case, compare Theorems 1.35 and 1.36.

Exercises for Section 1.4

1. Let $f : \mathbb{R} \to \mathbb{R}$ be defined as $f(x) = x^4$.
   (a) Compute $f(S)$, where $S = [-1, 2]$.
   (b) Compute $f^{-1}(T)$, where $T = [0, 16]$.
   (c) Compute $f^{-1}(T)$, where $(-\infty, 0)$.
   (d) If we restrict the domain and range so that $f : [0, \infty) \to [0, \infty)$, is the resulting function a bijection?
2. (a) Prove that if f is injective, then $f(S_1 \cap S_2) = f(S_1) \cap f(S_2)$. *[Hint: Note that one containment has already been proven in Proposition 1.26.]*
   (b) Find an example of a function $f : X \to Y$, and subsets $S_1, S_2 \subset X$ for which $f(S_1 \cap S_2) \neq f(S_1) \cap f(S_2)$. *[Hint: According to Proposition 1.26, $f(S_1 \cap S_2)$ is always a subset of $f(S_1) \cap f(S_2)$. According to the previous part of this problem, the other containment holds when f is injective. Thus, such an example would have to violate the latter containment and must not be one-to-one.]*
3. (a) Prove that if f is injective, then $f(X - S) = f(X) - f(S)$. *[Hint: Note that one containment has already been proven in Proposition 1.26.]*
   (b) Find an example of a function $f : X \to Y$, and a subset $S \subseteq X$ for which $f(X - S) \neq f(X) - f(S)$. *[Hint: According to Proposition 1.26, $f(X - S)$ always contains $f(X) - f(S)$. According to the previous part of this problem, the other containment holds when f is injective. Thus, such an example would have to violate the latter containment and must not be one-to-one.]*
4. Let $f, g : D \to \mathbb{R}$. Define

$$\begin{aligned} \max\{f, g\}(x) &= \max\{f(x), g(x)\}, \text{ and} \\ \min\{f, g\}(x) &= \min\{f(x), g(x)\}. \end{aligned}$$

   Prove that

$$\begin{aligned} \max\{f, g\}(x) &= \tfrac{1}{2}[(f + g) + |f - g|], \text{ and} \\ \min\{f, g\}(x) &= \tfrac{1}{2}[(f + g) - |f - g|]. \end{aligned}$$

5. Prove Assertion 2 of Proposition 1.28. That is, let $f : X \to Y$, and let $T_1, T_2 \subseteq Y$. Prove that $f^{-1}(T_1 \cap T_2) = f^{-1}(T_1) \cap f^{-1}(T_2)$.

6. Let $f : X \to Y$, and suppose $S \subseteq X$. Prove that $S \subseteq f^{-1}(f(S))$, and if f is injective, then these sets are equal.

7. Let $f : X \to Y$, and suppose $T \subseteq Y$. Prove that $f(f^{-1}(T)) \subseteq T$, and if f is surjective, then these sets are equal.

8. Prove Assertion 2 of Theorem 1.35. That is, suppose that $f : X \to Y$ and $g : Y \to Z$. Prove that if f and g are surjective, then $g \circ f$ is surjective.

9. Prove Assertion 1 of Theorem 1.36. That is, suppose that $f : X \to Y$ and $g : Y \to Z$. Prove that if $g \circ f$ is injective, then f is injective.

10. (a) Find a bijection from $(0, 1)$ to $(1, \infty)$.
    (b) Find a bijection from $(1, \infty)$ to $(0, \infty)$.
    (c) Find a bijection from $(0, 1)$ to $(0, \infty)$. *[Hint: You may find the first two parts helpful if you consider Assertion 3 of Theorem 1.35.]*

11. Suppose $f : X \to Y$. An *inverse* to f is a function $g : Y \to X$ satisfying the relations:

$$f(g(y)) = y \text{ for all } y \in Y, \text{ and } g(f(x)) = x \text{ for all } x \in X.$$

In the event that there exists such a function, we usually denote $f^{-1} = g$. Prove that an inverse to f exists if and only if f is a bijection.

1.5 Cardinality

This section introduces the notion of *cardinality*: the measure of how many elements a set has. For finite sets, this turns out to be nothing more than counting its elements. The situation is more detailed when it comes to infinite sets.

First, we will give a formal definition of *set equivalence* and define finite and infinite sets. After proving the intuitive fact that $\mathbb{N}$ is infinite, we go on to present some important facts about finite and infinite sets in terms of set equivalence and containment, and we give a characterization of infinite sets. Next, we study infinite sets more closely and define countable, countably infinite, and uncountable sets. After some more preliminary material, we prove that the countable union of countable sets is countable (Theorem 1.56). We close the section by proving that $[0, 1]$ is uncountable (Theorem 1.58) using Cantor's famous diagonalization argument. The organization of the material in this section is similar to Dangello and Seyfried [8].

1.5.1 Set equivalence and cardinality

When presented with two sets, it is natural to ask "which set is bigger?" Or rather, which set has "more elements?" In the case of what we think of as finite sets, one could simply determine the number of elements each set has, and do a comparison that way. But in comparing infinite sets, this method no longer accurately determines which set has "more elements." In fact, we shall see in Theorem 1.58 that there are different "sizes" of infinite sets, and indeed, that more is true (see Exercise 6).

The solution, then, is to compare two sets by way of a bijection: If there is a bijection between the sets, then there is a one-to-one pairing of each of their elements, and through this pairing one could have some confidence in saying that the sets had the "same number of elements." When a bijection exists between sets, we shall call them *equivalent*.

Definition 1.37. Let S and T be sets. We say that S and T are *equivalent*, or that S and T have the same *cardinality* if there exists a bijection $f : S \to T$. In this case, we write $S \sim T$.

Here are two brief but illuminating examples that we will refer to later in this section.

Example 1.38. These examples illustrate the notion of cardinality.

(a) Let $S_1 = \{\clubsuit, \blacklozenge, \heartsuit, \spadesuit\}$ and let $T_1 = \{1, 2, 3, 4\}$. Define a function $f : S_1 \to T_1$ by $f(\clubsuit) = 1, f(\blacklozenge) = 2, f(\heartsuit) = 3$, and $f(\spadesuit) = 4$. This is clearly a bijection (although it may be a good idea to convince yourself that this function is both one-to-one and onto), and illustrates that $S_1 \sim T_1$.

(b) Let $S_2 = 2\mathbb{N} = \{2n | n \in \mathbb{N}\}$, and let $T_2 = \mathbb{N}$. Let $f : S_2 \to T_2$ be given by $f(n) = \frac{1}{2}n$. This is clearly a bijection (again, it may be a good idea to check this), and so this demonstrates that $2\mathbb{N} \sim \mathbb{N}$. This is somewhat disturbing, since $2\mathbb{N}$ is a proper subset of $\mathbb{N}$, and yet this example illustrates that they have the same cardinality. ■

There are three important properties of set equivalence that we shall frequently use. Namely, the symmetric property (Assertion 2) allows us to choose a bijection from $S \to T$ or from $T \to S$ in the event that $S \sim T$.

Proposition 1.39. *Set equivalence is an equivalence relation. That is:*

1. *It is reflexive:* $S \sim S$.
2. *It is symmetric: If* $S \sim T$*, then* $T \sim S$.
3. *It is transitive: If* $S \sim T$ *and* $T \sim X$*, then* $S \sim X$.

Idea of proof. *We are proving statements about set equivalence, so by Definition 1.37, we will need to produce a bijection in each circumstance. For the first part, the identity function* $f(s) = s$ *will do. For the second part, since* $S \sim T$ *there is a bijection from* S *to* T*, and so its inverse will be a bijection from* T *to* S*. For the third part, since* $S \sim T$ *and* $T \sim X$*, there are bijections* f *and* g *from* $S \to T$ *and from* $T \to X$*, respectively. So their composition* $g \circ f$ *is a bijection from* $S \to X$.

Proof of Proposition 1.39. To prove the first assertion, define $f : S \to S$ as $f(s) = s$. This is clearly a bijection, and shows that $S \sim S$.

In proving Assertion 2, suppose $S \sim T$. Then there exists $f : S \to T$ that is a bijection. According to Exercise 11 of Section 1.4, the inverse of f exists, $f^{-1} : T \to S$, and is a bijection (since its inverse, f, exists). So, $T \sim S$.

Finally, to prove Assertion 3, suppose that $S \sim T$ and $T \sim X$. Then there exist bijections $f : S \to T$, and $g : T \to X$. According to Assertion 3 of Theorem 1.35 in Section 1.4, the composition $g \circ f : S \to X$ is a bijection. So, $S \sim X$. □

The remainder of this section is devoted to two objectives. First, we endeavor to classify certain sets in terms of their cardinality. In this way, we will distinguish between finite and infinite sets (Subsection 1.5.2), and then distinguish between certain types of infinite sets (Subsection 1.5.3). The second objective is to give an example illustrating that it is indeed the case that it is possible to have two infinite sets that have different cardinalities (Theorem 1.58).

1.5.2 Finite and infinite sets

We begin with a very intuitive definition of what it means for a set to be either *finite* or *infinite*.

Definition 1.40. Let S be a set.

1. S is *finite* if $S = \emptyset$, or if $S \sim \{1, 2, \ldots, n\}$ for some $n \in \mathbb{N}$. The positive integer $|S|$ is defined to be 0 if $S = \emptyset$, and n if $S \sim \{1, 2, \ldots, n\}$. The number $|S|$ is therefore defined for finite sets, and is called the *cardinality* or *cardinal number* of S.

2. S is *infinite* if S is not finite. In the event S is infinite, we say that the cardinality of S is infinite.

Example 1.38 illustrates the notions in Definition 1.40. In Part (a) of this example, we see that if $S_1 = \{\clubsuit, \blacklozenge, \heartsuit, \spadesuit\}$, then $S_1 \sim \{1, 2, 3, 4\}$. Therefore by Definition 1.40, S_1 is finite, and the cardinality of S_1 is 4. Part (b) of Example 1.38 considers the set $S_2 = 2\mathbb{N} = \{2n | n \in \mathbb{N}\}$, and shows that $2\mathbb{N} \sim \mathbb{N}$. It is tempting to say that as a result $2\mathbb{N}$ is infinite: not only is this intuitive, but it should seem that $\mathbb{N}$ is infinite, and that anything equivalent to $\mathbb{N}$ should also be infinite. Neither of these statements has been proven yet, and we do so now to make this line of reasoning rigorous. We begin with Lemma 1.41, which will be of help to us both now and later.

Lemma 1.41. *If S is a nonempty finite subset of $\mathbb{R}$, then S has a largest and smallest element.*

Idea of proof. *We will induct on the cardinality of S. Clearly if $|S| = 1$, then the only element of S is both the largest and smallest element. To prove the inductive step, we will consider a set where $|S| = n$, and remove an element from it, creating a set with one fewer element. By assumption, this smaller set has a largest and smallest element, and so when we put the previously removed element back we can compare it to the largest and smallest element from the smaller set.*

Proof of Lemma 1.41. We induct on the cardinality of S. Since $S \neq \emptyset$, we first consider the possibility that $|S| = 1$, so that $S = \{x\}$. Clearly x is the largest and smallest element of S. Now suppose that $|S| = n$ for some $n \geq 2$. Then $S \sim \{1, 2, \ldots, n\}$. Let $f : S \to \{1, 2, \ldots, n\}$ be a bijection, which is assumed to exist since $S \sim \{1, 2, \ldots, n\}$. Label the elements in S as $S = \{s_1, s_2, \ldots, s_n\}$ where $f(s_k) = k$ for $k = 1, 2, \ldots, n$. The restricted function

$$f|_{S - \{s_n\}} : S - \{s_n\} \to \{1, 2, \ldots, n-1\}$$

is a bijection illustrating $|S - \{s_n\}| = n - 1$. So, by our induction hypothesis, there exists a largest element $s_K \in S - \{s_n\}$ and a smallest element $s_k \in S - \{s_n\}$. If $s_n \geq s_K$, then the largest element of S is s_n, otherwise the largest element of S remains a_K. Similarly, if $s_n \leq s_k$, then the smallest element of S is s_n, otherwise the smallest element of S remains a_k. $\square$

We can now rigorously prove that $\mathbb{N}$ is infinite, which agrees with our intuition.

Theorem 1.42. *$\mathbb{N}$ is infinite.*

Idea of proof. *If $\mathbb{N}$ were finite, then there would be a largest element x of $\mathbb{N}$ according to Lemma 1.41. Adding one to x would produce another natural number $x+1$ which is bigger, contradicting the assumption that x was the largest natural number.*

Proof of Theorem 1.42. Suppose by way of contradiction that $\mathbb{N}$ were finite. By Lemma 1.41, there exists a largest element $x \in \mathbb{N}$. But $x+1 > x$, and $x+1 \in \mathbb{N}$. This contradicts the fact that x is the largest element of $\mathbb{N}$. □

The following proposition demonstrates that finite and infinite sets can be distinguished according to their cardinality.

Proposition 1.43. *Let S and T be sets.*

1. *If S is finite and $S \sim T$, then T is finite. In addition, $|S| = |T|$.*
2. *If T is infinite and $S \sim T$, then S is infinite.*

Idea of proof. *To prove the first assertion, we have to show that $T \sim \{1, 2, \ldots, n\}$ for some n. We use the symmetry and transitivity of the equivalence relation $\sim$: if $S \sim \{1, 2, \ldots, n\}$ for some n, and $S \sim T$, then $T \sim S \sim \{1, 2, \ldots, n\}$, so $T \sim \{1, 2, \ldots, n\}$ by transitivity. The second assertion is the contrapositive of the first.*

Proof of Proposition 1.43. We prove the first assertion and note that the second is the contrapositive of the first statement in the first assertion. Suppose S is finite. By Definition 1.40, there exists a bijection $f : S \to \{1, 2, \ldots, n\}$. Since $S \sim T$, we can conclude that $T \sim S$ since $\sim$ is symmetric (see Assertion 2 of Proposition 1.39). So, there exists a bijection $g : T \to S$. The composition $f \circ g$ is a bijection from T to $\{1, 2, \ldots, n\}$, showing $T \sim \{1, 2, \ldots, n\}$. Consequently, T is finite, and $|T| = n$. □

We can now complete a rigorous proof that $2\mathbb{N}$ is infinite as described in the paragraph above Lemma 1.41: By Theorem 1.42, $\mathbb{N}$ is infinite. Example 1.38 shows that $2\mathbb{N} \sim \mathbb{N}$. Using $T = \mathbb{N}$ and $S = 2\mathbb{N}$ in the second assertion of Proposition 1.43 shows that $S = 2\mathbb{N}$ is infinite.

There are two more intuitive facts concerning finite and infinite sets that are important. The first assertion of Proposition 1.44 proves that any subset of a finite set is still finite, while the second assertion shows that any superset of an infinite set is infinite.

Proposition 1.44. *Let S and T be sets.*

1. *If S is finite and $T \subseteq S$, then T is finite. In addition, $|T| \leq |S|$.*
2. *If T is infinite and $T \subseteq S$, then S is infinite.*

Idea of proof. *We prove the first assertion by inducting on the cardinality of S, and note that the second assertion is the contrapositive of the first. Clearly if $|S| = 0$, then $S = \emptyset$, and so $T = \emptyset$, which is finite, and $|T| = 0 \leq 0 = |S|$. Otherwise, if $|S| = n$ and $T \subseteq S$, then there are two cases, $T = S$ (in which case we are done), or $T \subsetneq S$. In the latter case, there is an element we can remove from S that makes T still a subset of this smaller set. The induction hypothesis will apply and make the rest of the argument straightforward.*

Proof of Proposition 1.44. We prove the first assertion and note that the second is the contrapositive of the first statement in the first assertion. We induct on the cardinality of S. If $|S| = 0$, then $S = \emptyset$, and since $T \subseteq S$, we must have $T = \emptyset$ as well. So, T is finite by definition, and $|T| = 0 \leq 0 = |S|$.

Now suppose that $|S| = n$ for some $n \geq 1$. Inductively assume that the result holds for every set that has a cardinality of $n - 1$ (or less). Suppose $T \subseteq S$. If $T = S$, then T is finite (any bijection from S to $\{1, 2, \ldots, n\}$ is also a bijection from T to $\{1, 2, \ldots, n\}$ if $T = S$), and $|T| \leq |S|$ since in this case $|T| = |S|$. On the other hand, if $T \subsetneq S$, then there is some element $x \in S - T$, and so $T \subseteq S - \{x\}$. Using the same argument as in the proof of Lemma 1.41, $|S - \{x\}| = n - 1$. Our induction hypothesis applies to $S - \{x\}$, and so we conclude that since $T \subseteq S - \{x\}$, that T is finite, and that $|T| \leq |S - \{x\}| = n - 1 \leq n = |S|$. □

There are two characterizations of infinite sets that will sometimes be useful that we present in Theorem 1.46. To prepare for this, we remind the reader of the intuitive *Pigeonhole Principle* in the next lemma.

Lemma 1.45 (Pigeonhole Principle). *If $n, m \in \mathbb{N}$, and each of n objects are to go into one of m holes with $n > m$, then at least one hole must have more than one object.*

Idea of proof. *This proof by contradiction could perhaps best be described with the use of an example–this example is of course not a proof of the general situation, this proof is given below. Suppose there were $n = 6$ objects to be placed into $m = 4$ holes, and we were actually able to do this by placing no more than one object into each hole. Then we can count all 6 objects by noticing that since there are only 4 holes, and by assumption there is no more than one object in each hole, there must be no more than 4 objects. This is a contradiction to our assumption that we could indeed distribute the objects in this way.*

Proof of Lemma 1.45. Suppose that these n objects could be placed into the m holes so that no more than one object was in each hole. Then, the total number of n objects can be recovered by counting the number of objects in each hole. Since there are m holes, and each hole contains no more than one object, there are at most m objects. But $m < n$, a contradiction. □

We can now prove the following result concerning infinite sets.

Theorem 1.46. *The following statements are equivalent for the set S:*

1. S is infinite.

2. For each $i \in \mathbb{N}$, there exists an element $s_j \in S$ so that $s_i \neq s_j$ for $i \neq j$.

3. There exists an injection from S onto a proper subset of S.

Idea of proof. *We prove that each statement implies any of the others by proving three assertions: $1 \Rightarrow 2 \Rightarrow 3 \Rightarrow 1$. To prove $1 \Rightarrow 2$, we build the elements s_i inductively: choose s_1 to be any element in S, and find the next element $s_2 \in S - \{s_1\}$. One can continue in this way since an infinite set without a finite number of elements is always nonempty. To prove $2 \Rightarrow 3$, we consider a function f that sends every element of S to itself, except that it advances the numbers $s_i \mapsto s_{i+1}$. This function is injective, but does not have the number s_1 in its range. Finally, we prove $3 \Rightarrow 1$ by contradiction: the Pigeonhole Principle disallows an injection from a finite set onto a proper subset of itself.*

Proof of Theorem 1.46. We prove these three statements are equivalent by demonstrating $1 \Rightarrow 2 \Rightarrow 3 \Rightarrow 1$. We begin by proving $1 \Rightarrow 2$. Suppose S is infinite. Since $\emptyset$ is finite, and S is not finite, $S \neq \emptyset$. Choose s_1 to be any element of S. If the set $S - \{s_1\}$ were empty, then $S = \{s_1\}$. In this event, S would be finite, whereas by assumption S is infinite. So we conclude $S - \{s_1\}$ is nonempty, and that we may choose $s_2 \in S - \{s_1\}$. Notice that by construction $s_2 \neq s_1$. We have just proven the base case (and the $n = 2$ case) of the following statement to be proven by Mathematical Induction: for every $n \in \mathbb{N}$, there exists $s_1, \ldots, s_n \in S$ with $s_i \neq s_j$ if $i \neq j$. To prove the inductive step, assume that such a collection $s_1, \ldots, s_n \in S$ has been chosen in this way. We aim to select $s_{n+1} \in S$ with $s_{n+1} \neq s_i$ for every $i \neq n+1$, and we shall do so similarly to how we selected s_2. The set $S - \{s_1, \ldots, s_n\}$ must not be empty, since if it were, then $S = \{s_1, \ldots, s_n\}$, which is clearly finite (the function $k \mapsto s_k$ is a bijection from $\{1, \ldots, n\}$ to $\{s_1, \ldots, s_n\}$) and contradicts our assumption. We then select s_{n+1} to be any element of $S - \{s_1, \ldots, s_n\}$, and by construction, $s_{n+1} \neq s_i$ for $i \neq n+1$. The existence of the elements $s_1, s_2, \ldots \in S$ with $i \neq j$ has now been shown.

To prove $2 \Rightarrow 3$, we begin by finding elements $s_1, s_2, \ldots \in S$, with $s_i \neq s_j$ for $i \neq j$. Define the function $f : S \to S$ as follows:

$$f(s) = \begin{cases} s & \text{if } s \neq s_i \text{ for any } i \in \mathbb{N}, \\ s_{i+1} & \text{if } s = s_i. \end{cases}$$

This function is injective. If $x, y \in S$ and $x \neq y$, then either exactly one, both, or neither of x and y are equal to one of the s_i for some i. If $x = s_i$ and $y \neq s_j$ for every j, then $f(x) = s_{i+1}$ and $f(y) = y$, and $y \neq s_j$ for every j implies

$y \neq s_{i+1}$. If $x = s_i$ and $y = s_j$, then $f(x) = s_{i+1} \neq s_{j+1} = f(y)$ since $i \neq j$ implies $i + 1 \neq j + 1$. If neither x nor y is equal to s_i for any i, then $f(x) = x$ and $f(y) = y$, and $x \neq y$ by assumption. Finally, there is no element in S that f maps to s_1: every element either maps to itself or to s_i for $i \geq 2$. So, f is not surjective.

To prove $3 \Rightarrow 1$, suppose there were an injective function f from S onto a proper subset of S, but that S were finite. If $|S| = n$, and the proper subset of S that f maps onto has cardinality $m < n$, then the injection f may be viewed as placing n objects into m holes, where no hole contains more than one object. The Pigeonhole Principle (Lemma 1.45) forbids this. □

Remark 1.47. The statements in Theorem 1.46 can be difficult to remember at an intuitive level. To help, we suggest thinking of Statement 2 as "there exists a sequence of distinct members of S," and Statement 3 as "S is equivalent to a proper subset of itself." Part (b) of Example 1.38 illustrates a concrete example of each of these statements, and is worth reviewing in the context of Theorem 1.46.

1.5.3 Countable and uncountable sets

In this subsection, we distinguish between certain types of infinite sets, and give an example illustrating that indeed there exist two infinite sets having a different cardinality.

Definition 1.48. Let S be a set.

1. S is *countably infinite* if $S \sim \mathbb{N}$.
2. S is *countable* if it is finite or countably infinite.
3. S is *uncountable* if it is not countable.

The terminology *countable* comes from the following notion: if S is countable, by definition, it is either finite or equivalent to $\mathbb{N}$. If S is finite, then you can manually count its elements. If $S \sim \mathbb{N}$, then any bijection $f : \mathbb{N} \to S$ allows you to still count (or enumerate) the elements of S as $f(1), f(2), f(3), \ldots$, since f is a bijection, every element of S will be counted (f is a surjection) and no element of S will be counted twice (f is an injection). Thus, the term *countable* is an appropriate one for such sets. At this point, it seems that one could simply enumerate the elements of *any* set T (in particular, an infinite set T) in that way, but unless there is a bijection $f : \mathbb{N} \to T$, we are not assured that every element of T would be counted (or, that elements of T would be counted twice, although this turns out to not be the problem). So, without any way of enumerating the elements of such a set, it makes sense to call it *uncountable*. We shall show in Theorem 1.58 that the set $[0, 1]$ is uncountable after a quick example and some preliminary material concerning countable and uncountable sets.

Example 1.49. Here are some examples concerning countable sets.

1. Any finite set is countable, according to Definition 1.48. In addition, $\emptyset$ is countable since it is finite.
2. Any set equivalent to $\mathbb{N}$ is countably infinite, according to Definition 1.48. In particular, $\mathbb{N}$ itself is countable since $\mathbb{N} \sim \mathbb{N}$. For instance, from Part (b) of Example 1.38, $2\mathbb{N} \sim \mathbb{N}$, and so $2\mathbb{N}$ is countably infinite.

Propositions 1.50 and 1.51 should appear very similar to Propositions 1.43 and 1.44, in that they relate the notion of equivalence and set containment to the notion of countability.

Proposition 1.50. *Let S and T be sets.*

1. *If S is countable, and $S \sim T$, then T is countable.*
2. *If T is uncountable, and $T \sim S$, then S is uncountable.*

Idea of proof. *To prove the first assertion, we notice that if S is countable, then S is either finite or is equivalent to $\mathbb{N}$. In either situation, we can use previous results to prove that T is either finite or equivalent to $\mathbb{N}$–that is, T is countable. The second assertion is the contrapositive of the first.*

Proof of Proposition 1.50. Suppose S is countable. Then S is finite, or $S \sim \mathbb{N}$. If S is finite, and $S \sim T$, then according to Proposition 1.43 (Assertion 1), T is finite. Otherwise, suppose $S \sim \mathbb{N}$, and $S \sim T$. By the transitivity of $\sim$ (Assertion 3 of Proposition 1.39), it follows that $T \sim \mathbb{N}$ as well, and so Assertion 1 is proven. Assertion 2 is the contrapositive of Assertion 1. $\square$

The following proposition relates countability to set containment, and shows that countable sets are, in some sense, the "smallest" in terms of cardinality.

Proposition 1.51. *Let S and T be sets.*

1. *If S is countable and $T \subseteq S$, then T is countable.*
2. *If T is uncountable and $T \subseteq S$, then S is uncountable.*

Idea of proof. *We approach the proof of Assertion 1 by considering whether or not T is infinite. If T is finite, then it is countable by definition 1.48. If T is infinite, then S must be infinite (Assertion 2 of Proposition 1.44). In this case, we can enumerate the elements of S, and we may use this enumeration to subsequently enumerate the elements of T, showing that $T \sim \mathbb{N}$. Assertion 2 is the contrapositive of Assertion 1.*

Proof of Proposition 1.51. To prove Assertion 1, we point out that the set T is either finite or infinite. If T is finite, then it is countable, and we are done. If T is infinite, then S must also be infinite (Assertion 2 of Proposition 1.44).

Since S is countable and not finite, $S \sim \mathbb{N}$. Thus, there exists a bijection $f : \mathbb{N} \to S$, and we use this bijection to enumerate the elements of S as $S = \{f(1), f(2), f(3), \ldots\}$. Since T is infinite, it is nonempty. Since $T \subseteq S$ and T is nonempty, by the Well-Ordering Principle (Axiom 1.14) there exists a least number $n_1 \in \mathbb{N}$ for which $f(n_1) \in T$ (specifically, the nonempty subset $f^{-1}(T) \subseteq \mathbb{N}$ has a least element). The set $T - \{f(n_1)\}$ is nonempty since T is infinite, and so in the same way as before there exists a least element $n_2 > n_1$ for which $f(n_2) \in T - \{f(n_1)\}$. In this way we enumerate the set $T = \{f(n_1), f(n_2), f(n_3), \ldots\}$, and we show that the function $g : \mathbb{N} \to T$ given by $g(k) = f(n_k)$ is a bijection. By definition, if $k < \bar{k}$, then $n_k < n_{\bar{k}}$, and so $f(n_k) \neq f(n_{\bar{k}})$ since f is injective. Thus, g is injective. By construction,

$$T = \{f(n_1), f(n_2), f(n_3), \ldots\} = g(\mathbb{N}).$$

Thus, g is surjective. We have shown that if T is infinite, then $T \sim \mathbb{N}$, completing the proof. □

In practice, it can sometimes be difficult to determine whether or not a set is countable or uncountable. Consider being given an infinite set S, and being asked whether or not S is countable or uncountable. If S were countable, since it is already infinite, it seems that you would have to find an explicit bijection $f : S \to \mathbb{N}$. If S were uncountable, you would somehow have to demonstrate that no such bijection exists. Depending on how complicated S is, either option would be unpalatable. After proving a simple but useful lemma, the following proposition gives sufficient conditions to more easily deduce whether or not a set is countable or uncountable–this proposition will be used subsequently in this section and illustrates its usefulness.

Lemma 1.52. *If $f : S \to T$ is an injection, then $S \sim f(S)$.*

Idea of proof. *To prove this lemma, we observe that the function with the restricted range $f : S \to f(S)$ is injective (by assumption), and surjective considering its possibly smaller range.*

Proof of Lemma 1.52. Consider the restricted function $f : S \to f(S)$. Since f is injective, this restricted function is as well. Since the new range of f is $f(S)$, it is surjective. Thus, the restricted $f : S \to f(S)$ is a bijection. □

Proposition 1.53. *Let S and T be sets.*

1. *If S is countable and there exists a surjection $f : S \to T$, then T is also countable.*
2. *If S is uncountable and there exists an injection $g : S \to T$, then T is also uncountable.*

Remark 1.54. Some nontechnical commentary may help to put Proposition 1.53 in context. If $f : S \to T$ is surjective, then from the point of view of cardinality, the cardinality of S should only be larger than or equal to the cardinality of T. Thus, Assertion 1 of Proposition 1.53 makes sense: if S were also countable, it makes sense that T must also be countable. If $f : S \to T$ is injective, then the opposite should be true: the cardinality of S should not exceed the cardinality of T. Thus, Assertion 2 of Proposition 1.53 makes sense: if S were uncountable, then T should be uncountable as well.

Idea of proof. *To prove Assertion 1, we consider the case that T is infinite or finite. If T is finite, then T is countable and we are done. If T is infinite, our goal will be to find a subset $\tilde{T} \subseteq S$ that is equivalent to T– we will do this using the surjectivity of f and Lemma 1.52. By Assertion 1 of Proposition 1.51, follows that $\tilde{T}$ is countable. Since $T \sim \tilde{T}$ and $\tilde{T}$ is countable, Assertion 1 of Proposition 1.50, T must be countable. The proof of Assertion 2 is perhaps more straightforward. If $g : S \to T$ is an injection, then $S \sim g(S)$ by Lemma 1.52. So, since S is uncountable and $S \sim g(S)$, by Assertion 2 of Proposition 1.50, $g(S)$ is uncountable. Since $g(S)$ is uncountable and $g(S) \subseteq T$, by Assertion 2 of Proposition 1.51, T must be uncountable as well.*

Proof of Proposition 1.53. We begin by proving Assertion 1. Suppose S is countable, and $f : S \to T$ is surjective. If T is finite, then T is countable and we are done. Suppose then that T is infinite. Since f is surjective, for each $t \in T$, there exists an $s_t \in S$ for which $f(s_t) = t$. For each $t \in T$, choose[3] one s_t for which $f(s_t) = t$, and define $F : T \to S$ by $F(t) = s_t$. The function F is injective, since if $F(t_1) = F(t_2) = s_t$, then $f(s_t) = t_1$ and $f(s_t) = t_2$ by construction, and so $t_1 = t_2$. Since F is injective, $T \sim F(T) \subseteq S$ by Lemma 1.52. Since S is countable and $F(T) \subseteq S$, $F(T)$ is countable by Assertion 1 of Proposition 1.51. Since $T \sim F(T)$, it follows that T is countable as well by Assertion 1 of Proposition 1.50.

We now prove Assertion 2. Suppose S is uncountable and $g : S \to T$ is injective. Then $S \sim g(S)$ by Lemma 1.52. Since S is uncountable, $g(S)$ is uncountable by Assertion 2 of Proposition 1.50. Since $g(S) \subseteq T$ and $g(S)$ is uncountable, it follows by Assertion 2 of Proposition 1.51 that T is uncountable as well. □

We shall use Proposition 1.53 to help us prove Theorem 1.56, which can be stated informally as: "the countable union of countable sets is countable." This fact will be very helpful in determining whether or not a set is countable, and one may use this to deduce that very many common subsets of $\mathbb{R}$ are in fact countable–see Example 1.57. We shall also need the following lemma.

Lemma 1.55. *$\mathbb{N} \times \mathbb{N}$ is countable.*

Idea of proof. *The idea is to find an injection $f : \mathbb{N} \times \mathbb{N} \to \mathbb{N}$. Once we have such a function, $\mathbb{N} \times \mathbb{N} \sim f(\mathbb{N} \times \mathbb{N})$ by Lemma 1.52. Since $f(\mathbb{N} \times \mathbb{N}) \subseteq \mathbb{N}$, by Proposition 1.51, $f(\mathbb{N} \times \mathbb{N})$ is countable. Finally, by Proposition 1.50, we conclude $\mathbb{N} \times \mathbb{N}$ is countable.*

[3]Here, we are forced to use the Axiom of Choice (see page 27).

Proof of Theorem 1.55. Define the function[4] $f : \mathbb{N} \times \mathbb{N} \to \mathbb{N}$ as $f(n, m) = 2^n 3^m$. To prove that f is injective, suppose that

$$f(n_1, m_1) = 2^{n_1} 3^{m_1} = 2^{n_2} 3^{m_2} = f(n_2, m_2). \tag{1.4}$$

Suppose without any loss in generality that $n_1 \geq n^2$. Then we may rewrite Equation (1.4) as

$$2^{n_1 - n_2} 3^{m_1} = 3^{m_2}. \tag{1.5}$$

The right-hand side of Equation (1.5) is not divisible by 2. Thus, there must be no powers of 2 on the left-hand side of the same equation. So, $n_1 - n_2 = 0$, or, $n_1 = n_2$. Using this fact and rewriting Equation 1.5, we conclude that $3^{m_1} = 3^{m_2}$. Using similar reasoning, we can show that $m_1 = m_2$. We have shown that if $f(n_1, m_1) = f(n_2, m_2)$, then $(n_1, m_1) = (n_2, m_2)$. That is, f is injective.

The remainder of the proof is straightforward. Since f is injective, $\mathbb{N} \times \mathbb{N} \sim f(\mathbb{N} \times \mathbb{N})$ by Lemma 1.52. Since $\mathbb{N}$ is countable and $f(\mathbb{N} \times \mathbb{N}) \subseteq \mathbb{N}$, $f(\mathbb{N} \times \mathbb{N})$ is countable by Proposition 1.51. Finally, since $\mathbb{N} \times \mathbb{N} \sim f(\mathbb{N} \times \mathbb{N})$, and $f(\mathbb{N} \times \mathbb{N})$ is countable, it follows by Proposition 1.50 that $\mathbb{N} \times \mathbb{N}$ is countable. □

We can now prove Theorem 1.56, a fundamental result in the area of cardinality and set theory:

Theorem 1.56. *The countable union of countable sets is countable. That is, if I is a countable indexing set, and each of the sets S_α is countable for every $\alpha \in I$, then the union $\cup_{\alpha \in I} S_\alpha$ is countable.*

Idea of proof. *There are some trivialities to avoid before really getting started: we can assume that the indexing set I is nonempty. We can also assume that all of the S_α are nonempty, since if any of them were, it would not contribute anything to the union. With these trivialities aside, the idea of the remainder of the proof is to find a surjective map $F : \mathbb{N} \times \mathbb{N} \to \cup_{\alpha \in I} S_\alpha$. Then, since $\mathbb{N} \times \mathbb{N}$ is countable (Lemma 1.55), Assertion 1 of Proposition 1.53 shows that $\cup_{\alpha \in I} S_\alpha$ is countable. In order to find this surjective map, since each S_α is countable, we may find surjections $f_\alpha : \mathbb{N} \to S_\alpha$ and $g : \mathbb{N} \to I$ that enumerate each of these sets. Then, we define $F(n, m)$ to be the mth element in $S_{g(n)}$.*

Proof of Theorem 1.56. If the indexing set I is empty, then the union is empty and we are done. So, we assume $I \neq \emptyset$. If any of the S_α are empty, then they contribute nothing to the union, and may be disregarded. So we assume S_α is nonempty for every $\alpha \in I$.

We proceed by repeatedly using the following fact: for every countable nonempty set, there exists a surjection from $\mathbb{N}$ onto that set (Exercise 1).

[4]The numbers 2 and 3 are not special here, any two distinct prime numbers will do.

Let $g : \mathbb{N} \to I$ and $f_\alpha : \mathbb{N} \to S_\alpha$ be surjections. We define the function $F : \mathbb{N} \times \mathbb{N} \to \cup_{\alpha \in I} S_\alpha$ by

$$F(n, m) = f_{g(n)}(m).$$

If we can prove that F is a surjection, we are done: $\mathbb{N} \times \mathbb{N}$ is countable (Lemma 1.55), and so Assertion 1 of Proposition 1.53 demonstrates that $\cup_{\alpha \in I} S_\alpha$ is countable as a result. So let $x \in \cup_{\alpha \in I} S_\alpha$. There exists at least one $\alpha \in I$ so that $x \in S_\alpha$. Since $g : \mathbb{N} \to I$ is surjective, $\alpha = g(n)$ for some n. Also for this same α, $f_\alpha : \mathbb{N} \to S_\alpha$ is surjective. So, since $x \in S_\alpha$ there exists an m for which $f_\alpha(m) = x$. So,

$$x = f_\alpha(m) = f_{g(n)}(m) = F(n, m).$$

□

Example 1.57 illustrates how to use Theorem 1.56 for several familiar sets.

Example 1.57. We show that each of the following sets is countable by expressing it as the countable union of countable sets, according to Theorem 1.56.

1. $\mathbb{Z} = \{\ldots, -3, -2, -1\} \cup \{0\} \cup \{1, 2, 3, \ldots\}$. Each of the sets in this finite (and hence countable) union is countable. Thus, $\mathbb{Z}$ is countable.

2. To show $\mathbb{Z} \times \mathbb{Z}$ is countable, we express it as the following countable union of countable sets. For each $n \in \mathbb{Z}$, let $T_n = \{(n, m) | m \in \mathbb{Z}\}$. The function $m \mapsto (n, m)$ is a bijection from $\mathbb{Z} \to T_n$, so since $\mathbb{Z}$ is countable and $\mathbb{Z} \sim T_n$, T_n is countable as well by Proposition 1.50. Now, we see that $\mathbb{Z} \times \mathbb{Z} = \cup_{n \in \mathbb{Z}} T_n$. Since the indexing set $\mathbb{Z}$ is countable, we have expressed $\mathbb{Z} \times \mathbb{Z}$ as the countable union of countable sets. Thus, $\mathbb{Z} \times \mathbb{Z}$ is countable by Theorem 1.56.

3. The set of rational numbers $\mathbb{Q}$ is countable by the following argument. Define $f : \mathbb{Z} \times (\mathbb{Z} - \{0\}) \to \mathbb{Q}$ as $f(n, m) = \frac{n}{m}$. This map is surjective (but not injective, $f(1, 2) = f(2, 4)$ for example). Since $\mathbb{Z} \times (\mathbb{Z} - \{0\}) \subseteq \mathbb{Z} \times \mathbb{Z}$, and $\mathbb{Z} \times \mathbb{Z}$ is countable, by Proposition 1.51 $(\mathbb{Z} - \{0\}) \times \mathbb{Z}$ is countable. Now f is a surjection from a countable set onto $\mathbb{Q}$, so according to Assertion 1 of Proposition 1.53, it follows that $\mathbb{Q}$ is countable.

We complete this section by presenting a set that is uncountable. The proof of Theorem 1.58 is a famous argument attributed to Cantor in the late 19th century, and is called *Cantor's diagonalization argument.*

Theorem 1.58. $[0, 1]$ *is uncountable.*

Idea of proof. *The proof is by contradiction: suppose $[0,1]$ were countable. It is clearly not finite, so if $[0,1]$ were countable, there would be a bijection $f : \mathbb{N} \to [0,1]$ that we could use to enumerate $[0,1]$ and to express them all in some sort of list: $f(1), f(2), f(3), \ldots$. We prove that f could not be a surjection by finding some element in $[0,1]$ that is not on this list. Cantor's idea is to express those elements targeted by f as decimal expansions. Then, by changing the kth decimal value of the number $f(k)$, one can find a number that is not on this list. In other words, f is not surjective.*

Proof of Theorem 1.58. Suppose $[0,1]$ were countable. Since it is infinite, this would mean that $[0,1]$ is countably infinite, and so $[0,1] \sim \mathbb{N}$. Let $f : \mathbb{N} \to [0,1]$ be a bijection. We express the numbers $[0,1]$ in a list and in decimal form. Since $.\bar{9} = 1$, and any terminating decimal can be completed with an infinite string of 0s, we may express

$$\begin{array}{rcl} f(1) & = & .d_{11}d_{12}d_{13}d_{14}\ldots \\ f(2) & = & .d_{21}d_{22}d_{23}d_{24}\ldots \\ f(3) & = & .d_{31}d_{32}d_{33}d_{34}\ldots \\ f(4) & = & .d_{41}d_{42}d_{43}d_{44}\ldots \\ & \vdots & \end{array}$$

where the numbers $d_{ij} \in \{0, 1, \ldots, 9\}$ are the digits in the decimal expansions of these numbers. We find a number $x \in [0,1]$ not on this list by altering the kth number to the right of the decimal of the number $f(k)$. Thus, by creating $x = .d_1d_2d_3\ldots$, where $d_k \neq d_{kk}$ above, $x \neq f(k)$ for any k since the two numbers have decimal expansions that differ at the kth decimal location. As a result, f is not a surjection. So choose[5]

$$d_k = \left\{ \begin{array}{ll} 2 & \text{if } d_{kk} \neq 2, \\ 8 & \text{if } d_{kk} = 2 \end{array} \right. .$$

Now $x = .d_1d_2d_3\ldots \neq f(k)$ for any k, and so f is not a surjection. This is a contradiction to the assumption that f is a bijection. □

Practice 1.59. Use the results from this section to prove the following statements.

(a) Prove that $\mathbb{R}$ is uncountable.

(b) Prove that $[0,1)$ is uncountable.

Methodology. *For the first statement, we simply note that $[0,1]$ is an uncountable subset of $\mathbb{R}$ (Theorem 1.58), and so by Assertion 2 of Proposition 1.51, $\mathbb{R}$ is uncountable. There are a variety of ways to approach the proof*

[5]The numbers 2 and 8 only need to be chosen as to not inadvertently create a number with a nonunique decimal expansion, for example, $.4\bar{9} = .5\bar{0}$. This cannot happen with the numbers 2 and 8.

of the second statement, and we present two such ways below. The first way would be to find an equivalent copy of $[0,1]$ *within* $[0,1)$*, perhaps* $[0, \frac{1}{2}]$*. Then, the same reasoning as in the proof of the first statement would show* $[0,1)$ *is uncountable. Another approach would be to assume that there is a bijection* $f : \mathbb{N} \to [0,1)$*. One could extend this function (by including one element in the domain of* f *mapping to the number* 1*) to create a bijection from* $\mathbb{N}$ *onto* $[0,1]$*, something we know to not be possible, according to Theorem 1.58.*

Solution.

(a) According to Theorem 1.58, $[0,1]$ is uncountable. Since $[0,1] \subseteq \mathbb{R}$, by Assertion 2 of Proposition 1.51, $\mathbb{R}$ is uncountable as well.

(b) *(Solution 1)* Consider the function $f : [0,1] \to [0, \frac{1}{2}]$, defined by $f(x) = \frac{1}{2}x$. This is clearly a bijection, showing $[0,1] \sim [0, \frac{1}{2}]$. Since $[0,1]$ is uncountable, by Assertion 2 of Proposition 1.50, $[0, \frac{1}{2}]$ is uncountable as well. Since $[0, \frac{1}{2}] \subseteq [0,1)$, by Assertion 2 of Proposition 1.51, $[0,1)$ is uncountable.

(Solution 2) Suppose $[0,1)$ were countable. Since it is infinite, $[0,1) \sim \mathbb{N}$ so that there exists a bijection $f : \mathbb{N} \to [0,1)$. Extend the function f to the domain $\mathbb{N} \cup \{0\}$ by defining $g(n) = f(n)$ if $n \in \mathbb{N}$, and $g(0) = 1$. Thus $g : \mathbb{N} \cup \{0\} \to [0,1]$ is a bijection. Define the new function $F : \mathbb{N} \to [0,1]$ as $F(n) = g(n-1)$. The function $n \mapsto n-1$ is a bijection, and since g is a bijection, F is the composition of bijections. Thus, F is a bijection by Assertion 3 of Theorem 1.35. But $[0,1]$ is uncountable (Theorem 1.58), and so no such bijection exists. Therefore $[0,1)$ is not countable. ■

Before ending the section, it makes sense to briefly discuss cardinalities of infinite sets. Notice that in Definition 1.40, a value is given for $|S|$ only when S is finite. Mathematicians have also given the value[6] of $\aleph_0$ to $|\mathbb{N}|$. This is not a real number, since if one were to consider $|\mathbb{N} \cup \{0\}|$, this should be $\aleph_0 + 1$. But $\mathbb{N} \cup \{0\} \sim \mathbb{N}$, so it should be the case that $|\mathbb{N} \cup \{0\}| = |\mathbb{N}|$, so that $\aleph_0 + 1 = \aleph_0$.

It can be shown that $\mathbb{R} \sim \mathcal{P}(\mathbb{N})$, the set of all subsets of $\mathbb{N}$ (see a related question in Exercise 6 in this section, and Exercise 6 of Section 1.2). In general, if $|S| = n$, then $|\mathcal{P}(S)| = 2^n$, and so it is sometimes written that $|\mathbb{R}| = 2^{\aleph_0}$. Sometimes people use the symbol $\aleph_1$ or $\mathfrak{c}$ for $|\mathbb{R}|$, and refer to it as the *cardinality of the continuum*. This process can be continued to define $\aleph_2 = |\mathcal{P}(\mathcal{P}(\mathbb{N}))|$, and so on. These $\aleph_k$ are examples of what are called *cardinal numbers*. Exercise 6 shows that $\aleph_0 < \aleph_1$, and the *continuum hypothesis* asserts that there is no set S with $\aleph_0 < |S| < \aleph_1$. It was shown that the continuum hypothesis is independent of the Zermelo-Fraenkel axioms of set theory (see the discussion on page 11). There is a lot of literature on the theory of cardinal numbers and the Continuum Hypothesis, the interested reader can find a brief summary and some reading suggestions on pages 40–41 of [27].

[6] $\aleph$ is the first letter of the Hebrew alphabet.

1.5.4 Review of important concepts from Section 1.5

1. The notion of set equivalence is the way we measure the relative size of a set. Set equivalence is defined in Definition 1.37. Set equivalence is an equivalence relation (Proposition 1.39).

2. Finite and infinite sets are defined in Definition 1.40. Countable, countably infinite, and uncountable sets are defined in Definition 1.48.

3. $\mathbb{N}$ is infinite (Theorem 1.42).

4. Propositions 1.43 and 1.50 demonstrate the property of being finite, infinite, countable, or uncountable is preserved by equivalence.

5. Propositions 1.44 and 1.51 uses set containment as a way of comparing sets in terms of cardinality. In particular, a subset of a finite set is finite, and a superset of an infinite set is infinite (Proposition 1.44). A subset of a countable set is countable, and a superset of an uncountable set is uncountable (Proposition 1.51).

6. There are a number of ways to identify an infinite set, see a characterization in Theorem 1.46.

7. Any surjection from a countable set must be mapped onto another countable set. Any injection from an uncountable set must be mapped into another uncountable set. See Proposition 1.53.

8. The countable union of countable sets is countable (Theorem 1.56).

9. $[0,1]$ is uncountable (Theorem 1.58).

Exercises for Section 1.5

1. If S is countable and nonempty, prove there exists a surjection $g : \mathbb{N} \to S$.

2. Suppose S and T are finite sets. Prove that $S \cup T$ is finite.

3. Find a bijection from $(0,1]$ to $(0,1)$. *[Hint: The construction given in the second part of the proof of Theorem 1.46 will perhaps be helpful.]*

4. Let S and T be sets. Prove or disprove each of the following statements[7]

   (a) If S and T are finite, then $S \sim T$.
   (b) If S and T are infinite, then $S \sim T$.
   (c) If S and T are countable, then $S \sim T$.
   (d) If S and T are countably infinite, then $S \sim T$.

[7] A question of this sort is asking you to prove the statement if it is true, or to provide an example illustrating it is not always true.

(e) If S and T are uncountable, then $S \sim T$. *[Hint: Consider using Exercise 6.]*

5. Prove that for any real numbers $a < b$, $[0,1] \sim [a,b]$. Use this to prove that $[a,b], (a,b], [a,b)$, and (a,b) are all uncountable. *[Hint: The methods in Practice 1.59 could be of some help.]*

6. Prove that the power set on X has a cardinality strictly greater than that of X (recall the power set $\mathcal{P}(X)$ is the set of all subsets of X, as defined in Exercise 6 of Section 1.2). That is, prove there is no surjection $f : X \to \mathcal{P}(X)$, but there is an injection $g : X \to \mathcal{P}(X)$. *[Hint: If f were a surjection, then consider the set $T = \{x \in X | x \notin f(x)\} = f(y)$ for some y. Show that you reach a contradiction whether $y \in T$ or $y \notin T$.]*

7. There are two problems with the "proof" below that if S and T are countable, then $S \times T$ is countable (this is true, see Exercise 8). One issue is a statement that is false. The other is something that requires more explanation. Find both of these issues:

   "Since S and T are countable, $S \sim \mathbb{N}$ and $T \sim \mathbb{N}$. So, $S \times T \sim \mathbb{N} \times \mathbb{N}$, which is countable by Lemma 1.55."

8. Prove that if S and T are countable, then $S \times T$ is countable. *[Hint: You may find a construction in Example 1.57 helpful, along with Exercise 5 of Section 1.2.]*

9. Prove that the set of irrationals are uncountable.

10. Let $S_1, S_2, S_3 \ldots$ be countable sets. Is $S_1 \times S_2 \times S_3 \times \cdots$ countable as well? *[Hint: Be careful not to use Theorem 1.56 incorrectly.]*

1.6 Troubleshooting Guide for Sets, Functions, and Proofs

Type of difficulty:	*Suggestion for help:*
Section 1.1: Logic and an Introduction to Proof	
"I still really don't know how to prove things, or what constitutes a proof."	Try writing down a proof to any one of the exercises in Section 1.1, and then review the proof writing suggestions on page 8. Does your proof incorporate those suggestions as appropriate? These suggestions are pursuant to the main goal of clearly expressing why a certain statement is true–do you think that your proof accomplishes this? If so, then your proof has a very good chance of being just fine. To be sure, however, you could share your proof with others, such as your professor, for clarity.
"How do I know if my proof is 'correct'?"	This is a tricky question, since a certain reader may be fine with your proof, while another may think you have left something out. Generally, if you adequately justify each of the steps in your proof, you should be okay. Also, review those suggestions found on page 8 as they may help as well.
"What level of rigor do I use?"	See suggestion 2 on page 8 concerning your audience.
Section 1.2: Sets and Their Operations	
"What is a set relation, and how do I know which one is appropriate?"	A set relation is a symbol that relates one set with another. These symbols include $\subseteq, =, \neq$, etc., and are described in Subsection 1.2.2. It is generally not obvious which of these to use until one has a full understanding of the sets involved. See Practice 1.7 for some examples.

Type of difficulty:	***Suggestion for help:***
"How many elements of a set must I list before the pattern is clear?"	This depends on what pattern you are trying to exhibit. Generally, listing the first 3 or 4 elements should be fine, although for more complicated patterns that are harder to discern, you may list out 5 or even 6 elements. Finding the general form of the pattern makes it more clear, and expressing elements of your set using set notation would be preferable, see Example 1.6 for some examples of set notation.
"I'm confused about how exactly to show two sets are equal."	Generally, one needs to use a double-inclusion argument: $S = T$ if and only if $S \subseteq T$ and $T \subseteq S$. See Practice 1.7 (part (b)), and Practice 1.9 for examples of this (the latter is an example in the context of unions and intersections of sets).
"Do I always have to use a double-inclusion argument to show that two sets are equal?"	Generally, the answer is yes, especially when it is the focus of the question as in Exercises 2 and 5. In Practice 1.11 a double-inclusion argument is not used since the sets involved were uncomplicated, and the point of that practice problem is to illustrate DeMorgan's Laws. See also the discussion on page 20.
Section 1.3: Mathematical Induction	
"Generally, I have no idea how to start these proofs."	Remember that for the proofs of some results, it would be inappropriate to use Mathematical Induction. For those where Mathematical Induction is appropriate, try writing out the first 4 or 5 statements and convince yourself that you understand the statements in general. Then to prove the inductive step, think about how to incorporate the previous statement into the next statement.

Type of difficulty:	***Suggestion for help:***
"If I plug in, say, 4 or 5 different values for n, I notice that my statement always holds. As a result, I'm convinced it's true for every n. Does this work as a proof?"	Just plugging in a few values certainly does not count as a proof, and misses the point of attempting a question of this sort. Remember, these questions normally ask you to prove something holds for *every* $n \in \mathbb{N}$, not just a few $n \in \mathbb{N}$. Review the second paragraph in Subsection 1.3.1 on page 22 for a detailed description of this and Mathematical Induction in general.
"I have a hard time determining what I'm supposed to do in a Mathematical Induction proof."	There could be several reasons for this. First, make sure you understand what Mathematical Induction actually is, see the discussion at the beginning of this section. Second, it is sometimes the case that the student is unclear on exactly what is the $(n+1)^{\text{st}}$ statement. See Practices 1.16, 1.17, and 1.18 for examples. It may also help to put the statements you are trying to prove into plain english to help you determine what the "next" statement is.
"It seems as though you're assuming what you're trying to prove with Mathematical Induction."	This is a common impression. In a Mathematical Induction proof, the induction hypothesis asks you to assume that the statement is true for one particular n. Overall, what you are trying to prove is that these statements are true for *all* n. Consider reviewing Theorem 1.15.
"Is it okay to use k instead of n?"	Yes, these letters simply indicate what statement one is talking about. It does not matter what you name them, provided you stick with that name throughout the proof.
Section 1.4: Functions	
"I don't understand how the inverse image $f^{-1}(T)$ has nothing to do with the inverse of a function, as it is described beneath Definition 1.23. I thought f^{-1} was the inverse of f."	This is an instance where the same notation is used for two different things. Yes, f^{-1} typically stands for "the inverse of f" (see Exercise 11) whenever it exists. But, the f^{-1} notation is also used to denote a certain set, see Definition 1.23.

Type of difficulty:	***Suggestion for help:***
"I don't really get the proof of $f(S_1 \cup S_2) = f(S_1) \cup f(S_2)$, and the proofs of the related facts in Propositions 1.26 and 1.28."	It is likely that a review of what this notation means will be of help, since the proofs of all of these facts really only disentangle what this notation really means. See Definition 1.23.
"I feel as though I understand the material in this section, but am having a hard time starting on the exercises."	Almost all of the homework questions are directly related to something covered in this section. A good suggestion would be to try to determine what material covered in the section relates to each exercise. Then, read over that material before attempting each exercise.
Section 1.5: Cardinality	
"I'm having a hard time grasping the concept of set equivalence and what it has to do with cardinality."	Equivalent sets are those that have a bijection between them (Definition 1.37). We use this bijection to match up the elements in the sets considered, and in this way, it could be said that the sets have the "same size," motivating the definition. See also the discussion on page 39 at the beginning of Subsection 1.5.1.
"Why can't a set's cardinality just be how many elements it has? Why all the complicated definitions?"	Defining a cardinality in this way is not precise enough, as it would lead to several unintended outcomes. See the discussion on page 39 at the beginning of Subsection 1.5.1.
"Why is the proof of Lemma 1.41 so complicated? In particular, if $\|S\| = n$ and you remove any element, it seems obvious that the cardinality of the resulting set would be $n - 1$."	Technically, to demonstrate a set has cardinality $n - 1$, one needs to provide a bijection from the set to $\{1, 2, \ldots, n-1\}$. In the proof, we identified the element that is mapped to n by the given bijection, so that the restricted bijection would have a range of $\{1, 2, \ldots, n - 1\}$, accomplishing this.
"I don't see how there could be infinite sets of different cardinalities. Doesn't 'infinite' just mean 'infinite', so that they should all have the same cardinality?"	Remember that two sets have the same cardinality if there is a bijection between them (Definition 1.37). Everything about cardinality follows from this definition. As such, it can be proven that there are some infinite sets that are not in set bijection with $\mathbb{N}$. Theorem 1.58 shows that $[0, 1]$ is such a set.

2

The Real Numbers

Chapter Summary:

This chapter introduces the real numbers axiomatically, and proceeds to deduce its important properties, including the Completeness Axiom, and the density of $\mathbb{Q}$ in $\mathbb{R}$.

Section 2.1: The Real and Extended Real Number System. This section introduces the field and order axioms for the real numbers (Axioms 2.1 and 2.5). Basic intuitive properties of the real numbers are then deduced. The absolute value (Definition 2.7), the Triangle Inequality (Theorem 2.9), the Extended Real Numbers (Definition 2.10), and neighborhoods (Definition 2.13) are then discussed.

Section 2.2: The Supremum and Infimum. This section's main goal is to define the supremum and infimum of a set (Definitions 2.18 and 2.22). The section first introduces the concept of a bounded set Definition 2.16, then defines the supremum of a set as the smallest upper bound of the set. The infimum is introduced subsequently, followed by several examples.

Section 2.3: The Completeness Axiom. This section covers three main concepts: The Completeness Axiom (Axiom 2.28), the Archimedean Property (Theorem 2.34), and the density of $\mathbb{Q}$ in $\mathbb{R}$ (Theorem 2.39).

Section 2.4: Troubleshooting Guide for The Real Numbers. A list of frequently asked questions and their answers, arranged by section. Come here if you are struggling with any material from this chapter.

2.1 The Real and Extended Real Number System

We begin our study of the real numbers in this section by introducing a collection of intuitive axioms that we assume they must obey: the "field axioms" (Axiom 2.1), and the "order axioms" (Axiom 2.5). Generally, when one begins studying a new object by starting with its axioms it can be a rather tedious process to finally deduce some basic and familiar facts. Since we do exactly that in this section, it is important to remember that our goal is to carefully deduce and prove these facts from the axioms, and is an unavoidable part of carefully exploring mathematical structures.

We want to heavily emphasize that in this section we do not construct or attempt to define the real numbers, but rather assume you have a familiarity with the set. We then assume that the real numbers satisfy Axioms 2.1 and 2.5. It is actually the case that the real numbers can be constructed, and shown to satisfy these axioms. This is a rather tedious process, and we do not do that here. We refer the reader instead to [24] where a careful construction of $\mathbb{R}$ is done, and these basic axioms are established. In addition to Rudin, the author was influenced by Lay [19] in the preparation of this section, and is another excellent reference.

In addition to proving these basic intuitive facts from Axioms 2.1 and 2.5, we introduce the absolute value function (Definition 2.7), and the Triangle Inequality (Theorem 2.9). These will be of central importance throughout the book. We finish the section by defining a certain type of set in Definition 2.13 that will also be commonly used: a "neighborhood" of a real number.

2.1.1 The Real numbers

Without an explicit construction of the real numbers, it is difficult to formally answer the question "What is a real number?" In addition, constructions of the real numbers unfortunately do not give a very authentic intuition of the answer, either. For the purposes of this book, we shall consider the set of real numbers $\mathbb{R}$ to be the set of all values x so that $-\infty < x < \infty$. This could include, for instance, numbers such as $0, \frac{7}{8}, \frac{-2}{4}, \sqrt{2}, \frac{-\pi}{6}$, where we notice that $\frac{-2}{4} = -\frac{1}{2}$, and $\sqrt{2}$ is defined to be the real number that squares to 2. Notice that we are specifically *not* including quantities such as $\sqrt{-1}$, or $\frac{1}{0}$: the former is not considered a real number (rather, it is called a *complex* number), and the latter is simply not a quantity that can be adequately defined (see Exercise 1).

We begin with a study of the important properties of the real numbers that we use freely throughout this book. In fact, the properties we list here are enjoyed by number systems other than just the real numbers (for example, the set of all quotients of whole numbers, known as the rational numbers and denoted by $\mathbb{Q}$, satisfy all of the below as well). We take as axiom that the real numbers $\mathbb{R}$ satisfy the following properties.

Axiom 2.1. The set of real numbers with the usual addition and multiplication operations make it a *field*. That is, the following properties hold:

1. (Closure) The real numbers are closed under the operations of addition and multiplication. That is, for every $x, y \in \mathbb{R}$, we have $x + y \in \mathbb{R}$ and $xy \in \mathbb{R}$.

2. (Commutativity) The operations of addition and multiplication are commutative. That is, for all $x, y \in \mathbb{R}$, we have $x + y = y + x$ and $xy = yx$.

3. (Associativity) The operations of addition and multiplication are associative. That is, for all $x, y, z \in \mathbb{R}$, we have $(x + y) + z = x + (y + z)$ and $(xy)z = x(yz)$.

4. (Identities) The operations of addition and multiplication each have their own identity. That is, there exists an element $0 \in \mathbb{R}$ so that $x + 0 = x$ for every $x \in \mathbb{R}$. There also exists an element $1 \in \mathbb{R}$ so that $x \cdot 1 = x$ for every $x \in \mathbb{R}$. We also have $0 \neq 1$.

5. (Inverses) Every $x \in \mathbb{R}$ has an additive inverse, and every nonzero $x \in \mathbb{R}$ has a multiplicative inverse. That is, for all $x \in \mathbb{R}$, there exists an additive inverse $y \in \mathbb{R}$ so that $x + y = 0$. For every nonzero $x \in \mathbb{R}$, there exists a multiplicative inverse $y \in \mathbb{R}$ so that $xy = 1$.

6. (Distributivity) The distributive law holds. That is, for every $x, y, z \in \mathbb{R}$, we have $x(y + z) = xy + xz$.

These properties of the real numbers should be very familiar. Note that the real number 0 is called the *additive identity*, while the number 1 is called the *multiplicative identity*. The additive inverse of x is usually written as $-x$, while the multiplicative inverse of a nonzero x is usually written as $\frac{1}{x}$. Finally, the expression $x + (-y)$ is sometimes expressed as $x - y$.

Assuming that the real numbers satisfy Axiom 2.1, we may draw many useful conclusions that we present in several groups. The first group of conclusions concerns the operation of addition.

Proposition 2.2. *Suppose* $x, y, z \in \mathbb{R}$.

1. *If* $x + y = x + z$, *then* $y = z$.

2. *The additive identity in* $\mathbb{R}$ *is unique.*

3. *Each real number has a unique additive inverse.*

4. *The additive inverse of* $-x$ *is* x. *In other words,* $-(-x) = x$.

5. $0 \cdot x = 0$.

Idea of proof. *The proofs of all of these assertions take advantage of one or more of the properties in Axiom 2.1, and are quite straightforward.*

Proof of Proposition 2.2. We begin by proving Assertion (1), and we suppose that $x+y = x+z$ for some real numbers x, y, and z. The number $-x \in \mathbb{R}$ exists according to Property (5) of Axiom 2.1. By adding $-x$ (this is the additive inverse of x) to both sides, we conclude that $(-x)+(x+y) = (-x)+(x+z)$. Using the fact that addition is associative (Property (3) of Axiom 2.1), we can rewrite this equation as $(-x+x)+y = (-x+x)+z$. Since $-x$ is the additive inverse of x, $-x+x = 0$, and according to Property (4) of Axiom 2.1, we have

$$\begin{aligned}(-x+x)+y &= 0+y = y, \text{and} \\ (-x+x)+z &= 0+z = z.\end{aligned}$$

Thus, $y = z$.

To prove Assertion (2), we assume that there are two additive identities in $\mathbb{R}$; suppose that $0, 0' \in \mathbb{R}$ satisfy the following property: for all $x \in \mathbb{R}$, $x+0 = x$ and $x+0' = x$. Consider using $x = 0'$ in the first equality, and $x = 0$ in the second. This would conclude that $0'+0 = 0'$, and $0+0' = 0$. According to Property (2) of Axiom 2.1, addition of real numbers is commutative. Thus,

$$0' = 0'+0 = 0+0' = 0,$$

so $0 = 0'$.

To prove Assertion (3), we let $x \in \mathbb{R}$, and assume that $y, y' \in \mathbb{R}$ so that $x+y = 0$ and $x+y' = 0$. In that case, $x+y = x+y'$, and by applying Assertion (1) of this proposition to this situation, we conclude that $y = y'$.

To establish Assertion (4), we need to show that the additive inverse of $-x$ is x. But x is a candidate for this, since $-x+x = 0$. According to Assertion (3), additive inverses are unique, and so the additive inverse of $-x$ can therefore be nothing other than x.

Finally, we can prove Assertion (5) by using the distributive rule and observing that

$$\begin{aligned}0 \cdot x &= (x-x) \cdot x \\ &= x^2 - x^2 \\ &= 0.\end{aligned}$$

□

This next group of conclusions is analogous to those in Proposition 2.2, and concerns the operation of multiplication. Remember that according to Property (5) of Axiom 2.1, only *nonzero* real numbers are expected to have multiplicative inverses.

Proposition 2.3. *Suppose $x, y, z \in \mathbb{R}$.*

1. *If $x \neq 0$ and $xy = xz$, then $y = z$.*
2. *The multiplicative identity of 1 is unique.*

3. *Every nonzero real number has a unique multiplicative inverse.*
4. *The multiplicative inverse of* $\frac{1}{x}$ *is* x. *In other words,* $\frac{1}{1/x} = x$.

Idea of proof. *As in the proof of Proposition 2.2, the proofs of all of these assertions again take advantage of one or more of the properties in Axiom 2.1, and are similarly straightforward. One might look at the similarity in the sort of argument that is used by comparing the proofs of Propositions 2.2 and 2.3.*

Proof of Proposition 2.3. Suppose that $x \neq 0$, and that $xy = xz$. Since $x \neq 0$, $\frac{1}{x} \in \mathbb{R}$ exists according to Property (5) of Axiom 2.1. Multiplying by $\frac{1}{x}$ (this is the multiplicative inverse of x), we conclude that $\frac{1}{x}(xy) = \frac{1}{x}(xz)$. Since multiplication is associative (Property (3) of Axiom 2.1), we have $\frac{1}{x}(xy) = \left(\frac{1}{x}x\right)y$, and $\frac{1}{x}(xz) = \left(\frac{1}{x}x\right)z$. Since $\frac{1}{x}$ is the multiplicative inverse of x, we have $\frac{1}{x}x = 1$, and according to Property (4) of Axiom 2.1, we have

$$\begin{aligned} \left(\tfrac{1}{x}x\right)y &= 1y = y, \text{and} \\ \left(\tfrac{1}{x}x\right)z &= 1z = z, \end{aligned}$$

so $y = z$.

To prove Assertion (2), we assume that there are two multiplicative identities in $\mathbb{R}$; suppose that $1, 1' \in \mathbb{R}$ satisfy the following property: for all $x \in \mathbb{R}$, $1x = x$ and $1'x = x$. Consider using $x = 1'$ in the first equality, and $x = 1$ in the second. This would conclude that $(1)(1') = 1'$, and $(1')(1) = 1$. According to Property (2) of Axiom 2.1, multiplication of real numbers is commutative. Thus,

$$1' = (1)(1') = (1')(1) = 1,$$

So $1' = 1$.

To prove Assertion (3), we let $x \in \mathbb{R}$ with $x \neq 0$, and assume that $y, y' \in \mathbb{R}$ so that $xy = 1$ and $xy' = 1$. In that case, $xy = xy'$, and by applying Assertion (1) of this proposition to this situation, we conclude that $y = y'$.

To establish Assertion (4), we need to show that the multiplicative inverse of $\frac{1}{x}$ is x. But x is a candidate for this, since $\frac{1}{x}x = 1$. According to Assertion (3), multiplicative inverses are unique, and so the multiplicative inverse of $\frac{1}{x}$ can therefore be nothing other than x.

□

This next group of conclusions concerns the distributive rule (Property (6) of Axiom 2.1), which demonstrates how the two operations of addition and multiplication interact.

Proposition 2.4. *Suppose* $x, y \in \mathbb{R}$.

1. $0x = 0$.
2. $-x = (-1)x$. *Here,* -1 *is the additive inverse of* 1.
3. $(-x)y = -(xy) = x(-y)$.
4. $(-x)(-y) = xy$. *In particular, if* $x = y = 1$, *then* $(-1)^2 = 1$.

5. $xy = 0$ if and only if one of x or y is 0.

Idea of proof. *The proofs of all of the above are again basic applications of Axiom 2.1, along with the use of some facts we have already proven.*

Proof of Proposition 2.4. We can prove the first assertion by noticing that

$$0x = (0+0)x = 0x + 0x.$$

Here, we have used the Distributive Rule (Property (6) of Axiom 2.1) in the second equality. By adding the quantity $-0x$ to both sides, we have $-0x+0x = -0x + 0x + 0x$, or, $0 = 0x$.

To prove Assertion (2), we demonstrate that $(-1)x$ is the additive inverse of x using the Distributive Rule and Assertion (1) of this proposition:

$$(-1)x + x = (-1)x + 1x = ((-1)+1)x = 0x = 0.$$

Since additive inverses are unique (Assertion (3) of Proposition 2.2), we will conclude that $(-1)x = -x$.

Assertion (3) can be proved using Assertion (2): $(-x)y = (-1)xy$, $-(xy) = (-1)xy$, and $x(-y) = x(-1)y = (-1)xy$ (the last of these because multiplication is commutative). Thus, they are all equal to the same number $(-1)xy$.

To prove Assertion (4), we use Assertion (3) twice: $(-x)(-y) = -x(-y) = -(-xy)$, and $-(-xy) = xy$ according to Assertion (4) of Proposition 2.2.

To prove the final Assertion (5), we suppose $xy = 0$, but that the conclusion is false: both $x \neq 0$ and $y \neq 0$. In that case, each of x and y has a multiplicative inverse $\frac{1}{x}$ and $\frac{1}{y}$, respectively. In this case, we have

$$1 = \frac{1}{x}x\frac{1}{y}y = \frac{1}{x}\frac{1}{y}xy = \frac{1}{x}\frac{1}{y}0 = 0,$$

which is a contradiction to the condition that $0 \neq 1$ in Property (4) of Axiom 2.1. Note that in the above sequence of equalities we have used the defining property of multiplicative inverses and the multiplicative identity in the first equality, the commutativity of multiplication in the second equality, the hypothesis that $xy = 0$ in the third equality, and Assertion (1) of this proposition in the fourth equality. The converse of Assertion (5) has already been proven as Assertion (1).

□

2.1.2 The order axioms of the Real numbers

The set of real numbers satisfies another important property that will be of critical use throughout the book: the real numbers are an *ordered set*. That is, there is a way to compare two real numbers according to which one of them is "larger."[1] The following is the order properties of $\mathbb{R}$, which we take as axiom.

[1] In other situations, an ordering on a set does not need to have any reference to "largeness" or "size" in the same way; however, in this case, the ordering on the real numbers is consistent with that intuition.

Axiom 2.5. *The set of real numbers is ordered. That is, there exists a relation $<$ on $\mathbb{R}$ that satisfies the following for every $x, y, z \in \mathbb{R}$:*

1. *If $x < y$, then $x + z < y + z$.*
2. *If $x < y$, and $z > 0$, then $xz < yz$.*
3. *(Transitivity) If $x < y$ and $y < z$, then $x < z$.*
4. *(Trichotomy) Exactly one of the following holds: $x < y$, $y < x$, or $x = y$.*

It should be noted that the common symbols $>$, $\leq$, and $\geq$ are defined as usual: $x > y$ if and only if $y < x$, $x \leq y$ if and only if $x < y$ or $x = y$, and $x \geq y$ if and only if $y < x$ or $x = y$. Similarly, strings of these relations are meant as combinations of statements. For example, the statement $x < y < z$ is meant to be a short way of saying the two statements $x < y$ and $y < z$.

When we introduced the field axioms of the real numbers (Axiom 2.1), we were able to draw some immediate and intuitive results (Propositions 2.2, 2.3, and 2.4). Similarly, we now present some immediate and intuitive results concerning the order axioms. As an example of some of these intuitive facts, notice Assertion (2) in Proposition 2.6 below: this assertion justifies the fact that one changes the order of the inequality when one multiplies both sides by a negative number. Further, by choosing x or y to be 0 there, one recovers the regular rules for multiplying negative numbers together: a negative times a negative equals a positive, and a positive times a negative equals a negative; if we set $x = 0$ in Assertion (2) of Axiom 2.5, we conclude that a positive times a positive equals another positive.

Proposition 2.6. *Let $x, y, z \in \mathbb{R}$.*

1. *If $x > 0$, then $-x < 0$. If $x < 0$, then $-x > 0$.*
2. *If $x < y$ and $z < 0$, then $xz > yz$.*
3. *If $x \neq 0$, then $x^2 > 0$. Also, $x^2 = 0$ if and only if $x = 0$.*
4. *$1 > 0$.*
5. *If $x \neq 0$, then x and $\frac{1}{x}$ have the same sign.*
6. *If $0 < x < y$, then $0 < \frac{1}{y} < \frac{1}{x}$.*
7. *Suppose x and y are positive. Then $x \leq y$ if and only if $x^2 \leq y^2$.*

Idea of proof. *These proofs are relatively straightforward, and follow mostly from the order axioms, Axiom 2.5. We only remark that it may be helpful to disentangle what the conclusion of Assertion (5) means: if $x > 0$ then $\frac{1}{x} > 0$, and if $x < 0$ then $\frac{1}{x} < 0$.*

Proof of Proposition 2.6. We begin with Assertion (1), and assume that $x > 0$. Using $z = -x$ as in part (1) of Axiom 2.5, and adding z to both sides (effectively, this is simply subtracting x from both sides), we have $0 - x = -x < 0 = x - x$.

Continuing with the second part of Assertion (1) and assuming $x < 0$, we again subtract x from both sides as we did above to conclude $0 < -x$.

Now suppose $x < y$ and $z < 0$ to prove Assertion (2). Since $z < 0$, according to Assertion (1) $-z > 0$. Now using part (2) of Axiom 2.5, we conclude $x(-z) < y(-z)$. But $x(-z) = -xz$, and $y(-z) = -yz$, so $-xz < -yz$. Now we add xz and yz to both sides to conclude $yz < xz$.

To prove Assertion (3), we note that if $x \neq 0$, then according to the Trichotomy axiom (part (4) of Axiom 2.5), it must be the case that $x > 0$ or $x < 0$ and we will show that $x^2 > 0$ in each case. If $x > 0$, setting $z = x$ in part (2) of Axiom 2.5, we have $0 < x \cdot x = x^2$. If, on the other hand, $x < 0$, then setting $y = 0$ and $z = x$ in Assertion (2), we conclude that $x \cdot x > 0 \cdot x$, in other words, $x^2 > 0$.

To prove the final part of Assertion (3), we suppose that $x^2 = 0$. Using $y = x$ in Assertion (5) of Proposition 2.4, we conclude that $x = 0$. Conversely, if $x = 0$, then by Assertion (1) of Proposition 2.4 we have $x^2 = 0$.

We now prove Assertion (4). By Trichotomy (part (4) of Axiom 2.5), exactly one of the following is true: $1 = 0$, $1 < 0$, or $1 > 0$. According to part (4) of the field axioms (Axiom 2.1), it cannot be the case that $1 = 0$. If it were the case that $1 < 0$, then by using $x = 1, y = 0$, and $z = 1$ in Assertion (2) above, we would have $1 \cdot 1 > 0 \cdot 1$, or in other words $1 > 0$. In conclusion, we have proven here that if $1 < 0$, then it is also the case that $1 > 0$, and this violates the Trichotomy axiom. Therefore, we must conclude $1 > 0$.

To prove Assertion (5), we again use Trichotomy to point out that if $x \neq 0$, then it must be the case that either $x < 0$ or $x > 0$. If $x < 0$ but $\frac{1}{x}$ were positive, then using $y = 0$ and $z = \frac{1}{x}$ in part (2) of Axiom 2.5, we would conclude that $x \cdot \frac{1}{x} = 1 < 0 = 0 \cdot \frac{1}{x}$, which violates Assertion (4). We conclude that if $x < 0$, then $\frac{1}{x}$ cannot be positive. By Trichotomy, $\frac{1}{x}$ must therefore be either 0 or negative. But $\frac{1}{x} \neq 0$, since otherwise

$$1 = x \cdot \frac{1}{x} = x \cdot 0 = 0,$$

a contradiction to part (4) of Axiom 2.1. Therefore, if $x < 0$ it must be the case that $\frac{1}{x} < 0$ as well. If, on the other hand $x > 0$, but $\frac{1}{x} < 0$, then we may use a similar argument to find the same contradiction (see Exercise 2).

To prove Assertion (6), we first assume that $0 < x < y$. According to Assertion (5), $\frac{1}{x} > 0$ and $\frac{1}{y} > 0$. Repeatedly using part (2) of Axiom 2.5, we prove this assertion by the following argument:

$$\begin{array}{ccccccl}
0 & < & x & < & y & & \text{implies} \\
0 \cdot \frac{1}{x} & < & x \cdot \frac{1}{x} & < & y \cdot \frac{1}{x}, & & \text{or} \\
0 & < & 1 & < & y \cdot \frac{1}{x}, & & \text{so} \\
0 \cdot \frac{1}{y} & < & 1 \cdot \frac{1}{y} & < & y \cdot \frac{1}{x} \cdot \frac{1}{y}, & & \text{or} \\
0 & < & \frac{1}{y} & < & \frac{1}{x}. & &
\end{array}$$

Finally, to prove Assertion (7), we assume first that $x \leq y$. Then

$$x^2 = x \cdot x \leq x \cdot y \leq y \cdot y = y^2. \tag{2.1}$$

To prove the converse, suppose that $x^2 \leq y^2$, but that $y < x$. Then using a similar argument as in Equation (2.1), we conclude $y^2 < x^2$, a contradiction to our hypothesis that $x^2 \leq y^2$. □

2.1.3 Absolute value and the triangle inequality

One of the foundational objects in Analysis is a way to measure distance, and from a certain perspective the absolute value function does exactly that: it measures the distance from any point on the real number line to 0. The formal definition of the absolute value of a real number is as follows.

Definition 2.7. If $x \in \mathbb{R}$, then the *absolute value* of x, denoted $|x|$, is defined as

$$|x| = \begin{cases} x & \text{if } x \geq 0 \\ -x & \text{if } x < 0 \end{cases}.$$

At first, this definition seems a little strange, but it is exactly what we would want the distance from x to 0 to be. For example, if $x = 5$, then the distance from 5 to 0 is clearly 5, and since $5 \geq 0$, according to the first line of the definition above, $|5| = 5$, and recovers that same answer. If, on the other hand, $x = -5$, then the distance from -5 to 0 is still 5. Since $-5 < 0$, looking at the second line of the definition above, we would conclude that $|-5| = -(-5) = 5$ again. The definition of $|x|$ is set up to ensure that you will leave x unchanged if $x \geq 0$, and to change the sign of x to make it positive if x were negative.

We can now prove some very useful facts about the absolute value function in Proposition 2.8 below. We remark that an almost identical proof to the one given below will allow us to replace $\leq$ with $<$ and $\geq$ with $>$ in Assertions (3) and (4) (see Exercise 4).

Proposition 2.8. *Let $x, y \in \mathbb{R}$.*

1. $|x| \geq 0$.
2. $|x| = 0$ *if and only if* $x = 0$.
3. *Suppose* $a \in \mathbb{R}$ *and* $a \geq 0$. *Then* $|x| \leq a$ *if and only if* $-a \leq x \leq a$.
4. *Suppose* $a \in \mathbb{R}$ *and* $a \geq 0$. *Then* $|x| \geq a$ *if and only if* $x \leq -a$ *or* $x \geq a$.
5. $|xy| = |x||y|$.
6. $|x|^2 = x^2$, *so that* $|x| = \sqrt{x^2}$.

Idea of proof. *Each of these assertions will be proven straightforwardly by using the order axioms (Axiom 2.5) and Proposition 2.6, and the definition of the absolute value (Definition 2.7).*

Proof of Proposition 2.8. To prove Assertion (1), we consider the two cases $x \geq 0$ and $x < 0$. If $x \geq 0$, then $|x| = x \geq 0$. If $x < 0$, then $|x| = -x > 0$ by Assertion (1) of Proposition 2.6.

To prove Assertion (2), we prove the logically equivalent statement: $|x| \neq 0$ if and only if $x \neq 0$. Since $|x|$ is either x or $-x$, $|x| \neq 0$ means either $x \neq 0$ or $-x \neq 0$. In either case, $x \neq 0$. Conversely, if $x \neq 0$, then neither of x or $-x$ is equal to 0, so $|x| \neq 0$.

We now prove Assertion (3). If $|x| \leq a$, and $x \geq 0$, then $|x| = x$, and so $0 \leq x \leq a$. If $|x| \leq a$ but $x < 0$, then $|x| = -x \leq a$, and so by Assertion (2) of Proposition 2.6, we must have $0 > x \geq -a$. In conclusion, we must have $-a \leq x \leq a$.

We prove Assertion (4) in a very similar way. If $|x| \geq a$ and $x \geq 0$, then $|x| = x \geq a$. If $|x| \geq a$ and $x < 0$, then $|x| = -x \geq a$, so again by Assertion (2) of Proposition 2.6, we must have $x \leq -a$.

Assertion (5) can be proven by considering all four possibilities of x or y being negative or nonnegative. See Exercise 6.

We can prove Assertion (6) by observing that $|x|^2 = |x| \cdot |x| = |x \cdot x| = |x^2|$ by the previous assertion. Also, $x^2 \geq 0$ by Assertion (3) of Proposition 2.6, so $|x|^2 = |x^2| = x^2$. The result now follows. □

We can now introduce a fact known as the *Triangle Inequality* in Theorem 2.9. Its name comes from the geometric intuition that the shortest route from one vertex of a triangle to any other is along the straight line that connects them, as opposed to the path through the other vertex. It is arguably one of the most commonly used theorems throughout this entire book and in Analysis in general.

Theorem 2.9 (Triangle Inequality). *For any real numbers a and b, we have*

$$|a + b| \leq |a| + |b|.$$

Idea of proof. *We algebraically manipulate $|x+y|^2$ to conclude $|x+y|^2 \leq (|x| + |y|)^2$, and the result will follow from there. Part of that manipulation uses the fact that $x \leq |x|$ (see Exercise 7).*

Proof of Theorem 2.9. We manipulate $|a + b|^2$ and repeatedly use Assertion (6) of Proposition 2.8:

$$\begin{aligned}
|a+b|^2 &= (a+b)^2 && \text{(Assertion (6) of Proposition 2.8)} \\
&= a^2 + b^2 + 2ab \\
&= |a|^2 + |b|^2 + 2ab && \text{(Assertion (6) of Proposition 2.8)} \\
&\leq |a|^2 + |b|^2 + 2|a||b| && \text{(Exercise 7)} \\
&= (|a| + |b|)^2.
\end{aligned}$$

Now that $|a+b|^2 \leq (|a|+|b|)^2$, by Assertion (7) of Proposition 2.6, we can conclude $|a+b| \leq |a|+|b|$. □

Before continuing, it is worth briefly expounding on a somewhat advanced perspective on the absolute value and the triangle inequality. There are different sorts of sets to do analysis on other than $\mathbb{R}$. For example, one could want to study the $xy-$plane instead (sometimes called $\mathbb{R}^2$). What is required, however, for this study to be a study of analysis is a notion of distance on that set, and that this notion of distance must satisfy the triangle inequality. On $\mathbb{R}$, this notion of distance is the absolute value, and one could define the "distance" between any two points (x_1, y_1) and (x_2, y_2) according to the Pythagorean Theorem as

$$\sqrt{(x_2-x_1)^2+(y_2-y_1)^2}.$$

These sets that are endowed with a compatible notion of distance are called *metric spaces*, and many of the analytical properties of $\mathbb{R}$ that we study in this book extend to these more abstract metric spaces with very little change.

2.1.4 The Extended Real numbers

It will be sometimes convenient to preform certain arithmetic operations with the symbols ∞ and $-\infty$. The *extended real numbers* is the set of real numbers with these two symbols included, and is defined below.

Definition 2.10 (The Extended Real Numbers)**.** Let $\mathbb{R}$ be the set of real numbers, and define the *extended real numbers*, $\mathbb{R}^\sharp$, as

$$\mathbb{R}^\sharp = \mathbb{R} \cup \{\infty\} \cup \{-\infty\}.$$

The symbols ∞ (sometimes written as "$+\infty$") and $-\infty$ are not real numbers, and we define them to satisfy $-\infty < \infty$, and for all $x \in \mathbb{R}$ we have $-\infty < x < \infty$. This declaration makes $\mathbb{R}^\sharp$ an ordered set, where the symbol $\leq$ has its usual meaning: for all $a, b \in \mathbb{R}^\sharp$, then $a \leq b$ means either $a < b$ or $a = b$. Using this ordering, one notices that Trichotomy also holds in $\mathbb{R}^\sharp$: for every $x, y \in \mathbb{R}^\sharp$, exactly one of $x < y, y < x$, or $x = y$ is true. For convenience, if $x = \infty$ or $x = -\infty$, then we may call x **infinite**. Accordingly, if $x \in \mathbb{R}$, we may call x **finite**, or simply **real**.

There are four arithmetic properties of $\mathbb{R}^\sharp$ that we define as well:

Definition 2.11. Arithmetic and order are defined in $\mathbb{R}^\sharp$ according to the following rules:

1. Any arithmetic involving only members of $\mathbb{R}$ is carried out as before, and the operations of $+$ and $\cdot$ in $\mathbb{R}^\sharp$ are commutative and associative wherever defined.
2. If $x \in \mathbb{R}$, then $x \pm \infty = \pm\infty$. Also, $\infty + \infty = \infty$, and $-\infty - \infty = -\infty$.

3. Suppose $x \in \mathbb{R}$. If $x > 0$, then $x \cdot (\pm\infty) = \pm\infty$. If $x < 0$, then $x \cdot (\pm\infty) = \mp\infty$. In particular, $(-1) \cdot \infty = -\infty$. Also, $(\infty)(\infty) = \infty$.

4. If $x \in \mathbb{R}$, then $\frac{x}{\pm\infty} = 0$.

We cannot simply include $\pm\infty$ as real numbers since, if they were, then they would also have to obey the rest of the field axioms (Axiom 2.1), and this is not possible to do in a mathematically consistent way. For instance, the arithmetic operations $\infty - \infty$, $0 \cdot (\pm\infty)$, and $\frac{\pm\infty}{\pm\infty}$ are not defined.

Example 2.12. This example is designed to illustrate what goes wrong if we were to try to assign an extended real number to $\infty - \infty$. Suppose that $\infty - \infty = x \in \mathbb{R}^\sharp$, and that the field axioms (Axiom 2.1) held true for the extended real numbers. Since $1 + \infty = \infty$ (Definition 2.11, part (2)), we must have

$$x = \infty - \infty = (\infty + 1) - \infty = \infty - \infty + 1 = x + 1.$$

We conclude that $x = x + 1$, and when we subtract x, it leads to the erroneous conclusion that $0 = 1$ (see Axiom 2.1, part (4)). The other operations of $0 \cdot (\pm\infty)$ and $\frac{\pm\infty}{\pm\infty}$ are similarly impossible to define in a consistent way. See Exercises (10) and (11). ■

Since the extended real numbers are the set of all objects x which satisfy $-\infty \leq x \leq \infty$, sometimes people write $[-\infty, \infty] = \mathbb{R}^\sharp$.

2.1.5 Neighborhoods

We complete this section by defining a type of set, called a *neighborhood* that will be of great use to us throughout the book. To help understand the terminology used, consider the neighborhood you live in. It is roughly defined as the locations that are "close by". A neighborhood of a real number x is treated in the same way: it is the set of locations (real numbers) that are close to x. Sometimes we wish to quantify just how close this is by assigning a positive real number as the distance (or radius) we would consider above and below x. This is formalized in the following definition.

Definition 2.13. Let $x \in \mathbb{R}$. A *neighborhood*, U, of x is the set

$$U = (x - \epsilon, x + \epsilon) \quad \text{for some } \epsilon.$$

A neighborhood U of ∞ is of the form

$$U = (M, \infty) \quad \text{for some } M \in \mathbb{R},$$

and a neighborhood of $-\infty$ is of the form

$$U = (-\infty, M) \quad \text{for some } M \in \mathbb{R}.$$

If $U = (x - \epsilon, x + \epsilon)$ is a neighborhood of x, then we will sometimes want to include ϵ in our discussion. In this event, we may refer to U as an $\epsilon-$*neighborhood* of x, or even refer to U as U_ϵ (when we wish to emphasize the radius ϵ), or as $U_\epsilon(x)$ (when we wish to emphasize the center x). In addition, sometimes ϵ is called the *radius* of the neighborhood. Note that if a number z is in the $\epsilon-$neighborhood U of x, then

$$\begin{aligned} z \in U & \iff z \in (x - \epsilon, x + \epsilon) \\ & \iff x - \epsilon < z < x + \epsilon \\ & \iff -\epsilon < z - x < \epsilon \\ & \iff |z - x| < \epsilon. \end{aligned} \tag{2.2}$$

If $S \subseteq \mathbb{R}$ and $x \in S$, then x is an *interior point of* S if there exists a neighborhood U of x so that $U \subseteq S$. For example, if $S = [0, 1]$, then the set of interior points of S is $(0, 1)$, and $0, 1 \in [0, 1]$ are not interior points.

For neighborhoods of ∞ or $-\infty$, we do not generally refer to M as in Definition 2.13 as some sort of "radius", and the analogy of a neighborhood being the set of things that are "close by" breaks down: no real number could legitimately be called "close" to ∞. All the same, these sorts of sets will play a central and unifying role in the theory to come.

We close this section with an important fact that will be of use repeatedly throughout the book: for every distinct pair of real numbers, there are neighborhoods around them that are disjoint, "separating" the points. This property is more commonly referred to as being *Hausdorff*.

Theorem 2.14. *Suppose $x, y \in \mathbb{R}$, with $x \neq y$. There exists a neighborhood U of x and a neighborhood V of y with $U \cap V = \emptyset$.*

Idea of proof. *We need to construct the required neighborhoods of x and y in this proof, which amounts to finding a radius of each neighborhood that ensures that they will not intersect. We could use the same radius for both neighborhoods U and V, and half of the distance from x to y will do, see Figure 2.1.1. We will find a contradiction to the assumption that there exists a real number in $U \cap V$ by measuring the distance from x to y and using the Triangle Inequality (Theorem 2.9). If the following proof does not make sense, read Example 2.15 and the discussion preceding it, then try reading the proof again.*

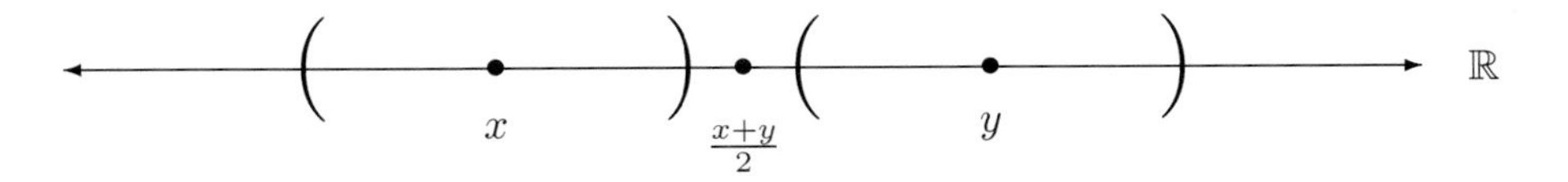

FIGURE 2.1.1
Disjoint neighborhoods of x and y.

Proof of Theorem 2.14. Let $\epsilon = \frac{1}{2}|x - y|$, and define U to be the ϵ−neighborhood $U = (x - \epsilon, x + \epsilon)$. Using the same ϵ, define V to be the ϵ−neighborhood $V = (y - \epsilon, y + \epsilon)$. We claim that $U \cap V = \emptyset$. Suppose there did exist a $z \in U \cap V$. Since $z \in U$ we must have $|z - x| < \epsilon$, and since $z \in V$ we must have $|z - y| < \epsilon$ (see the discussion following Definition 2.13, and Equation (2.2)). Then, using the Triangle Inequality (Theorem 2.9),

$$\begin{aligned} |x - y| &= |(x - z) + (z - y)| && \\ &\leq |x - z| + |z - y| && \text{(Triangle Inequality, Theorem 2.9)} \\ &< \epsilon + \epsilon && (z \in U \text{ and } z \in V) \\ &= 2\epsilon && \\ &= |x - y|. && \text{(definition of } \epsilon) \end{aligned}$$

We conclude that $|x - y| < |x - y|$, which is not the case. □

In the proof of Theorem 2.14, we chose ϵ, the radius of both neighborhoods, to be half of the distance between the two real numbers x and y. There is no one "correct" way to choose this ϵ: we could have also chosen ϵ to be one-third the distance from x to y (or any fraction of that distance smaller than one half), and the neighborhoods constructed will still be disjoint. There is, however, a wrong way to choose ϵ, as is illustrated in the next example.

Example 2.15. Suppose $x = 1$, and $y = 13$. The goal of this example is to demonstrate that there are many ways to find neighborhoods of x and y that are disjoint, but that there are also choices of ϵ that do not make it so.

1. First, choose ϵ to be one-half of the distance from $x = 1$ to $y = 13$: $\epsilon = \frac{1}{2}|x - y| = \frac{1}{2}|1 - 13| = 6$. The neighborhoods U and V defined in the proof above are

   $$U = (1 - 6, 1 + 6) = (-5, 7), \text{ and } \quad V = (13 - 6, 13 + 6) = (7, 19).$$

   One can clearly see that $U \cap V = \emptyset$, so that this choice of ϵ "works", although just barely: if we were to include only 7 into both sets, they would no longer be disjoint.

2. Now choose ϵ to be one-third the distance from $x = 1$ to $y = 13$: $\epsilon = \frac{1}{3}|1 - 13| = 4$. Now

   $$U = (1 - 4, 1 + 4) = (-3, 5) \text{ and } V = (13 - 4, 13 + 4) = (9, 17).$$

   U and V are still disjoint in this case, and there is a little more breathing room: the two sets are separated by all numbers from 5 to 9. This is not any "better" than before (when $\epsilon = 6$), since having more breathing room was never a consideration in Theorem 2.14, rather, we only cared that the neighborhoods were disjoint. In this way, there are very many different choices for our neighborhoods that are acceptable.

3. This time, choose ϵ to be exactly the distance between x and y: $\epsilon = |x - y| = 12$. Now

$$U = (1-12, 1+12) = (-11, 13) \text{ and } V = (13-12, 13+12) = (1, 25).$$

These neighborhoods are not suitable for the purposes of demonstrating Theorem 2.14 in this case, since all of the numbers from 1 to 13 are contained in both sets.

Thus, the choice of the radius of these neighborhoods must be done with care: some choices will produce the desired result, some will not. ■

2.1.6 Review of important concepts from Section 2.1

1. We begin our study of the real numbers by taking as axiom some basic properties, known as the field axioms (Axiom 2.1) and the order axioms (Axiom 2.5). All of what we deduce about the real numbers can be traced back to these axioms.
2. The absolute value function on $\mathbb{R}$ is extremely important and is defined in Definition 2.7. The Triangle Inequality (Theorem 2.9) is an extremely important fact used throughout the book.
3. The extended real number system $\mathbb{R}^\sharp = \mathbb{R} \cup \{\infty\} \cup \{-\infty\}$ is sometimes used as a convenient way of stating interesting facts. Beware: although some intuitive arithmetic operations are defined that involve $\pm\infty$, there are some others that are not defined. See Example 2.12 and Exercises (10) and (11).
4. A neighborhood of a real number or $\pm\infty$ is defined in Definition 2.13. These neighborhoods will be a unifying way to consider much of the upcoming theory. One important fact is that for any two distinct real numbers, there is a neighborhood of each that does not intersect. See Theorem 2.14.

Exercises for Section 2.1

1. This exercise is designed to convince you that $\frac{1}{0}$ and $\frac{0}{0}$ could not ever adequately define a real number.
   (a) Suppose $\frac{1}{0}$ were a real number, say $\frac{1}{0} = x \in \mathbb{R}$. Deduce the contradiction $0 = 1$. (*Hint: See part (4) of Axiom 2.1, and part (1) of Proposition 2.4.*)
   (b) Suppose that $\frac{0}{0}$ were a real number, say $\frac{0}{0} = x \in \mathbb{R}$. Let y be *any* other real number. Prove that $\frac{0}{0} = y$ as well, so that $\frac{0}{0}$ does not uniquely determine one real number.
2. Prove that if $x > 0$, then $\frac{1}{x} > 0$. (This completes the proof of Assertion (5) of Proposition 2.6.)

3. Prove that if x and y are nonequal real numbers, then their average lies strictly between them.

4. * Suppose $x, a \in \mathbb{R}$, with $a > 0$.

   (a) Prove $|x| < a$ if and only if $-a < x < a$.

   (b) Prove $|x| > a$ if and only if $x < -a$ or $x > a$.

5. Prove that if $x \in \mathbb{R}$, then $|(|x|)| = |x|$.

6. If $x, y \in \mathbb{R}$, then $|xy| = |x||y|$.

7. Prove that for every $x \in \mathbb{R}$, we have $x \leq |x|$.

8. The Triangle Inequality (Theorem 2.9) states that for every real numbers a and b, we have $|a + b| \leq |a| + |b|$. What additional conditions on a and b will force $|a + b| = |a| + |b|$?

9. Let a and c be real numbers. This exercise is designed to help you prove that $||a| - |c|| \leq |a - c|$ for any $a, b, c \in \mathbb{R}$.

   (a) Show that $|c| - |a| \leq |c - a|$. *[Hint: Write $c = a + b$ for some number b, apply the Triangle Inequality to $|a + b|$, and express your result in terms of a and c.]*

   (b) Show that $-|a - c| \leq |a| - |c|$. *[Hint: Interchange the roles of a and c, and express $a = c + \tilde{b}$ for some $\tilde{b}$. Now apply the Triangle Inquality to $|c + \tilde{b}|$ and express your result in terms of a and c. Notice that $|a - c| = |c - a|$.]*

   (c) Use the previous parts and Assertion (3) of Proposition 2.8 to deduce $||a| - |c|| \leq |a - c|$.

10. Demonstrate that it is not possible to define $0 \cdot \infty$ in a way that is consistent with the field axioms (Axiom 2.1). (*Hint: See Example 2.12.*)

11. Demonstrate that it is not possible to define $\frac{\infty}{\infty}$ in a way that is consistent with the field axioms (Axiom 2.1). (*Hint: See Example 2.12.*)

12. (This exercise proves that $\mathbb{R}^\sharp$ is Hausdorff, compare it with Theorem 2.14. See the proof of Theorem 3.111.) Suppose $x, y \in \mathbb{R}^\sharp$, with $x \neq y$. There exists a neighborhood U of x and a neighborhood V of y with $U \cap V = \emptyset$. *(Hint: Consider different cases where either x or y (or both) are infinite, and use Theorem 2.14 for the case that both x and y are finite.)*

2.2 The Supremum and Infimum

In this section, we define the supremum and infimum of a set (Definitions 2.18 and 2.22). Before these definitions, we discuss *bounded sets* in Definition 2.16, and after these definitions we give several practice problems illustrating how to determine the supremum and infimum of a set and some theoretical examples that further illustrate this process.

2.2.1 Bounded sets

In order to properly define the supremum and infimum of a set, we must define an *upper* and *lower bound* for a set.

Definition 2.16. Let $S \subseteq \mathbb{R}^\sharp$. An element $b \in \mathbb{R}^\sharp$ is

1. an *upper bound* for S if $b \geq x$ for all $x \in S$.
2. a *lower bound* for S if $b \leq x$ for all $x \in S$.

If a real upper bound for S exists, then S is said to be *bounded above.* If a real lower bound for S exists, then S is said to be *bounded below.* The set S is *bounded* if there exists an upper and lower bound for S, and *unbounded* if it is not bounded.

The following example illustrates some common features of bounds on a set, and that bounds for a set need not be unique.

Example 2.17. Let $S = (-\infty, 6)$. The number 6 is an upper bound for the set S, and so according to Definition 2.16, S is bounded above. The number 47 is also an upper bound for S, although for some reason it does not seem quite as good of an upper bound: the smaller upper bound of 6 seems to better describe how large the values in the set are, whereas 47 does not seem to quite as well. All the same, according to Definition 2.16, 47 is still an upper bound for S.

The set S is not bounded below, since there is no real number b for which $b \leq x$ for all $x \in S$. The element $-\infty$ is a lower bound for S (as it is for every subset of $\mathbb{R}^\sharp$), although $-\infty$ is not real. Therefore, although there exists a real upper bound for S, there does not exist a real lower bound for S, so S is unbounded. ■

Before moving on, we pause to establish a bit of terminology. The elements $\pm\infty \in \mathbb{R}^\sharp$ are not best described as *numbers*, since it seems to suggest that they are *real numbers.* Instead, we may refer to elements of $\mathbb{R}^\sharp$ as *quantities* to allow for $\pm\infty$ as part of the discussion. In reality, $\pm\infty$ are not numbers, but more appropriately described as bounds as in Definition 2.16.

2.2.2 The Supremum and Infimum of a set

Example 2.17 pointed out that a set can have many upper bounds, and that the smaller the upper bound of a set is, the more descriptive that number is in terms of describing how large the numbers are in the set. The *supremum* of a set is meant to be just that: the smallest upper bound for a set.

Definition 2.18. Let $S \subseteq \mathbb{R}$. The *supremum* of S, written $\sup(S)$, is the smallest upper bound for S.

Remark 2.19. Sometimes the supremum of a set is referred to as the *least upper bound* of S, or the $lub(S)$.

The following example illustrates the thought process behind how one could find the supremum of a set.

Example 2.20. Let $S = (-\infty, 6)$ as in Example 2.17. We noted there that both 6 and 47 are upper bounds, and in fact, there are many upper bounds. The supremum of $(-\infty, 6)$ is defined in Definition 2.18 as the smallest upper bound for S. Given that the set is made up of all real numbers less than 6, it seems natural to think that 6 is that smallest upper bound. But how do we *prove* that $6 = \sup(S)$?

Since it is clear that 6 is already an upper bound for S, we only need to show that it is the smallest of all of the upper bounds. We can prove this by contradiction: suppose that b is an upper bound for S, and that $b < 6$. To reach a contradiction, we could show that b is not an upper bound for S. Since $b < 6$, we can find the element $x = \frac{b+6}{2}$ (the average of b and 6), and according to Exercise 2.1.3, $b < x < 6$. Therefore, $x \in (-\infty, 6) = S$ but $b < x$. This contradicts the assumption that b is an upper bound for S. We have proven that 6 is an upper bound and that any number smaller than 6 is *not* an upper bound, therefore 6 is the smallest upper bound. ■

Example 2.20 presents an excellent method for proving that a certain number is the supremum of a set S. First, find an upper bound s that you think is the supremum of your set, and illustrate that it is indeed an upper bound (this is usually obvious). Next, suppose that there were a smaller upper bound $b < s$ for S, and proceed to find an element $x \in S$ for which $b < x$. This demonstrates that any number smaller than s is not an upper bound for S.

We now practice the process of finding the supremum of a set. Notice that in what follows and Example 2.20, the supremum of a set or any upper bound for a set need not be an element of the set.

Practice 2.21. In the following, find $\sup(S)$.

(a) $S = (-1, 1]$.

(b) $S = \mathbb{R}$.

Methodology. *We shall follow the idea given in the discussion following Example 2.20: for each set we find an upper bound that we think is the supremum, and then show that no smaller number could be an upper bound.*

Solution. (a) We claim that $\sup(S) = 1$. Notice that 1 is clearly an upper bound for S. If b is an upper bound for S and $b < 1$, then their average $x = \frac{b+1}{2}$ has $b < x < 1$. If $b \leq -1$, then it is clearly not an upper bound for S, since $1 \in S$ and $b < 1$, which contradicts our assumption that b is an upper bound for S. If $-1 < b$, then the element $x \in S$, and since $b < x$, this also contradicts the assumption that b is an upper bound for S. Therefore $1 = \sup(S)$.

(b) We claim that $\sup(S) = \infty$. Clearly, ∞ is an upper bound for S (as it is for every subset of the extended real numbers). But if b is an upper bound for S where $b < \infty$ (clearly $b = -\infty$ is not an upper bound for S), then since b is real, the element $b + 1 > b$, and $b + 1 \in S$. This illustrates that b is not an upper bound for S. So, $\sup(S) = \infty$. ■

The infimum of a set functions in the same way that the supremum does, except that instead of trying to find a descriptive upper bound, it is a descriptive lower bound for the set.

Definition 2.22. Let $S \subseteq \mathbb{R}$. The *infimum* of S, written $\inf(S)$, is the largest lower bound for S.

Remark 2.23. Sometimes the infimum of a set is referred to as the *greatest lower bound* of S, or the $glb(S)$.

Remark 2.24. It is quite clear from the definition that if $S \neq \emptyset$, then for any element $s \in S$, $\inf(S) \leq s \leq \sup(S)$, so that $\inf(S) \leq \sup(S)$, which agrees with our intuition.

Before practicing this concept, we pause to make a helpful remark that will assist us in finding the infimum of a set that is similar to the discussion following Example 2.20. In order to find the infimum of a set S, we first find a *lower bound* s for S that we think is the infimum. Next, suppose that there were a *larger* lower bound $b > s$, and proceed to find an element $x \in S$ for which $b > x$. This demonstrates that any number larger than s is not a lower bound for S.

We now practice the process of finding the infimum of a set. Compare the answers to Example 2.20 and Practices 2.21 and 2.25: the sup or inf of a set need not be in the set, and is almost never a consideration in determining what $\sup(S)$ or $\inf(S)$ actually is.

Practice 2.25. In the following, find $\inf(S)$.

(a) $S = (-1, 1]$.

(b) $S = \mathbb{R}$.

Methodology. *We shall follow the idea given in the discussion above following Definition 2.22: for each set we find a lower bound that we think is the infimum, and then show that no larger number could be a lower bound.*

Solution. (a) We claim that $\inf(S) = -1$. Notice that -1 is clearly a lower bound for S. If b is a lower bound for S and $b > -1$, then their average $x = \frac{b+1}{2}$ has $-1 < x < b$. If $b \geq 1$, then it is clearly not an upper bound for S, since $1 \in S$ and $b > 1$, which contradicts our assumption that b is a lower bound for S. If $b < 1$, then the element $x \in S$, and since $b > x$, this also contradicts the assumption that b is a lower bound for S. Therefore, $-1 = \inf(S)$.

(b) We claim that $\inf(S) = -\infty$. Clearly, $-\infty$ is a lower bound for S (as it is for every subset of the extended real numbers). But if b is a lower bound for S where $b > -\infty$ (clearly $b = \infty$ is not a lower bound for S), then since b is real, the element $b - 1 < b$, and $b - 1 \in S$. This illustrates that b is not a lower bound for S. So, $\inf(S) = -\infty$. ■

We pause for a moment to point something out: **we have not demonstrated that any set actually has a supremum or infimum. Nor have we illustrated that these quantities are unique should they exist.** This is a very important issue that we discuss in the next section, and we shall find these issues reasonably resolved in the *Completeness Axiom* (Axiom 2.28).

2.2.3 More practice with the sup and inf

We close this section with some illustrations of how to prove statements involving the sup and inf in situations where you may not explicitly know what set you are given. These facts are more theoretical than those presented in Example 2.20 and Practices 2.21 and 2.25.

Practice 2.26. Let $S \subseteq \mathbb{R}$, and suppose S is nonempty. Let $a \in \mathbb{R}$. Define the set

$$a + S = \{a + s | s \in S\}.$$

Prove that $\sup(a + S) = a + \sup(S)$, and that $\inf(a + S) = a + \inf(S)$. (For this problem, you may assume that $\sup(S)$ and $\inf(S) \in \mathbb{R}$.)

Methodology. *We tackle each question separately, although the solutions will be very similar. To verify* $\sup(a + S) = a + \sup(S)$*, we simply check that* $a + \sup(S)$ *is an upper bound for* $a + S$*, and that any number smaller than* $a + \sup(S)$ *is not an upper bound for* S*. This is the approach outlined in the discussion following Example 2.20. The inf question is handled similarly, according to the discussion following Definition 2.22.*

Solution. For convenience, let $\alpha = \sup(S)$, and $\beta = \inf(S)$.

We first show that $a + \alpha$ is an upper bound for $a + S$. Since $\alpha = \sup(S)$, α is already an upper bound for S. Therefore, if $s \in S$, then $s \leq \alpha$. It follows that if $s \in S$, then $a + s \leq a + \alpha$. Therefore $a + \alpha$ is an upper bound for $a + S$.

Next we show that if b is an upper bound for $a + S$, and $b < a + \alpha$, then there exists an element $a + x \in a + S$ so that $b < a + x$, showing that b could not be an upper bound for $a + S$. Since $b < a + \alpha$, we have $b - a < \alpha$. Therefore,

$b - a$ is not an upper bound for S since α is the smallest upper bound for S (this statement is crucial). Therefore, there exists an $x \in S$ so that $b - a < x$, or, $b < a + x \in a + S$. This completes the proof that $a + \alpha = \sup(a + S)$.

We prove that $\inf(a + S) = a + \beta$ in a similar manner. We first show that $a + \beta$ is a lower bound for $a + S$. Since β is a lower bound for S, $\beta \leq s$ for every $s \in S$. So, $a + \beta \leq a + s$ for every $s \in S$. Therefore, $a + \beta$ is a lower bound for $a + S$.

Next we show that if b is a lower bound for $a+S$, and $b > a+\beta$, then there exists an element $a + x \in a + S$ so that $b > a + x$, showing that b could not be a lower bound for $a + S$. Since $b > a + \beta$, we have $b - a > \beta$. Therefore, $b - a$ is not a lower bound for S since β is the largest lower bound for S (again, this statement is crucial). So, there exists an $x \in S$ so that $b - a > x$, or, $b > a + x \in a + S$. This completes the proof that $a + \beta = \inf(a + S)$. ■

Practice 2.27. Let A and B be nonempty subsets of $\mathbb{R}$, and $A \subseteq B$. Prove that

$$\inf(B) \leq \inf(A) \leq \sup(A) \leq \sup(B).$$

Methodology. *The middle inequality* $\inf(A) \leq \sup(A)$ *is true since A is nonempty (see Remark 2.24). We prove* $\sup(A) \leq \sup(B)$ *by noticing that every upper bound for B is also an upper bound for A, since $A \subseteq B$. Thus, since* $\sup(A)$ *is the smallest of the upper bounds of A, it follows that* $\sup(A) \leq \sup(B)$. *The other inequality regarding infs is proven similarly.*

Solution. The middle inequality is obvious, see Remark 2.24. To prove $\sup(A) \leq \sup(B)$, we note that every upper bound for B is an upper bound for A since $A \subseteq B$. Since $\sup(A)$ is the smallest upper bound for A, and $\sup(B)$ is an upper bound for A (since it is an upper bound for B), it follows that $\sup(A) \leq \sup(B)$.

We now prove that $\inf(B) \leq \inf(A)$ in a similar way. Every lower bound for B is a lower bound for A since $A \subseteq B$. Since $\inf(A)$ is the largest upper bound for A, and $\inf(B)$ is a lower bound for A (since it is a lower bound for B), it follows that $\inf(B) \leq \inf(A)$. ■

2.2.4 Review of important concepts from Section 2.2

1. An upper bound of a set is a quantity that is larger than (or equal to) any of the elements in the set. A set is bounded above if it has a real upper bound. Similarly, a lower bound for a set is a quantity that is smaller than (or equal to) any of the elements in the set. A set is bounded below if it has a real lower bound. See Definition 2.16.

2. The supremum of a set is its smallest upper bound, and the infimum of a set is its largest lower bound (Definitions 2.18 and 2.22). These are meant to be descriptive quantities describing something about the set.

3. To prove that a certain quantity is the supremum of a set, first show that it is an upper bound for the set, and then show that any smaller number is not an upper bound (see the discussion on page 76 after Example 2.20). To prove that a certain quantity is the infimum of a set, first show that it is a lower bound for the set, and then show that any larger number is not a lower bound (see the discussion on page 77 after Definition 2.22).

Exercises for Section 2.2

For the problems in this section, you may assume that all supremums and infimums exist in $\mathbb{R}^\sharp$.

1. Determine $\sup(S)$ and $\inf(S)$ for the given sets S below.
   (a) $S = (-7, 4]$.
   (b) $S = (0, \infty)$.
   (c) $S = \mathbb{N}$.
   (d) $S = \mathbb{Q}$.
2. Prove that the set S is bounded if and only if $|S|$ is bounded, where
$$|S| = \{|s| | s \in S\}.$$
3. Let S be a nonempty subset of $[a, b]$. Prove that $\sup(S) \in [a, b]$.
4. Suppose $a > 0$, $S \subseteq \mathbb{R}$, and $\sup(S) \in \mathbb{R}$. Define the set $aS = \{as | s \in S\}$ (this is a similar construction of a set to Practice 2.26).
   (a) Prove $\sup(aS) = a\sup(S)$.
   (b) Prove $\inf(aS) = a\inf(S)$.
5. Suppose $a < 0$, and define aS as in Exercise 4. Prove that $\inf(aS) = a\sup(S)$. Deduce from this that $\inf(-S) = -\sup(S)$.
6. Suppose A and B are nonempty subsets of $\mathbb{R}$, and that $\sup(A), \sup(B) \in \mathbb{R}$. Prove that $\sup(A \cup B) = \max\{\sup(A), \sup(B)\}$.
7. Suppose A and B are nonempty subsets of $\mathbb{R}$, and that $\sup(A), \sup(B) \in \mathbb{R}$. Prove that $\inf(A \cup B) = \min\{\inf(A), \inf(B)\}$.
8. Suppose A and B are bounded nonempty subsets of $\mathbb{R}$. Define the set
$$A + B = \{a + b | a \in A, b \in B\}.$$
   (a) Prove that $A + B$ is bounded.
   (b) Prove that $\sup(A + B) = \sup(A) + \sup(B)$. (You may assume that $\sup(A + B), \sup(A), \sup(B) \in \mathbb{R}$.)

9. Suppose A and B are bounded nonempty subsets of $\mathbb{R}$. Define the set
$$AB = \{ab | a \in A, b \in B\}.$$
Find sets A and B where $\sup(A), \sup(B) \in \mathbb{R}$, but $\sup(AB) \neq \sup(A)\sup(B)$. (Compare this result with Exercise 8.)

10. Suppose S is a nonempty subset of $\mathbb{R}$, and that $\sup(S), \inf(S) \in \mathbb{R}$.
    (a) Prove that for every $\epsilon > 0$, there exists an $s \in S$ so that $\sup(S) - \epsilon < s \leq \sup(S)$.
    (b) Prove that for every $\epsilon > 0$, there exists an $s \in S$ so that $\inf(S) \leq s < \inf(S) + \epsilon$.

11. Let S be a nonempty set which is bounded above, and let U be the set of upper bounds of S:
$$U = \{x \in \mathbb{R} | x \text{ is an upper bound for } S\}.$$
Prove that $\sup(S) = \inf(U)$. (For this problem, you may assume that $\sup(S)$ and $\inf(S)$ are real.)

12. Suppose $f, g : A \to \mathbb{R}$, where A is a nonempty subset of $\mathbb{R}$. Let
$$f(A) = \{f(x) | x \in A\}, \quad g(A) = \{g(x) | x \in A\}, \text{ and}$$
$$(f + g)(A) = \{(f + g)(x) | x \in A\}.$$
Prove that $\sup((f + g)(A)) \leq \sup(f(A)) + \sup(g(A))$. Give an example to show equality is not always possible.

13. Suppose S is a finite, nonempty set. Prove that $\sup(S) \in S$, and so therefore such a set has a "largest element."

2.3 The Completeness Axiom

In this section we introduce the last axiom of the real numbers we shall assume: The Completeness Axiom (Axiom 2.28). This is actually a fact that can be proven after a careful construction of $\mathbb{R}$, but since we do not present that here, we will simply assume The Completeness Axiom holds. This very important fact guarantees the existence of the supremum of a set in the event the set is bounded above, and that this supremum is real (as opposed to infinite). With this and other observations, we prove that the infimum of a set which is bounded below exists (and is real), and that overall the supremum and infimum of a set can be found for *any* subset of $\mathbb{R}^\sharp$.

We then prove that $\mathbb{N}$ is not bounded above (Theorem 2.33) and use that to establish the Archimedean Property (Theorem 2.34), which states that for any positive real number, there exists a natural number whose reciprocal is smaller than it. This fact will be crucial in later sections.

We close the section by proving that between any two distinct real numbers, there exists a rational number strictly between them. This can be stated as: $\mathbb{Q}$ is *dense* in $\mathbb{R}$. See Definition 2.36 and Theorem 2.39.

2.3.1 The Completeness Axiom and suprema and infima

A quick scan of the last section will reveal one important observation: We have not yet demonstrated that any subset of $\mathbb{R}$ actually has a supremum or infimum. This leads us to the final axiom concerning the real numbers that we shall use, known as The Completeness Axiom.

Axiom 2.28 (The Completeness Axiom). *Every nonempty subset of* $\mathbb{R}$ *that is bounded above has a supremum in* $\mathbb{R}$.

Remark 2.29. Recall from Section 2.1 that in this book we had assumed the real numbers existed and satisfied certain axioms (Axioms 2.1, 2.5, and now, The Completeness Axiom (Axiom 2.28)). We pointed out that the real numbers can be constructed and shown to satisfy those axioms, and indeed the same is true of The Completeness Axiom: in any construction of the real numbers, one can show The Completeness Axiom holds. We assume it here as an axiom since we do not present a construction of the real numbers. Again, the construction found in Rudin [24] is an excellent resource for the interested reader.

Immediately, several observations can be made.

Proposition 2.30. *Every nonempty subset of* $\mathbb{R}$ *that is bounded below has an infimum in* $\mathbb{R}$.

Idea of proof. *We want to relate our set which is bounded below (call it* S*), to one which is bounded above, and to then use The Completeness Axiom*

(Axiom 2.28) on that set which is bounded above and relate the supremum of that set with the infimum of A. To do this, we use Exercise 2.2.5.

Proof of Proposition 2.30. Let S be a nonempty subset of $\mathbb{R}$ that is bounded below. Then $-S$ is a nonempty subset of $\mathbb{R}$ that is bounded above (see Exercise 2.2.5). Therefore, $\sup(-S)$ exists according to Axiom 2.28, and is in $\mathbb{R}$. According to Exercise 2.2.5 (replacing $-S$ for S), we have $\inf(S) = -\sup(-S)$. $\square$

Proposition 2.31. *Suppose $S \subseteq \mathbb{R}$.*

1. *If S is nonempty, and* not *bounded above, then* $\sup(S) = \infty$.
2. *If S is nonempty, and* not *bounded below, then* $\inf(S) = -\infty$.
3. *If $S = \emptyset$, then* $\sup(S) = -\infty$, *and* $\inf(S) = \infty$. *This is the only circumstance where* $\inf(S) \not\leq \sup(S)$.

Idea of proof. *For the first assertion, we simply use the definition of what it means to not be bounded above to conclude that there is no real upper bound. As a result, ∞ is the only upper bound, and is therefore the least upper bound. The proof of the second assertion is very similar, and the last assertion follows by vacuous reasoning, and is best described as pathological.*

Proof of Proposition 2.31. To prove Assertion (1), note that by Definition 2.16 if S is not bounded above, then there exists no real upper bound. Therefore, ∞ is the only upper bound for S, and so as the smallest upper bound, $\sup(S) = \infty$.

The proof of Assertion (2) is similar: since S is assumed to be unbounded below, there exists no real lower bound. Therefore, $-\infty$ is the only lower bound for S, and so as the largest lower bound, $\inf(S) = -\infty$.

To prove Assertion (3), let $b \in \mathbb{R}^\sharp$. We claim that b is an upper and lower bound for the empty set. To prove b is an upper bound, we need to check the claim: if $x \in S$, then $x \leq b$. Since x belongs to the empty set, the hypothesis of this implication is never met, and therefore the statement is true. Similar reasoning shows that b is a lower bound for S. Therefore every element of $\mathbb{R}^\sharp$ is an upper and lower bound for S. The smallest of these is the supremum, so $\sup(S) = -\infty$. The largest of these is the infimum, so $\inf(S) = \infty$. In any other situation, S is nonempty, and by the argument in Remark 2.24, we see that $\inf(S) \leq \sup(S)$. $\square$

We can now state an essential aspect of suprema and infima using The Completeness Axiom (Axiom 2.28), and Propositions 2.30 and 2.31: **If S is any subset of the real numbers, then the supremum and infimum of S exist in $\mathbb{R}^\sharp$.** Furthermore, we show in the next proposition that the supremum and infimum of a given subset of $\mathbb{R}$ are unique.

Proposition 2.32. *If $S \subseteq \mathbb{R}^\sharp$, then* $\sup(S)$ *and* $\inf(S)$ *are uniquely determined elements in* $\mathbb{R}^\sharp$.

Idea of proof. *Depending on whether or not S is bounded above, The Completeness Axiom 2.28 or Proposition 2.31 declare that* $\sup(S)$ *exists in* $\mathbb{R}^\sharp$. *Insofar as proving that* $\sup(S)$ *is uniquely determined, we argue by contradiction: if* $\sup(S)$ *were equal to two quantities at once, then one of these quantities would have to be larger, and would not be the smallest upper bound. Similarly for the infimum, which is left as Exercise 2.*

Proof of Proposition 2.32. Let $S \subseteq \mathbb{R}^\sharp$. According to The Completeness Axiom (Axiom 2.28) or Proposition 2.31, $\sup(S)$ exists and is an element of $\mathbb{R}^\sharp$. Suppose that $\sup(S) = \alpha$ and $\sup(S) = \beta$, with $\alpha < \beta$. Then α and β are both upper bounds for S, and they are both smallest upper bounds for S. Since $\alpha < \beta$ and β is a smallest upper bound for S, α must not be an upper bound for S, a contradiction. We leave the remainder of the proof (establishing the result for $\inf(S)$) for Exercise 2. □

2.3.2 The Archimedean Property

As an important application of The Completeness Axiom, we prove The Archimedean Property, which essentially says that there is a fraction of an integer that can be made as close to zero as we like. We will prove this fact in Theorem 2.34 after demonstrating that the natural numbers are not bounded above in Theorem 2.33.

Theorem 2.33. *The natural numbers are not bounded above.*

Idea of proof: *We prove this by contradiction: if* $\mathbb{N}$ *were bounded above, then The Completeness Axiom would give us a least upper bound for* $\mathbb{N}$. *We can exceed this supposed bound, providing a contradiction.*

Proof of Theorem 2.33. Suppose that $\mathbb{N}$ were bounded above. Then since $\mathbb{N}$ is nonempty, The Completeness Axiom (Axiom 2.28) applies, and $\sup(\mathbb{N}) = \alpha \in \mathbb{R}$. By definition, $\alpha - 1$ is *not* an upper bound of $\mathbb{N}$, and so there exists an $N \in \mathbb{N}$ with $\alpha - 1 < N$. Adding 1, we see that $\alpha < N + 1$, and $N + 1 \in \mathbb{N}$. Therefore α is not an upper bound for $\mathbb{N}$, a contradiction. □

Theorem 2.34 (Archimedean Property). *For every* $\epsilon > 0$, *there exists an* $N \in \mathbb{N}$ *so that* $\frac{1}{N} < \epsilon$.

Idea of proof: *We will prove the equivalent fact: for every* $\epsilon > 0$, *there exists an* $N \in \mathbb{N}$ *so that* $\frac{1}{\epsilon} < N$. *Such an* N *would have to exist, otherwise* $\frac{1}{\epsilon}$ *would be a real upper bound for* $\mathbb{N}$, *and no such bound exists according to Theorem 2.33.*

Proof of Theorem 2.34. Let $\epsilon > 0$ be given. According to Theorem 2.33, $\frac{1}{\epsilon}$ cannot be an upper bound for $\mathbb{N}$. Therefore, there exists an $N \in \mathbb{N}$ so that $\frac{1}{\epsilon} < N$, or rather, $\frac{1}{N} < \epsilon$. □

We can use The Archimedean Property (Theorem 2.34) to establish a rather intuitive fact about the real numbers that will be of great use in the upcoming proof of Theorem 2.39: for every real number x, there is an integer between $x-1$ and x; this is the content of the next Theorem 2.35.

Theorem 2.35. *If $x \in \mathbb{R}$, then there exists an $m \in \mathbb{Z}$ so that $x - 1 < m \leq x$.*

Idea of proof: *The basic idea is to find a set that could capture this integer m that we are looking for, and to use the defining properties of the supremum of that set to demonstrate that it satisfies the result of the theorem.*

Proof of Theorem 2.35. Let $S = (-\infty, x] \cap \mathbb{Z}$. We first show that S is nonempty and bounded above. To show that S is nonempty, note that the natural numbers are not bounded above, and so there is some natural number n for which $-x \leq n$. That is, the integer $-n \leq x$, so $-n \in S$. By definition, S is bounded above by x.

Since S is nonempty and bounded above, The Completeness Axiom (Axiom 2.28) applies, so $\sup(S)$ exists and is real. For convenience, call $\sup(S) = \alpha$. Since α is the least upper bound of S, $\alpha - 1$ is not an upper bound for S. So, there exists an $m \in S$ with

$$\alpha - 1 < m \leq \alpha.$$

By the definition of S, $m \in S$ means that both $m \in \mathbb{Z}$ and $m \leq x$. By adding 1 to the above equation, we see that $\alpha < m + 1$. If it were the case that $m + 1 \leq x$, then since $m + 1 \in \mathbb{Z}$ we would conclude $m + 1 \in S$, which contradicts α being an upper bound for S and $\alpha < m + 1$. This contradiction forces us to conclude $x < m + 1$, or, that $x - 1 < m$. When we combine this with our earlier observation that $m \leq x$ and $m \in \mathbb{Z}$, the proof is complete. □

2.3.3 The Density of $\mathbb{Q}$ in $\mathbb{R}$

As an application of The Archimedean Property, we can deduce another highly significant fact about the real numbers in Theorem 2.39: any real number is arbitrarily close to a rational number. This concept is more commonly referred to using the term "dense", which is described in the following definition.

Definition 2.36. Let $S \subseteq \mathbb{R}$. S is *dense* in $\mathbb{R}$ if every neighborhood contains an element of S.

Definition 2.36 seems a bit strange at first, but it is meant to apply to other areas of Analysis as well. We can specialize this definition to characterize the property of being dense in $\mathbb{R}$ in the following theorem, which we shall mainly use.

Theorem 2.37. *S is dense in $\mathbb{R}$ if and only if for every $x, y \in \mathbb{R}$ with $x < y$, there exists $s \in S$ with $x < s < y$.*

Idea of proof: *We shall prove each implication separately. If S is dense in $\mathbb{R}$, and $x < y$ are given, then we can find a neighborhood existing entirely between x and y, for which an element of S must exist according to Definition 2.36. Conversely, if a neighborhood U is given, we can choose two different elements in U and find an element of S between them that lies in the neighborhood.*

Proof of Theorem 2.37. Suppose first that S is dense in $\mathbb{R}$, and let $x < y$ be given. Find $z = \frac{x+y}{2}$, so that $x < z < y$, and choose $\epsilon = |z-x| = |y-z| = \frac{y-x}{2}$. Let U be the ϵ−neighborhood about z, and notice that $U \subseteq (x, y)$. Since S is dense in $\mathbb{R}$, there exists an element $s \in U \cap S$. So, $x < s < y$.

Conversely, suppose that for every real number $x < y$, there exists an $s \in S$ with $x < s < y$, and let U be a neighborhood of $\mathbb{R}$ about the real number z. The set U contains infinitely many elements, so choose $z_1 < z_2$ with $z_1, z_2 \in U$. By assumption, there exists an $s \in S$ with $z_1 < s < z_2$, so that $s \in U$. □

The following presents some examples of dense subsets of $\mathbb{R}$, and suggests why the word "dense" is used to describe such a set.

Example 2.38. For each of the following, we show that S is dense in $\mathbb{R}$, and we do this by using the characterization in Theorem 2.37.

(a) Let $S = \mathbb{R} - \{0\}$. Let $x < y$ be given. Consider $z = \frac{x+y}{2}$, the average of x and y. Then $x < z < y$. In the event that $z = 0$ so that $z \notin S$, then we have $x = -y < 0 < y$, and instead we could choose $z = \frac{y}{2} \in S$.

(b) Let $S = \mathbb{R} - \{\frac{1}{n} | n \in \mathbb{N}\}$. Let $x < y$ be given. If there are no elements of the form $\frac{1}{n}$ between x and y, then any element z between them is in S. In the event that there is only one element of the form $\frac{1}{n}$ between x and y, suppose $\frac{1}{N}$ is this element, and consider the average $z = \frac{1}{2}(x + \frac{1}{N})$. Then z is still clearly between x and y (since it is between x and $\frac{1}{N}$), and not equal to $\frac{1}{N}$. By assumption, there are no other elements of this sort between x and y, and so z is not of this form, and so $z \in S$. The last possibility is that there are two or more elements of the form $\frac{1}{n}$ between x and y. Let $\frac{1}{N}$ and $\frac{1}{M}$ be two such elements, where $M > N$. Then it is also the case that $\frac{1}{N+1} \in (x, y)$ since $\frac{1}{M} < \frac{1}{N+1} < \frac{1}{N}$. Now the average $z = \frac{1}{2}(\frac{1}{N+1} + \frac{1}{N})$ is not of the form $\frac{1}{n}$ for any $n \in \mathbb{N}$, and so $z \in S$, and

$$x < \frac{1}{M} < \frac{1}{N+1} < z < \frac{1}{N} < y.$$

(c) Let $S = \mathbb{R} - C$, where C is any countable set. If $x < y$ are given, then the set (x, y) is uncountable (see Exercise 1.5.5). We are looking for an element $s \in S$ with $s \in (x, y)$. Since C is countable, there must still be an uncountable number of elements in $(x, y) - C$, so there must exist such an s. Notice that the argument given here could have also worked for the other parts of this example. ■

We are now ready to present the result that the rational numbers are dense in $\mathbb{R}$. This fact is used repeatedly throughout this and other subjects in mathematics.

Theorem 2.39. *$\mathbb{Q}$ is dense in $\mathbb{R}$.*

Idea of proof: *We wish to find a real number of the form $\frac{n}{m}$ (for integers m, n) between two given numbers x and y. To find such a number between x and y, we dilate the distance from x to y by multiplying by some large $n \in \mathbb{N}$. The point of this is to spread nx far enough from ny until there is an $m \in \mathbb{N}$ with $nx < m < ny$, so that $x < \frac{n}{m} < y$. This proof is very similar to the one found in Rudin [24].*

Proof of Theorem 2.39. Let $x < y$ be given, so that $\frac{y-x}{2} > 0$. Using The Archimedean Property (Theorem 2.34), find an $n \in \mathbb{N}$ so that $\frac{1}{n} < \frac{y-x}{2}$. Then

$$\begin{aligned} 2 &< n(y-x) = ny - nx, \text{ so} \\ nx &< ny - 2. \end{aligned}$$

We use Theorem 2.35 to find an $m \in \mathbb{Z}$ with

$$ny - 2 < m \leq ny - 1.$$

Now we use the inequality $nx < ny - 2$ from above and the obvious inequality $ny - 1 < ny$ to conclude

$$nx < ny - 2 < m \leq ny - 1 < ny.$$

In other words, $nx < m < ny$, and so $x < \frac{m}{n} < y$. □

Knowing Theorem 2.39, there are a variety of interesting sets of which we may now compute the supremum and infimum.

Practice 2.40. Compute the supremum of the following sets.

(a) $S = (0, 1) \cap \mathbb{Q}$.

(b) $S = \{x \in \mathbb{Q} | x^2 < 2\}$.

Methodology. *The method for determining these suprema is the same as in Practice 2.21: we find an upper bound for S that we think could be the supremum, and then show that any smaller number is not an upper bound. It is in the second of these steps that we shall need to use the density of $\mathbb{Q}$ in $\mathbb{R}$ (Theorem 2.39).*

Solution. (a) We prove that $\sup(S) = 1$. Clearly, 1 is an upper bound for S. We now suppose $b < 1$ and show that b is not an upper bound for S. Since $\mathbb{Q}$ is dense in $\mathbb{R}$ (Theorem 2.39), there exists a rational number r between b

and 1: $b < r < 1$. If $b < 0$, then this r may be chosen to be $\frac{1}{2}$. If $0 \leq b < r < 1$, then $r \in (0,1)$, and so since r is also rational, $r \in S$. In either case, this contradicts the fact that b is an upper bound for S, so that $1 = \sup(S)$.

(b) We prove that $\sup(S) = \sqrt{2}$. Clearly, $\sqrt{2}$ is an upper bound for S, since if $x \in S$, $x^2 < 2$ so that $|x| < \sqrt{2}$, and so $-\sqrt{2} < x < \sqrt{2}$. Now suppose that $b < \sqrt{2}$. If $b < 0$, then we may find $0 \in S$ with $b < 0$. If $0 \leq b < \sqrt{2}$, since $\mathbb{Q}$ is dense in $\mathbb{R}$ (Theorem 2.39), there exists a rational number r between b and $\sqrt{2}$: $b < r < \sqrt{2}$. Since $0 < r < \sqrt{2}$, we have $b^2 < 2$, and so since r is rational, $r \in S$. In either case, this contradicts the fact that b is an upper bound for S, so that $\sqrt{2} = \sup(S)$. ■

2.3.4 Review of important concepts from Section 2.3

1. The Completeness Axiom (Axiom 2.28) guarantees that if S is any subset of $\mathbb{R}$ that is bounded above, then $\sup(S)$ exists and is real. One can use this axiom to show that if S is bounded below, then $\inf(S)$ exists and is real. Propositions 2.31 and 2.32 consider the case of unbounded sets, and demonstrate that if S is *any* subset of $\mathbb{R}$, then $\sup(S)$ and $\inf(S)$ exist and are unique quantities.
2. The natural numbers are not bounded above (see Theorem 2.33).
3. The Archimedean Property says that there is a fraction of a natural number that can be made as close to zero as desired (see Theorem 2.34). This fact will be repeatedly used later on.
4. We prove that $\mathbb{Q}$ is dense in $\mathbb{R}$ in Theorem 2.39. An equivalent formulation of this fact is to say that between any two distinct real numbers, there exists a rational number strictly between them.

Exercises for Section 2.3

1. Compute the following:
   (a) $\inf\{\frac{1}{n} | n \in \mathbb{N}\}$.
   (b) $\sup\{1 - \frac{1}{n^2} | n \in \mathbb{N}\}$.
   (c) $\sup\{r \in \mathbb{Q} | 0 < r < 1\}$.
2. Prove the remainder of Proposition 2.32. That is, prove that if $S \subseteq \mathbb{R}$, then $\inf(S)$ is unique.
3. Prove that if The Archimedean Property holds, then $\mathbb{N}$ is not bounded above. More specifically, prove that if for every $\epsilon > 0$ there exists an $N \in \mathbb{N}$ where $\frac{1}{N} < \epsilon$, then $\mathbb{N}$ is not bounded above.
4. Prove the following generalization of the Archimedean Property: Suppose $S \subseteq \mathbb{R}$ that is nonempty and not bounded above. Prove that for every $\epsilon > 0$, there exists an $s \in S$ so that $\frac{1}{s} < \epsilon$.
5. Show that if $S \subseteq \mathbb{R}$ is finite, then S is not dense in $\mathbb{R}$.

6. Find an example of a countably infinite subset of $\mathbb{R}$ that is not dense in $\mathbb{R}$.

7. Let α be any real number. Prove that for every $n \in \mathbb{N}$, there exists a rational number r_n with $\alpha - \frac{1}{n} < r_n < \alpha$. Using this, deduce that for every real number there is a rational number arbitrarily close to it.

8. Show that between any two distinct real numbers, there exist infinitely many rational numbers.

9. Prove that the set of irrational numbers is dense in $\mathbb{R}$. *[Hint: it may be helpful to use the fact that $\sqrt{2}$ is irrational (see Practice 1.4), that the rational numbers are dense in $\mathbb{R}$, and that rational numbers and irrational numbers often combine in certain ways to irrational numbers, as in Exercise 1.1.7.]*

2.4 Troubleshooting Guide for The Real Numbers

Type of difficulty:	*Suggestion for help:*
Section 2.1: The Real and Extended Real Number System	
"This section does not seem to contain any mathematical content. I also don't really understand what the point of all of this is."	The purpose of this section is to introduce the real numbers. We proceed very carefully, starting with Axioms 2.1 and 2.5 defining how the real numbers are supposed to behave. From there, we carefully and painstakingly derive from these axioms all of the familiar rules about the real numbers. Read the introduction to this section on page 60 for a perspective on what is designed to be accomplished in this section.
"Why can't we just consider ∞ and $-\infty$ as real numbers?"	If we did, then we would expect them to obey the field axioms (Axiom 2.1). Example 2.12 illustrates what goes wrong when trying to consider $\infty - \infty$, and Exercises (10) and (11) illustrate two other problems.
"Why is there so much discussion after Theorem 2.14 about the choices of ϵ that are suitable? I'm not even sure what 'suitable' means here."	The number ϵ is the common radius of the neighborhoods U and V, and a "suitable" choice of ϵ is one where the resulting neighborhoods U and V are disjoint. For more explanation about this, see Example 2.15, and then the proof of Theorem 2.14. Very many aspects of our further study of Analysis will involve choosing a "suitable" neighborhood (or neighborhoods), and exploring exactly what that means in this one situation will be helpful later on.
Section 2.2: The Supremum and Infimum	
"I don't really understand what the supremum of a set is."	Consider reading Example 2.17, followed by Definition 2.18. Example 2.20 carefully finds the supremum of a set, and the discussion that follows on page 76 gives a method for finding the supremum of a set.

Type of difficulty:	***Suggestion for help:***
"How do I determine $\sup(S)$, the supremum of a set?"	Find an upper bound for the set, and prove that any smaller number is not an upper bound. See Example 2.20, and the discussion following it on page 76. Then, try reading Practice 2.21.
"I don't understand what the infimum of a set is or how to compute it."	If you understand the supremum of a set and how to compute it, then see Definition 2.22 and the subsequent discussion. If you do not understand the supremum, then see the above difficulties regarding the supremum.
Section 2.3: The Completeness Axiom	
"I don't understand why The Completeness Axiom is such a big deal."	Up to this point, there is no reason why the supremum of a set should exist. In fact, one can prove that it must through a construction of the real numbers, although since we do not construct the real numbers here, we assume The Completeness Axiom is true and move forward. See Remark 2.29, the subsequent propositions, and finally the discussion afterward on page 83.
"Why is $\sup(\emptyset) = -\infty$? This seems to go against my intuition."	The issue is that there is no element in the empty set to illustrate that $-\infty$ is not an upper bound. See the proof of Proposition 2.31.
"It seems rather obvious that $\mathbb{N}$ is not bounded above (Theorem 2.33), why do we even need to prove that?"	There is nothing in the definition of $\mathbb{N}$ that automatically deduces this, and it therefore must be proven. In addition, this fact is crucial in the proof of The Archimedean Property (Theorem 2.34).

3

Sequences and Their Limits

Chapter Summary:

The theory of sequences and their limits is the starting point of modern analysis. With a firm background in our knowledge of the real numbers from Chapter 2, we can develop results in this chapter that are both interesting in their own right, and useful for almost every remaining chapter in the text.

Section 3.1: Sequences. This is an introduction to sequences, with emphasis on later concepts and ideas.

Section 3.2: Limits and Convergence. This section rigorously develops the idea of *convergence* in Definition 3.13: the tendency of a sequence of numbers to approach one particular number (the *limit* of the sequence).

Section 3.3: Limit Theorems. Since the definition of convergence is somewhat technical, we present a number of results in this section that will allow us to deal with limits of sequences a little more easily; we call these results "supporting results."

Section 3.4: Subsequences. Here, we introduce the concept of a *subsequence*, and relate the convergence of a sequence to the convergence of any of its subsequences.

Section 3.5: Monotone Sequences. This section introduces monotone sequences, a special kind of sequence. The Monotone Convergence Theorem (Theorem 3.74) and The Nested Intervals Theorem (Theorem 3.81) are established here in preparation for the Bolzano-Weierstrass Theorems (Theorems 3.84 and 3.91) in the next section.

Section 3.6: The Bolzano-Weierstrass Theorems. This section presents two of the most important theorems in real analysis, known as the Bolzano-Weierstrass Theorems.

Section 3.7: Cauchy Sequences. This section presents an equivalence between the set of convergent sequences and a set of sequences known as *Cauchy sequences.* These sequences are shown to be convergent, but with the added benefit that this can be done without explicit knowledge of the limit of the sequence.

Section 3.8: Infinite Limits. This section extends the notion of convergence to allow for the situation a list of numbers could tend to plus or minus infinity. Very many of the results up to this point that concern finite limits are extended to the infinite setting.

Section 3.9: The $\limsup$ **and** $\liminf$**.** A generalization of convergence is developed here, giving us an ability to study sequences that are not as well behaved as those previously encountered.

Section 3.10: Troubleshooting Guide for Sequences and Their Limits. A list of frequently asked questions and their answers, arranged by section. Come here if you are struggling with any material from this chapter.

3.1 Sequences

We begin the chapter by defining the objects that are central to the study of analysis: sequences. We will start with an intuitive definition of what a sequence should be, and progress toward a formal mathematical definition of a sequence that will be of use. After establishing the (standard) notation for sequences that we will be using, we will discuss two types of ways to define sequences, and how to construct new sequences from old ones. We conclude the section with a discussion of inequalities of sequences.

3.1.1 How to express sequences

Intuitively, a sequence is just some list of objects. In addition, the order of such a list matters as much as the objects themselves. For example, if you are driving somewhere unfamiliar, you may construct a sequence of directions that will take you to where you would like to go. If you perform these directions in a scrambled order, there is no telling where you would end up!

One way of constructing a sequence is to simply list out the members of it in some pre-determined order. For example, the sequence $2, 4, 6, 8, 10, \ldots$ has 2 as its first member, 4 as its second, 6 as its third, and so on. In this way, the sequence $2, 4, 6, 8, 10, \ldots$ is a function $f : \mathbb{N} \to \mathbb{R}$:

"the first member of my sequence is 2"	means	$f(1) = 2$,
"the second member of my sequence is 4"	means	$f(2) = 4$,
"the third member of my sequence is 6"	means	$f(3) = 6$,
$\vdots$		$\vdots$

Note also that this sequence is **infinite**, in that it is not meant to cease. It is sometimes the case (for example, computer science, cryptography, and any number of other subjects) that one might only be concerned with **finite** sequences, although, in this course, we will only be studying infinite sequences of real numbers.

Definition 3.1. A *sequence* is a function from $\mathbb{N}$ to $\mathbb{R}$.

In the above paragraph, we considered the sequence $2, 4, 6, \ldots$, and it would be convenient to have a better way of expressing this data than to always have to list it out. **We let a_n denote the nth term of the sequence. In addition, the sequence as a whole will usually be denoted as $(a_n)_{n\in\mathbb{N}}$.** In this case, the symbol n is what we call an index variable, or dummy variable. The numbers a_n are the individual members of the sequence (one for each $n \in \mathbb{N}$), while $(a_n)_{n\in\mathbb{N}}$ is the sequence itself.

The reason the "$n \in \mathbb{N}$" appears in the subscript of $(a_n)_{n\in\mathbb{N}}$ is because it tells us that as n increases in $\mathbb{N}$ from 1 forward, the numbers a_n generate the list of numbers that define the sequence. For this reason, we may encounter the notation such as $(a_n)_{n\geq 1}$, or $(a_k)_{k\in\mathbb{N}}$ (here k is the indexing variable). Sometimes it is the case that our sequence begins with a zeroth element a_0, and our notation can accommodate this as well: such a sequence could be denoted as $(a_n)_{n\in\mathbb{N}\cup\{0\}}$. Admittedly, this is a bit awkward, and one could instead denote this sequence as $(a_n)_{n\in A}$ (where $A = \mathbb{N}\cup\{0\}$), or as $(a_n)_n$ when the context is clear in which set the sequence is indexed. We usually denote a sequence as (a_n) when it is clear that n is the index variable, and in what set it is indexed. One could also use the notation (a_n) when the indexing set is irrelevant, to make the mathematical statement easier to read. In this way, we may talk about a property holding true "for all n" or "for every n," and we would mean that such a property would hold for whatever values of n are appropriate from context. Whenever the context does not clearly indicate what values of n this could be for (or when it would be important to note something specific about the values of n one considers), the specific values of n one considers will be made very clear.

Before considering some examples of sequences, we consider how we will actually encounter them in practice. **Although there are many ways to define sequences, the most common are to give a rule for a_n (known as a sequence given in closed form), or to recursively define them.**

3.1.2 Closed form sequences

Consider the sequence we began our section with: $2, 4, 6, \ldots$. If you were to try to describe this sequence to someone, you would probably have to continue to give them enough terms in the sequence before they could continue the pattern themselves–this is not at all a rigorous or precise method of discussing sequences. Instead, you could give them the formula $a_n = 2n$. One can check that indeed, $a_1 = 2, a_2 = 4, a_3 = 6, \ldots$, and that this formula actually does accurately describe the sequence. To define a_n by giving its rule is to define this sequence in **closed form**.

Let us practice a bit with these types of sequences. Part (1) is used in Practice 3.17.

Practice 3.2. List out the first 5 members of the following sequences:

1. $a_n = n^2$.
2. $a_n = \frac{1}{n}$.
3. $a_n = \frac{1}{n^2+1}$.

Methodology. *Listing out members of a sequence is straightforward: simply plug in the value of n into both sides of the defining equation.*

Solution. We plug in the values $n = 1$ through $n = 5$ for each of the sequences. For part (1), $a_n = n^2$, so

$$a_1 = 1^2 = 1, a_2 = 2^2 = 4, a_3 = 3^2 = 9, a_4 = 4^2 = 16, a_5 = 5^2 = 25.$$

For part (2), $a_n = \frac{1}{n}$, so

$$a_1 = \frac{1}{1} = 1, a_2 = \frac{1}{2}, a_3 = \frac{1}{3}, a_4 = \frac{1}{4}, a_5 = \frac{1}{5}.$$

For part (3), $a_n = \frac{1}{n^2+1}$, so

$$a_1 = \frac{1}{1^2+1} = \frac{1}{2}, a_2 = \frac{1}{2^2+1} = \frac{1}{5}, a_3 = \frac{1}{3^2+1} = \frac{1}{10},$$

$$a_4 = \frac{1}{4^2+1} = \frac{1}{17}, a_5 = \frac{1}{5^2+1} = \frac{1}{26}.$$

■

Practice 3.3. Find a closed form expression for the sequence b_n of positive odd numbers: $1, 3, 5, 7, 9, \ldots$.

Methodology. *We are looking for a function that sends 1 to 1, 2 to 3, 3 to 5, 4 to 7, 5 to 9, etc. We observe that these outputs are one less than the outputs of the sequence considered above, and work from there.*

Solution. We notice that the sequence a_n of positive even numbers 2, 4, 6, ..., has closed form expression $a_n = 2n$. The positive odd numbers are, term by term, one less. So we suspect the sequence $b_n = a_n - 1 = 2n - 1$ will suffice. Indeed,

$$\begin{array}{rcl} b_1 & = & 2(1) - 1 = 1, \\ b_2 & = & 2(2) - 1 = 3, \\ b_3 & = & 2(3) - 1 = 5, \\ \vdots & & \vdots \end{array}$$

■

Solution 2. We give an answer where the indexing set includes 0, and begin the sequence there. Consider the sequence $b_n = 2n + 1$. We have

$$\begin{array}{rcl} b_0 & = & 2(0) + 1 = 1, \\ b_1 & = & 2(1) + 1 = 3, \\ b_2 & = & 2(2) + 1 = 5, \\ \vdots & & \vdots \end{array}$$

■

These two different solutions to the same problem above show that questions of this sort can have many different solutions.

Practice 3.4. Find a closed form expression for the sequence a_n given by the pattern $1, -1, 1, -1, \ldots$.

Methodology. *We are looking for a function that sends 1 to 1, 2 to -1, 3 to 1, etc. Two concepts come to mind: repeated multiplication of negative* *numbers alternates sign, and several trigonometric functions oscillate between 1 and -1. A deeper and more precise consideration of these concepts will point us in the right direction.*

Solution. We consider the sequence $a_n = (-1)^{n+1}$. This apparently works, as

$$\begin{array}{rcl} a_1 & = & (-1)^{1+1} = (-1)^2 = 1, \\ a_2 & = & (-1)^{2+1} = (-1)^3 = -1, \\ a_3 & = & (-1)^{3+1} = (-1)^4 = 1, \\ \vdots & & \vdots \end{array}$$

■

Solution 2. We notice that the sine function oscillates between 1 and -1, and it does so at odd multiples of $\frac{\pi}{2}$. Using the sequence for odd numbers we found in the first solution to the above example, we can build the sequence $a_n = \sin\left((2n-1)\frac{\pi}{2}\right)$. Then,

$$\begin{array}{rcl} a_1 & = & \sin\left((2(1)-1)\frac{\pi}{2}\right) = \sin\left(\frac{\pi}{2}\right) = 1, \\ a_2 & = & \sin\left((2(2)-1)\frac{\pi}{2}\right) = \sin\left(\frac{3\pi}{2}\right) = -1, \\ a_3 & = & \sin\left((2(3)-1)\frac{\pi}{2}\right) = \sin\left(\frac{5\pi}{2}\right) = 1, \\ \vdots & & \vdots \end{array}$$

■

The following example introduces geometric sequences. These sequences are common and useful, in particular, in the study of infinite series and sequences of real numbers and of functions in Chapter 7.

Example 3.5 (Geometric Sequences)**.** Let $r \in \mathbb{R}$. **A sequence of the form** $a_n = r^n$ **is called a geometric sequence**. Notice that if $r = 0$, then the sequence $a_n = 0$ for all $n \geq 1$. If $r = 1$, then $a_n = 1$ for all $n \geq 0$, and if $r = -1$, then we more or less recover the same sequence as in Practice 3.4. If $r > 1$, then the sequence $a_n = r^n$ grows larger without bound, while if $r < -1$, the magnitude $|a_n|$ of the sequence grows without bound, while r^n alternates sign. If $r \in (-1, 1)$, then the sequence seems to tend toward 0 (see Exercise 14 of Section 3.2). For example, if $r = \frac{1}{2}$, then the first five members of the sequence $a_n = \left(\frac{1}{2}\right)^n$ are

$$\frac{1}{2}, \frac{1}{4}, \frac{1}{8}, \frac{1}{16}, \frac{1}{32}.$$

■

The concept of a sequence approaching one number is called *convergence*, and is discussed in the next section, and a great deal throughout the remainder of the book.

There is no general process that one could follow to determine the closed form expression for any given sequence from a collection; in the previous examples, it was fortunate that we could relate the problem to one that had already been solved, or one that was already intuitive to us. Generally speaking, one must be creative. Thinking about the answers to these questions will sometimes help: how quickly or slowly is the sequence changing? Does the sequence change sign? Is the sequence combined from already previously known sequences?

3.1.3 Recursively defined sequences

There is another way to describe a sequence that is quite common, and that is to describe how previous members of the sequence combine to produce the next member of the sequence. Such a sequence is called a *recursive sequence* or is said to be defined *recursively*. The example at the beginning of the section could be defined as follows:

$$a_1 = 2, \text{ and } a_{n+1} = a_n + 2 \text{ for } n \geq 1. \tag{3.1}$$

One checks again, perhaps more laboriously, that

a_1	$=$	2	(by definition)
$a_2 = a_{1+1}$	$=$	$a_1 + 2 = 2 + 2 = 4$	(from Equation 3.1, with $n = 1$)
$a_3 = a_{2+1}$	$=$	$a_2 + 2 = 4 + 2 = 6$	(from Equation 3.1, with $n = 2$)
$\vdots$		$\vdots$	

Here are some more examples of sequences that are recursively defined.

Practice 3.6. We define the sequence (a_n) recursively as follows: $a_1 = 2$, and for $n \geq 1$ we define $a_{n+1} = \sqrt{a_n} + 2$. List out the first 4 terms of the sequence (a_n).

Methodology. *Similar to the method in Practice 3.2, we look at the defining formula for a_n that is given to us, and plug in the values of $n = 1, 2, 3$, and 4 into the formula $a_{n+1} = \sqrt{a_n} + 2$. In contrast to the closed form situation, there is nothing to do but observe the definition that a_1 is given to us already as 2. The work will begin when $n = 2$.*

Solution. We observe that by definition, $a_1 = 2$. Using this, and the recurrence relation $a_{n+1} = \sqrt{a_n} + 2$, we conclude

$$\begin{aligned} a_2 = a_{1+1} &= a_1 + 2 = \sqrt{2} + 2 && (n = 1) \\ a_3 = a_{2+1} &= a_2 + 2 = \sqrt{\sqrt{2} + 2} + 2 && (n = 2) \\ a_4 = a_{3+1} &= a_3 + 2 = \sqrt{\sqrt{\sqrt{2} + 2} + 2} + 2 && (n = 3) \end{aligned}$$

We will consider the behavior of this sequence as n gets large (the *convergence* of this sequence) in Practice 3.75 in Section 3.5. ■

We point out that it would be quite difficult and awkward to construct or use a closed-form expression for the sequence defined in Practice 3.6. Here is another such example that is very well known.

Example 3.7 (The Fibonacci Sequence). (The following sequence $(F_n)_{n\geq 0})$ was originally discovered by Fibonacci (Leonardo de Pisa) in the early thirteenth century as a solution of a problem regarding rabbit breeding. It is known today as the *Fibonacci Sequence.*) Set $F_0 = 0$, and $F_1 = 1$. Then, for $n \geq 2$, define $F_n = F_{n-1} + F_{n-2}$. The first 6 members of this sequence begin with the already defined $F_0 = 0$ and $F_1 = 1$, and continue as:

$$\begin{aligned} F_2 = F_{2-1} + F_{2-2} &= F_1 + F_0 = 1 + 0 = 1 \quad (n = 2), \\ F_3 = F_{3-1} + F_{3-2} &= F_2 + F_1 = 1 + 1 = 2 \quad (n = 3), \\ F_4 = F_{4-1} + F_{4-2} &= F_3 + F_2 = 2 + 1 = 3 \quad (n = 4), \\ F_5 = F_{5-1} + F_{5-2} &= F_4 + F_3 = 3 + 2 = 5 \quad (n = 5). \end{aligned}$$

Again, it would seem to be quite difficult to discover a closed form for this sequence (see Exercise 13 for such a closed form). ■

Example 3.8 (Geometric sequences, again). In Example 3.5, we defined geometric sequences to be those of the form $a_n = r^n$ for some $r \in \mathbb{R}$. Here is a recursive expression for geometric sequences. Let $a_0 = 1$, and for $n \geq 1$, set $a_n = ra_{n-1}$. This apparently generates the sequence $1, r, r^2, r^3, \ldots$. Next, in Practice 3.9, we use Mathematical Induction to prove explicitly that this is indeed the case.

Practice 3.9. Let $r \in \mathbb{R}$ with $r \neq 0$. Let a_n be the recursively defined sequence above in Example 3.8: $a_0 = 1$, and for $n \geq 1$, set $a_n = ra_{n-1}$. Let b_n be the sequence from Example 3.5, given by $b_n = r^n$ for $n \geq 0$. Prove that $a_n = b_n$ for all $n \geq 0$.

Methodology. *We will use a straightforward Mathematical Induction argument. The only other observation is that the number 0^0 is not defined, so if $r = 0$ and $n = 0$, the sequence b_n would not be defined at 0. This is the justification for the extra hypothesis that $r \neq 0$.*

Solution. We use Mathematical Induction. The base case is easy to check: since $r \neq 0$, we have $b_0 = r^0 = 1 = a_0$. Now assume that for some $n \geq 0$, we have $a_n = b_n$; we demonstrate $a_{n+1} = b_{n+1}$. Since $a_{n+1} = ra_n$, and $b_{n+1} = r^{n+1}$, and since our induction hypothesis is the statement $a_n = b_n = r^n$, we have

$$a_{n+1} = ra_n = rb_n = r \cdot r^n = r^{n+1} = b_{n+1}.$$

Therefore, by Mathematical Induction, $a_n = b_n$ for all integers $n \geq 0$. ■

3.1.4 Constructing new sequences from old ones

Given two sequences (a_n) and (b_n) and a real number c, there are a variety of ways to combine them to form new sequences. For example, we define new sequences

$$(a_n \pm b_n), (a_n b_n), (ca_n), \text{ and } \left(\frac{a_n}{b_n}\right),$$

where in the final expression, $b_n \neq 0$ for every n. We close the section with some examples of this concept.

Practice 3.10. Find a closed-form expression for the sequence x_n given by the pattern $1, -3, 5, -7, 9 \ldots$.

Methodology. *We notice that this sequence seems to be built by multiplying the sequence $a_n = 2n - 1$ of odd numbers with the oscillating sequence $b_n = (-1)^{n+1}$.*

Solution. Using $a_n = 2n-1$ and $b_n = (-1)^{n+1}$, we construct their product $x_n = a_n b_n = (-1)^{n+1}(2n - 1)$. Practices 3.3 and 3.4 demonstrate that $x_1 = a_1 b_1 = 1, x_2 = a_2 b_2 = -3$, and so forth. ■

3.1.5 Inequalities of sequences

As you may have already seen from Chapter 2, inequalities play a central role in the study of analysis. This subsection is here to provide several examples of common inequalities among sequences, and to better prepare the student for what lies ahead.

Practice 3.11. Let $a_n = \frac{n-1}{n}$ for $n \geq 1$.

(a) List out the first 5 members of this sequence.

(b) Prove that $a_n \leq a_{n+1}$ for all $n \geq 1$.

(c) Does there exist a real number M so that $|a_n| \leq M$ for every n? If so, find such an M.

(d) Consider the behavior of the numbers a_n as n gets very large. Does it appear that a_n approaches some quantity as n approaches ∞? If so, what is that quantity, and what justification can you give that this is the case?

Methodology. *The first part follows as we have done in the other practice problems. The next part we will prove directly, although very often one uses Mathematical Induction to prove such a statement. The answer to part (c) is yes, there does exist such an M: we are looking for an overall bound on the magnitude of each member of the sequence. Since the sequence is an increasing one by part (b), a lower bound is furnished by the absolute value of the first number a_1. We will simply notice that this sequence is a fraction of nonnegative quantities where the numerator is smaller than the denominator. In other words, 1 is an upper bound, and so we may put $M = 1$. There is not meant to be a right or wrong answer to part (d), but to simply reflect on what the question asks for this sequence. Using the very same sequence considered here, the next section explores this concept in much further and more rigorous detail.*

Solution. (a) The first 5 members of the sequence are

$$\begin{aligned} a_1 &= \tfrac{1-1}{1} = 0, \\ a_2 &= \tfrac{2-1}{2} = \tfrac{1}{2}, \\ a_3 &= \tfrac{3-1}{3} = \tfrac{2}{3}, \\ a_4 &= \tfrac{4-1}{4} = \tfrac{3}{4}, \\ a_5 &= \tfrac{5-1}{5} = \tfrac{4}{5}. \end{aligned}$$

(b) We must show that $\frac{n-1}{n} = a_n \leq a_{n+1} = \frac{n}{n+1}$. To do so, we follow the reversible statements of inequality. Notice that we clear denominators in the first step, and since $n \geq 1$ (and so $n+1 \geq 0$), there is no resulting change of inequality:

$$\begin{aligned} \tfrac{n-1}{n} &\leq \tfrac{n}{n+1}, \\ (n-1)(n+1) &\leq n^2, \\ n^2 - 1 &\leq n^2, \\ -1 &\leq 0. \end{aligned}$$

(c) We notice that each of $a_n \geq 0$, so a lower bound for a_n is 0. Moreover, since $n - 1 \leq n$, we have $\frac{n-1}{n} \leq 1$, so 1 is an upper bound. The larger of the two numbers $|0| = 0$ and $|1| = 1$ is 1, so we set $M = 1$.

(d) It appears that the a_n are approaching the number 1. As n grows to be very large, the numerator and denominator are nearly indistinguishable, and so their quotient will be very nearly equal to 1 for large values of n. ■

The next practice problem is a result named after Jacob Bernoulli (1654–1705).

Practice 3.12 (Bernoulli's Inequality)**.** For $h > -1$, prove that $(1+h)^n \geq 1 + nh$ for $n \geq 0$.

Methodology. *We use Mathematical Induction to prove this. Notice that since h is strictly greater than -1, the number $1 + h$ is never 0, and so when $n = 0$ there is no problem with not being well-defined.*

Solution. We prove this by Mathematical Induction. When $n = 0$, there is nothing to check, since $(1+h)^0 = 1$, and $1 + 0 \cdot h = 1$ as well. Assume then, for some $n \geq 0$, we have $(1+h)^n \geq 1 + nh$. Then since $1 + h > 0$, we multiply both sides of $(1+h)^n \geq 1 + nh$ by it and see that

$$\begin{aligned} (1+h)^{n+1} &= (1+h)^n(1+h) \\ &\geq (1+nh)(1+h) \\ &= 1 + h + nh + nh^2 \\ &= 1 + (n+1)h + nh^2. \end{aligned}$$

Since $n \geq 1$ and $h^2 \geq 0$, we have $nh^2 \geq 0$, so we complete the Mathematical Induction proof by noting that

$$(1+h)^{n+1} \geq 1 + (n+1)h + nh^2 \geq 1 + (n+1)h.$$

■

3.1.6 Review of important concepts from Section 3.1

1. A sequence is a function from $\mathbb{N}$ to $\mathbb{R}$. We will consider infinite sequences only, unless otherwise stated.
2. We write a_n for the nth term of the sequence, and write $(a_n)_{n\in\mathbb{N}}$ or simply (a_n) for the sequence itself.
3. Sequences given in closed form are those for which a rule for a_n is given.
4. Recursive sequences are those whose previous terms are used to define the next terms.
5. Generally speaking, it is difficult to discover a pattern for a given sequence of numbers and find either a closed form or recursive formula for it. One must draw on intuition and experience.
6. One can build new sequences from existing ones and/or scalars through addition, subtraction, multiplication, division, or scaling.
7. Inequalities are central to the study of Analysis.

Exercises for Section 3.1

1. Let $a_n = \frac{n!}{(n+1)!}$.
   (a) List out the first 5 terms of this sequence (and reduce into lowest terms).
   (b) Is there a more simple closed form expression for this sequence?
2. Find a closed form expression for the sequence $0, 2, 0, 2, 0, 2, \ldots$. *[Hint: there can be many solutions to questions of this type.]*
3. Find a closed form expression for the sequence $0, 1, 0, 1, 0, 1, \ldots$.
4. Find a closed form expression for the sequence $0, 1, 0, -1, 0, 1, 0, -1, \ldots$.
5. Find a closed form expression for the sequence $1, \frac{1}{6}, \frac{1}{120}, \frac{1}{5040}, \frac{1}{362{,}880}, \ldots$.
6. For $n \geq 1$, show that $n^2 + 1 \geq n$.
7. For $n \geq 3$, show that $9n^2 - 51 \geq n$.
8. Find an integer K so that $n! \geq 2^n$ for every $n \geq K$. Then, prove that this is indeed the case using Mathematical Induction. Conclude that $\frac{1}{n!} \leq \frac{1}{2^n}$ for every $n \geq K$.
9. Let $a > 1$ be a real number. Find an integer K so that $n! \geq a^n$ for every $n \geq K$. Then, prove that this is indeed the case using Mathematical Induction. Conclude that $\frac{1}{n!} \leq \frac{1}{a^n}$ for every $n \geq K$.
10. Let $a_n = \frac{n+1}{n}$.
    (a) List out the first 5 members of this sequence.
    (b) Prove that $a_n \geq a_{n+1}$ for all $n \geq 1$.

(c) Does there exist a real number M so that $|a_n| \leq M$ for every n? If so, find such an M.

(d) Consider the behavior of the numbers a_n as n gets very large. Does it appear that a_n approaches some quantity as n approaches ∞? If so, what is that quantity, and what justification can you give that this is the case?

11. This exercise is designed to compare the eventual behavior of sequences of the form $((-1)^n a_n)$ with that of (a_n).

(a) Let $a_n = \frac{(-1)^n}{1+n^4}$.

i. List out the first 5 members of this sequence.

ii. Does it appear a_n approaches some quantity as n approaches ∞? If so, what is that quantity, and what justification can you give that this is the case? *[Hint: $a_n \to 0$.]*

iii. Find an easier expression for $|a_n|$. Does it appear $|a_n|$ approaches some quantity as n approaches ∞? If so, what is that quantity, and what justification can you give that this is the case?

(b) Let $b_n = \frac{(-1)^n(n-1)}{n+1}$.

i. List out the first 5 members of this sequence.

ii. Does it appear b_n approaches some quantity as n approaches ∞? If so, what is that quantity, and what justification can you give that this is the case? *[Hint: b_n does not approach one single quantity.]*

iii. Find an easier expression for $|b_n|$. Does it appear $|b_n|$ approaches some quantity as n approaches ∞? If so, what is that quantity, and what justification can you give that this is the case? *[See also Exercise 1c of the next section.]*

(c) If you know a sequence c_n tends to c as n gets large, must it always be the case that $|c_n|$ tends to $|c|$? If you know $|c_n|$ tends to $|c|$ as n gets large, must it be the case that c_n tends to c as n gets large? *[Hint: See Exercises 9 and 10 of Section 3.2.]*

12. Recursively define the sequence (g_n) as follows: Let $g_0 = 1$, and for $n \geq 0$, set $g_{n+1} = 1 + \frac{1}{g_n}$.

(a) List out the first 5 members of this sequence.

(b) There are two solutions $\varphi > 0$ and $\bar{\varphi} < 0$ to the equation $g = 1 + \frac{1}{g}$. Find them. (φ is known as the *golden ratio*.)

(c) Does it seem that the sequence g_n approaches φ as n approaches ∞?

13. (This exercise references Exercise 12, and gives a closed form for the Fibonacci sequence from Example 3.7. It is known as Binet's Formula.) Let φ and $\bar{\varphi}$ be the solutions of the equation $g = 1 + \frac{1}{g}$ as

in Exercise 12. Prove that the Fibonacci sequence F_n from Example 3.7 has the closed form

$$F_n = \frac{1}{\sqrt{5}}[\varphi^n - \bar{\varphi}^n] \quad \text{for} \quad n \geq 0.$$

3.2 Limits and Convergence

In this section we introduce a foundational concept in Analysis: convergence of sequences. We introduce the concept through a familiar example, and then formally define convergence in Definition 3.13. Since this is such an important topic, we list several considerations to help the reader become more quickly familiarized with it.

We begin by considering the sequence $a_n = \frac{n-1}{n}$ in Practice 3.11 of the previous section (see page 101). The exercise ultimately asks you the question: "Does it appear that a_n approaches some quantity as n approaches ∞? If so, what is that quantity, and what justification can you give that this is the case?" This question is an example of what is known as *convergence* of sequences, and we consider this question in detail. We list some values of a_n for very large values of n in Figure 3.2.1.

n	a_n	
5	$\frac{4}{5}$	$= .8$
10	$\frac{9}{10}$	$= .9$
100	$\frac{99}{100}$	$= .99$
1000	$\frac{999}{1000}$	$= .999$
1 million	$\frac{999999}{1000000}$	$= .999999$
10^{12}		$= .999999999999$

FIGURE 3.2.1
The behavior of $a_n = \frac{n-1}{n}$ for some large values of n.

It certainly seems that the sequence $\frac{n-1}{n}$ is approaching 1 as n gets larger. **How can we rigorously define the intuitive concept of convergence?** Recall that the distance between two real numbers a and b is the absolute value of their difference: $|a - b|$. So to say

"The quantities $\frac{n-1}{n}$ tend to 1 as n gets large."

is the same as saying

"The distance between $\frac{n-1}{n}$ and 1 is negligible as n gets large."

Since the distance between the numbers $\frac{n-1}{n}$ and 1 is $\left|\frac{n-1}{n} - 1\right|$, we could rephrase the above as

"The quantities $\left|\frac{n-1}{n} - 1\right|$ are negligible as n gets large."

Let us consider exactly what we mean by the word "negligible" in the above statement. A quantity that is negligible is one that is not only insignificantly small, but also nonnegative. So, if we were to estimate such a number, we might use a comparison to another small but positive number. Let ϵ be such a number: we intend for ϵ to be something we may consider very small, but also $\epsilon > 0$. Moreover, we should have some freedom in declaring that no matter how small such an ϵ could be, the quantity that becomes negligible would eventually become smaller than ϵ. So, we could rephrase the above as

"For every $\epsilon > 0$, the quantities $\left|\frac{n-1}{n} - 1\right| < \epsilon$, as n gets large."

Now let us examine exactly what we mean in this context by the phrase "as n gets large." It seems that if you were to measure the distance between $\frac{n-1}{n}$ and 1, and compare it to some overtly small quantity ϵ, it might not be the case that for *every* n, that distance would be less than ϵ. Figure 3.2.2 is a list of quantities that compare the distance from $\frac{n-1}{n}$ to 1 for various values of n. So, if we decided that $\epsilon = 14$, then it would be that case that, for every

n	$\left\|\frac{n-1}{n} - 1\right\|$
1	1
2	.5
5	.2
10	.1
100	.01
1000	.001
1 million	.000001
10^{12}	.000000000001

FIGURE 3.2.2
The distance from $\frac{n-1}{n}$ to 1 for various values of n.

value of $n \geq 1$, the quantity $\left|\frac{n-1}{n} - 1\right| < \epsilon$. If instead, $\epsilon = .4$, then it appears from Figure 3.2.2 that there are some values of n for which $\left|\frac{n-1}{n} - 1\right|$ is *not* less than ϵ ($n = 1$ and 2, for example); however, there is a point at which these numbers stay between 0 and ϵ. By looking at Figure 3.2.2, we can see that this happens at least by the time $n = 5$. If ϵ instead were .00002, then again, there are very many values of n for which the distance in question is *greater* than .00002. But the same phenomenon occurs: there is some value of n at which these distances dip below .00002, and stay below .00002. According again to Figure 3.2.2, it seems that this happens at least by $n =$ one million. So, **to say that a certain distance becomes negligible as n gets large is to say that no matter what positive value of ϵ you supply, there is *always* a value of n so that from this point on, these distances are less than that value of ϵ. Let us use N for this important value.** Based on this discussion, it seems as though N **depends on ϵ.** This will be made more clear in Example 3.15. For now, we complete our mathematical formulation of the convergence of sequences. Notice that this is more or less the way Definition 3.13 reads upon application to the sequence $a_n = \frac{n-1}{n}$ that converges to 1:

> "For every $\epsilon > 0$, there exists N so that $|\frac{n-1}{n} - 1| < \epsilon$ for every $n \geq N$."

The above statement surely sounds good, but did we define the concept that we were aiming for? Thinking of this concept from the opposite perspective for an arbitrary sequence a_n, and a supposed quantity that it tends to L: what if we could not always demonstrate the above phenomenon occurred? That is, what if there was some positive number ϵ, and that no matter what values of n you consider, there was *always* some n for which the distance between a_n and L was larger than this ϵ. In such a situation, it could fairly be said that the numbers a_n stayed at least a distance of ϵ from L, and this should be exactly what it means for a sequence to *not* converge to L. We will practice this concept in Practice 3.25.

3.2.1 Convergence

We may now, finally, define what it means for a sequence to converge to a real number.

Definition 3.13. Let (a_n) be a sequence, and let $L \in \mathbb{R}$. The sequence (a_n) *converges* to L if, for every $\epsilon > 0$, there exists N so that $|a_n - L| < \epsilon$ for every $n \geq N$. If the sequence does not converge, then we say that it *diverges.*

There are several different ways we could symbolically express that the sequence (a_n) converges to L. These include $\lim_{n\to\infty} a_n = L$ (the limit of a_n, as n tends toward infinity, is L), or simply $a_n \overset{n}{\to} L$ as $n \to \infty$, or, when the index variable is clear from context, simply $a_n \to L$.

Let us consider the same example $a_n = \frac{n-1}{n}$. Using Definition 3.13, we formally prove that the sequence $a_n = \frac{n-1}{n}$ converges to 1. The reader should note that in this practice and throughout the section we will make heavy use of The Archimedean Property of the real numbers and its equivalent forms. It may be of help to go back to Theorem 2.34 on page 84 to review this before proceeding.

Practice 3.14. Let $a_n = \frac{n-1}{n}$, and let $L = 1$. Prove that the sequence (a_n) converges to L.

Methodology. *Unless there are some other supporting results (which we do not yet have), we must demonstrate that the sequence a_n converges to L as described in Definition 3.13. So, we assume some $\epsilon > 0$ has been given to us, and we proceed to find an N so that whenever $n \geq N$, it is the case that $|a_n - L| < \epsilon$. We will find such an N by using the Archimedean Property.*

Solution. Let $\epsilon > 0$ be given. We must find N so that $|a_n - L| < \epsilon$

whenever $n \geq N$. Notice that

$$\begin{aligned} |a_n - L| &= \left|\frac{n-1}{n} - 1\right| \\ &= \left|\frac{n}{n} - \frac{1}{n} - 1\right| \\ &= \left|\frac{1}{n}\right| \\ &= \frac{1}{|n|} \\ &= \frac{1}{n}, \text{ since } n \geq 1. \end{aligned}$$

According to The Archimedean Property, there exists a N so that $\frac{1}{N} < \epsilon$.
Suppose that $n \geq N$. It follows that $\frac{1}{n} \leq \frac{1}{N}$, and so

$$|a_n - L| = \frac{1}{n} \leq \frac{1}{N} < \epsilon.$$

■

As the first formal proof of convergence of a sequence, Practice 3.14 is an important example. Let us go a little bit further into the mechanics of the proof with Example 3.15; the goal of Example 3.15 is to demonstrate how N depends on ϵ. In particular, if you did not understand the solution to Practice 3.14, read Example 3.15, and then try reading Practice 3.14 again.

Example 3.15. In the above proof of convergence, notice that once ϵ was given, only then can one go searching for N. **It is generally the case that N depends on ϵ,** and this example is designed to demonstrate this with a familiar example, in addition to further familiarizing the reader with Definition 3.13. (There is really only one situation where N does not depend on ϵ, see Practice 3.16.)

Consider again $a_n = \frac{n-1}{n}$. In the solution to Practice 3.14, ϵ was given to us, and we proceeded to find an N so that $|a_n - N| < \epsilon$ for every $n \geq N$. The crux of the solution is this: no matter what ϵ we are given, we can find a number N that satisfies a certain property: $\left|\frac{n-1}{n} - 1\right| < \epsilon$ for every $n \geq N$. The solution to Practice 3.14 proceeds to find this N for an arbitrary ϵ, but let us attempt to do this for various given values of ϵ. (This example follows closely the discussion following Figure 3.2.2, but is more rigorous.)

Suppose first that the value of ϵ given to us is $\epsilon = 14$. The discussion near Figure 3.2.2 suggests that one could have $N = 1$. That is, for every $n \geq N = 1$, we wish to know if $\left|\frac{n-1}{n} - 1\right| < \epsilon = 14$. Since $n \geq 1$ (i.e., not negative), then $\left|\frac{n-1}{n} - 1\right| = \frac{1}{n}$, as in the solution to Practice 3.14. Indeed, this is the case: if $n \geq 1$, then $\frac{1}{n} < 14$, since, at the very least, $\frac{1}{n} \leq 1$ for such $n \geq 1$. In conclusion, we have verified the convergence condition given in Definition 3.13 in the one case that $\epsilon = 14$. We have found that, in this case, $N = 1$. This analysis is *not* sufficient to conclude that $a_n \to 1$, since Definition 3.13 requires that one find an N for *any* given (positive) value of ϵ, not just the one we have just studied.

Now suppose that ϵ is chosen to be .00002. Earlier, it seemed plausible that an N that "works" is $N =$ one million. Let us check that this is indeed the case. Since $\left|\frac{n-1}{n} - 1\right| = \frac{1}{n}$, we simply note that if $n \geq 1000000$, then $\frac{1}{n} \leq \frac{1}{1000000} < .00002$. Again, based only on the work found in this example, we have not yet proven that $a_n \to 1$. Only a proof where an arbitrary ϵ is considered, such as the solution found in Practice 3.14, is sufficient. Later (in Section 3.3 and in many other places), we will learn certain additional methods for demonstrating convergence, but for now, we have only Definition 3.13. ■

There is one situation in which N does not depend on ϵ, we consider this in the next practice problem.

Practice 3.16. Let $a_n = a$ be a constant sequence. Prove (a_n) converges, and discuss the relationship between ϵ and N from your solution.

Methodology. *As in Practice 3.14, we demonstrate this using only the definition of convergence (Definition 3.13). If we are to show that this sequence converges, we first have to determine what it converges to. A good guess would be the number a: after all, we are trying to show that this sequence tends toward a certain number, and this sequence is always the same number, a. Since a_n is constant, the quantity we need to study, $|a_n - a|$, will always be zero. The determination of N is pretty easy: N can be anything!*

Solution. We show that the sequence $a_n = a \to L$, where $L = a$. Let $\epsilon > 0$. We must find an N so that for every $n \geq N$, the quantity $|a_n - L| = |a - a| = 0 < \epsilon$ for *every* n. Therefore, the choice of N is arbitary, and select $N = 1$.

In previous discussions about how N depends on ϵ, we have pointed out that the sequence may not always be within a distance of ϵ away from its limit, but that at some point and beyond, it is. Since this sequence never changes from a, the distance from it to a remains at 0. So, the point at which it stays within ϵ away from its limit is at the very beginning. Exercise 8 explores this concept in greater detail. ■

Since the notion of convergence is such an important one, we now provide a number of practice problems to (1) help the reader see just how to approach problems of convergence of sequences, and (2) provide formal proofs to model your own off of to help you get started with the exercises at the end of the section.

Practice 3.17. Let $a_n = \frac{1}{n}$. Prove that the sequence (a_n) converges to 0.

Methodology. *According to the definition of convergence, we will have to study the distances $|a_n - L| = \left|\frac{1}{n} - 0\right| = \frac{1}{n}$. Given an $\epsilon > 0$, the task will be to try to find an N so that these distances are less than ϵ. The Archimedean Property does exactly this.*

Solution. Let $\epsilon > 0$ be given. We must find an N so that

$$|a_n - L| = \left|\frac{1}{n} - 0\right| = \frac{1}{n} < \epsilon \quad \text{for} \quad n \geq N.$$

According to The Archimedean Property, there exists an N so that $\frac{1}{N} < \epsilon$. Then, for $n \geq N$, we have $\frac{1}{n} \leq \frac{1}{N}$, and

$$|a_n - L| = \left|\frac{1}{n} - 0\right| = \frac{1}{n} \leq \frac{1}{N} < \epsilon.$$

■

Practice 3.18. Prove that $\frac{2n-3}{n} \to 2$.

Methodology. *This the same type of question as in Practice 3.17, but asked using a different notation. We will have to study the distances*

$$|a_n - L| = \left|\frac{2n-3}{n} - 2\right| = \left|\frac{-3}{n}\right| = \frac{3}{n}.$$

Unlike the previous examples, $\frac{3}{n}$ is not $\frac{1}{n}$, where a direct application of The Archimedean Property was possible. All the same, $\frac{3}{n}$ is 3 times $\frac{1}{n}$, and if $\frac{1}{n}$ can be made arbitrarily small, so can $\frac{3}{n}$ by applying The Archimedean Property to the value $\frac{\epsilon}{3}$ instead of ϵ.

Solution. Let $\epsilon > 0$ be given. We must find an N so that

$$|a_n - L| = \left|\frac{2n-3}{n} - 2\right| = \left|\frac{2n-3}{n} - \frac{2n}{n}\right| = \left|\frac{-3}{n}\right| = \frac{3}{n}.$$

By The Archimedean Property, there exists an N so that $\frac{1}{N} < \frac{\epsilon}{3}$, and so $\frac{3}{N} < \epsilon$. Then, for $n \geq N$, we have $\frac{3}{n} \leq \frac{3}{N}$, and

$$|a_n - L| = \frac{3}{n} \leq \frac{3}{N} < \epsilon.$$

■

In the previous practice, notice that The Archimedean Property did not apply in the same direct way as Practices 3.14 and 3.17. It turned out to not be a problem: we just had to adjust ϵ ahead of time for our use of The Archimedean Property. Generally speaking, such an adjustment is commonplace, and the difficulty in this subject is to identify exactly how to find that $|a_n - L| < \epsilon$. One usually assumes that he or she has found a suitable N ahead of time, and then works backwards to determine how he or she could justify such a choice. We give two solutions to the following practice problem: the first illustrates the "working backward" mentality, while the second uses a helpful inequality that will allow us to approach the problem more directly.

Practice 3.19. Show that $\lim_{n\to\infty} \frac{1}{n^2} = 0$.

Methodology. *Asked again using a different notation, this is the same sort of question as the other practices. We will use the definition of convergence to prove the claim $\frac{1}{n^2} \to 0$. In the spirit of working backward from our goal, let* *us assume that an $\epsilon > 0$ has been given, and we have determined a suitable N already. This N would have to satisfy, for $n \geq N$,*

$$\left|\frac{1}{n^2} - 0\right| = \frac{1}{n^2} \leq \frac{1}{N^2} < \epsilon.$$

Again, The Archimedean Property does not apply directly. But, if N is to satisfy $\frac{1}{N^2} < \epsilon$, then equivalently, $\frac{1}{N} < \sqrt{\epsilon}$. This is an expression with which we can use The Archimedean Property.

Another line of reasoning for this problem (that we give as a seperate solution below) is to notice that for every $n \geq 1$, the quantity $\frac{1}{n^2} \leq \frac{1}{n}$. So with this observation, we could use The Archimedean Property directly.

Solution 1. Let $\epsilon > 0$ be given. We must find an N so that

$$\left|\frac{1}{n^2} - 0\right| = \frac{1}{n^2} \leq \frac{1}{N^2} < \epsilon \quad \text{for} \quad n \geq N.$$

According to The Archimedean Property, there exists an N so that $\frac{1}{N} < \sqrt{\epsilon}$. It follows that $\frac{1}{N^2} < \epsilon$. Using this N, for $n \geq N$, we have

$$\left|\frac{1}{n^2} - 0\right| = \frac{1}{n^2} \leq \frac{1}{N^2} < \epsilon.$$

■

Solution 2. Let $\epsilon > 0$ be given. We must find an N so that if $n \geq N$, then

$$\left|\frac{1}{n^2} - 0\right| = \frac{1}{n^2} < \epsilon.$$

Notice that $\frac{1}{n^2} \leq \frac{1}{n}$ for every $n \geq 1$. Choose N so that $\frac{1}{N} < \epsilon$. Then, for every $n \geq N$, we have

$$\left|\frac{1}{n^2} - 0\right| = \frac{1}{n^2} \leq \frac{1}{n} < \epsilon.$$

■

Sometimes this "working backward" approach is one that could be easier done with the use of *inequality* rather than *equality*, as Practice 3.20 shows. Generally, the idea is to make a fraction bigger (but not too big!) by making the denominator smaller.

Practice 3.20. Show that $\frac{n}{2n+1} \to \frac{1}{2}$.

Methodology. *Just like the other practices, we use Definition 3.13 and study*

$$\left|\frac{n}{2n+1} - \frac{1}{2}\right| = \left|\frac{-1}{2(2n+1)}\right| = \frac{1}{4n+2}.$$

In the end, we are simply looking for a way to make $\frac{1}{4n+2}$ larger–but still less than ϵ–in a sensible way. A general observation of fractions of positive numbers *is that there is an inverse relationship between the size of the denominator and the size of the fraction: the smaller the denominator, the larger the fraction. So, we could make* $4n+2$ *smaller by replacing it with* $4n$ *and notice that the fraction gets bigger:* $\frac{1}{4n+2} \leq \frac{1}{4n}$*. And we could make* $4n$ *smaller by replacing it with* n *and notice that* $\frac{1}{4n} \leq \frac{1}{n}$*. Then, we could use The Archimedean Property.*

Solution. Let $\epsilon > 0$ be given. We must find an N so that

$$\left|\frac{n}{2n+1} - \frac{1}{2}\right| = \left|\frac{-1}{2(2n+1)}\right| = \frac{1}{4n+2} < \epsilon \quad \text{for} \quad n \geq N.$$

But

$$\frac{1}{4n+2} \leq \frac{1}{4n} \leq \frac{1}{n}.$$

Using The Archimedean Property, there exists an N so that $\frac{1}{N} < \epsilon$. Now, for $n \geq N$, we have

$$\left|\frac{n}{2n+1} - \frac{1}{2}\right| = \frac{1}{4n+2} \leq \frac{1}{n} \leq \frac{1}{N} < \epsilon.$$

■

There are a number of important initial concepts regarding convergence, and we complete the section by discussing some of them.

3.2.2 Basic concepts about convergence

As with any new mathematical definition, there are a variety of concepts that would help the student become more familiar with a new definition. In this case, we wish to become more familiar with the definition of convergence.

Unless there are other helpful supporting results, you must use Definition 3.13 for any formal proof or demonstration regarding convergence. For a moment, compare the discussion near Figure 3.2.2 with that of Example 3.15. The discussion near Figure 3.2.2 made claims that were unsubstantiated, but plausible. Example 3.15 made the same claims, but backed them up with explicit calculations. Until there are supporting results concerning convergence available to us, every statement in a formal proof must be substantiated with some justification (unless, of course, such a statement is completely obvious). The examples and practice in this section not only serve to deepen your understanding and facility with demonstrating convergence, but also serve as models to help you understand what an acceptable solution is. Section 3.3 is devoted to providing different sorts of supporting results, so as to deepen our understanding of this concept of convergence, so will our proofs take different forms. For this section, however, we only have the definition of convergence to work with, and so our proofs refer only to that.

Definition 3.13 is awkward. There is an equivalent formulation

that might be more intuitive. We would like to reword Definition 3.13 in an attempt to more intuitively understand it; we note that this is not a type of "supporting result" mentioned above, but rather, a restatement of the same definition–we will prove in a moment in Theorem 3.21 that they are equivalent.

The definition of convergence is arguably not as concerned about the beginning of a sequence, but rather, how the sequence behaves beyond a certain N. We say that a sequence (a_n) *eventually* has a property if there exists an N so that (a_n) satisfies that property for every $n \geq N$ (and not necessarily for values of n between 1 and N). One might refer to the numbers $a_1, \ldots, a_{N-1}$ as the beginning (or start) of a sequence, and we refer to $(a_n)_{n \geq N}$ as the *tail* of a sequence.

Finally, before we can more intuitively rephrase Definition 3.13, recall from Definition 2.13 in Chapter 2 that if L is a real number, then an $\epsilon-$*neighborhood* U_ϵ (or, just a *neighborhood* U) of L is the interval $(L - \epsilon, L + \epsilon)$. That is,

$$U = \{x \in \mathbb{R} | |x - L| < \epsilon\}.$$

Theorem 3.21. *Let (a_n) be a sequence, and let $L \in \mathbb{R}$. The following are equivalent:*

1. *(a_n) converges to L.*
2. *For every neighborhood U of L, (a_n) is eventually in U.*

Idea of proof: *This theorem is an equivalence of two statements, we will show (1) $\Rightarrow$ (2), then show (2) $\Rightarrow$ (1). It seems that there is nothing much to show here, except to unravel what exactly the words "neighborhood" and "eventually" mean.*

Proof of Theorem 3.21. ((1) $\Rightarrow$ (2)) Suppose $a_n \to L$, and let U be a neighborhood of L. We wish to show that (a_n) is eventually in U. The neighborhood $U = (L - \epsilon, L + \epsilon)$ for some ϵ. Since $a_n \to L$, for this ϵ, there exists an N so that $|a_n - L| < \epsilon$ for all $n \geq N$. In other words, $a_n \in U$ for $n \geq N$, or, that the sequence (a_n) is eventually in U.

((2) $\Rightarrow$ (1)) Suppose that (2) holds; we wish to show $a_n \to L$. Let $\epsilon > 0$ be given. We must find an N so that $|a_n - L| < \epsilon$ for every $n \geq N$. Let U be the neighborhood $(L - \epsilon, L + \epsilon)$. By assumption, (a_n) is eventually in U, which means that there exists an N so that $a_n \in U$ for every $n \geq N$. By the definition of U, another way to say this is that $|a_n - L| < \epsilon$ for every $n \geq N$, which is the definition of what it means for $a_n \to L$. □

In demonstrating (a_n) ***converges, one usually must determine*** L ***for themselves in advance. This limit*** L ***is unique, provided*** (a_n) ***converges.*** In all of the practice that follows Definition 3.13, the limit L is explicitly given. Very often, you will have to determine L for yourself, and *then* demonstrate that your choice of L is, indeed, the correct one. The following theorem demonstrates that, if a sequence converges, there is only one correct choice for its limit.

Theorem 3.22. *If a sequence converges, its limit is unique. In other words, if $a_n \to L_1$ and $a_n \to L_2$, then $L_1 = L_2$.*

Idea of proof. *Intuitively, if the numbers a_n really are approaching some number L_1, and different number L_2, they must not be approaching them in the way Definition 3.13 describes it. Using the language in our reformulated (equivalent) definition of convergence (Theorem 3.21), we could find a neighborhood of L_1 and a neighborhood of L_2 that do not intersect (this is possible since $\mathbb{R}$ is Hausdorff, see Theorem 2.14). Then, how could the tail of the sequence eventually be in* both *of these neighborhoods if the neighborhoods do not have any numbers in common? In our proof below, we will be sure to consider the terms a_n for large enough n that are assuredly in both of these neighborhoods–this will be our contradiction.*

Proof of Theorem 3.22. Suppose that $a_n \to L_1$, and that $a_n \to L_2$. Suppose also that $L_1 \neq L_2$. According to Theorem 2.14, there exist neighborhoods U_1 of L_1 and U_2 of L_2 so that $U_1 \cap U_2 = \emptyset$. By the assumption that $a_n \to L_1$, the sequence is eventually in U_1. That is, there exists an N_1 so that $a_n \in U_1$ for $n \geq N_1$. By the assumption that $a_n \to L_2$, the sequence is eventually in U_2 as well. In the same way, there exists an N_2 so that $a_n \in U_2$ for all $n \geq N_2$. Let $N = \max\{N_1, N_2\}$. Since $N \geq N_1$, the number $a_N \in U_1$. Since $N \geq N_2$, the number $a_N \in U_2$ as well. This is a contradiction, since U_1 and U_2 were chosen to have no elements in common. So, the assumption that $L_1 \neq L_2$ is false. □

Generally speaking, determining what a sequence converges to can be quite difficult. Although there are some common tools we sometimes employ (such as those found in the next section), there is no one method that will always lead you to the correct limit. In fact, there are many examples of sequences of real numbers that are known to converge, but what exactly they converge to is unknown (infinite series form a large class of these, see Chapter 7). In Section 3.7, we show that there are ways to demonstrate that a sequence converges without knowing what it converges to. For now, we are forced to initially make an educated guess as to what a sequence's limit could be, then demonstrate that our guess is correct by supplying a proof.

Practice 3.23. Let $a_n = \frac{n^2+1}{3n^2-17}$ for $n \geq 0$. Show that a_n converges, and compute $\lim_{n\to\infty} a_n$.

Methodology. *In contrast to the other practices, we need to determine what $\lim_{n\to\infty} a_n$ is first, and then go ahead and demonstrate that (a_n) converges to it. In order to determine $\lim_{n\to\infty} a_n$ we will employ a common trick in analysis. Not until next section will we give a rigorous justification for its use. That is okay here, since we only use this trick to get an educated guess as to what the limit could be, and we will be giving a rigorous proof using*

Definition 3.13. We factor an n^2 out of the numerator and denominator of a_n:

$$\frac{n^2+1}{3n^2-17} = \frac{n^2(1+\frac{1}{n^2})}{n^2(3-\frac{17}{n^2})} = \frac{1+\frac{1}{n^2}}{3-\frac{17}{n^2}}.$$

We proved in Practice 3.19 that $\frac{1}{n^2} \to 0$, and so it stands to reason that $1+\frac{1}{n^2} \to 1$ and $3-\frac{17}{n^2} \to 3$, and that $\frac{1+\frac{1}{n^2}}{3-\frac{17}{n^2}} \to \frac{1}{3}$ as well. We emphasize that this reasoning is not a proof that this sequence converges to $\frac{1}{3}$, our proof that this is, indeed, the limit will be according to Definition 3.13. As part of that proof, we will need to consider the quantity

$$\left|\frac{n^2+1}{3n^2-17} - \frac{1}{3}\right| = \frac{20}{|9n^2-51|}.$$

For large enough n, $9n^2-51>0$ and so the absolute value bars will become unnecessary. The main problem is that we cannot give the same direct line of reasoning as we did in Practice 3.20, since the resulting inequality would go the wrong way. On the other hand, $9n^2-51 \geq n$ eventually (see Exercise (7) of Section 3.1), and this, along with The Archimedean Property will find the N we are looking for.

Solution. We claim that $\lim_{n\to\infty} \frac{n^2+1}{3n^2-17} = \frac{1}{3}$. Let $\epsilon > 0$ be given. We must find an N so that if $n \geq N$, then

$$\left|\frac{n^2+1}{3n^2-17} - \frac{1}{3}\right| = \frac{20}{|9n^2-51|} < \epsilon.$$

Notice that $9n^2-51>0$ for $n \geq 3$ (so, in our estimates we may remove the absolute value signs provided $N \geq 3$), and that $9n^2-51 \geq n$ for $n \geq 3$ (see Exercise 7 of Section 3.1). Then, for $n \geq 3$, we have

$$\frac{20}{|9n^2-51|} = \frac{20}{9n^2-51} \leq \frac{20}{n}. \tag{3.2}$$

We use The Archimedean Property to find an N_1 satisfying

$$\begin{array}{rcl} \frac{1}{N_1} & < & \frac{\epsilon}{20}, \text{ so that} \\ \frac{20}{n} \leq \frac{20}{N_1} & < & \epsilon. \end{array} \tag{3.3}$$

We choose $N = \max\{N_1, 3\}$ so that when $n \geq N$, we have $n \geq 3$, and $n \geq N_1$. Then, for $n \geq N$,

$$\begin{array}{rcll} \left|\frac{n^2+1}{3n^2-17} - \frac{1}{3}\right| & = & \frac{20}{|9n^2-51|} & \\ & = & \frac{20}{9n^2-51} \leq \frac{20}{n} & \text{(Equation (3.2))} \\ & \leq & \frac{20}{N_1} & \text{(Equation (3.4))} \\ & < & \epsilon. & \end{array} \tag{3.4}$$

■

Remark 3.24. In the solution of Practice 3.23, we chose our N to be the maximum of two numbers: 3 and N_1. The reason why such a declaration was advantageous in this practice problem is so that every number $n \geq N$ will simultaneously satisfy $n \geq 3$ and $n \geq N_1$. Generally speaking, declaring N to be the maximum of several numbers is designed to force any $n \geq N$ to be greater than or equal to all of these numbers, and usually, several useful inequalities will simultaneously be true to assist you. For example, in Practice 3.23, that $n \geq N \geq 3$ means that we may use the inequality in Equation (3.2). That $n \geq N \geq N_1$ means that we may use the inequality in Equation (3.3). This idea of choosing N to be the maximum of two numbers is quite common, and is next used in the proof of Theorem 3.31.

The process of finding N given ϵ is one that needs to be approached with care and creativity. In Practice 3.23, we were fortunate to remember the inequality $9n^2 - 3 \geq n$ for $n \geq 3$. That leads the student to wonder **"How am I supposed to know what inequalities will help me?"** Unfortunately, there is no golden list of inequalities that will always help you in your analysis. This question is, in fact, the wrong one to ask. Instead, **try asking the question: "To what place am I trying to take this expression?"** For example, in Practice 3.23 we were not concerned too much with what inequality would help, rather, we were concerned with a possible method to relate $\frac{20}{|9n^2-51|}$ to $\frac{1}{n}$ in order to use The Archimedean Property. It is advisable to consider this as you practice this concept.

Not every sequence converges, and very many diverge. It is important to know how to show a sequence diverges. Very similarly to the earlier suggestion: unless there are any supporting results that help you in this endeavor, you must use Definition 3.13 to show that a sequence diverges. The negation of Definition 3.13 reads as follows:

"The sequence (a_n) does *not* converge to L if there exists an $\epsilon > 0$, so that for every N, there exists an $n \geq N$ with $|a_n - L| \geq \epsilon$. "

This has an equivalent formulation using the language of neighborhoods, as in the equivalent formulation of Definition 3.13 in Theorem 3.21:

The sequence (a_n) does *not* converge to L if there exists a neighborhood of L that does not contain any tail of the sequence.

It is worth pointing out that the phrase "does not contain any tail of the sequence" means "for no N does U contain a_n for every $n \geq N$". Let us practice this concept of divergence.

Practice 3.25. Let $a_n = (-1)^{n+1}$. Show that the sequence (a_n) diverges.

Methodology. *(This is the same sequence that is in Practice 3.4.) Let L be arbitrary: we want to show that (a_n) does not converge to L. We demonstrate* *this according to the first boxed statement above. As such, we are looking for an $\epsilon > 0$ that is small enough so that for every N, there is some member of the sequence beyond a_N that stays at least a distance of ϵ away from L. Since this sequence oscillates between 1 and -1, it is feasible to choose ϵ much smaller than the distance between them to eliminate the inadvertent possibility that our neighborhood of L be so large that it could possibly contain the tail of a_n. So, since the distance from -1 to 1 is 2, try $\epsilon = \frac{1}{2}$.*

Solution. Let L be arbitrary; we show that $(-1)^{n+1}$ does not converge to L. Let $\epsilon = \frac{1}{2}$. Since the distance between 1 and -1 is 2, and the distance between any two numbers in the set $(L - \frac{1}{2}, L + \frac{1}{2})$ must be less than 1, it is not the case that both 1 and -1 are within a distance of ϵ from L, so that either $|1 - L| \geq \epsilon$, or $|-1-L| \geq \epsilon$. Now, let N be arbitrary. We need to show that there exists an $n \geq N$ with $|a_n - L| \geq \epsilon$. Assume first that $|1 - L| \geq \epsilon$. Then there exists an odd number $n \geq N$, and for this n, we have $a_n = 1$, and so $|a_n - L| = |1 - L| \geq \epsilon$. Now assume that $|-1-L| \geq \epsilon$. Then there exists an even number $n \geq N$, and for this n, we have $a_n = -1$, and so $|a_n - L| = |-1-L| \geq \epsilon$. Putting these cases together, we have shown that for $\epsilon = \frac{1}{2}$, for every N there exists an $n \geq N$ so that $|a_n - L| \geq \epsilon$. ∎

Briefly and nonrigorously put, the reason $a_n = (-1)^{n+1}$ diverges is because it oscillates between two quantities. The sequence $a_n = n^2 + 1$ below will not converge to a real number because it is not bounded. Practice 3.26 demonstrates this through a contradiction argument: our solution will assume that the sequence does converge to a real number L, and a contradiction will ensue. This demonstrates that one is not forced to use the boxed statements above that negate Definition 3.13 to demonstrate the divergence of a sequence. In Section 3.8, we further consider sequences that tend toward either ∞ or $-\infty$.

Practice 3.26. Let $a_n = n^2 + 1$. Show that a_n does not converge to L for any $L \in \mathbb{R}$.

Methodology. *It seems that a_n is increasing without bound, and it is for this reason that it seems impossible that a_n be approaching one real number in particular. We will exploit this unbounded behavior of a_n in order to reach a natural contradiction to the assumption that a_n converge to L.*

Solution. Suppose that $a_n \to L$. Choose $\epsilon = 1$. By the definition of convergence, there exists an N with $|a_n - L| < 1$ whenever $n \geq N$. Rewriting this expression, for $n \geq N$:

$$\begin{array}{rrcll} & |a_n - L| & < & 1 & \\ \iff & -1 & < & a_n - L & < 1 \\ \iff & L - 1 & < & a_n & < 1 + L. \end{array}$$

One checks that $a_n = n^2 + 1 \geq n$ for every $n \geq 1$ (see Exercise 6 of Section 3.1). Therefore

$$n \leq n^2 + 1 = a_n \leq 1 + L \text{ for every } n \geq N.$$

This contradicts the fact that the natural numbers are bounded above (Theorem 2.33). ■

In the previous practice problem, we noted that $a_n = n^2 + 1$ must not converge, otherwise the natural numbers would be bounded. This contradiction was found through the observation that the sequence itself was not bounded. In fact, there is a relationship between convergent and bounded sequences.

Definition 3.27. A sequence (a_n) is *bounded* if there exists a real number $M \geq 0$ so that $|a_n| \leq M$ for all n. The number M is called a bound for (a_n).

The reader has already been introduced to bounded sequences, see Practice 3.11 and Exercise 10 of Section 3.1. Notice that a bounded sequence a_n must have $M \geq 0$ as its bound, and in every case other than the constant sequence $a_n = 0$, that bound $M > 0$. The following theorem proves that all convergent sequences must be bounded.

Theorem 3.28. *If (a_n) converges, then (a_n) is bounded.*

Idea of proof. *The numbers a_n are approaching some number L by assumption. The definition of convergence will force a bound on the tail of the sequence so that not any of the tail will be too far away from the limit L. The largest magnitude of that number and the beginning values of the sequence will be our bound M.*

Proof of Theorem 3.28. Suppose $a_n \to L$. Let $\epsilon = 1$. Then there exists an N so that $|a_n - L| < 1$ for $n \geq N$. Using the triangle inequality (Exercise 9 of Section 2.1), we have

$$\begin{aligned} ||a_n| - |L|| \leq |a_n - L| &< 1, \text{ so} \\ -1 < |a_n| - |L| &< 1, \text{ so} \\ |a_n| &< 1 + |L| \text{ for } n \geq N. \end{aligned}$$

Now set $M = \max\{|a_1|, \ldots, |a_{N-1}|, L+1\}$. By definition of M, if $n \leq N-1$, then $|a_n| \leq M$, and if $n \geq N$ then $|a_n| \leq M$ as well. Therefore, (a_n) is bounded. □

The contrapositive of Theorem 3.28 is given as Corollary 3.29.

Corollary 3.29. *If (a_n) is not bounded, then (a_n) diverges.*

We apply this idea in the following practice.

Practice 3.30. Let $a_n = \frac{n^2+1}{n+1}$. Determine whether or not (a_n) converges.

Methodology. *It appears that this sequence is unbounded, since this sequence is defined as a quotient of polynomials, and the degree of the numerator is strictly larger. Then we will use Corollary 3.29 to conclude (a_n) does not converge.*

Solution. We show that (a_n) is unbounded by contradiction. Suppose there is a number M so that $|a_n| \leq M$ for all n. Since $a_n \geq 0$, $|a_n| = a_n$, and as a result,

$$\frac{n^2+1}{n+1} \leq M, \quad \text{or,} \quad n^2 - Mn \leq M - 1.$$

Since $n^2 - Mn$ is a polynomial with positive leading term (of 1), it is not possible that it is less than or equal to any fixed number for every n. This contradiction demonstrates that (a_n) is not bounded. Therefore, by Corollary 3.29, the sequence (a_n) must diverge. ■

3.2.3 Review of important concepts from Section 3.2

1. Convergence of sequences is defined formally in Definition 3.13. There is an equivalent and more intuitive formulation of this in Theorem 3.21. Unless there are supporting results that apply, you must use this definition to prove a sequence converges.
2. In proving a sequence converges using Definition 3.13, one typically considers an arbitrary $\epsilon > 0$, and searches for an N that will satisfy Definition 3.13. Generally speaking, this N depends on ϵ.
3. If you are asked to prove a sequence converges, then you must determine what the sequence converges to unless it is given to you. There are exceptions to this (but not until Section 3.7 about Cauchy Sequences).
4. If you are proving that a sequence diverges, you must also use (the negated form of) Definition 3.13, or its equivalent. One could also assume such a sequence converges to some arbitrary limit, and find a contradiction. In this way, you would be proving that a sequence does not converge to L for every real number L.
5. Convergent sequences are bounded (Theorem 3.28). Therefore, an unbounded sequence must not converge (Corollary 3.29).

Exercises for Section 3.2

1. Show the following sequences converge to the given limit using only Definition 3.13. Reviewing Practices 3.14 and 3.17–3.20 may be helpful.
   (a) $\frac{1}{n^3} \to 0$.
   (b) $\frac{(-1)^n}{1+n^5} \to 0$.
   (c) $\frac{n-1}{n+1} \to 1$.
   (d) $\frac{-7}{n-4} \to 0$ (here, the sequence is defined for $n \geq 5$).
2. In the following, sequences (a_n) are given. Determine whether or not the sequences converge, and if so, determine $\lim_{n\to\infty} a_n$ with proof. Reviewing Practices 3.23, 3.25, and 3.30 may be helpful.

(a) $a_n = \frac{n^3-7}{n^3-15}$.
(b) $a_n = \frac{n^2}{n+1}$.
(c) $a_n = \frac{-4n^3}{(2n+1)^3}$.
(d) $a_n = (-1)^n \frac{n}{2n-1}$. *[Hint: See Exercise 11.]*
(e) $a_n = \sqrt{n+1} - \sqrt{n}$. *[Hint: Rationalize.]*
(f) $a_n = \begin{cases} \frac{1}{n} & \text{for odd } n, \\ 0 & \text{for even } n. \end{cases}$
(g) $a_n = \begin{cases} \frac{1}{n} & \text{for odd } n, \\ 1 & \text{for even } n. \end{cases}$
(h) $a_n = \begin{cases} n^2 & \text{for odd } n, \\ -n & \text{for even } n. \end{cases}$

3. Prove that $\frac{1}{\sqrt{n}} \to 0$.

4. Suppose that $a_n \geq 0$, and that $a_n \to a$.

   (a) Show that $a \geq 0$.
   (b) Show that $\sqrt{a_n} \to \sqrt{a}$. *[Hint: Consider two cases, $a = 0$, and then $a > 0$.]*
   (c) Show that $a_n^{\frac{1}{p}} \to a^{\frac{1}{p}}$ for $p \in \mathbb{N}$. *[Hint: Find a polynomial $\Phi_p(x, y)$ so that $x^p - y^p = (x - y)\Phi_p(x, y)$. Then, proceed as in the last part.]*

5. Let $h > 0$. Prove that $\lim_{n\to\infty} \frac{1}{1+nh} = 0$.

6. Let $(a_n)_{n\geq 1}$ be a sequence. Define the sequence $(b_n)_{n\geq 1}$ as $b_n = a_{n+1}$. That is, the sequence (b_n) is the same as (a_n), except it begins with a_2 instead of a_1.

   (a) Prove that (a_n) converges to x if and only if (b_n) converges to x.
   (b) Define $(c_n)_{n\geq 1}$ as $c_n = a_{n+p}$ for some $p \geq 1$. Use Mathematical Induction on p to show that (a_n) converges to x if and only if (c_n) converges to x. (This exercise fully demonstrates the fact that convergence is only concerned with the eventual behavior of a sequence.)

7. This problem is an example of how to use the tool developed in Exercise 6.

   (a) Prove that if $a, b \geq 0$, then $\frac{1}{n^2+an+b} \leq \frac{1}{n}$ for $n \geq 1$.
   (b) Let $a_n = \frac{1}{n^2-20}$. Find a closed-form expression for $b_n = a_{n+5}$.
   (c) Show $b_n \to 0$. Use Exercise 6 to conclude $a_n \to 0$ as well.

8. Let (a_n) be a sequence that is eventually constant. Show that (a_n) converges. Then, discuss the relationship between ϵ and N from your solution, and compare and contrast it to the discussion from Practice 3.16.

9. Suppose that $a_n \to L$. Prove that $|a_n| \to |L|$. *[Hint: Use Exercise 9 of Section 2.1.]*

10. This exercise demonstrates that the converse of Exercise 9 is almost always false.
    (a) Show that if $|a_n| \to 0$, then $a_n \to 0$.
    (b) Construct an example of a sequence a_n so that $|a_n| \to L \neq 0$, but (a_n) does not converge.
11. Suppose $a_n \to L$, where $L \neq 0$. Prove that $((-1)^n a_n)$ diverges.
12. Suppose that $x_n \in \mathbb{Z}$ for every n, and that $x_n \to x$. What can you say about the sequence x_n? What can you say about x? *[Hint: Consider the property in Exercise 8.]*
13. Suppose that $x_n \in \mathbb{Q}$ for every n, and that $x_n \to x$. Must $x \in \mathbb{Q}$ as well? *[Hint: See Practice 3.45.]*
14. This exercise will study the convergence of geometric sequences. Let $r \in \mathbb{R}$ be given.
    (a) $(r = 1)$ Show that the sequence $(1^n)_{n \geq 0}$ converges to 1.
    (b) $(r = 0)$ Show that the sequence $(0^n)_{n \geq 1}$ converges to 0.
    (c) $(r = -1)$ Show that the sequence $((-1)^n)_{n \geq 0}$ diverges. *[Hint: See Practice 3.25.]*
    (d) $(0 < r < 1)$ Suppose $0 < r < 1$. Let $x = \frac{1}{r}$.
        i. Note that $x > 1$ if $0 < r < 1$.
        ii. Write $x = 1+h$ for $h > 0$. Show that $x^n \geq 1+nh$. *[Hint: Consider the Binomial Theorem or Bernoulli's Inequality (Practice 3.12).]*
        iii. Use the inequality above and Exercise 5 to show that $r^n \to 0$.
    (e) $(r > 1)$ Suppose $r > 1$.
        i. Write $r = 1 + h$ for some $h > 0$. Use a method similar to the last part to show $r^n \geq 1 + nh$.
        ii. Show that $(r^n)_{n \geq 0}$ diverges. *[Hint: Use the method of Practice 3.26.]*
    (f) $(-1 < r < 0)$. Suppose $-1 < r < 0$. Use previous work and Exercise 10 to show $r^n \to 0$.
    (g) $(r < -1)$ Write $|r| = 1 + h$ for some $h > 0$, and apply the methodology of part (e) to show that $(r^n)_{n \geq 0}$ diverges.
15. Construct a divergent sequence that is bounded. Conclude that the converse to Theorem 3.28 is false.
16. Show that (a_n) is eventually bounded if and only if it is bounded.
17. Suppose $a_n \to L$. Let $M \neq L$. Show that (a_n) is eventually bounded away from M. In other words, show that there exists an $\epsilon > 0$ and N so that $|a_n - M| \geq \epsilon$ for all $n \geq N$.
18. Suppose $M \in \mathbb{R}$. Show that if (a_n) is bounded away from M, then (a_n) does not converge to M.

19. Suppose that $b_n \neq 0$ for every n, and that $b_n \to b \neq 0$. Show that $\frac{1}{b_n} \to \frac{1}{b}$. *[Hint: You will, in some manner, need to use the result in Exercise 17.]*

20. Let $f : A \to \mathbb{R}$ be a function, where $A, B \subseteq \mathbb{R}$. Suppose that $x_n \to x$, with $x_n, x \in A$ for every n, and that for every $x, y \in A$, there exists a number $c \in (0, 1)$ with $|f(x) - f(y)| < c|x - y|$. Show that $f(x_n) \to f(x)$.

3.3 Limit Theorems

There were several places in the last section where we spoke of "supporting results" to aid us in determining convergence, and in the absence of such results, we were forced to use Definition 3.13 to demonstrate that a sequence converged. In this section, we develop our first "supporting results" so that we may not always have to use Definition 3.13 to demonstrate convergence. After these supporting results have been proven, it will be another tool to help us both justify a sequence converges, and to help us find its limit.

There are three types of limit theorems given in this section. The first type in Subsection 3.3.1 are basic results to assist us in determining that certain new sequences (built from old ones, as in Section 3.1) will converge if built from already convergent sequences. In addition, the limits of these new sequences are given as well.

The second sort of results are given in Subsection 3.3.2; these results deal with inequalities of sequences, and how that inequality is preserved through convergence. For example, Theorem 3.48 shows that if there is an inequality of convergent sequences, then there is an inequality of their limits.

The final collection of results is more of a dictionary of special sequences and their limits, and this is found in Subsection 3.3.3. These are sequences that are quite common, and knowing the limits of these special sequences will be of great use later, in particular, when used in conjunction with any one of the other results from this section. In addition to some other important results, the exercises at the end of the section also have you show that, among other things, exponentiation and taking logarithms are preserved by convergence. All combined, the results in this section will be used repeatedly throughout the text.

3.3.1 Computing limits

Recall from Section 3.1 that we can build new sequences from old ones using familiar operations. **The following result illustrates how, if the old sequences converge, then it is a straightforward task to show that the new sequence converges and to determine the limit of the new sequence.** Theorem 3.31 will be the first of our "supporting results" that will allow us to determine convergence of a sequence without specifically referencing Definition 3.13. Be sure to read the remarks following the proof as well.

Theorem 3.31. *Suppose $a_n \to a$ and $b_n \to b$.*

1. $a_n \pm b_n \to a \pm b$.
2. $a_n b_n \to ab$.
3. *If c is a real number, then* $ca_n \to ca$.

4. $|a_n| \to |a|$.

5. If $b \neq 0$, *then* $\frac{a_n}{b_n} \to \frac{a}{b}$.

Idea of proof. *This is the first of the theorems that we will consider a "supporting result". So, since we do not actually have such a supporting result yet, we will establish each of the assertions by using Definition 3.13. We note that part (4) is Exercise 9 of the last section. In part (5), there is an additional requirement that the limit of the denominator is nonzero. To complete the proof of part (5) we will have to use Exercise 17 of the last section to conclude that* b_n *is eventually bounded away from* 0.
Since we are using Definition 3.13, we will have to approximate the distance between each of these sequences and their supposed limits. For example, in part 1, we will have to study $|(a_n + b_n) - (a + b)|$, *and this could be rewritten and compared to* $|a_n - a| + |b_n - b|$ *using the triangle inequality. Each separate part will require its own method for approximating these distances.*

Proof of Theorem 3.31. We prove each of the assertions using the definition of convergence (Definition 3.13).

(1) We prove that $a_n + b_n \to a + b$, and leave the proof that $a_n - b_n \to a - b$ for Exercise 9. Let $\epsilon > 0$ be given. We notice that, by the triangle inequality,

$$|(a_n + b_n) - (a + b)| = |(a_n - a) + (b_n - b)| \leq |a_n - a| + |b_n - b|.$$

Since $a_n \to a$, there exists an N_1 so that $|a_n - a| < \frac{\epsilon}{2}$ for every $n \geq N_1$. Similarly, since $b_n \to b$, there exists an N_2 so that $|b_n - b| < \frac{\epsilon}{2}$ for every $n \geq N_2$. Set $N = \max\{N_1, N_2\}$. Then, for every $n \geq N$, we have

$$|(a_n + b_n) - (a + b)| \leq |a_n - a| + |b_n - b| < \frac{\epsilon}{2} + \frac{\epsilon}{2} = \epsilon.$$

(2) Let $\epsilon > 0$ be given. Notice that, by subtracting and adding the same quantity $a_n b$, and then by the triangle inequality, we have

$$\begin{aligned} |a_n b_n - ab| &= |a_n b_n - a_n b + a_n b - ab| \\ &\leq |a_n b_n - a_n b| + |a_n b - ab| \\ &= |a_n||b_n - b| + |a_n - a||b|. \end{aligned} \tag{3.5}$$

We will approximate each of $|a_n||b_n - b|$ and $|a_n - a||b|$ separately. Since (a_n) is a convergent sequence, it must be bounded (Theorem 3.28). So, there exists a real number $M \geq 0$ so that $|a_n| \leq M$ for every n. In that case,

$$|a_n||b_n - b| \leq M|b_n - b|. \tag{3.6}$$

Since $b_n \to b$, there exists an N_1 so that $|b_n - b| < \frac{\epsilon}{2(M+1)}$. So, for $n \geq N_1$, since $0 \leq \frac{M}{M+1} < 1$ we have

$$|a_n||b_n - b| < M\frac{\epsilon}{2(M+1)} = \left(\frac{M}{M+1}\right)\frac{\epsilon}{2} < \frac{\epsilon}{2}. \tag{3.7}$$

To approximate $|a_n - a||b|$, we notice that $0 \leq \frac{|b|}{|b|+1} < 1$, and that since $a_n \to a$, there exists an N_2 so that for every $n \geq N_2$, we have $|a_n - a| < \frac{\epsilon}{2(|b|+1)}$. In that case, for $n \geq N_2$, we have

$$|a_n - a||b| < \frac{\epsilon}{2(|b|+1)}|b| = \frac{\epsilon}{2}\left(\frac{|b|}{|b|+1}\right) < \frac{\epsilon}{2}. \tag{3.8}$$

Set $N = \max\{N_1, N_2\}$. So, for $n \geq N$, we have

$$\begin{array}{rcll} |a_n b_n - ab| & \leq & |a_n||b_n - b| + |a_n - a||b| & \text{(Equation (3.5))} \\ & < & \frac{\epsilon}{2} + \frac{\epsilon}{2} & \text{(Equations (3.7) and (3.8))} \\ & = & \epsilon. & \end{array} \tag{3.9}$$

(3) Let $\epsilon > 0$ be given. Notice that $0 \leq \frac{|c|}{|c|+1} < 1$. Since $a_n \to a$, there exists an N so that, for every $n \geq N$, we have

$$|a_n - a| < \frac{\epsilon}{|c|+1}. \tag{3.10}$$

In that case, for $n \geq N$, we have

$$\begin{array}{rcl} |ca_n - ca| & = & |c||a_n - a| \\ & < & |c|\frac{\epsilon}{|c|+1} \\ & = & \epsilon\left(\frac{|c|}{|c|+1}\right) \\ & < & \epsilon. \end{array} \tag{3.11}$$

(5) Let $\epsilon > 0$ be given. We wish to approximate

$$\left|\frac{a_n}{b_n} - \frac{a}{b}\right| = \frac{|a_n b - ab_n|}{|b_n||b|}. \tag{3.12}$$

Since $b_n \to b \neq 0$, according to Exercise 17 of Section 3.2, the sequence (b_n) is eventually bounded away from 0. As such, there exists $\gamma > 0$ and an N_1 so that $|b_n| \geq \gamma > 0$ for every $n \geq N_1$. By assumption that the sequence $\left(\frac{a_n}{b_n}\right)$ is defined, we have $b_n \neq 0$ for every n. Set

$$\alpha = \min\{|b_1|, \ldots, |b_{N_1-1}|, \gamma\}.$$

Since each of $|b_i| > 0$ for $i = 1, \ldots, N_1 - 1$, and $\gamma > 0$, the minimum of them $\alpha > 0$. Therefore, from Equation (3.12), for every n we have

$$\frac{|a_n b - ab_n|}{|b_n||b|} \leq \frac{|a_n b - ab_n|}{\alpha|b|}. \tag{3.13}$$

We can now proceed to estimate the numerator in Equation (3.13) similarly to Equations (3.5) to (3.9) from the previous part.

We subtract and add the quantity ab, and use the triangle inequality to conclude

$$\begin{aligned}\frac{|a_nb-ab_n|}{\alpha|b|} &= \frac{|a_nb-ab+ab-ab_n|}{\alpha|b|}\\ &\leq \frac{1}{\alpha|b|}|a_n-a||b|+\frac{1}{\alpha|b|}|a||b-b_n|\\ &= \frac{1}{\alpha}|a_n-a|+\frac{|a|}{\alpha|b|}|b_n-b|. \qquad \text{(Note } |b-b_n|=|b_n-b|\text{)}\end{aligned} \tag{3.14}$$

Since $a_n \to a$, there exists an N_2 so that, for $n \geq N_2$, we have

$$|a_n-a| < \frac{\alpha\epsilon}{2}, \text{ so } \frac{1}{\alpha}|a_n-a| < \frac{\epsilon}{2}. \tag{3.15}$$

Since $b_n \to b$, there exists an N_3 so that, for $n \geq N_3$, we have

$$|b_n-b| < \frac{\alpha|b|\epsilon}{2(|a|+1)}, \text{ so } \frac{|a|}{\alpha|b|}|b_n-b| < \frac{|a|}{\alpha|b|}\frac{\alpha|b|\epsilon}{2(|a|+1)} < \frac{\epsilon}{2}. \tag{3.16}$$

Let $N = \max\{N_2, N_3\}$. Then for every $n \geq N$, we have

$$\begin{aligned}\left|\frac{a_n}{b_n}-\frac{a}{b}\right| &= \frac{|a_nb-ab_n|}{|b_n||b|}\\ &\leq \frac{|a_nb-ab_n|}{\alpha|b|} && \text{(Equation (3.13))}\\ &\leq \frac{1}{\alpha}|a_n-a|+\frac{|a|}{\alpha|b|}|b_n-b| && \text{(Equation(3.14))}\\ &< \frac{\epsilon}{2}+\frac{\epsilon}{2} && \text{(Equations (3.15) and (3.16))}\\ &= \epsilon.\end{aligned} \tag{3.17}$$

(4) See Exercise 9 of the last section.

□

There are several remarks about the previous proof that are in order.

Remark 3.32. There is the possibility, however remote, that $M = 0$ in Equation (3.6) (see also the paragraph following Definition 3.27 on page 119). The reason why we fuss with the quantity $\frac{M}{M+1}$ in Equation (3.7) rather than just to simply bound $|b_n - b|$ by $\frac{\epsilon}{2M}$ is to avoid the consideration that M could be 0. Of course, if $M = 0$ and is a bound for the sequence (a_n), then $a_n = 0$ for every n, and the product sequence $a_nb_n = 0$, and is a constant sequence. Since constant sequences are already known to converge (see Practice 3.16), there would be nothing new to prove in this case anyway. All the same, once this consideration is taken into account, the proof is perfectly valid without splitting it up into cases. The same sort of considerations are made about $|c|$ in Equation (3.10), and about $|a|$ in Equation (3.16). These considerations are *not* required concerning α and $|b|$ in Equation (3.16), since these quantities are already known to be nonzero.

Remark 3.33. There is a more straightforward proof of part (3) that uses part (2), and the fact that a constant sequence converges: If the sequence $b_n = c$, a constant sequence, then by Practice 3.16, $b_n \to c$. In that case, by part (2), the sequence $a_n b_n = c a_n \to ca$. This is an example of how to use supporting results (part (2) and Practice 3.16) to deduce convergence without the need to resort to a proof involving an ϵ.

Remark 3.34. There is also a more straightforward proof of part (5) that uses part (2) and Exercise 19 of Section 3.2. If $a_n \to a$, and $b_n \to b$ with b_n, and $b \neq 0$, then we may use Exercise 19 to conclude $\frac{1}{b_n} \to \frac{1}{b}$. Then we note that by part (2), the sequence

$$\frac{a_n}{b_n} = a_n \cdot \frac{1}{b_n} \to a \cdot \frac{1}{b} = \frac{a}{b}.$$

The most important remark to be made about Theorem 3.31 is that it concludes that, under certain circumstances, new sequences fashioned from convergent old ones are convergent! In addition, this theorem computes the limits of these new sequences. For example, part (2) concludes that, if both (a_n) and (b_n) are convergent sequences, then $(a_n b_n)$ converges as well. In addition, it concludes that $a_n b_n \to ab$. We now practice applications of this result, and then demonstrate through the use of examples and exercises that each of the hypotheses is necessary in Theorem 3.31.

Practice 3.35. Let $a_n = \frac{n-1}{n}$. Show a_n converges and find its limit.

Methodology. *This is the same sequence that we studied in detail at the beginning of Section 3.2. In Practice 3.14 we proved directly that $a_n \to 1$ using the definition of convergence of sequences (Definition 3.13). The point of this practice problem is to use our new supporting results to determine the limit of a_n and to demonstrate convergence simultaneously with an application of Theorem 3.31.*

To that end, we notice two things. The first is that we may not use part (5) of Theorem 3.31: although $a_n = \frac{n-1}{n}$ is given as the quotient of two sequences ($n-1$ and n), neither of those sequences converge since neither is bounded. The hypotheses of Theorem 3.31 insists that, in order to use this part, both the numerator and denominator must be convergent sequences (and the denominator must not converge to 0).

The second observation is the more helpful one, and will allow us to complete the problem quite easily, and without the use of Definition 3.13. We notice that

$$a_n = \frac{n-1}{n} = \frac{n}{n} - \frac{1}{n} = 1 - \frac{1}{n}.$$

So a_n is the difference of two sequences. These sequences are the constant sequence 1, and the sequence $\frac{1}{n}$. Both of these are known to converge, $1 \to 1$, and $\frac{1}{n} \to 0$ from Practice 3.17. So, according to part (1) of Theorem 3.31, we may conclude that $a_n = 1 - \frac{1}{n}$ converges, and that it converges to $1 - 0 = 1$.

Solution. We notice that $a_n = \frac{n-1}{n} = 1 - \frac{1}{n}$. Since the constant sequence 1 converges to 1, and $\frac{1}{n} \to 0$ (see Practice 3.17), a_n is the difference of two convergent sequences. By part (1) of Theorem 3.31, $a_n \to 1 - 0 = 1$. ■

Practice 3.36. Let $a_n = \frac{-7}{n}$. Determine if a_n converges, and find its limit if it does.

Methodology. *Just as in Practice 3.35, this sequence is the quotient of two sequences, although we still cannot use part (5) of Theorem 3.31, since the denominator is not a convergent sequence. Just because this sequence does not fit the hypotheses of that particular theorem does* **not** *imply that the sequence is divergent. (In fact, the last practice problem demonstrates that as well.) On the other hand,* $\frac{-7}{n} = (-7)\frac{1}{n}$. *The sequence* $\frac{1}{n} \to 0$, *and* -7 *is a scalar. So, according to part (3) of Theorem 3.31, the sequence* $\frac{-7}{n} = (-7)\frac{1}{n} \to (-7) \cdot 0 = 0$.

Solution. Notice that $a_n = \frac{-7}{n} = (-7)\frac{1}{n}$, and that $\frac{1}{n} \to 0$. So, by part (3) of Theorem 3.31, a_n converges to $(-7) \cdot 0 = 0$. ■

Practice 3.37. Let $a_n = \frac{n^2+1}{3n^2-17}$. Determine if a_n converges, and find its limit if it does.

Methodology. *This is the same sequence that we considered in Practice 3.23. In that problem, we deduced the limit of this sequence should be* $\frac{1}{3}$ *because of the following equality:*

$$\frac{n^2+1}{3n^2-17} = \frac{n^2(1+\frac{1}{n^2})}{n^2(3-\frac{17}{n^2})} = \frac{1+\frac{1}{n^2}}{3-\frac{17}{n^2}}.$$

At the time, we had no supporting results to justify that this methodology was valid, and we were forced to demonstrate that $a_n \to \frac{1}{3}$ *using Definition 3.13. But now, we have found the sequence* a_n *to be equal to the quotient of two convergent sequences, and Theorem 3.31 will make the problem much easier to do.*

Solution. Notice that

$$a_n = \frac{n^2+1}{3n^2-17} = \frac{n^2(1+\frac{1}{n^2})}{n^2(3-\frac{17}{n^2})} = \frac{1+\frac{1}{n^2}}{3-\frac{17}{n^2}}.$$

We demonstrate that the numerator is a convergent sequence, and that the denominator is a sequence that converges to a nonzero number. Then, we will use part (5) of Theorem 3.31 to compute the limit.

In the numerator, the sequence $1 + \frac{1}{n^2}$ is the sum of two convergent sequences: the constant sequence 1, and the sequence $\frac{1}{n^2} \to 0$ (see Practice 3.19). So, by part (1) of Theorem 3.31, the numerator is a sequence that converges to $1 + 0 = 1$.

In the denominator, the sequence $3 - \frac{17}{n^2}$ is also convergent by the following reasoning. The constant sequence 3 converges (to 3), and the sequence $\frac{17}{n^2} = (17)\left(\frac{1}{n^2}\right)$ converges to $(17)(0) = 0$ according to part (3) of Theorem 3.31. The sequence $3 - \frac{17}{n^2}$ is then the difference of convergent sequences, and converges to $3 - 0 = 3$ by part (1) of Theorem 3.31.

We have demonstrated that $1 + \frac{1}{n^2} \to 1$, and $3 - \frac{17}{n^2} \to 3 \neq 0$. So according to part (5) of Theorem 3.31,

$$a_n = \frac{1 + \frac{1}{n^2}}{3 - \frac{17}{n^2}} \to \frac{1}{3}.$$

■

Generally speaking, after one has become accustomed to using the various parts of Theorem 3.31, one rarely needs to reference it. We have included these references in the solutions of these previous practice problems to emphasize exactly what part(s) of Theorem 3.31 were used, how they were used, and where. After this section, we will rarely reference Theorem 3.31 unless its reference is an important part of our work.

3.3.2 Convergence inequalities

Although there are countless results about inequalities in Analysis, there are three useful and basic inequalities that we shall repeatedly use. The first is known as the Squeeze Theorem.

Theorem 3.38 (Squeeze Theorem)**.** *Suppose that $a_n \leq b_n \leq c_n$. Suppose also that $a_n \to x$, and that $c_n \to x$. Then $b_n \to x$ as well.*

Idea of proof. *If b_n is squeezed between a_n and c_n, then the distance from b_n to x must be squeezed between the distances $|a_n - x|$ and $|c_n - x|$ as well. We will use this in order to approximate $|b_n - x|$, and we will bound $b_n - x$ above by $c_n - x$, and below by $a_n - x$.*

Proof of the Squeeze Theorem (Theorem 3.38). Let $\epsilon > 0$ be given. Since $a_n \leq b_n \leq c_n$, it follows that

$$a_n - x \leq b_n - x \leq c_n - x. \tag{3.18}$$

Since $c_n \to x$, there exists an N_1 so that $|c_n - x| < \epsilon$ for all $n \geq N_1$. In other words,

$$-\epsilon < c_n - x < \epsilon \text{ for } n \geq N_1. \tag{3.19}$$

Since $a_n \to x$, there exists an N_2 so that $|a_n - x| < \epsilon$ for all $n \geq N_2$. In other words,

$$-\epsilon < a_n - x < \epsilon \text{ for } n \geq N_2. \tag{3.20}$$

Let $N = \max\{N_1, N_2\}$. We start with Equation (3.18), extend it to the right using the second inequality of Equation (3.19), and extend it to the left using the first inequality from Equation (3.20) to obtain, for $n \geq N$,

$$-\epsilon < a_n - x \leq b_n - x \leq c_n - x < \epsilon.$$

In other words, for $n \geq N$, $|b_n - x| < \epsilon$. □

The Squeeze Theorem can help to establish the convergence of certain limits that otherwise would be difficult to demonstrate.

Practice 3.39. Let $a_n = \frac{\cos(n)}{n}$. Prove $a_n \to 0$.

Methodology. *Considering the behavior of this sequence as a whole is more helpful than considering any of a_n individually, especially since* $\cos(n)$ *is relatively bizarre for $n \in \mathbb{N}$. Generally, the denominator of $\frac{\cos(n)}{n}$ seems to grow without bound, while the numerator is always between -1 and 1.*

Solution. Notice that for every n, we have $-1 \leq \cos(n) \leq 1$, and so

$$\frac{-1}{n} \leq \frac{\cos(n)}{n} \leq \frac{1}{n}.$$

Since $\frac{1}{n} \to 0$ and $\frac{-1}{n} \to 0$ as well, by the Squeeze Theorem we can conclude $\frac{\cos(n)}{n} \to 0$. ■

Practice 3.40. Let $a_n = \frac{n+(-1)^n}{n+1}$ for $n \geq 1$. Determine if (a_n) converges, and if it does, compute $\lim_{n\to\infty} a_n$.

Methodology. *By observation, we can see that $(-1)^n \leq 1$, and that $-1 \leq (-1)^n$ for $n \geq 1$. Therefore*

$$n-1 \leq n+(-1)^n \leq n+1, \quad \text{so} \quad \frac{n-1}{n+1} \leq \frac{n+(-1)^n}{n+1} \leq \frac{n+1}{n+1} = 1.$$

If we could show that $\frac{n-1}{n+1} \to 1$, then the Squeeze Theorem would do the rest of the job.

Solution. Notice that

$$\frac{n-1}{n+1} \leq \frac{n+(-1)^n}{n+1} \leq \frac{n+1}{n+1} = 1.$$

If we can show that $\frac{n-1}{n+1} \to 1$, then the Squeeze Theorem applies, and it would conclude that $a_n \to 1$. But

$$\frac{n-1}{n+1} = \frac{n\left(1-\frac{1}{n}\right)}{n\left(1+\frac{1}{n}\right)} = \frac{1-\frac{1}{n}}{1+\frac{1}{n}}.$$

But $1-\frac{1}{n} \to 1$ and $1+\frac{1}{n} \to 1 \neq 0$ as well. So $\frac{n-1}{n+1} \to \frac{1}{1} = 1$. ■

Remark 3.41. The estimate $-1 \leq (-1)^n \leq 1$ that we give in Practice 3.40 is a fairly good one. If we were to instead use the estimate $-n \leq (-1)^n \leq n$, then we would arrive at the inequality

$$\frac{n-n}{n+1} \leq \frac{n+(-1)^n}{n+1} \leq \frac{n+n}{n+1}, \text{ or } 0 \leq \frac{n+(-1)^n}{n+1} \leq \frac{2n}{n+1}.$$

Since $\frac{2n}{n+1} \to 2$, this estimate–although valid–would not help us demonstrate that $\frac{n+(-1)^n}{n+1}$ converges since $0 \neq 2$. This observation should tell the student that there is no one right way of going about your estimations in order to arrive finally at an estimate that will be of help. See Remark 3.44 for another example of different helpful estimates for the same problem. ■

Practice 3.42. Show $a_n \to 0$, where we define a_n as follows:

$$a_n = \begin{cases} \frac{1}{n^2} & \text{for odd } n, \\ \frac{-1}{n} & \text{for even } n. \end{cases}$$

Methodology. *This sequence's rule is pasted together from two others:* $\frac{1}{n^2}$ *and* $\frac{-1}{n}$*. Both of these sequences tend to 0, and so the Squeeze Theorem should help us out.*

Solution. Notice that $a_n \leq \frac{1}{n^2}$ for every n since it is equal to that for every odd n, and $a_n < 0 \leq \frac{1}{n^2}$ for every even n. A similar argument demonstrates that $a_n \geq \frac{-1}{n}$ for every n as well. Since $\frac{1}{n^2} \to 0$ and $\frac{-1}{n} \to 0$, the Squeeze Theorem applies and $a_n \to 0$ as well. ■

Practice 3.43. Suppose that $|r| > 1$. Show that $a_n = \frac{1}{nr^n}$ converges.

Methodology. *The sequence* $\frac{1}{nr^n}$ *seems to be quite a bit smaller (in magnitude) than the sequence* $\frac{1}{n}$*. So, we may trap* $|a_n|$ *between* 0 *and* $\frac{1}{n}$*, and use the Squeeze Theorem.*

Solution. Notice that since $|r| > 1$, the number $\frac{1}{|r|} < 1$, and so $\frac{1}{|r|^n} < 1$. We then notice that for every n

$$0 \leq \left| \frac{1}{nr^n} \right| = \frac{1}{n} \frac{1}{|r|^n} < \frac{1}{n}.$$

So, since $\frac{1}{n} \to 0$, the Squeeze Theorem applies, and so $|a_n| \to 0$. By Exercise 10 of Section 3.2, we conclude $a_n \to 0$ as well. ■

Remark 3.44. In Practice 3.43, we could have made a different estimate to aid us in demonstrating $a_n \to 0$, again showing that the inequalities one employs in an effort to use the Squeeze Theorem are not at all unique. Consider the fact that if $|r| > 1$, then $\left|\frac{1}{r}\right| < 1$, so $\frac{1}{|r|^n} \to 0$ by Exercise 14 of Section 3.2. Also, since $\frac{1}{n} \leq 1$ for every $n \geq 1$, we have the estimate

$$0 \leq |a_n| = \frac{1}{n} \frac{1}{|r|^n} \leq \frac{1}{|r|^n}.$$

Now the Squeeze Theorem applies, and $|a_n| \to 0$, so again by Exercise 9 of Section 3.2, $a_n \to 0$. ■

The following practice problem demonstrates that every real number is the limit of a sequence of rational numbers. Exercise 10 asks you to show that every real number is also the limit of a sequence of irrational numbers.

Practice 3.45. Let α be any real number. Prove there exists a sequence of rational numbers (r_n) with $r_n \to \alpha$.

Methodology. *Since the rational numbers are dense in* $\mathbb{R}$*, every interval* $(\alpha - \frac{1}{n}, \alpha)$ *will contain a rational number* r_n *for every* n*. The idea is to always find a rational number* r_n *between* α *and some sequence* a_n *that will approach* α*; the sequence* $a_n = \alpha - \frac{1}{n}$ *will do. According to part (1) of Theorem 3.31,*

$\alpha - \frac{1}{n} \to \alpha$, and $\alpha - \frac{1}{n} < r_n < \alpha$. So, r_n is squeezed between $\alpha - \frac{1}{n}$ and α. By the Squeeze Theorem, $r_n \to \alpha$.

Solution. Let $a_n = \alpha - \frac{1}{n}$. Since a_n is the difference of the two convergent sequences $\alpha \to \alpha$, and $\frac{1}{n} \to 0$, a_n converges to $\alpha - 0 = \alpha$ by part (1) of Theorem 3.31. The intervals $I_n = (a_n, \alpha)$ contain infinitely many rational numbers since $\mathbb{Q}$ is dense in $\mathbb{R}$. For each n, choose a rational $r_n \in I_n$. By construction, $a_n < r_n < \alpha$, and so by the Squeeze Theorem, $r_n \to \alpha$. ■

Before moving on to the next useful inequality, we pause to establish a useful result. The proof of this theorem will follow closely to the solution of Practice 3.45.

Theorem 3.46. *Let $S \subseteq \mathbb{R}$ be nonempty.*

1. *If S is bounded above, then there exists a sequence $x_n \in S$ that converges to* $\sup(S)$.
2. *If S is bounded below, then there exists a sequence $x_n \in S$ that converges to* $\inf(S)$.

Idea of proof. *Whatever subset of $\mathbb{R}$ you have, imagine the part of it that has the largest elements. Since S is bounded above, the Completeness Axiom states that $\sup(S) \in \mathbb{R}$. There are two possibilities: either $\sup(S) \in S$, or $\sup(S) \notin S$. In the event that $\sup(S) \in S$, then the constant sequence $x_n = \sup(S)$ will do, even if that feels like cheating (after all, what if S contains only one element?). If $\sup(S) \notin S$, then we shall construct a sequence in S converging to $\sup(S)$ in the same manner as Practice 3.45, using the fact that $\sup(S)$ is the least upper bound of S.*

Proof of Theorem 3.46. (We prove part (1), and leave part (2) for Exercise 25.) Since S is bounded above, $\sup(S) \in \mathbb{R}$. For convenience, denote $\sup(S)$ as s. We consider two cases: $s \in S$, and $s \notin S$. If $s \in S$, then define $x_n = s$ for every n. Since it is a constant sequence, it converges, and it converges to s.

Now suppose $s \notin S$. Since s is the *least* upper bound of S, any number less than s is *not* an upper bound of S. So, for every $\epsilon > 0$, there exists an $x \in S$ (depending on that ϵ) with

$$s - \epsilon \leq x < s \qquad (x \neq s \text{ since we assumed } s \notin S, \text{ and } x \in S.)$$

We use this fact repeatedly. For every $n \geq 1$, set $\epsilon = \frac{1}{n}$. Then there exists an $x_n \in S$ satisfying

$$s - \frac{1}{n} \leq x_n < s.$$

Now we use the Squeeze Theorem: $s - \frac{1}{n} \to s$, and so $x_n \to s$. □

Remark 3.47. It is important to note that the sequence x_n we constructed in the proof of Theorem 3.46 need not satisfy any other conditions beyond just converging to the correct number. In Exercise 3 of Section 3.5, we ask the reader to show that such a sequence can be constructed subject also to the condition that $x_n \leq x_{n+1}$. ■

The second useful inequality states that an inequality of convergent sequences gives rise to an inequality of their limits.

Theorem 3.48. *Suppose $a_n \leq b_n$ for every n, and that $a_n \to a$ and $b_n \to b$. Then $a \leq b$.*

Idea of proof. *Intuitively, something must go wrong if the opposite conclusion were true. That is, we could attempt to prove this theorem by contradiction. If $a > b$, then the sequence (a_n) is eventually bounded away from (and larger than) b, whereas the tail of (b_n) will be eventually within any distance we choose from b. This will contradict the assumption $a_n \leq b_n$.*

Proof of Theorem 3.48. Suppose that $a > b$, and set $\epsilon = \frac{1}{2}(a - b)$. By assumption, $\epsilon > 0$. Since $a_n \to a$, there exists an N_1 so that, for every $n \geq N_1$, $|a_n - a| < \epsilon$. In other words,

$$-\epsilon < a_n - a < \epsilon \text{ for } n \geq N_1. \tag{3.21}$$

Since $b_n \to b$, there exists an N_2 so that $|b_n - b| < \epsilon$ for every $n \geq N_2$. In other words,

$$-\epsilon < b_n - b < \epsilon \text{ for } n \geq N_2. \tag{3.22}$$

Let $N = \max\{N_1, N_2\}$. Then, for every $n \geq N$, we may rewrite Equation (3.21) to read $a - \epsilon < a_n$, and Equation (3.22) to read $b_n < b + \epsilon$. Putting these together, along with our original hypothesis that $a_n \leq b_n$ for each n, we have

$$a - \epsilon < a_n \leq b_n < b + \epsilon. \tag{3.23}$$

But adding ϵ and subtracting b in Equation (3.23) gives us

$$a - b < 2\epsilon.$$

But, this is impossible, since $\epsilon = \frac{1}{2}(a - b)$, so $2\epsilon = a - b$, and it is not the case that $a - b < a - b$. This contradiction completes the proof. □

Notice that a special case of Theorem 3.48 states that if $b_n \geq 0$ for every n, and $b_n \to b$, then $b \geq 0$.

The last useful inequality is a corollary of Theorem 3.48.

Corollary 3.49. *Suppose $x_n \in [a, b]$ for every n. If $x_n \to x$, then $x \in [a, b]$.*

Idea of proof. *We may prove that $x \leq b$ by using x_n for a_n and the constant sequence b for b_n in Theorem 3.48. Since the constant sequence $b \to b$, Theorem 3.48 will apply and give us our result. A similar argument will establish $a \leq x$.*

Proof of Corollary 3.49. Notice that the constant sequence $b_n = b \to b$, and that since $x_n \in [a, b]$ for every n, it follows that $x_n \leq b_n = b$. Therefore, by Theorem 3.48, $x \leq b$.

Similarly, let $a_n = a$ be a constant sequence. Then, $a_n \to a$, and by assumption, $a_n = a \leq x_n$ for each n. Therefore, by Theorem 3.48, $a \leq x$. So, combining these observations gives us $a \leq x \leq b$, or, $x \in [a, b]$. □

The following example is designed to show you that the hypotheses in Corollary 3.49 are necessary: there exist examples of convergent sequences in (a, b) that do not converge within (a, b).

Example 3.50. Let $x_n = \frac{1}{n}$. Notice that the sequence $x_n \in (0, 2)$ for every $n \geq 1$. However, $\frac{1}{n} \to 0$, and $0 \notin (0, 2)$. Therefore, the hypothesis in Corollary 3.49 that the interval that x_n is in contain its endpoints is necessary. ∎

Since convergence is only really concerned with the eventual behavior of sequences (not necessarily the beginning of them), it makes sense that the three results in this subsection have unnecessarily strict hypotheses. We rephrase each of Theorems 3.38, and 3.48 and Corollary 3.49, and leave the proofs of these results to the reader in the exercises (see Exercises 17, 18, and 19).

Theorem 3.51. *Suppose that $a_n \leq b_n \leq c_n$ eventually. Suppose also that $a_n \to x$ and $c_n \to x$. Then $b_n \to x$.*

Theorem 3.52. *Suppose that $a_n \leq b_n$ eventually. Suppose also that $a_n \to a$ and $b_n \to b$. Then $a \leq b$.*

Corollary 3.53. *Suppose $x_n \in [a, b]$ eventually, and suppose $x_n \to x$. Then $x \in [a, b]$.*

To close this subsection, we give a useful example of the modified Squeeze Theorem, Theorem 3.51. In this example, we use Theorem 3.51 to deduce an interesting way to show a sequence converges to 0.

Example 3.54. Suppose that $a_n > 0$ for every n, and that $\frac{a_{n+1}}{a_n} \to \beta < 1$. We show that $a_n \to 0$. This result is sometimes referred to as "the Ratio Test for sequences," and is not to be confused with the Ratio Test (Theorem 4.31 on Page 239).

Since $\beta < 1$, there exists an α so that $\beta < \alpha < 1$. Since $\frac{a_{n+1}}{a_n} \to \beta < \alpha$, it is eventually bounded above by α. More specifically, there exists an N so that

$$\frac{a_{n+1}}{a_n} \leq \alpha, \quad \text{for } n \geq N. \tag{3.24}$$

We prove that $a_n \leq \alpha^{n-N} a_N$ for every $n \geq N$ by Mathematical Induction. If $n = N$, then the result is trivial. Assume that it holds for some $n \geq N$, and notice that by Equation (3.24) and our induction hypothesis,

$$a_{n+1} \leq \alpha a_n \leq \alpha \alpha^{n-N} a_N = \alpha^{(n+1)-N} a_N.$$

Now that we have shown $a_n \leq \alpha^{n-N} a_N$ for every $n \geq N$, we may complete the proof that $a_n \to 0$. Since $a_n > 0$, we have

$$0 \leq a_n \leq \alpha^{n-N} a_N = \left(\frac{a_N}{\alpha^N}\right) \alpha^n \text{ eventually.}$$

Since $\alpha < 1$, we have $\alpha^n \to 0$ (Exercise 14 of Section 3.2). By part (3) of Theorem 3.31, $\frac{a_N}{\alpha^N} \alpha^n \to 0$. Finally, by the modified Squeeze Theorem (Theorem 3.51), we may conclude $a_n \to 0$. ∎

3.3.3 Limits of special sequences

In this section we exhibit the limits of some common sequences.

Theorem 3.55. *The following sequences converge, and their limits are as given.*

1. *Let* $p > 0$. *Then* $\frac{1}{n^p} \to 0$.
2. *Let* $|r| < 1$. *Then* $r^n \to 0$.
3. *Let* $p > 0$. *Then* $\sqrt[n]{p} \to 1$.
4. $\sqrt[n]{n} \to 1$.

Idea of proof. *Each of these sequences escapes the limit theorems that we have developed in this section, and so we will prove each of these using Definition 3.13. Each of these proofs will require a different estimate, although the second part has already been done in Exercise 14 of Section 3.2. In part (1), we simply use The Archimedean Property. In part (3), we break it up into three cases:* $p = 1$ *(easy, since the sequence is constant),* $p > 1$ *and* $0 < p < 1$. *If* $p > 1$, *then we use the Squeeze Theorem and bound* $\sqrt[n]{p} - 1$ *by 0 below and* $\frac{p-1}{n}$ *above. The* $0 < p < 1$ *case can be related to the* $p > 1$ *case by considering reciprocals. Finally, in part (4), we use a similar approach as in part (3) and the Binomial Theorem.*

Proof of Theorem 3.55. (1) Let $\epsilon > 0$ be given. We find an N so that $\frac{1}{N} < \epsilon^{1/p}$. In that case, for $n \geq N$, we have

$$\frac{1}{n^p} \leq \frac{1}{N^p} < [\epsilon^{1/p}]^p = \epsilon.$$

(2) This is Exercise 14 of Section 3.2.

(3) First, if $p = 1$, the sequence $\sqrt[n]{p} = 1$ is constant and converges to 1. If $p > 1$, then we set $a_n = \sqrt[n]{p} - 1$, so $(1 + a_n)^n = p$. Since $p > 1$, $\sqrt[n]{p} > 1$, so $a_n > 0$. By applying Bernoulli's Inequality (Practice 3.12), we have

$$1 + na_n \leq (1 + a_n)^n = p, \quad \text{so} \quad a_n \leq \frac{p-1}{n}.$$

So since $0 \leq a_n \leq \frac{p-1}{n}$, and $\frac{p-1}{n} \to 0$, it must be the case that $a_n \to 0$. Since $\sqrt[n]{p} = a_n + 1$, by part (1) of Theorem 3.31, we have shown $\sqrt[n]{p} \to 1$ in the event $p > 1$.

If $0 < p < 1$, then set $q = \frac{1}{p}$. Then $q > 1$, so $\sqrt[n]{q} \to 1$. But then

$$\frac{1}{\sqrt[n]{q}} \to \frac{1}{1} = 1, \quad \text{and} \quad \frac{1}{\sqrt[n]{q}} = \sqrt[n]{\frac{1}{q}} = \sqrt[n]{p}.$$

So, $\sqrt[n]{p} \to 1$ in the case $0 < p < 1$.

(4) Set $a_n = \sqrt[n]{n} - 1$, so that $(a_n + 1)^n = n$. It is the case that $\sqrt[n]{n} \geq 1$, so that $a_n \geq 0$. The Binomial Theorem states that, since $a_n \geq 0$,

$$(a_n + 1)^n = 1 + na_n + \frac{n(n-1)}{2}a_n^2 + \cdots \geq 1 + \frac{n(n-1)}{2}a_n^2. \tag{3.25}$$

Since $(a_n + 1)^n = n$, we deduce from Equation (3.25) that for $n \geq 2$,

$$(n-1) \cdot \frac{2}{n(n-1)} \geq a_n^2, \quad \text{so} \quad a_n \leq \sqrt{\frac{2}{n}}. \tag{3.26}$$

We may now use the Squeeze Theorem (actually, Theorem 3.51). We already know that $a_n \geq 0$, and $a_n \leq \sqrt{\frac{2}{n}}$ for $n \geq 2$. But $\sqrt{\frac{2}{n}} = \sqrt{2}\frac{1}{n^{1/2}}$. By part (1) with $p = \frac{1}{2}$, we know $\frac{1}{n^{1/2}} \to 0$, and so by part (3) of Theorem 3.31, $\sqrt{2}\frac{1}{n^{1/2}} \to 0$ as well. So the Squeeze Theorem applies, and $a_n \to 0$. Since $a_n + 1 = \sqrt[n]{n}$, by part (1) of Theorem 3.31, $\sqrt[n]{n} \to 1$. □

There is another famous sequence that we establish the limit of here. Although the proof will require some tools from elsewhere in the book (namely, Chapter 6), it is important to have access to the limit of this sequence now.

Example 3.56. In this example, we prove that

$$e = \lim_{n\to\infty} \left(1 + \frac{1}{n}\right)^n.$$

We use several results to help us. First, we assume e^x and $\ln(x)$ are inverse functions, as in Chapter 1. Next, we assume that the sequence $e_n = \left(1 + \frac{1}{n}\right)^n$ (sometimes referred to as "Euler's Sequence") converges; this fact will be shown in Section 3.5, see Example 3.77. We also use Exercise 11 and Theorem 3.31 of this section. Finally, we assume the reader is familiar with the Mean Value Theorem and the derivative of the ln function from basic calculus, so that they may understand Practice 6.15 in Chapter 6. We treat the Mean Value Theorem in detail in Chapter 6 (see Theorem 6.14).

We assume that $\left(1 + \frac{1}{n}\right)^n$ converges to some number y. By Exercise 11, this implies that

$$\ln\left(1 + \frac{1}{n}\right)^n = n\ln\left(1 + \frac{1}{n}\right) \to \ln y.$$

We wish to show $n\ln\left(1 + \frac{1}{n}\right) \to 1$, so that by the uniqueness of limits, $\ln y = 1$, or, $y = e$. We begin with the inequality from Practice 6.15 in Chapter 6, which states that

$$\frac{x}{1+x} \leq \ln(1+x) \leq x \quad \text{for } x \geq 0.$$

Using $x = \frac{1}{n} \geq 0$, and multiplying by n, we have

$$\frac{1}{1 + \frac{1}{n}} \leq n\ln\left(1 + \frac{1}{n}\right) \leq 1.$$

By part 5 of Theorem 3.31,

$$\frac{1}{1+\frac{1}{n}} \to 1.$$

Since the constant sequence $1 \to 1$ as well, by the Squeeze Theorem, we conclude

$$n \ln\left(1 + \frac{1}{n}\right) \to 1.$$

This completes the proof that $\left(1 + \frac{1}{n}\right)^n \to e$. ■

We give an example of how the known quantities of these special sequences can be used in conjunction with Theorem 3.31 to compute limits.

Practice 3.57. For each sequence a_n, determine if (a_n) converges. If so, determine $\lim_{n\to\infty} a_n$.

1. $a_n = \frac{\sqrt[n]{6}}{n^6}$.
2. $a_n = \sqrt[2n]{n}$.
3. $a_n = \sqrt[n]{4n^2 + n + 1}$.

Methodology. *Parts (1) and (2) are just combinations of some of the sequences we have encountered already, and Theorem 3.31 will apply. Part (3), however, is a little more complicated. If the $n+1$ were not there, this sequence would amount to product of sequences whose limit is known. And so, we will use the Squeeze Theorem to trap the sequence in part (3) between these known sequences.*

Solution. (1) Notice that this sequence $a_n = \sqrt[n]{6} \cdot \frac{1}{n^6}$. We know from Part 3 of Theorem 3.55 that $\sqrt[n]{6} \to 1$, and by Part 1 of Theorem 3.55 that $\frac{1}{n^6} \to 0$. So, by Part 2 of Theorem 3.31, $a_n \to 1 \cdot 0 = 0$.

(2) We notice that $\sqrt[2n]{n} = n^{\frac{1}{2n}} = \sqrt{n^{\frac{1}{n}}}$. Since $n^{\frac{1}{n}} \to 1$, by Exercise 4 of Section 3.2, $\sqrt[2n]{n} = \sqrt{n^{\frac{1}{n}}} \to \sqrt{1} = 1$.

(3) Notice that $n+1 \leq 4n^2$ for all $n \geq 1$. Then $4n^2 + n + 1 \leq 4n^2 + 4n^2 = 8n^2$, and

$$\sqrt[n]{4n^2+n+1} \leq \sqrt[n]{8n^2} = 8^{\frac{1}{n}}[n^{\frac{1}{n}}]^2.$$

Now $8^{\frac{1}{n}} \to 1$, and $n^{\frac{1}{n}} \to 1$ by Parts 3 and 4, respectively, of Theorem 3.55, so $[n^{\frac{1}{n}}]^2 \to 1^2 = 1$ and $8^{\frac{1}{n}}[n^{\frac{1}{n}}]^2 \to 1 \cdot 1^2 = 1$ by repeated use of Part 2 of Theorem 3.31.

On the other hand, notice that $n + 1 \geq 0$ and $4n^2 \geq 1$ for every $n \geq 1$. Then $4n^2 \leq 4n^2 + n + 1$, and

$$\sqrt[n]{4n^2} \leq \sqrt[n]{4n^2+n+1}.$$

By a reasoning similar to above, $\sqrt[n]{4n^2} = 4^{\frac{1}{n}}[n^{\frac{1}{n}}]^2 \to 1 \cdot 1^2 = 1$ as well. So the Squeeze Theorem applies, and $a_n \to 1$. ■

3.3.4 Review of important concepts from Section 3.2

1. Theorem 3.31 can both establish the convergence of certain sequences, and determine limits. Make sure you meet the hypotheses of the theorem before you use it.
2. The Squeeze Theorem (Theorem 3.38) is a powerful tool to establish the convergence of some complicated looking sequences. In addition, there are other results that show that generally, inequality is preserved by convergence (Theorem 3.48) and that limits of convergent sequences cannot be too far from the members of the sequence themselves (Corollary 3.49). There are versions of these results to account for situations where the hypotheses are eventually satisfied (see Theorems 3.51 and 3.52, and Corollary 3.53).
3. A variety of common sequences and their limits are given in this section (Theorem 3.55 and Example 3.56). One can use these facts and Theorem 3.31 together to evaluate a great deal of limits.
4. In the exercises below, you will show that certain basic functions are also preserved by convergence. These include taking logarithms (Exercise 11), exponentiation (Exercises 13 and 14), and powers (Exercises 5 for integer exponents and/or 15 otherwise).

3.3.5 Exercises for Section 3.3

1. Please determine if the following sequences converge, and if so, determine $\lim_{n\to\infty} a_n$.

   (a) $a_n = \frac{n-1}{n+1}$.
   (b) $a_n = 7 + \frac{1}{2n-1}$.
   (c) $a_n = \frac{n^3}{1+n^2}$.
   (d) $a_n = \frac{n^{\frac{1}{3}} - 7n^{\frac{1}{5}}}{3n^{\frac{1}{6}} - 2n^{\frac{1}{3}}}$.
   (e) $a_n = \frac{n^e}{n^\pi}$.
   (f) $a_n = \frac{n}{e^n}$.
   (g) $a_n = \frac{3^n - 2^n}{6^n}$.
   (h) $a_n = \frac{(14n)^{\frac{1}{n}}}{n!}$.
   (i) $a_n = \left(\frac{n+1}{n}\right)^{2n}$.
   (j) $a_n = \left(1 + \frac{1}{n}\right)^{\frac{1}{n}}$.
   (k) $a_n = \left(\frac{n2^{\frac{1}{n}} + 2^{\frac{1}{n}}}{n}\right)^n$.

2. Prove $\frac{1}{n!} \to 0$. *[Hint: You may find Exercise 8 of Section 3.1 and Theorem 3.51 helpful.]*
3. Verify the following sequences converge to their given limit. You may find Example 3.54 helpful.

   (a) $\frac{a^n}{n!} \to 0$ for $a > 0$.
   (b) $\frac{a^n}{n!} \to 0$ for $a < 0$. (Use the above result and Exercise 10 of Section 3.2.)
   (c) $\frac{n!}{n^n} \to 0$. (Compute $\beta = \frac{1}{e} < 1$ in Example 3.54.)

4. Let $a > 0$. Show that there exists an $N \in \mathbb{N}$ so that for every $n \geq N$, we have $n! \geq a^n$. *[Hint: Use part (a) of the previous problem.]*

5. Suppose $a_n \to a$, $b_n \to b$, and $c_n \to c$.

   (a) Show that $a_n + b_n + c_n \to a + b + c$. Generalize this result to the finite sum of any (finite number of) convergent sequences.

   (b) Show that $a_n b_n c_n \to abc$. Generalize this result to the finite product of any (finite number of) convergent sequences.

6. Use the last exercise to prove that if $x_n \to x$, then $x_n^3 \to x^3$. Could your line of reasoning deduce that $x_n^{\frac{1}{3}} \to x^{\frac{1}{3}}$? Why or why not? (For more on this, see Exercise 15.)

7. This exercise investigates what happens when you remove a hypothesis of convergence from Theorem 3.31.

   (a) Construct divergent sequences (a_n) and (b_n) where

      i. $(a_n + b_n)$ converges.
      ii. $(a_n + b_n)$ diverges.
      iii. $(a_n b_n)$ converges.
      iv. $(a_n b_n)$ diverges.

   (b) Suppose that (a_n) converges, but (b_n) diverges. What can you say about the convergence or divergence of

      i. $(a_n + b_n)$?
      ii. $(a_n b_n)$?
      iii. $(|b_n|)$?

8. Show $\frac{\sin n}{n} \to 0$ through the following argument.

   (a) Recall that $-1 \leq \sin n \leq 1$, and demonstrate that $\frac{-1}{n} \leq \frac{\sin n}{n} \leq \frac{1}{n}$.

   (b) Using the fact that $\pm\frac{1}{n} \to 0$, use the Squeeze Theorem (Theorem 3.38) to prove that $\frac{\sin n}{n} \to 0$.

9. Suppose that $a_n \to a$ and $b_n \to b$. Prove that $a_n - b_n \to a - b$. *[Hint: Mimic the proof of Theorem 3.31, and note that $|b - b_n| = |b_n - b|$.]*

10. Let α be a real number. Prove there exists a sequence of irrational numbers (ζ_n) with $\zeta_n \to \alpha$. *[Hint: Adopt the methodology of Practice 3.45 and use the density of the irrationals in $\mathbb{R}$.]*

11. This exercise is designed to show that natural logarithms are preserved by convergence. That is, if $x_n, x > 0$, and $x_n \to x > 0$, then $\ln(x_n) \to \ln(x)$.

   (a) Show that if $y \geq 1$, then $|\ln(y)| \leq |y - 1|$. *[Hint: Use the fact that $\ln(y)$ and e^y are inverse functions, that when $y \geq 1$ both of $\ln(y) > 0$ and $y - 1 > 0$, and inequality is preserved by raising both sides to be powers of e. Or, you could use the Mean Value Theorem 6.14.]*

(b) Show that if $0 < y < 1$, then $|\ln(y)| \leq \frac{1}{y}|y-1|$. *[Hint: Rewrite $z = \frac{1}{y}$, where now $z > 1$. Then use the last part and rewrite your conclusions in terms of y.]*

(c) Suppose now for the remainder of the problem that $x_n, x > 0$, and that $x_n \to x$. Prove that $\frac{x_n}{x} \to 1$. Conclude that $\frac{x_n}{x}$ eventually satisfies $\left|\frac{x_n}{x}\right| \geq \frac{1}{2}$.

(d) Prove that eventually $|\ln(x_n) - \ln(x)| \leq \frac{2}{x}|x_n - x|$.

(e) Prove that $\ln(x_n) \to \ln(x)$.

12. Let $b > 0$. Show that if $x_n \to x$ with $x_n, x > 0$, then $\log_b(x_n) \to \log_b(x)$. *[Hint: Use the so-called "change of base formula" $\log_b(a) = \frac{\ln a}{\ln b}$, and Exercise 11.]*

13. This exercise is designed to show that the transformation e^x is preserved by convergence. That is, if $x_n \to x$, then $e^{x_n} \to e^x$. Let $\epsilon > 0$ be given.

    (a) Show that $|e^y - 1| \leq e^{|y|} - 1$ for any real y.

    (b) Show $|e^{x_n} - e^x| \leq e^x \left(e^{|x_n - x|} - 1\right)$.

    (c) Use part (3) of Theorem 3.55 to find an N so that $e^{|x_n - x|} - 1 < \frac{\epsilon}{e^x}$.

    (d) Show that if $x_n \to x$, then $e^{x_n} \to e^x$.

14. Let $a > 0$. Suppose that $x_n \to x$. Prove that $a^{x_n} \to a^x$. *[Hint: Write $a^{x_n} = e^{x_n \ln a}$, and use Exercise 13.]*

15. Suppose $x_n \to x$, that $x_n, x > 0$, and $\alpha \in \mathbb{R}$. Show $x_n^\alpha \to x^\alpha$. *[Hint: Write $x_n^\alpha = e^{\alpha \ln x_n}$, and use Exercises 11 and 13.]*

16. Exercises 11, 13, and 15 demonstrate that certain common mathematical transformations are preserved by convergence. This exercise exhibits one that is not: you cannot evaluate separate limits inside of an expression at different times.

    (a) Verify $1 + \frac{1}{n} \to 1$.

    (b) Verify $1^n \to 1$.

    (c) Does it make sense to say $\left(1 + \frac{1}{n}\right)^n \to 1^n \to 1$? (See Example 3.56.)

17. Prove Theorem 3.51.

18. Prove Theorem 3.52.

19. Prove Corollary 3.53.

20. Let $r > 1$ be a real number, and let q be any integer.

    (a) Show $\frac{n^q}{r^n} \to 0$. *[Hint: Use Example 3.54.]*

    (b) Let $f(n)$ be any polynomial. Show that $\frac{f(n)}{r^n} \to 0$.

21. (This exercise lays the groundwork for the study of Geometric Series in Chapter 7.) Suppose $r \neq 1$.

(a) Show that $1 + r + r^2 + \cdots + r^n = \frac{1-r^{n+1}}{1-r}$.

(b) Suppose that $|r| < 1$. Show that $\lim_{n\to\infty}(1+r+r^2+\cdots+r^n) = \frac{1}{1-r}$.

22. Let $a_n = \frac{p(n)}{q(n)}$, where p and q are polynomials.

(a) Suppose the degree of p is strictly smaller than the degree of q. Show that $a_n \to 0$.

(b) Suppose the degree of p is equal to the degree of q. Show that $a_n \to \frac{c}{d}$, where c and d are the leading coefficients of p and q, respectively.

(c) Suppose the degree of p is strictly greater than the degree of q. Show that (a_n) diverges. (This question is studied in more detail in Exercise 4 of Section 3.8.)

23. Suppose that $a_n \to 0$, and (b_n) is bounded. Show that $a_n b_n \to 0$ (notice that we do *not* assume b_n is a convergent sequence, otherwise one could use Theorem 3.31). Is the same result true if (b_n) is bounded, but $a_n \to a \neq 0$? Is the same result true if $a_n \to 0$ but (b_n) is unbounded?

24. This exercise further studies the technique in Example 3.54, and you may find it helpful as you answer the following questions. Let $a_n > 0$ for every n, and suppose that $\frac{a_{n+1}}{a_n} \to \beta$.

(a) Suppose $\beta > 1$. Show that (a_n) is unbounded. Conclude using Corollary 3.29 that (a_n) diverges.

(b) Construct a convergent sequence (a_n) so that $a_n > 0$, and $\lim_{n\to\infty} \frac{a_{n+1}}{a_n} = 1$.

(c) Construct a divergent sequence (a_n) so that $a_n > 0$, and $\lim_{n\to\infty} \frac{a_{n+1}}{a_n} = 1$. (As a result of parts (b) and (c), this test is inconclusive should you ever compute $\beta = 1$, since there are examples of convergent *and* divergent sequences that share that property.)

25. Prove part (2) of Theorem 3.46. That is, show that if S is a nonempty subset of $\mathbb{R}$ that is bounded below, then there is a sequence $x_n \in S$ that converges to $\inf(S)$.

3.4 Subsequences

Subsequences will play a pivotal role in this book, and we introduce them here. We first discuss the (somewhat awkward, at first) definition of a subsequence before going on to study the close link between the convergence of sequences and the convergence of any associated subsequence.

3.4.1 Defining subsequences

Intuitively, if (a_n) is a sequence, then a *subsequence* of (a_n) should be a sequence (b_k) where every element in the sequence (b_k) is also an element of the sequence (a_n). The only other condition that we should insist upon is that our subsequence (b_k) be ordered in the same manner as (a_n). The following example paves the way for how we will formally define subsequences, and how we will express them mathematically.

Example 3.58. Let $a_n = 2n$. An example of a subsequence of a_n would be to choose some subcollection of the numbers $a_n = 2n$, and to order them in the same manner. So, for this example, let us choose every other member of a_n to create the subsequence b_k, as demonstrated in Figure 3.4.1. Notice that the index variable k of (b_k) is different than the index variable n of (a_n) to avoid confusion: these are different sequences, even if one is a subsequence of another.

a_n	b_k
$a_1 = 2$	$b_1 = 2$
$a_2 = 4$	
$a_3 = 6$	$b_2 = 6$
$a_4 = 8$	
$a_5 = 10$	$b_3 = 10$
$a_6 = 12$	
$a_7 = 14$	$b_4 = 14$
$a_8 = 16$	
$a_9 = 18$	$b_5 = 18$
$\vdots$	
$\vdots$	$\vdots$

FIGURE 3.4.1
The sequence (b_k) is a subsequence of (a_n).

There are two immediate ways to interpret this data. The first would be to come up with a closed form for (b_k), and by inspection, it appears that we could have $b_k = 4k - 2$. But this description lacks reference to the original sequence (a_n) that it was conjured from, and there should be a better way to describe (b_k) in terms of (a_n).

Perhaps the best way to describe this relationship is to indicate just which subscripts of (a_n) are used in the subsequence (b_k). Notice in Figure 3.4.1 that

$$\begin{array}{rcl} b_1 & = & a_1, \\ b_2 & = & a_3, \\ b_3 & = & a_5, \\ b_4 & = & a_7, \\ b_5 & = & a_9, \\ \vdots & & \vdots \end{array}$$

The relationship in the subscripts defined by $1 \to 1$, $2 \to 3$, $3 \to 5$, $4 \to 7$, $5 \to 9$, etc., can be summarized by the function $k \to 2k - 1$. So define the function $n_k = 2k-1$, and now the reference to the original sequence is complete when we write

$$b_k = a_{n_k}.$$

Let us check for the first 5 values of k that $b_k = a_{n_k}$:

k	$n_k = 2k - 1$	b_k	a_{n_k}
1	$n_1 = 1$	$b_1 = 2$	$a_{n_1} = 2$
2	$n_2 = 3$	$b_2 = 6$	$a_{n_2} = 6$
3	$n_3 = 5$	$b_3 = 10$	$a_{n_3} = 10$
4	$n_4 = 7$	$b_4 = 14$	$a_{n_4} = 14$
5	$n_5 = 9$	$b_5 = 18$	$a_{n_5} = 18$

So the assignment $k \to n_k = 2k - 1$ recovers the equation $b_k = a_{n_k}$. In fact, we could rediscover the closed-form expression for b_k that we found earlier:

$$n_k = 2k - 1, \text{ so } b_k = a_{n_k} = a_{2k-1} = 2(2k - 1) = 4k - 2.$$

Now that we have developed a notation for subsequences, only one thing remains: Although the collection of numbers $\{b_k | k \in \mathbb{N}\}$ is a subset of $\{a_n | n \in \mathbb{N}\}$, why must the ordering of (a_n) be preserved in the subsequence (b_k)? The trick is to look at the function n_k: if it is the case that $n_k < n_{k+1}$ for every k, then the ordering will be preserved; we call such functions *strictly increasing*. We make this more formal in Definitions 3.59 and 3.60 below. ■

Definition 3.59. Let $f : \mathbb{N} \to \mathbb{N}$. If $f(k) < f(k + 1)$ for every k, then we say f is *strictly increasing*.

Notice that if $k \to n_k$ is a strictly increasing function, then $n_k \geq k$ (see Exercise 3). We can now give the formal definition of a subsequence.

Definition 3.60. Let (a_n) be a sequence. Any sequence (b_k) of the form $b_k = a_{n_k}$ for a strictly increasing n_k is called a subsequence of (a_n).

Before considering the relationship between subsequences and convergence in the next subsection, we pause to give a few more examples of subsequences.

Example 3.61. If (a_n) is a sequence, then setting $n_k = k$ demonstrates that every sequence is a subsequence of itself.

Example 3.62. This example is designed to demonstrate that a subsequence of a subsequence is again a subsequence. It follows from the fact that the composition of strictly increasing functions is again strictly increasing. Suppose $b_k = a_{n_k}$ is a subsequence of (a_n), and that $c_s = b_{k_s}$ is a subsequence of (b_k). We claim that $c_s = a_{n_{k_s}}$ is a subsequence of (a_n). All that we need to show is that the assignment $s \to n_{k_s}$ is a strictly increasing one. But this is a straightforward check: $k_s < k_{s+1}$, and so $n_{k_s} < n_{k_{s+1}}$ since both of $s \to k_s$ and $k \to n_k$ are strictly increasing. ∎

3.4.2 Subsequences and convergence

There are several important relationships between the possible convergence of a sequence, and the possible convergence of any of its subsequences. In the following examples, we investigate this relationship. Afterwards, we give two theorems that formalize our observations.

Example 3.63. Let $a_n = (-1)^n$. There are two constant subsequences that are worth pointing out. These are $a_{2k} = (-1)^{2k} = 1$ (here, $n_k = 2k$), and $a_{2k+1} = (-1)^{2k+1} = -1$ (here, $n_k = 2k+1$). Notice that $(-1)^n$ is a divergent sequence, but it has two convergent subsequences that converge to two different limits: $a_{2k} = 1 \to 1$, and $a_{2k+1} = -1 \to -1$. ∎

Example 3.64. Define

$$a_n = \frac{(n+1)\cos\left(\frac{n\pi}{2}\right)}{n+2} \quad \text{for } n \geq 0.$$

Since the sequence $\cos\left(\frac{n\pi}{2}\right)$ proceeds as $1, 0, -1, 0, 1, 0, -1, 0 \ldots$, the sequence

$$a_n \text{ proceeds as } \frac{1}{2}, 0, -\frac{3}{4}, 0, \frac{5}{6}, 0, -\frac{7}{8}, 0, \frac{9}{10}, 0, -\frac{11}{12} \ldots.$$

It appears that there are at least 3 convergent subsequences. First and most obviously, the constant subsequence $c_k = 0$. A formula for this could be $c_k = a_{2k-1}$ for $k \geq 1$. There are at least two other subsequences that converge, and those would be the subsequence $b_k = a_{4k}$ $(k \geq 0)$ made from the strictly positive terms, and the subsequence $d_k = a_{4k+2}$ $(k \geq 0)$ made from the strictly negative terms. It appears, however, that the original sequence (a_n) is divergent, since for every N, there exists an $n \geq N$ with $|a_n| > \frac{1}{2}$, and another $n \geq N$ with $a_n = 0$. Therefore, the tail of this sequence would not stay within any neighborhood of a potential limit of diameter less than, say, $\frac{1}{4}$ (see page 117 to recall the formulation of what a divergent sequence is). ∎

In the previous examples, we considered subsequences of (a_n) where (a_n) was divergent. We saw that in these cases, there were still be convergent subsequences, and that the limits of these convergent subsequences were not all the same. In the next practice problem, we shall study a convergent $a_n \to L$, and demonstrate that *every* subsequence is a convergent one, and that it converges to L as well. This, in fact, is always the case, and is proven in Theorem 3.66.

Practice 3.65. Let $a_n = \frac{1}{n}$. Notice $a_n \to 0$. Prove that every subsequence a_{n_k} converges to 0 as well.

Methodology. *Let (a_{n_k}) be any subsequence of (a_n). Since n_k could be anything, $a_{n_k} = \frac{1}{n_k}$ could be anything, forbidding us from directly using our supporting results in Section 3.3. Although, there is one more piece of information we may use: n_k is strictly increasing (by Definition 3.60), and we remarked that $n_k \geq k$ (see Exercise 3). With this information, we will use the original definition of convergence to demonstrate $a_{n_k} \to 0$.*

Solution. Let $\epsilon > 0$ be given, and let (a_{n_k}) be any subsequence of (a_n) where $a_n = \frac{1}{n}$. Since $n_k \geq k \geq 1$, it follows that $\frac{1}{n_k} \leq \frac{1}{k}$. So choose an N so that $\frac{1}{N} < \epsilon$. Then for every $k \geq N$, we have

$$|a_{n_k} - 0| = \frac{1}{n_k} \leq \frac{1}{k} \leq \frac{1}{N} < \epsilon.$$

■

The following theorem demonstrates that the result in Practice 3.65 is not a coincidence: every subsequence (a_{n_k}) of a convergent sequence (a_n) converges to the same limit L if $a_n \to L$.

Theorem 3.66. *Suppose $a_n \to a$. Then every subsequence $a_{n_k} \to a$ as well.*

Idea of proof. *The idea is the same as in Practice 3.65, in that we will use the definition of convergence (Definition 3.13), and use the fact that $k \leq n_k$. The N that ensures $|a_k - L| < \epsilon$ for $k \geq N$ will be the same N that ensures $|a_{n_k} - L| < \epsilon$ for $k \geq N$. This makes intuitive sense: since $n_k \geq k$, the tail of (a_k) arrives slower than the tail of (a_{n_k}), so it should take no longer for the tail of (a_{n_k}) to be within ϵ of L than it would take for the tail of (a_n) to be within ϵ of L.*

Proof of Theorem 3.66. Let $\epsilon > 0$ be given. Since $a_n \to L$, there exists an N so that for every number $k \geq N$, we have $|a_k - L| < \epsilon$. Now consider an arbitrary subsequence $b_k = a_{n_k}$, and let $k \geq N$. Since $n_k \geq k$ (Exercise 3), and $k \geq N$ (by assumption), it follows that $n_k \geq N$, and so $|a_{n_k} - L| = |b_k - L| < \epsilon$. □

The theory of subsequences offers another tool in the study of the convergence of sequences.

Theorem 3.67. *The sequence (a_n) does* not *converge to a if and only if (a_n) has a subsequence that is bounded away from a.*

Idea of proof. *This is an "if and only if" theorem, and we prove both implications separately. We use the characterization of divergence on page 117 for the $\Rightarrow$ implication. To prove the $\Leftarrow$ implication, we proceed by contradiction. If there is a subsequence that is bounded away from a, but that $a_n \to a$, then Theorem 3.66 says that such a subsequence would also have to converge to a, and it is not possible that a (sub)sequence both converge to a number and be bounded away from it.*

Proof of Theorem 3.67. ($\Rightarrow$) Suppose (a_n) does not converge to a. Then (quoting the discussion of divergence on page 117), there exists an $\epsilon > 0$ so that for every N, there exists an $n \geq N$ with $|a_n - a| \geq \epsilon$. We repeatedly use this fact to construct a subsequence that is bounded away from a. Let $N = 1$. There exists an $n_1 \geq 1$ so that $|a_{n_1} - a| \geq \epsilon$. Now let $N = n_1 + 1$. There exists an $n_2 \geq n_1 + 1 > n_1$ so that $|a_{n_2} - a| \geq \epsilon$. Note that $n_2 > n_1$ so that we are indeed constructing a subsequence. Continuing in this fashion, we have constructed a subsequence (a_{n_k}), and every member of this sequence satisfies $|a_{n_k} - a| \geq \epsilon$. Therefore, (a_{n_k}) is bounded away from a.

($\Leftarrow$) Suppose that (a_n) has a subsequence (a_{n_k}) that is bounded away from a by a distance of $\epsilon > 0$, but that $a_n \to a$. Since $a_n \to a$, Theorem 3.66 ensures that $a_{n_k} \to a$ as well. Therefore (a_{n_k}) is eventually within $\frac{\epsilon}{2}$ from a. But this is not possible, as (a_{n_k}) is always at a distance greater than ϵ from a, and $\epsilon > \frac{\epsilon}{2}$. This contradicts the assumption that $a_n \to a$. □

3.4.3 Review of important concepts from Section 3.4

1. A subsequence (a_{n_k}) of (a_n) is a sequence of numbers that respects the same ordering as (a_n). This is accomplished by insisting that the function $k \to n_k$ be strictly increasing.

2. If $a_n \to a$, then $a_{n_k} \to a$ for any subsequence (a_{n_k}) of (a_n) (Theorem 3.66).

3. There is a strong link between the possible convergence of a sequence and the possible convergence of any of its subsequences (Theorems 3.66 and 3.67).

Exercises for Section 3.4

1. Let $a_n = \frac{n}{n+1}$ for $n \geq 1$.

   (a) Write out the first 5 terms of the subsequence a_{2k+1} for $k \geq 0$

   (b) Write out the first 5 terms of the subsequence $a_{k!}$ for $k \geq 0$.

   (c) Write out the first 5 terms of the subsequence a_{2k}. Why must this subsequence proceed from $k = 1$ instead of $k = 0$?

2. Let (a_n) be a sequence. Is $(a_{\sqrt{k}})$ considered a subsequence of (a_n)? What about $(a_{k/2})$? Why or why not?

3. Suppose that $k \to n_k$ is a strictly increasing function from $\mathbb{N}$ to $\mathbb{N}$. Prove that $n_k \geq k$.

4. Recall that the sequence $e_n = \left(1 + \frac{1}{n}\right)^n \to e$ from Example 3.56 of Section 3.3.

   (a) Show $\left(1 + \frac{1}{2n}\right)^{2n} \to e$ as well.

   (b) Show $\left(1 + \frac{1}{2n}\right)^n \to \sqrt{e}$.

   (c) Let $p \in \mathbb{N}$. Show $\left(1 + \frac{1}{pn}\right)^n \to \sqrt[p]{e}$.

5. Show that if (a_n) is bounded, then any subsequence of (a_n) is also bounded.

6. Suppose that (a_n) is a sequence with two subsequences $a_{n_k} \to x$ and $a_{m_k} \to y$. If $x \neq y$, prove that (a_n) diverges.

7. Suppose (a_n) has a divergent subsequence. Show that (a_n) diverges.

8. Let (q_n) be any enumeration of the rationals, and set

$$a_n = \begin{cases} q_n & \text{for odd } n \\ \cos\left(\frac{\pi}{2}n\right) & \text{for even } n. \end{cases}$$

   Show that (a_n) diverges. *[Hint: Use Exercise 7.]*

9. Suppose that (a_n) is a sequence, and that $a_{2n} \to a$ and $a_{2n+1} \to a$ as well. Show that $a_n \to a$. (See Exercise 2 of Section 3.6 for a generalization of this problem.)

3.5 Monotone Sequences

In this section, we consider an important class of sequences, known as *monotone sequences.* We will discuss some basic facts about these sequences, and then go on to share several results that concern the convergence of monotone sequences (The Monotone Convergence Theorem, Theorem 3.74), and the famous Nested Intervals Theorem (Theorem 3.81).

3.5.1 Definition and examples of monotone sequences

We begin with the definition of a monotone sequence.

Definition 3.68. Let (a_n) be a sequence.

1. If $a_n \leq a_{n+1}$ for all n, then (a_n) is *monotone increasing.* If $a_n < a_{n+1}$ for all n, then (a_n) is *strictly monotone increasing.*
2. If $a_n \geq a_{n+1}$ for all n, then (a_n) is *monotone decreasing.* If $a_n > a_{n+1}$ for all n, then (a_n) is *strictly monotone decreasing.*

If (a_n) is either monotone increasing or monotone decreasing, we can refer to (a_n) simply as being *monotone.* If (a_n) is either strictly monotone increasing or strictly monotone decreasing, we can refer to (a_n) simply as being *strictly monotone.*

In the next subsection, we consider some useful results about the convergence of monotone sequences, but for now, it would be helpful to have some practice with demonstrating that a sequence is monotone. Generally speaking, there is no one way to approach such a problem, although there are two common methods: (1) demonstrate the required inequality directly, perhaps using Mathematical Induction, or (2) if the sequence (a_n) can be extended to a differentiable function f (so that $f(n) = a_n$), then one can demonstrate monotonicity by showing the derivative f' has constant sign. We illustrate these methods in the following practice problems.

Practice 3.69. Let $a_n = \frac{n-1}{n}$. Show that (a_n) is monotone increasing.

Methodology. *There are a number of ways to demonstrate a sequence is monotone. We could approach this problem directly and apply "common sense" sort of inequalities (Solution 1), or, we notice that if* $f(x) = \frac{x-1}{x}$, *then* $f(n) = a_n$. *Basic calculus tells us that a function is increasing if* $f'(x) > 0$, *and so we could also show that* $\frac{d}{dx}\frac{x-1}{x} \geq 0$ *(Solution 2).*

Solution 1. Notice that $a_n = 1 - \frac{1}{n}$. So, since $\frac{1}{n+1} < \frac{1}{n}$ for $n \geq 1$, we have $a_{n+1} = 1 - \frac{1}{n+1} > 1 - \frac{1}{n} = a_n$. (In fact, this solution shows that a_n is *strictly* monotone increasing.)

Solution 2. Let $f(x) = \frac{x-1}{x} = 1 - \frac{1}{x}$. Notice that $f(n) = a_n$ for every n. The derivative $f'(x) = \frac{1}{x^2} > 0$ for all $x > 0$. Therefore, $f(x)$ is increasing on the interval $(0, \infty)$. So, for every $n \in \mathbb{N}$, we have $f(n) < f(n+1)$, or, since $f(n) = a_n$, that $a_n < a_{n+1}$. (This solution also shows that (a_n) is *strictly* monotone increasing.) ■

Practice 3.70. Let $a_n = \frac{1}{n!}$. Determine if (a_n) is monotone, and prove your claim.

Methodology. *In contrast to Practice 3.69, there is no obvious differentiable function*[1] $f(x)$ *one can use to simply differentiate as in the second solution of Practice 3.69. By thinking just a little bit about the behavior of this sequence, we can see that each increase in* n *produces another (larger) factor in the denominator of* a_n.

Solution. We can see that $a_n = \frac{1}{n!}$ is monotone decreasing by noticing that $\frac{1}{n+1} \leq 1$ for $n \geq 1$, and so

$$a_{n+1} = \frac{1}{(n+1)!} = \left(\frac{1}{n+1}\right)\frac{1}{n!} \leq \frac{1}{n!} = a_n.$$

■

It will be the case that *eventual* monotonicity is just as important. A sequence (a_n) is *eventually monotone* if there exists an N so that (a_n) is monotone for $n \geq N$. When it is known whether or not the sequence is eventually monotone increasing or monotone decreasing, one could call such a sequence *eventually increasing* or *eventually decreasing.* The following practice problem studies a sequence that is eventually decreasing.

Practice 3.71. Let $a_n = \frac{n-4}{n^2-8n+17}$. Show that (a_n) is eventually decreasing, and determine at what point this occurs.

Methodology. *Thankfully we may use calculus for this problem, since there is a differentiable function* $f(x)$ *that models* (a_n), *in the sense that* $f(n) = a_n$. *We will identify an interval* $[a, \infty)$ *in which* f *is decreasing, and for* N *the next largest integer from* a, *the sequence* (a_n) *will be decreasing for* $n \geq N$.

Solution. Let $f(x) = \frac{x-4}{x^2-8x+17}$. We compute (check for yourself using the quotient rule)

$$f'(x) = \frac{-x^2+8x-15}{(x^2-8x+17)^2} = \frac{-(x-3)(x-5)}{(x^2-8x+17)^2}.$$

We notice that $x^2 - 8x + 17 \neq 0$ for any real x, so that indeed this function is differentiable on $\mathbb{R}$, and that $(x^2 - 8x + 17)^2 > 0$ for every x. Therefore, for $x \geq 5$, the product $(x-3)(x-5) \geq 0$, and so $-(x-3)(x-5) \leq 0$, and for $x \in [5, \infty)$, $f'(x) \leq 0$. So, the sequence (a_n) is decreasing for $n \geq 5$. ■

Remark 3.72. Our analysis in Practice 3.71 actually shares more information

[1]There is a function that models $n!$ known as the Gamma function.

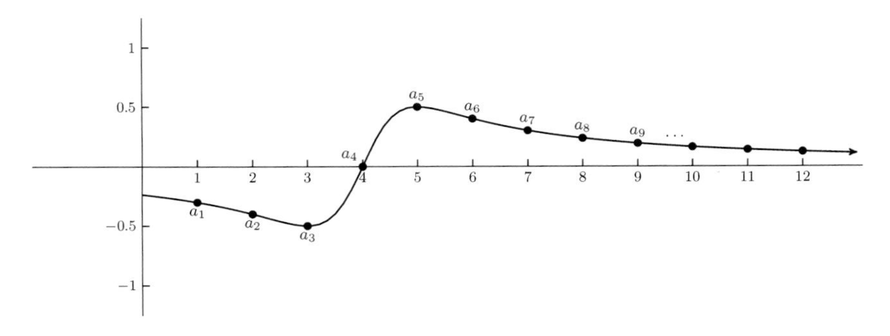

FIGURE 3.5.1
A graph of $f(x) = \frac{x-4}{x^2-8x+17}$.

which is not relevant to the problem, but worth mentioning to reinforce our methodology. Since $f'(x) \leq 0$ on the interval $(-\infty, 3]$, it is the case that

$$a_1 \geq a_2 \geq a_3.$$

Since $f'(x) \geq 0$ on $[3, 5]$, it is the case that

$$a_3 \leq a_4 \leq a_5.$$

Then, as the solution points out, since $f'(x) \leq 0$ on the interval $[5, \infty)$, it is the case that

$$a_5 \geq a_6 \geq a_7 \geq \cdots.$$

See Figure 3.5.1 for a graph of $f(x) = \frac{x-4}{x^2-8x+17}$ which graphically illustrates these conclusions. ■

Sometimes it can be difficult to determine if a sequence is monotone, or even eventually monotone.

Practice 3.73. Let $a_n = \frac{2+(-1)^n}{n}$. Determine if a_n is monotone, eventually monotone, or not eventually monotone, and prove your claim.

Methodology. *One sees almost immediately from listing out the first few terms of a_n that is is not monotone; the first 4 members of the sequence are $1, \frac{3}{2}, \frac{1}{3}, \frac{3}{5}$. The sequence could still be eventually monotone. Since the function $(-1)^x$ is not defined for very many real numbers (consider $(-1)^{1/2}$ or $(-1)^{1/10}$), we cannot expect to use calculus to demonstrate possible monotonicity. On more of a "common sense" approach, we see that there are two possibilities for the numerator of a_n: the numerator is 1 when n is odd, and it is 3 when n is even. It seems, then, that any problem of monotonicity would be in studying the two cases when n is even, and then when n is odd. If eventually monotone, it would have to be eventually monotone decreasing since the denominator seems to be growing without bound while the numerator is bounded.*

A careful analysis of the parity of n will demonstrate that this sequence is not *eventually monotone.*

Solution. Suppose that (a_n) eventually satisfied the condition $a_{n+1} \leq a_n$. In that case, for some even n that is large enough we would have

$$\frac{1}{n+1} \leq \frac{3}{n}.$$

But this is equivalent to $n \leq 3(n+1)$, or $0 \leq 2n+3$, which can certainly happen (in fact, precisely when $n \geq -\frac{3}{2}$). Although for an n large enough that is odd, we would have

$$\frac{3}{n+1} \leq \frac{1}{n}.$$

And this is equivalent to $3n \leq n+1$, or $0 \leq 1-2n$, or that $n \leq \frac{1}{2}$. For (a_n) to be eventually monotone decreasing, we would need both of the previous conditions to occur for all n larger than some N that we would need to produce. But the only n for which both conditions occur are when $-\frac{3}{2} \leq n \leq \frac{1}{2}$. In other words, there is no N so that for every $n \geq N$, we have $a_{n+1} \leq a_n$. ∎

3.5.2 Theorems about the convergence of monotone sequences

There are three important theorems that relate to monotone sequences and the property of being bounded. We prove two of them in this section: The Monotone Convergence Theorem (Theorem 3.74) and the Nested Intervals Theorem (Theorem 3.81). The third theorem is known as the Bolzano-Weierstrass Theorem (Theorem 3.84), and it is proven in the next section. Recall that a bounded sequence (a_n) is a sequence for which there is a real number M that satisfies $|a_n| \leq M$ for all n (see Definition 3.27).

The Monotone Convergence Theorem asserts that any bounded monotone sequence converges. It is a foundational result that will be heavily used throughout the book.

Theorem 3.74 (Monotone Convergence Theorem)**.** *Every bounded and monotone sequence converges.*

Idea of proof. *If (a_n) is our bounded and monotone sequence that is supposed to converge, we will first need to determine what it converges to before proceeding. If it is the case that (a_n) is increasing and bounded, then the set of numbers $\{a_n | n \in \mathbb{N}\}$ has an upper bound, and its least upper bound, intuitively, is a good candidate for the limit. See Figure 3.5.2. The defining properties of least upper bounds will demonstrate that (a_n) converges to that least upper bound. We will demonstrate convergence using the definition of convergence, Definition 3.13. If (a_n) is bounded and decreasing, then a similar proof will show that (a_n) converges to the greatest lower bound of $\{a_n | n \in \mathbb{N}\}$.*

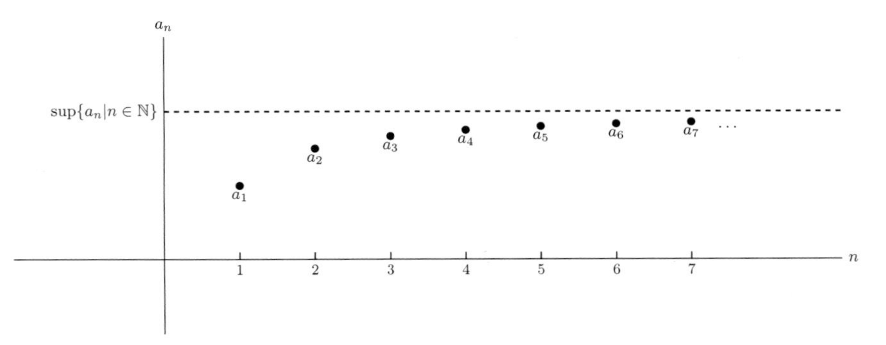

FIGURE 3.5.2
A bounded monotone increasing sequence converges to $\sup\{a_n|n \in \mathbb{N}\}$.

Proof of Theorem 3.74. Let $S = \{a_n|n \in \mathbb{N}\}$. Since (a_n) is bounded, the set S is bounded, and is obviously nonempty. We first consider the case that (a_n) is increasing, and prove that $a_n \to \sup(S)$.

Let $\epsilon > 0$ be given. By the definition of the supremum of S, $\sup(S) - \epsilon$ is not an upper bound of S, so there exists an element $a_N \in S$ satisfying

$$\sup(S) - \epsilon < a_N \leq \sup(S).$$

Since (a_n) is increasing, and $\sup(S)$ is an upper bound for S, it follows that

$$\sup(S) - \epsilon < a_N \leq a_n \leq \sup(S) \text{ for every } n \geq N.$$

So, for $n \geq N$, we have

$$\sup(S) - \epsilon < a_n \leq \sup(S) < \sup(S) + \epsilon,$$

and so $|a_n - \sup(S)| < \epsilon$ for $n \geq N$. Therefore, by the definition of convergence, $a_n \to \sup(S)$.

We now assume that (a_n) is decreasing, and proceed similarly in showing that $a_n \to \inf(S)$. Let $\epsilon > 0$ be given. By the definition of infimum of S, $\inf(S) + \epsilon$ is not a lower bound for S. So, there exists an element $a_N \in S$ satisfying

$$\inf(S) \leq a_N < \inf(S) + \epsilon.$$

Since (a_n) is decreasing, and $\inf(S)$ is a lower bound for S, it follows that

$$\inf(S) - \epsilon < \inf(S) \leq a_n \leq a_N < \inf(S) + \epsilon \text{ for every } n \geq N.$$

So, for $n \geq N$, we again have

$$\inf(S) - \epsilon < a_n < \inf(S) + \epsilon,$$

and so $|a_n - \inf(S)| < \epsilon$ for $n \geq N$. Therefore, by the definition of convergence, $a_n \to \inf(S)$. □

The Monotone Convergence Theorem is a powerful tool for demonstrating convergence of sequences.

Practice 3.75. Define a_n recursively as follows: Let $a_1 = 2$, and for $n \geq 1$, we define $a_{n+1} = \sqrt{a_n} + 2$. Show that this sequence converges, and find its limit. (This is the same sequence we considered in Practice 3.6).

Methodology. *We will show that the sequence above is bounded and monotone increasing. The Monotone Convergence Theorem will then conclude that the sequence indeed converges. In order to find the limit, we use Exercise 6 from Section 3.2:* $a_n \to a$ *if and only if* $a_{n+1} \to a$. *Then, we may use the limit theorems in Section 3.3 and the recursive definition of* (a_n) *in order to find the limit.*

Solution. We first show that (a_n) is monotone increasing, and we do so by Mathematical Induction. When $n = 1$, it is easy to see that $a_2 = \sqrt{2} + 2 \geq 2 = a_1$. Now we suppose that $a_n \geq a_{n-1}$ for some $n \geq 2$, and prove that $a_{n+1} \geq a_n$. For every k, we have $a_k = \sqrt{a_{k-1}} + 2 \geq 2 \geq 1$. So,

$$\begin{aligned} \sqrt{a_n} &\geq \sqrt{a_{n-1}}, \text{ and so} \\ a_{n+1} = \sqrt{a_n} + 2 &\geq \sqrt{a_{n-1}} + 2 = a_n. \end{aligned}$$

We now show that (a_n) is bounded. We have previously noted that $a_n \geq 2$ for every n. We show by Mathematical Induction that $a_n \leq 4$ for every n. The trivial case is obvious: $a_1 = 2 \leq 4$. If we now assume that $a_n \leq 4$ for some $n \geq 1$, then

$$a_{n+1} = \sqrt{a_n} + 2 \leq \sqrt{4} + 2 = 4.$$

This completes the proof that (a_n) is bounded and monotone. By the Monotone Convergence Theorem (Theorem 3.74), (a_n) converges to some number a. According to Exercise 6 of Section 3.2, $a_{n+1} \to a$ as well. Therefore

$$\begin{array}{ccc} a_{n+1} & = & \sqrt{a_n} + 2 \\ \downarrow & & \downarrow \\ a & = & \sqrt{a} + 2. \end{array} \tag{3.27}$$

To be more precise in Equation (3.27), we notice that

$$a_{n+1} = \sqrt{a_n} + 2, \text{ so } a = \lim_{n\to\infty} a_{n+1} = \lim_{n\to\infty} \sqrt{a_n} + 2 = \sqrt{a} + 2.$$

This provides a characterization of a. Namely, $a = \sqrt{a} + 2$, or

$$(a - 2)^2 = a, \text{ or } a^2 - 5a + 4 = 0.$$

The above has two solutions, $a = 1$ and $a = 4$, although $a = 4$ is the only one of those two that satisfies $a = \sqrt{a} + 2$. So, $a = 4$. ■

Remark 3.76. There are two subtle points to be made about the solution of Practice 3.75, once the equation $a = \sqrt{a} + 4$ is found. First, we recall that if a sequence converges, its limit is unique (Theorem 3.22), so it would not have been possible that the sequence (a_n) converge to *both* 1 and 4. Second, if the possibility $a = 1$ were not eliminated as a false solution of $a = \sqrt{a} + 4$, we could have (and should have) eliminated it another way: We determined that $a_n \geq 2$, and so in particular, it is bounded away from 1. By Exercise 18 of Section 3.2, it is not possible that $a_n \to 1$. ■

Using the Monotone Convergence Theorem, we can prove that the sequence $e_n = \left(1 + \frac{1}{n}\right)^n$ from Example 3.56 converges.

Example 3.77. In this example, we prove that the sequence $e_n = \left(1 + \frac{1}{n}\right)^n$ converges. This fact was used in Example 3.56 of Section 3.3, where we used the assumed convergence of (e_n) to show that $e_n \to e$. We shall show that the sequence (e_n) is monotone increasing, and bounded, and therefore according to the Monotone Convergence Theorem, (e_n) converges. In order to complete this goal, we follow the following steps:

1. We use calculus to show (e_n) is increasing. More specifically, we show $y' \geq 0$, where $y = \left(1 + \frac{1}{x}\right)^x$. Since $y(n) = e_n$, this will show that (e_n) is increasing.
2. We expand $\left(1 + \frac{1}{n}\right)^n$ using the binomial theorem to overestimate e_n. Since $e_n > 0$ for every n, this overestimation will be our bound for (e_n).

Let us begin with Step 1. If $y = \left(1 + \frac{1}{x}\right)^x$, for $x > 0$, then $\ln y = x \ln\left(1 + \frac{1}{x}\right)$, and so

$$\begin{aligned} \frac{y'}{y} &= \ln\left(1 + \frac{1}{x}\right) + x\frac{1}{\left(1+\frac{1}{x}\right)} \cdot \frac{d}{dx}\left[1 + \frac{1}{x}\right] \\ &= \ln\left(1 + \frac{1}{x}\right) - \frac{1}{1+x}. \end{aligned}$$

But $\ln\left(1 + \frac{1}{x}\right) \geq \frac{\frac{1}{x}}{1+\frac{1}{x}}$ (see Practice 6.15 of Chapter 6, using $\frac{1}{x}$ instead of x), and $\frac{\frac{1}{x}}{1+\frac{1}{x}} = \frac{1}{1+x}$. Since $y > 0$ for every $x > 0$, and $\ln\left(1 + \frac{1}{x}\right) - \frac{1}{1+x} \geq 0$ for $x > 0$, it follows that $y' \geq 0$, and so (e_n) is increasing.

We continue with Step 2, and show that (e_n) is bounded. First, we expand e_n using the binomial theorem, and rearrange the factors in each summand:

$$\begin{aligned} e_n &= \left(1 + \frac{1}{n}\right)^n \\ &= \sum_{k=0}^{n} \binom{n}{k} \left(\frac{1}{n}\right)^k \\ &= 1 + \frac{n}{1!} \cdot \left(\frac{1}{n}\right) + \frac{n(n-1)}{2!}\left(\frac{1}{n}\right)^2 + \frac{n(n-1)(n-2)}{3!}\left(\frac{1}{n}\right)^3 + \cdots \\ &\qquad \cdots + \frac{n(n-1)(n-2)\cdots 2}{(n-1)!}\left(\frac{1}{n}\right)^{n-1} + \frac{n!}{n!}\left(\frac{1}{n}\right)^n \\ &= 1 + 1 + \frac{n(n-1)}{n^2}\frac{1}{2!} + \frac{n(n-1)(n-2)}{n^3}\frac{1}{3!} + \cdots \\ &\qquad \cdots + \frac{n(n-1)(n-2)\cdots 2}{n^{n-1}}\frac{1}{(n-1)!} + \frac{n!}{n^n}\frac{1}{n!}. \end{aligned} \tag{3.28}$$

Notice that $\frac{n(n-1)}{n^2} = \frac{n}{n} \cdot \frac{n-1}{n} \leq 1$. Using similar inequalities in each of the summands of Equation (3.28), we can conclude

$$e_n \leq 1 + 1 + \frac{1}{2!} + \frac{1}{3!} + \cdots + \frac{1}{(n-1)!} + \frac{1}{n!}. \tag{3.29}$$

We now point out that by Exercise 8 of Section 3.1, for $n \geq 4$ it is the case that $n! \geq 2^n$, so that $\frac{1}{n!} \leq \frac{1}{2^n}$ for these n. In addition, by Exercise 21 of Section 3.3, the sum

$$1 + \frac{1}{2} + \frac{1}{2^2} + \cdots + \frac{1}{2^n} = \frac{1 - \left(\frac{1}{2}\right)^{n+1}}{1 - \frac{1}{2}} \leq \frac{1}{1 - \frac{1}{2}} = 2.$$

So,

$$\frac{1}{4!} + \frac{1}{5!} + \cdots + \frac{1}{n!} \leq \frac{1}{2^4} + \frac{1}{2^5} + \cdots + \frac{1}{2^n} \leq 2 - 1 - \frac{1}{2} - \frac{1}{2^2} - \frac{1}{2^3} = \frac{1}{8}.$$

In that case, from Equation (3.29),

$$e_n \leq 1 + 1 + \frac{1}{2!} + \frac{1}{3!} + \frac{1}{8} = \frac{67}{24} = 2.791\bar{6}.$$

So (e_n) is bounded. Since we have already shown it is increasing, we conclude that (e_n) converges. ■

Remark 3.78. In completing step 2 of Example 3.77, we showed that $e_n \leq 2.791\bar{6}$. In the proof of the Monotone Convergence Theorem (Theorem 3.74), we showed that a bounded and monotone increasing sequence (e_n) converges to $\sup(S)$, where $S = \{e_n | n \in \mathbb{N}\}$, and so **in general, a better bound for S will be a better approximation of the limit.** That being said, we could have been quite a bit more careless with our estimation above, and that would have completed our task of showing (e_n) is bounded, but the information our more careful efforts gave us is a better approximation of the number e, which is the limit of (e_n). It is known that

$$e = 2.71828182846\ldots,$$

and so it appears our estimates were quite close. In Chapter 7, we will discuss the accuracy of estimates as well. ■

The next result is really quite amazing: it states that every sequence in $\mathbb{R}$ has a monotone subsequence. We will use it to prove the Bolzano-Weierstrass Theorem (Theorem 3.84) in the next section, and in Section 3.8 we will again use it to prove Theorem 3.118, an analogue to the Bolzano-Weierstrass Theorem involving infinite limits.

Theorem 3.79. *Every sequence in $\mathbb{R}$ has a monotone subsequence.*

Idea of proof. *We will use an argument from Bartle and Sherbert's* Introduction to Real Analysis, *as it is quite elegant. Define a* peak *of the sequence (a_n) to be a member of the sequence a_k that is larger than the subsequent members of the sequence. More precisely, a_k is a peak if $a_k \geq a_\ell$ for every $\ell \geq k$. If there are infinitely many peaks, we may enumerate them into a subsequence, and this sequence will be decreasing. If there are only finitely many peaks, then we can construct an increasing subsequence that begins with some member of the sequence beyond the last peak. The remaining members of the sequence are found by the hypothesis that we are beyond this last peak.*

Proof of Theorem 3.79. Let (a_n) be our sequence. For the purposes of this proof, we define the element a_k to be a *peak* if $a_k \geq a_\ell$ for every $\ell \geq k$. We split up the proof into two cases: if there are infinitely many peaks, we will construct a monotone decreasing subsequence; if there are finitely many peaks, then we will construct a monotone increasing sequence.

Suppose first that there are infinitely many peaks. We enumerate them: suppose a_{n_1} is the first peak, a_{n_2} is the second peak, etc. This forms a subsequence (a_{n_k}). This sequence is decreasing: if a_{n_k} is a member of the subsequence, it must be a peak. Further, $n_{k+1} > n_k$, and so by the definition of peak, $a_{n_k} \geq a_{n_{k+1}}$.

Now suppose that there are finitely many peaks. Suppose a_N is the last of these peaks. Set $n_1 = N + 1$. Since a_N is the last peak, a_{n_1} is not a peak. Therefore, there exists a member of the sequence a_{n_2} with $n_2 > n_1$ with $a_{n_2} > a_{n_1}$. But a_{n_2} is not a peak either, so there exists an $n_3 > n_2$ with $a_{n_3} > a_{n_2}$. Continuing in this way, we have constructed a strictly increasing subsequence (a_{n_k}). □

The previous theorem provides a short argument to following fact.

Practice 3.80. Let (q_n) be an enumeration of the countable set $(0, 1) \cap \mathbb{Q}$. Prove that there exists a monotone subsequence (q_{n_k}).

Methodology. *This is immediate from Theorem 3.79.*

Solution. Theorem 3.79 guarantees the existence of such a subsequence.
■

We will employ the theory of monotone sequences in the proof of the following theorem. Theorem 3.81 is not only an interesting fact on its own, but the theorem and generalizations of it are useful in other areas of mathematics.

Theorem 3.81 (Nested Intervals Theorem)**.** *Suppose the intervals $I_n = [a_n, b_n]$ is a sequence of intervals satisfying the property $I_n \supseteq I_{n+1}$ for every n. Then*

$$\cap_{n=1}^{\infty} I_n \neq \emptyset.$$

Idea of proof. *We need to show that there is a number $a \in [a_n, b_n]$ for every n, or rather, that $a_n \leq a \leq b_n$ for every n. The sequence (a_n) of left endpoints of I_n is an increasing one since $[a_n, b_n] \supseteq [a_{n+1}, b_{n+1}]$, and the right-most endpoint b_1 is an upper bound for these numbers (while a_1 is a lower bound), so (a_n) is increasing and bounded, hence it converges to some number $a = \sup(S)$, where $S = \{a_n | n \in \mathbb{N}\}$. This number a is a good candidate to satisfy $a_n \leq a \leq b_n$. See Figure 3.5.3 for a picture of this.*

That $a_n \leq a$ for every n is clear since a is the supremum of the set $S = \{a_n | n \in \mathbb{N}\}$. To show that $a \leq b_n$ for every n we will point out that b_n is an upper bound for S for any n. In that case, since a is the ***least*** *upper bound of S, a will be less than or equal to b_n.*

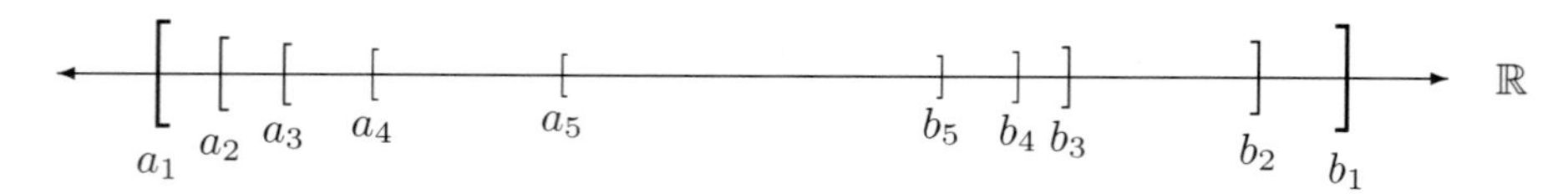

FIGURE 3.5.3
Nested intervals on a number line.

Proof of Theorem 3.81. We construct a number $a \in I_n$ for each n as follows. Notice that

$$I_n = [a_n, b_n] \supseteq [a_{n+1}, b_{n+1}] = I_{n+1}, \text{ so } a_n \leq a_{n+1} \text{ and } b_{n+1} \leq b_n.$$

The sequence (a_n) is therefore increasing. In addition, $a_1 \leq a_n \leq b_n \leq b_1$, so (a_n) is bounded (within $[a_1, b_1]$). According to the Monotone Convergence Theorem (Theorem 3.74), (a_n) converges to $a = \sup(S)$, where $S = \{a_n | n \in \mathbb{N}\}$. We claim that $a_n \leq a \leq b_n$ for every n, so that $a \in \cap_{n=1}^{\infty} I_n$.

It is clear that $a_n \leq a$ for every n, since a is the supremum of $S = \{a_n | n \in \mathbb{N}\}$. We show now that, for every n, the number b_n is an upper bound for S. Since a is the *least* upper bound of S, it follows that $a \leq b_n$ for each n, completing the proof.

Let n be given. We wish to show b_n is an upper bound for S, or put differently, that $a_k \leq b_n$ for every k. If it is the case that $k \leq n$, then

$$a_k \leq a_n \leq b_n.$$

If it is the case that $k \geq n$, then

$$a_k \leq b_k \leq b_n.$$

In any case, $a_k \leq b_n$, so the proof is complete. □

Remark 3.82. If $I_n \subset \mathbb{R}$ is *any* collection of sets satisfying the property $I_n \supseteq I_{n+1}$, then one usually refers to the collection (I_n) as being a *nested* sequence of sets. More specifically, in this case we would say these intervals are *nested downward* (the opposite condition that $I_n \subseteq I_{n+1}$ is referred to as being *nested upward*, or also just as being *nested*). As the following examples indicate, it is not always the case that a sequence of nested intervals will have at least one point in common: The Nested Intervals Theorem applies only to nested intervals of the form $I_n = [a_n, b_n]$. In fact, more is known about $\cap_{n=1}^{\infty} I_n$ in this setting–see Exercise 6. ■

If one were to remove the hypothesis that $I_n = [a_n, b_n]$ (that is, that I_n is a bounded interval that contains its endpoints), then the conclusion of the Nested Intervals Theorem would not always hold.

Example 3.83. In this example, we investigate what happens if one replaces $I_n = [a_n, b_n]$ with an interval of any other type. We shall see that the conclusion of Theorem 3.81 does not have to hold.

1. Let $I_n = [0, \frac{1}{n}]$. The sequence of intervals (I_n) is nested downward, and is of the form $I_n = [a_n, b_n]$. According to the Nested Intervals Theorem, $\cap_{n=1}^{\infty} I_n \neq \emptyset$. Indeed, this is the case, as 0 is an element of (the *only* element of) the infinite intersection (see Exercise 6 for the explanation of why 0 is the only element of this set).

2. Let $I_n = (0, \frac{1}{n})$, or $(0, \frac{1}{n}]$. It is clear that the sequence (I_n) of intervals is nested downward. But it is also clear that $\cap_{n=1}^{\infty} I_n = \emptyset$: Any element in the infinite intersection of these sets would have to be *strictly* greater than 0, but also be less than (or equal to, for $I_n = (0, \frac{1}{n}]$) the values $\frac{1}{n}$ for *every* n. Since there is no such number, the intersection is empty.

3. Suppose $I_n = (n, \infty)$, or $[n, \infty)$. This sequence of intervals is nested downward as well, although the infinite intersection $\cap_{n=1}^{\infty} I_n = \emptyset$: Any element in the infinite intersection would have to be greater than (or equal to, for $I_n = [n, \infty)$) the number n for *every* $n \in \mathbb{N}$. If there was such a number, this would be an upper bound on $\mathbb{N}$, and there is no such upper bound (see Theorem 2.33).

■

3.5.3 Review of important concepts from Section 3.5

1. Monotone sequences are an important class of sequences, and they will be used often.
2. The Monotone Convergence Theorem (Theorem 3.74) says that any bounded and monotone sequence must converge.
3. Every sequence has a monotone subsequence (Theorem 3.79).
4. The Nested Intervals Theorem (Theorem 3.81) says that any sequence of nested intervals of the form $I_n = [a_n, b_n]$ has at least one real number common to all of them.

Exercises for Section 3.5

1. Prove that a bounded sequence that is eventually monotone is convergent.
2. Let $a_n = 1 + \frac{1}{2} + \cdots + \frac{1}{n}$. Show that (a_n) is monotone, but not bounded.
3. (Compare this exercise with Theorem 3.46, and Exercise 25 of Section 3.3.) Suppose that S is a nonempty subset of $\mathbb{R}$.

   (a) Show that if S is bounded above, there exists a monotone increasing sequence x_n converging to $\sup(S)$.

(b) Show that if S is bounded below, there exists a monotone decreasing sequence x_n converging to $\inf(S)$.

4. Prove that every bounded sequence has a convergent subsequence. *[Hint: Use Theorem 3.79 and Exercise 5 of Section 3.4.]*

5. We define the element a_k to be a *valley* of (a_n) if $a_k \leq a_\ell$ for every $\ell \geq k$. That is, there are no smaller elements of the sequence (a_n) beyond a_k. Redo the proof of Theorem 3.79 using instead the idea of a "valley" of a sequence (a_n). *[Hint: This proof will follow very closely to the proof of Theorem 3.79. The point of this exercise is to test your understanding of this proof.]*

6. Suppose $I_n = [a_n, b_n]$, and that the intervals I_n are nested downward: $I_n \supseteq I_{n+1}$ for every n.

   (a) Prove that there exists real numbers $a \leq b$ so that
   $$\cap_{n=1}^{\infty} I_n = [a, b].$$
   (it is possible that $a = b$ in this case, so that $[a, a] = \{a\}$.) *[Hint: As in the proof of Theorem 3.81, set $a = \sup\{a_n | n \in \mathbb{N}\}$, and set $b = \inf\{b_n | n \in \mathbb{N}\}$.]*

   (b) Show that if $\lim_{n\to\infty}(b_n - a_n) = 0$, then there exists a real number x so that $\cap_{n=1}^{\infty} I_n = \{x\}$. *[Hint: Show that $a = b$ in the above part by contradiction. If $a < b$, then $b - a > 0$, and by assumption there exists an N so that $0 \leq b_N - a_N < b - a$, or rather, $a - a_N < b - b_N$. Then use the assumption that $a_N \leq a$ and $b \leq b_N$ to arrive at a contradiction.]*

7. Prove that the Nested Intervals Theorem implies the Completeness Axiom. *[Hint: Let A be a subset of $\mathbb{R}$ which is nonempty and bounded above. Let b_1 be an upper bound of A. Since A is nonempty, we can find a real number a_1 which is* not *an upper bound. Now the midpoint $m_1 = \frac{1}{2}(b_1 - a_1)$ is an upper bound for A, or it is not. If it is, then set $b_2 = m_1$, and $a_2 = a_1$. If it is not an upper bound, set $a_2 = m_1$ and $b_2 = b_1$. The interval $[a_2, b_2] \subseteq [a_1, b_1]$. Continue this process to construct a sequence of nested intervals whose widths tend to 0, then use Exercise 6 to find a number x and show that x is the least upper bound of A.]*

8. Let $I_n = (-\infty, a_n) \cup (b_n, \infty)$ be a sequence of intervals that is nested upward. Show that $\cup_{n=1}^{\infty} I_n \neq \mathbb{R}$. *[Hint: Consider De Morgan's Laws (Theorem 1.12) and the Nested Intervals Theorem (Theorem 3.81).]*

3.6 The Bolzano-Weierstrass Theorems

This section is devoted to the Bolzano-Weierstrass Theorems (Theorems 3.84 and 3.91), which were originally proved by Bernhard Bolzano, and then subsequently modified slightly by Karl Weierstrass in the 1860s [6]. We will use these results as part of our study of the continuity of functions in Chapter 5.

We will first state and prove the Bolzano-Weierstrass Theorem for sequences (Theorem 3.84), then go on to introduce limit points and isolated points (Definitions 3.86 and 3.87) before stating and proving the Bolzano-Weierstrass Theorem for sets (Theorem 3.91).

3.6.1 The Bolzano-Weierstrass Theorem for sequences

The Bolzano-Weierstrass Theorem (for sequences) is a famous result in the field of Real Analysis, and it can readily be extended to $\mathbb{R}^n$. There are two well-known proofs to the theorem: the first proof is quite short, relies on what have developed so far, and applies only to $\mathbb{R}$ (more specifically, it requires the ordering of the real numbers, as that is what the needed previous results are based upon). The second proof uses a more basic argument that is independent of those restrictions and can be adapted to prove the extension of Theorem 3.84 to $\mathbb{R}^n$.

Theorem 3.84 (Bolzano-Weierstrass Theorem for sequences). *Every bounded sequence in $\mathbb{R}$ has a convergent subsequence.*

Idea of first proof. *If (a_n) is a bounded sequence, then every subsequence of it is also bounded by Exercise 5 of Section 3.4. Theorem 3.79 guarantees the existence of a monotone subsequence. Such a sequence is monotone and bounded, and so it must converge.*

First proof of Theorem 3.84. Suppose (a_n) is a bounded sequence. There exists a monotone subsequence (a_{n_k}) by Theorem 3.79, and this monotone subsequence must be bounded as well (by Exercise 5 of Section 3.4). Since (a_{n_k}) is monotone and bounded, it converges by the Monotone Convergence Theorem. □

Idea of second proof. *The idea of the second proof is based on the idea of following where the tail of the sequence leads. More specifically, since (a_n) is bounded, it sits wholly inside of an interval $[a, b]$. Unless the sequence only has finitely many values, there are infinitely many values of a_n that are in the first half of the interval, or infinitely many values of a_n that are in the second half of the interval. Choose one member of the sequence from whatever half of $[a, b]$ contains infinitely many elements. Applying repeatedly this process to the half interval that contains infinitely many a_n, we have constructed a subsequence* *that we can show converges. For this proof, the* width *of an interval* $[a, b]$ *we define to be* $b - a$.

Second proof of Theorem 3.84. Let (a_n) be a bounded sequence. Since (a_n) is bounded, there exists real numbers (bounds) a, b so that $a_n \in [a, b]$ for every n. Let $S = \{a_n | n \in \mathbb{N}\}$. The set S is either finite or infinite. If S is finite, then there is one number $x \in S$ that is repeated infinitely many times, and we may then create a subsequence (a_{n_k}) so that $a_{n_k} = x$ for all k, and hence, $a_{n_k} \to x$ converges.

If S is infinite, there must be infinitely members of S in either $\left[a, \frac{b+a}{2}\right]$ or $\left[\frac{b+a}{2}, b\right]$, and define I_1 to be that subinterval that contains infinitely many elements of S. If there are infinitely many members of S in *both* of these subsets, then choose I_1 to be $\left[a, \frac{b+a}{2}\right]$. In the same way, there are infinitely many members of S in either the first half of I_1 or the second half of I_1. Choose I_2 to be the subinterval of I_1 containing infinitely many members, where again, if there are infinitely many members of S in both subintervals, choose the subinterval that contains the smaller real numbers (see Figure 3.6.1).

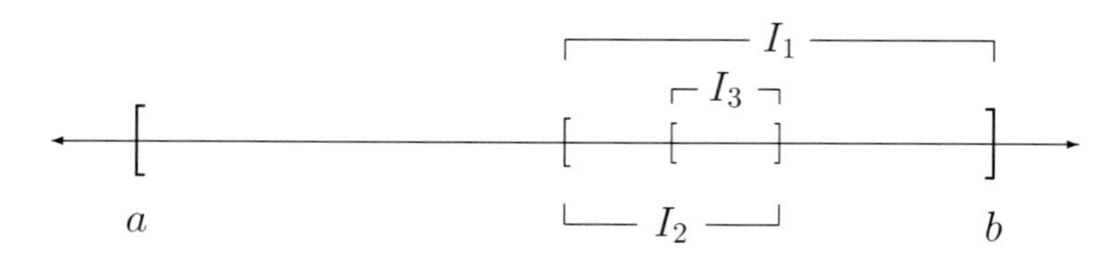

FIGURE 3.6.1
A possible subdivision of the interval $[a, b]$.

Continuing in this process, we have a sequence of intervals I_n that satisfy these properties:

1. $I_n \supseteq I_{n+1}$: This is clear from our definition of the I_n.

2. The width of I_n is $\frac{b-a}{2^n}$: This can be proven by Mathematical Induction. Clearly the width of I_1 is $\frac{b-a}{2}$, and the width of I_{n+1} is half of the width of I_n.

3. If $I_n = [a_n, b_n]$, then from (2), the width of I_n, which is $b_n - a_n = \frac{b-a}{2^n} \to 0$.

According to (1), the intervals (I_n) are nested downward. Since the widths of the I_n approach 0 by (3), according to the last part of Exercise 6 of Section 3.5 (ultimately from the Nested Intervals Theorem, Theorem 3.81), $\cap_{n=1}^{\infty} I_n = \{x\}$. Since I_k contains infinitely many elements of S, there exists at least one: choose $a_{n_k} \in I_k$ for each k. We claim $a_{n_k} \to x$, and we prove this by the definition of convergence after establishing one more fact.

Since the width of I_k is $\frac{b-a}{2^k}$, if $y, z \in I_k$, then the largest $|y-z|$ can be is $\frac{b-a}{2^k}$. So in particular, since $a_{n_k} \in I_k$ (for any given k), and $x \in I_k$ (for all k), it must be the case that

$$|a_{n_k} - x| \leq \frac{b-a}{2^k}.$$

We may now show that $a_{n_k} \to x$. Let $\epsilon > 0$ be given. Since $\frac{b-a}{2^k} \to 0$, there exists an N so that

$$\frac{b-a}{2^k} < \epsilon \text{ for all } k \geq N.$$

Then, for $k \geq N$, we have

$$|a_{n_k} - x| \leq \frac{b-a}{2^k} < \epsilon.$$

□

The Bolzano-Weierstrass Theorem will be used in subsequent chapters (namely, in the proof of the Extreme Value Theorem (Theorem 5.44) in Chapter 5). But for now, here is an elementary practice problem.

Practice 3.85. Suppose (a_n) is a sequence with $a_n \in (-5, 2)$. Prove that there is a convergent subsequence of (a_n). What are the possibilities of the limit of this convergent subsequence?

Methodology. *This is a direct application of the Bolzano-Weierstrass Theorem for sequences. The only question is: what number could such a convergent subsequence converge to? We recall Corollary 3.49: if $x_n \to x$ and $x_n \in [a,b]$, then $x \in [a,b]$ as well. The interval $(-5,2)$ is not of the type $[a,b]$, but $(-5,2) \subseteq [-5,2]$, and so at the very least we can say that our $a_n \in [-5,2]$. Our convergent subsequence is also contained in $[-5,2]$, and so its limit must be in $[-5,2]$ as well.*

Solution. Since (a_n) is bounded, by the Bolzano Weierstrass Theorem for sequences (Theorem 3.84), there exists a convergent subsequence $a_{n_k} \to a$. Since $a_n \in (-5,2)$ for all n, the subsequence $a_{n_k} \in [-5,2]$, and by Corollary 3.49, $a \in [-5,2]$ as well. ■

3.6.2 Limit points and isolated points

The Bolzano-Weierstrass Theorem for sets concerns a certain type of point known as a "limit point". In fact, given a subset $A \subseteq \mathbb{R}$, there are two types of points associated with A: limit points and isolated points. We discuss them here before we discuss the Bolzano-Weierstrass Theorem for sets in the next subsection.

Definition 3.86. Let A be a subset of $\mathbb{R}$, and let $x \in \mathbb{R}$. Suppose that for every neighborhood U of x, there is a point $y \in U$ with $y \neq x$. Then x is called a *limit point* of A.

The term *limit point* is given to those points satisfying Definition 3.86 because there is a relationship between limit points of a set and limits of sequences residing within that set (see Exercise 8). Moreover, in Definition 3.86, we insist that each neighborhood of a limit point contain a member of the set *other* than the limit point itself. This part of the definition is crucial to defining limits of functions (see Definition 5.1 on Page 255).

The opposite condition of limit point, more or less, is that of an *isolated point.*

Definition 3.87. Let $x \in A \subseteq \mathbb{R}$. If x is not a limit point of A, then x is called an *isolated point* of A.

Here is some practice to help familiarize you with the notions of limit points and isolated points. **Keep in mind that, although an isolated point of a set A must be a member of A, a limit point may fall outside of A. Further, since each neighborhood of a limit point of A must contain a member of A other than the limit point itself, the limit points of A cannot be too far away from A.** This idea will help to locate the possible limit points of a set (see Exercise 7 for a formal statement of this notion).

Practice 3.88. Find all of the limit points and isolated points of the following sets:

1. (a, b) with $a < b$.
2. $(a, b]$ with $a < b$.
3. $[a, b]$ with $a < b$.
4. $A = \{\frac{1}{n} | n \in \mathbb{N}\}$.
5. $B = (\mathbb{Q} \cap [0, 1]) \cup \mathbb{Z}$.

Methodology. *For each part, we refer to the definitions of limit points and isolated points. Remember to keep in mind that the limit points cannot be too far from the set in question, and that isolated points are members of the sets that are not limit points.*

Solution. Parts (1) through (3): The answers of parts (1), (2), and (3) are the same: The set of limit points is $[a, b]$, and the set of isolated points is empty. To see this, we show that every element of $[a, b]$ is a limit point point of A, where $A = (a, b), (a, b]$, or $[a, b]$. We will then show that every number outside $[a, b]$ is *not* a limit point. Let $x \in [a, b]$, and let U be a neighborhood of x. Since $a < b$ there are infinitely many elements $y \in U$ with $y \in A$ (in particular, there is one other than x).

Now if $x > b$, then there is a neighborhood $U = (x - \epsilon, x + \epsilon)$ for $\epsilon = \frac{1}{2}(x - b)$, and there are no points of U that intersect any of (a, b), $(a, b]$, or $[a, b]$. Similarly, if $x < a$ we will find the same by choosing $\epsilon = \frac{1}{2}(a - x)$. So, the set of limit points of A is $[a, b]$.

We now show the set of isolated points of $(a, b), (a, b]$, or $[a, b]$ is empty. We have already shown that the set of limit points of any of these sets is $[a, b]$, and an isolated point is a point in the set that is *not* a limit point. There are no such points that are in any of $(a, b), (a, b]$ or $[a, b]$ that are not already limit points, so the set of isolated points is empty.

(4) We claim that $x = 0$ is the only limit point of A, and that every member of the set A is isolated. Let U be a neighborhood of 0. Since $\frac{1}{n} \to 0$, the tail of the sequence $a_n = \frac{1}{n}$ is eventually in U, and so there are infinitely many members of A in U. Namely, since $0 \notin A$, there are points of A in U other than 0. If x is any other number, we claim that x is *not* a limit point. If $x < 0$ or $x > 1$, then one can find an $\epsilon-$neighborhood U that contains no elements of A. If $x < 0$, then choose $\epsilon = \frac{|x|}{2}$ (all elements of U are strictly negative and hence not in A), and if $x > 1$, then choose $\epsilon = \frac{x-1}{2}$ (all elements of U are strictly greater than 1, and hence not in A).

In order to show that all of the points in A are isolated, and that any other values of x are *not* limit points, we study the set A more closely. If $y \in A$, then there exists an n so that $y = \frac{1}{n}$, and

$$\frac{1}{n+1} < y < \frac{1}{n-1} \quad (n > 1), \quad \text{or} \quad \frac{1}{2} < y \quad (n = 1).$$

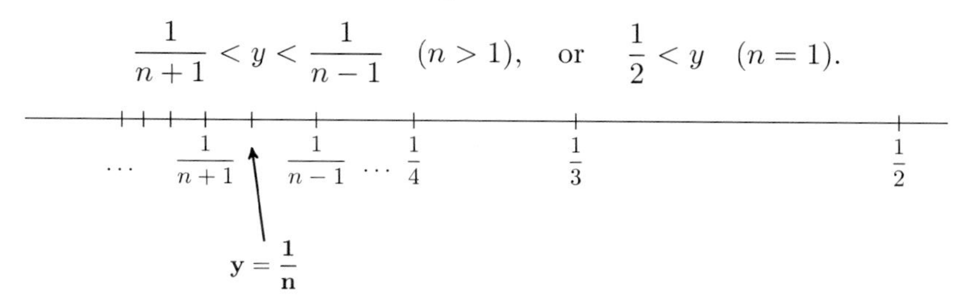

Therefore if $\epsilon = \frac{1}{2}\min\{y - \frac{1}{n+1}, \frac{1}{n-1} - y, \frac{1}{4}\}$, the $\epsilon-$neighborhood U will contain only the element y from the set A. If $y = 1$, then $\epsilon = \frac{1}{4}$ will do, while if $y = \frac{1}{n}$ for $n > 1$, then the smaller of $\frac{1}{2}(y - \frac{1}{n+1})$, and $\frac{1}{2}(\frac{1}{n-1} - y)$ will do. So choosing ϵ in the manner above guarantees that for every $y \in A$, the $\epsilon-$neighborhood U contains only y from A (notice here the subtle point that ϵ is chosen *after* y is). So, every element of A is an isolated point, which in turn means that no point of A is a limit point.

We complete the problem by showing that there are no other limit points of A other than $x = 0$. We have already considered all x with $x < 0$, and $x > 1$, and for every $x \in A$. We have shown $x = 0$ *is* a limit point, and that $x = 1 \in A$ is already not a limit point. So, the only points left to check are $x \in (0, 1)$ with $x \notin A$. Suppose that such an x is given. There exists an $n \geq 1$ with

$$\frac{1}{n+1} < x < \frac{1}{n}.$$

$$\cdots \quad \frac{1}{n+1} \quad \mathbf{x} \quad \frac{1}{n} \quad \cdots \quad \frac{1}{4} \quad \frac{1}{3} \quad \frac{1}{2}$$

Let $\epsilon = \frac{1}{2}\min\{x - \frac{1}{n+1}, \frac{1}{n} - x\}$. Similar to the above, the $\epsilon-$neighborhood U contains no elements from A, and so x is not a limit point of A.

(5) We claim that the set of limit points of B is $[0, 1]$, and the set of isolated points of B is $\mathbb{Z}\backslash\{0, 1\}$. We begin with showing that the set of limit points is $[0, 1]$.

We show that every element of the set $[0, 1]$ is a limit point, and that every real number *not* in $[0, 1]$ is *not* a limit point. First, let $x \in [0, 1]$, and let U be a neighborhood of x. Since the rational numbers are dense in the real numbers, there exists infinitely many rational numbers in U, and since $x \in [0, 1]$, there are infinitely many rational numbers in $U \cap B$. So x is a limit point.

Now, if $x > 1$ or $x < 0$, then the $\epsilon-$neighborhood U contains only finitely many points of B, where similar to our choice of ϵ above, if $x < 0$, then choose $\epsilon = \frac{|x|}{2}$ (all elements of U are strictly negative and hence can contain only finitely many elements (integers) from B), and if $x > 1$, then choose $\epsilon = \frac{x-1}{2}$ (all elements of U are strictly greater than 1, and hence again can only contain finitely many points of B). So, if $x \notin [0, 1]$, x is not a limit point.

Now we show that $\mathbb{Z}\backslash\{0, 1\}$ is the set of isolated points of B. Since isolated points of B are members of B which are not limit points, and $[0, 1]$ is the set of limit points of B, the only members of B left to be isolated points are the integers other than 0 and 1. If n is such an integer, then the $\epsilon-$neighborhood U of n will not contain any points of B other than n, where $\epsilon = \frac{1}{2}$. So, $\mathbb{Z}\backslash\{0, 1\}$ is the set of isolated points of B. ■

There is an alternate characterization of limit points (Theorem 3.89, see also Exercise 8) and isolated points (Theorem 3.90 and Exercise 12) that will be useful.

Theorem 3.89. *The real number x is a limit point of A if and only if for every neighborhood U of x, there are infinitely many points in $U \cap A$.*

Idea of proof. *We prove both implications seperately. For $\Rightarrow$, we proceed by contradiction. If there was a neighborhood of x that had only finitely many points (other than x) in $U \cap A$, then we could find a smaller neighborhood V of x that contained no points other than possibly x in $V \cap A$, which could contradict the assumption that x is a limit point of A. There is not much to prove in the $\Leftarrow$ implication: if every neighborhood of x contains infinitely many points of $U \cap A$, then there must be one that is not already x.*

Proof of Theorem 3.89. ($\Rightarrow$) Suppose x is a limit point of A, but there is a neighborhood U of x with only finitely many points in $U \cap A$. Suppose then, that $\{p_1, \ldots, p_n\} = U \cap A - \{x\}$. Let

$$\epsilon = \frac{1}{2} \min\{|x - p_1|, |x - p_2|, \ldots, |x - p_n|\},$$

and let $V = (x - \epsilon, x + \epsilon)$ be the neighborhood of x of radius ϵ. Since $V \subseteq U$, it follows that

$$V \cap A \subseteq U \cap A = \{p_1, \ldots, p_n\}.$$

But none of $p_1, \ldots, p_n$ could be an element of V, since they are too distant from x: $|x - p_i| > \epsilon$, and V is the set of all elements within a distance of ϵ from x.

($\Leftarrow$) Suppose that for every neighborhood U of x, there exists infinitely many points in $U \cap A$. Therefore, even if $x \in U \cap A$, there must still be (infinitely many) other $y \in U \cap A$ with $y \neq x$. So, by Definition 3.86, x is a limit point of A. □

Theorem 3.90. *The element x is an isolated point of A if and only if there exists a neighborhood U of x so that $U \cap A = \{x\}$.*

Idea of proof. *This requires two separate implications. The $\Rightarrow$ is actually quite straightforward, and will follow from the definition of isolated point as in Definition 3.87, and from an argument similar to the proof of Theorem 3.89: if there are only finitely many points in a neighborhood of x, then we can shrink the neighborhood to ensure that x is the only member of that smaller neighborhood. The $\Leftarrow$ implication follows immediately from the definition.*

Proof of Theorem 3.90. Exercise 12. □

3.6.3 The Bolzano-Weierstrass Theorem for sets

Now that we have the requisite background information, we can establish the Bolzano-Weierstrass Thoerem for sets.

Theorem 3.91 (Bolzano-Weierstrass Theorem for sets)**.** *Every bounded infinite subset of $\mathbb{R}$ has a limit point.*

Idea of proof. *If A is our infinite and bounded set, then there must be a sequence of distinct points $a_n \in A$ (that the a_n are distinct is crucial to the proof). This sequence is bounded, and according to the Bolzano-Weierstrass Theorem for sequences, this bounded sequence has a convergent subsequence. The limit of this subsequence will be our limit point.*

Proof of Theorem 3.91. Let A be a bounded and infinite subset of $\mathbb{R}$. Since A is infinite, it is at least countable, and so there exists a sequence of *distinct* numbers $a_n \in A$. Since A is bounded, the sequence (a_n) is bounded. The Bolzano-Weierstrass Theorem for sequences (Theorem 3.84) assures us that there is a convergent subsequence $a_{n_k} \to a$. We claim that a is a limit point of A.

Let U be a neighborhood of a. Since the sequence (a_{n_k}) converges to a, the tail of the subsequence is eventually in U. Since the numbers in the sequence (a_n) are distinct, the numbers in the sequence (a_{n_k}) are distinct. Therefore the tail of the subsequence is comprised of infinitely many numbers, and so $U \cap A$ contains infinitely many numbers. Therefore, by Theorem 3.89, x is a limit point of A. □

3.6.4 Review of important concepts from Section 3.6

1. There are two Bolzano-Weierstrass Theorems: one for sequences (every bounded sequence has a convergent subsequence, Theorem 3.84), and one for sets (every bounded infinite subset of $\mathbb{R}$ has a limit point, Theorem 3.91).
2. If $A \subseteq \mathbb{R}$, then the limit points of A and the isolated points of A are relevant mathematical objects. In particular, the Bolzano-Weierstrass Theorem for sets claims that every bounded and infinite subset of $\mathbb{R}$ contains a limit point.
3. There are alternate characterizations for limit points and isolated points that are sometimes useful. See Theorems 3.89 and 3.90.

Exercises for Section 3.6

1. Suppose (a_n) is a sequence with $a_n \in (0,1)$. Prove that there is a convergent subsequence of (a_n). What are the possibilities of the limit of this convergent subsequence?
2. (This problem generalizes Exercise 9 of Section 3.4, and is expanded upon in Section 3.9.) Suppose (x_n) is a bounded sequence. Let E be the set of *subsequential limits* of (x_n), defined by

$$E = \{x \in \mathbb{R} |\text{ there exists a subsequence } x_{n_k} \text{ converging to } x\}.$$

   (a) Show that E is nonempty.
   (b) Show that E is bounded.
   (c) Show that $x_n \to x$ if and only if $E = \{x\}$ contains only one element.
3. Find all of the limit points and isolated points of the following sets:
   (a) $[0,1)$.
   (b) $\{\frac{1}{n^2} | n \in \mathbb{N}\}$.
   (c) $\mathbb{Q}$.
   (d) $\mathbb{Q} \cap (0,1)$.
   (e) $\mathbb{Z} \cup \{2^{-n} | n \in \mathbb{N}\}$.
4. Construct a subset $A \subset \mathbb{R}$ with
   (a) no limit points.
   (b) exactly one limit point.
   (c) exactly two limit points.
   (d) countably many limit points.
   (e) uncountably many limit points.
5. Construct a subset $A \subset \mathbb{R}$ with

(a) no isolated points.

(b) exactly one isolated point.

(c) countably many isolated points.

(d) every element of A isolated and A is infinite.

6. Show that every finite set contains only isolated points.

7. (This exercise demonstrates that a limit point of a set cannot be too far away from the set.) Let $x \in \mathbb{R}$, and let $A \subseteq \mathbb{R}$. Define the *distance* from $\{x\}$ to A as

$$\inf\{|a - x| | a \in A\}.$$

Show that if the distance between x and A is strictly greater than zero, then x is not a limit point of A.

8. (This exercise equates the term "limit point" with the limit of a sequence.) Show that x is a limit point of A if and only if there exists a sequence $a_n \in A$ of distinct points that converges to x and $a_n \neq x$ for all n.

9. Let $A \subseteq \mathbb{R}$. A real number x is called a *condensation point* if every neighborhood U of x contains uncountably many points from A.

   (a) Show that every condensation point of a subset $A \subseteq \mathbb{R}$ is a limit point of A.

   (b) Is it true that every limit point of A is also a condensation point of A? *[Hint: What if A is countable?]*

10. Let A be a nonempty subset of $\mathbb{R}$.

   (a) Suppose A is bounded above. Show that if $\sup(A) \notin A$, then $\sup(A)$ is a limit point of A.

   (b) Suppose that A is bounded below. Show that if $\inf(A) \notin A$, then $\inf(A)$ is a limit point of A.

11. Suppose that x is an isolated point of A. What must be true of a sequence $a_n \in A$ that converges to x? Prove your claim.

12. Prove Theorem 3.90. That is, show that x is an isolated point of A if and only if there exists a neighborhood U of x so that $U \cap A = \{x\}$.

13. Suppose A has a limit point $x \in \mathbb{R}$. Prove A is infinite.

14. Let $A \subseteq \mathbb{R}$. The *derived set* A' of A is the set of all limit points of A.

   (a) Show that if A is infinite and bounded, then $A' \neq \emptyset$.

   (b) Show that $(A')' \subseteq A'$.

   (c) Construct an example that shows $(A')' \neq A'$.

   (d) Show that $(A \cup B)' = A' \cup B'$.

(e) Show that $(A \cup A')' \subseteq A \cup A'$. (The set $A \cup A'$ is known as the *closure* of A in $\mathbb{R}$, denoted as $\bar{A}$.)

15. Let A be dense in $\mathbb{R}$. Prove that every real number is a limit point of A.

3.7 Cauchy Sequences

In this section, we introduce Cauchy sequences (Definition 3.92). Nontechnically put, a Cauchy sequence is one where eventually, the distance between any two members of the sequence is negligible. After several practice problems that illustrate this condition, we will show in this section that every convergent sequence is Cauchy (Theorem 3.97), and that every Cauchy sequence converges (Theorem 3.99). The benefit of studying these Cauchy sequences is that, since Cauchy sequences and convergent sequences are one and the same (Corollary 3.101), it will provide for us a way to demonstrate that a sequence converges without specifically knowing its limit.

3.7.1 Cauchy sequences and examples

We begin with a definition of what it means for a sequence to be Cauchy.

Definition 3.92. The sequence (a_n) is *Cauchy* if, for every $\epsilon > 0$, there exists an N so that for every $n, m \geq N$, we have

$$|a_n - a_m| < \epsilon.$$

Practice 3.93. Let $a_n = \frac{1}{n}$. Show that (a_n) is Cauchy.

Methodology. *According to the definition of Cauchy sequence (Definition 3.92), we need to study the quantity*

$$|a_n - a_m| = \left|\frac{1}{n} - \frac{1}{m}\right|,$$

and find an N so that this quantity is less than any given ϵ. We endeavor to make $|a_n - a_m|$ bigger, but not too big, similar to our initial efforts to show a sequence converged by using only the definition of convergence (see, for example, Practice 3.20). Along those lines, we notice that in this case, the difference between $\frac{1}{n}$ and $\frac{1}{m}$ can be no bigger than the larger of the two. Since the larger of the two can be made to be less than ϵ, the smaller difference can be made to be less than ϵ.

Solution. Let $\epsilon > 0$ be given. By the Archimedean Property, there exists an N so that $\frac{1}{N} < \epsilon$. Now let $n, m \geq N$. One of n or m is larger, suppose it is n. Then $\frac{1}{n} \leq \frac{1}{m} \leq \frac{1}{N}$, and so

$$|a_n - a_m| = \left|\frac{1}{n} - \frac{1}{m}\right| = \frac{1}{m} - \frac{1}{n} \leq \frac{1}{m} \leq \frac{1}{N} < \epsilon.$$

So, by definition, the sequence $a_n = \frac{1}{n}$ is Cauchy. ■

The condition of (a_n) being Cauchy insists that the difference between *any* two members of (a_n) is eventually minuscule. It is important to point out that this is *not* the same as the distance between any two *consecutive* members being eventually minuscule. We give two practice problems to demonstrate this difference. The first practice problem (Practice 3.94) gives an example of a sequence where the difference between consecutive terms is eventually minuscule, but the sequence itself is not Cauchy. The second practice problem (Practice 3.95) gives an example of a sequence where the difference between consecutive terms is also eventually minuscule, but the sequence turns out to be Cauchy.

Practice 3.94. Let $a_n = \sqrt{n}$. Given any $\epsilon > 0$, show that there exists an N so that $|a_{n+1} - a_n| < \epsilon$ for every $n \geq N$, but that (a_n) is *not* Cauchy.

Methodology. *We study the difference* $|a_{n+1} - a_n|$*, and will rationalize (the numerator) to find*

$$|a_{n+1} - a_n| \leq \frac{1}{2\sqrt{n}}.$$

The quantity above is one that converges to 0*, and so it is eventually less than any given* ϵ*.*

So, why is this sequence not Cauchy? If it were, then for a given $\epsilon < 0$ *(say,* $\epsilon = 1$*), there exists an* N *so that the quantity*

$$|a_n - a_m| < 1 \text{ for all } n, m \geq N.$$

If there were such an N*, and we fixed* $N = m$ *and let* n *increase without bound, the statement* $|a_n - a_N| < 1$ *will (with a small amount of work) provide an upper bound for* $\mathbb{N}$*, which we know is not possible according to Theorem 2.33 in Chapter 2.*

Solution. We estimate $|a_{n+1} - a_n|$ as follows:

$$|a_{n+1} - a_n| = \sqrt{n+1} - \sqrt{n} = (\sqrt{n+1} - \sqrt{n})\frac{\sqrt{n+1} + \sqrt{n}}{\sqrt{n+1} + \sqrt{n}} = \frac{1}{\sqrt{n+1} + \sqrt{n}}.$$

We note that

$$\frac{1}{\sqrt{n+1} + \sqrt{n}} \leq \frac{1}{\sqrt{n} + \sqrt{n}} = \frac{1}{2\sqrt{n}}.$$

Let $\epsilon > 0$ be given. Since $\frac{1}{2\sqrt{n}} \to 0$, there exists an N so that $\frac{1}{2\sqrt{n}} < \epsilon$ for all $n \geq N$, so that

$$|a_{n+1} - a_n| \leq \frac{1}{2\sqrt{n}} < \epsilon \text{ for all } n \geq N.$$

We now show $a_n = \sqrt{n}$ is not Cauchy. Suppose it were, so that for $\epsilon = 1$, there exists an N so that $|a_n - a_m| < 1$ for all $n, m \geq N$. Then for $m = N$, and $n \geq m = N$, we have

$$\begin{aligned} |a_n - a_N| &< \epsilon, \text{ or differently put,} \\ -1 < \sqrt{n} - \sqrt{N} &< 1, \text{ and so} \\ n &< (1 + \sqrt{N})^2. \end{aligned}$$

Thus, the number $(1 + \sqrt{N})^2$ is an upper bound for $\mathbb{N}$, contradicting Theorem 2.33 in Chapter 2. ■

The next practice problem gives an example of a sequence whose consecutive differences $|a_{n+1} - a_n| \to 0$, and, in contrast to the above, turns out to be Cauchy. **The methodology in Practice 3.95 is a very common one to prove that a sequence–especially a recursively defined sequence–is Cauchy. Exercise 4 generalizes this approach.** See also Practice 3.103 for another demonstration of this technique.

Practice 3.95. Suppose that $0 < r < 1$, and that the sequence (a_n) satisfies

$$|a_{n+1} - a_n| \leq r^n. \tag{3.30}$$

Show that (a_n) is Cauchy.

Methodology. *We solve this problem in two parts. The first part is a careful estimate of $|a_n - a_m|$ using the defining condition found in Equation (3.30), and our previous work about sums of geometric sequences (Exercise 21 of Section 3.3). The second part uses that estimate to show how Definition 3.92 is satisfied.*

Solution. Let n, m be fixed. Suppose that $n > m$, and write $n = m + k$ for some $k \geq 1$. We intend to prove by Mathematical Induction that

$$|a_{m+k} - a_m| \leq r^m(1 + r + \cdots + r^{k-1}). \tag{3.31}$$

If $k = 1$, then $n = m + 1$, and $|a_n - a_m| \leq r^m$ by using $n = m$ in the defining condition given in Equation (3.30). So, the trivial case is established. Now suppose that the result is true for some $k \geq 1$. Then

$$\begin{aligned}
|a_{m+k+1} - a_m| &= |a_{m+k+1} - a_{m+k} + a_{m+k} - a_m| \\
&\leq |a_{m+k+1} - a_{m+k}| + |a_{m+k} - a_m| && \text{(Triangle Inequality)} \\
&\leq r^{m+k} + |a_{m+k} - a_m| && \text{(given condition)} \\
&\leq r^{m+k} + r^m(1 + r + \cdots + r^{k-1}) && \text{(induction hypothesis)} \\
&= r^m(r^k + 1 + r + \cdots + r^{k-1}) && \text{(factor out } r^m\text{)} \\
&= r^m(1 + r + \cdots + r^{k-1} + r^k).
\end{aligned}$$

We may now show that the sequence (a_n) is Cauchy using the estimate in Equation (3.31). Let $\epsilon > 0$ be given. Let n, m be given. If $n \geq m$, then there exists a $k \geq 0$ so that $n = m + k$, and using the estimate above, according to Exercise 21 of Section 3.3 we have

$$|a_n - a_m| \leq r^m(1 + r + \cdots + r^{k-1}) = r^m \cdot \frac{1 - r^k}{1 - r} \leq \frac{r^m}{1 - r}.$$

(Notice that if $k = 0$, then $n = m$, and the inequality above holds trivially.) Since $\frac{r^m}{1-r} \to 0$, there exists an N so that

$$\frac{r^m}{1 - r} < \epsilon \quad \text{for all } m \geq N.$$

Now suppose that $n, m \geq N$. Putting all of our estimates together, we have

$$|a_n - a_m| \leq \frac{r^m}{1 - r} < \epsilon.$$

■

Remark 3.96. It is worth mentioning how we found Equation (3.31) since it is so crucial to the solution of Practice 3.95. We illustrate it using $m = 4, n = 7$, which would make $k = 3$. A generalization of the following process is how one could think to stumble upon Equation (3.31):

$$\begin{aligned}
|a_7 - a_4| &= |a_7 - a_6 + a_6 - a_4| \\
&\leq |a_7 - a_6| + |a_6 - a_4| \\
&\leq r^6 + |a_6 - a_5 + a_5 - a_4| \\
&\leq r^6 + |a_6 - a_5| + |a_5 - a_4| \\
&\leq r^6 + r^5 + r^4 \\
&= r^4(1 + r + r^2).
\end{aligned}$$

Recall that $m = 4$, and $k = 3$, so $k - 1 = 2$. Exercise 4 asks you to proceed similarly as in Practice 3.95. ■

3.7.2 Cauchy sequences and convergence

There are two crucial ideas that this entire section is about. **The first crucial idea one should take away from this section is that every Cauchy sequence (of real numbers) is convergent, and every convergent sequence (of real numbers) is Cauchy.** In other words, we are trying to prove that the set of all convergent sequences is equal to the set of all Cauchy sequences. In this subsection, we demonstrate this by double-inclusion. We shall first show in Theorem 3.97 that

$$\boxed{(a_n) \text{ is convergent}} \overset{\text{Thm.3.97}}{\Longrightarrow} \boxed{(a_n) \text{ is Cauchy.}}$$

We will then prove

$$\boxed{(a_n) \text{ is Cauchy}} \overset{\text{Thm.3.98}}{\Longrightarrow} \boxed{(a_n) \text{ is bounded.}}$$

which will aid us in establishing one of the main results of this section:

$$\boxed{(a_n) \text{ is Cauchy}} \overset{\text{Thm.3.99}}{\Longrightarrow} \boxed{(a_n) \text{ is convergent.}}$$

Once we have established the above, we can point out that **the second crucial idea to take away from this section is that, by demonstrating that a sequence is Cauchy, we may now show that a sequence converges without knowing specifically what it converges to.** Nowhere in the definition of Cauchy sequence (Definition 3.92) does it mention any real number that is supposed to be the limit of the sequence. Below we prove that Cauchy sequences and convergent sequences are the same (Corollary 3.101). The only other place that we have considered a special type of sequence that converged without specifically knowing what it converged to was in Section 3.5; specifically, we showed in the Monotone Convergence Theorem (Theorem 3.74) that all bounded monotone sequences converge (although, in the proof of that theorem the limit of such a sequence is found). What is interesting, however, is that not every convergent sequence is bounded and monotone. So the set of convergent sequences is, in a sense, bigger than the set of bounded monotone sequences, whereas, in this section we prove that the set of convergent sequences is *exactly the same* as the set of Cauchy sequences, where again, we can show that a sequence is Cauchy (and thus convergent) without knowing the limit of the sequence in advance.

Theorem 3.97. *Every convergent sequence is Cauchy.*

Idea of Proof. *If (a_n) is our sequence, we are assuming that $a_n \to a$. We have to show (a_n) is Cauchy, and as such, we need to estimate $|a_n - a_m|$ for large enough n, m. Since $a_n \to a$, both of a_n and a_m are not too far away from a, and so a_n and a_m are not too far away from each other.*

Proof of Theorem 3.97. Suppose $a_n \to a$ is a convergent sequence. Let $\epsilon > 0$ be given. Since (a_n) converges, there exists an N so that

$$|a_n - a| < \frac{\epsilon}{2} \quad \text{for all } n \geq N.$$

So, if both of n and m are greater than or equal to N, we have

$$\begin{aligned} |a_n - a_m| &= |a_n - a + a - a_m| \\ &\leq |a_n - a| + |a - a_m| \\ &< \tfrac{\epsilon}{2} + \tfrac{\epsilon}{2} \\ &= \epsilon. \end{aligned}$$

□

Theorem 3.98. *Every Cauchy sequence is bounded.*

Idea of Proof. *This proof will be very similar to the proof of Theorem 3.28 (every convergent sequence is bounded). If (a_n) is our Cauchy sequence, then we will use the Cauchy condition to find a bound for the tail of the sequence, and then increase that bound if necessary in considering the start of the sequence.*

Proof of Theorem 3.98. Suppose (a_n) is Cauchy. Choosing $\epsilon = 1$, there exists an N so that

$$|a_n - a_m| < 1 \quad \text{for all } n, m \geq N.$$

Let $m = N$. For every $n \geq N$, the above condition states that

$$\begin{aligned} |a_n - a_N| &< 1, \text{and so since} \\ |a_n| - |a_N| &\leq |a_n - a_N| \text{ by the triangle inequality, we have} \\ |a_n| &\leq |a_N| + 1 \text{ for all } n \geq N. \end{aligned} \tag{3.32}$$

Now set

$$M = \max\{|a_1|, \ldots, |a_{N-1}|, |a_N| + 1\}.$$

We claim M is a bound for (a_n). If $n < N$, then $|a_n| \leq M$ since M is defined as being larger than any of $|a_1|, \ldots, |a_{N-1}|$. If $n \geq N$, then M is still larger than $|a_n|$, as this is exactly the content of Equation (3.32). So, M is a bound for (a_n). □

The last theorem also provides a way to show that a sequence is *not* Cauchy: prove it is unbounded. This observation would have helped in Practice 3.94, since the sequence $a_n = \sqrt{n}$ is unbounded.

We can now complete the equivalence of convergent and Cauchy sequences.

Theorem 3.99. *Every Cauchy sequence converges.*

Idea of Proof. *If (a_n) is Cauchy, it must be bounded, and so the Bolzano-Weierstrass Theorem for sequences (Theorem 3.84): there must exist a convergent subsequence $a_{n_k} \to a$. We will show that $a_n \to a$ by estimating the distance $|a_n - a|$ using our convergent subsequence, and the Cauchy condition.*

Proof of Theorem 3.99. Suppose (a_n) is Cauchy. By Theorem 3.98, (a_n) is bounded. Since (a_n) is bounded, the Bolzano-Weierstrass Theorem for sequences (Theorem 3.84), there exists a convergent subsequence, $a_{n_k} \to a$. We will prove that $a_n \to a$ by the definition of convergence (Definition 3.13).

Let $\epsilon > 0$ be given. Since $a_{n_k} \to a$, there exists an N_1 so that

$$|a_{n_k} - a| < \frac{\epsilon}{2} \quad \text{for all } k \geq N_1.$$

Since (a_n) is Cauchy, there exists an N_2 so that

$$|a_n - a_m| < \frac{\epsilon}{2} \quad \text{for all } n, m \geq N_2.$$

Set $N = \max\{N_1, N_2\}$. Let $n \geq N$, and choose $k \geq N \geq N_1$. Then for $m = n_k \geq N \geq N_2$, we have

$$\begin{aligned} |a_n - a| &= |a_n - a_{n_k} + a_{n_k} - a| \\ &\leq |a_n - a_{n_k}| + |a_{n_k} - a| \\ &< \tfrac{\epsilon}{2} + \tfrac{\epsilon}{2} \\ &= \epsilon. \end{aligned}$$

□

Remark 3.100. In the proof above, we find ourselves with a sequence (a_n), and a subsequence (a_{n_k}) which converges. We are trying to prove (a_n) converges, and it is incorrect to assume that just because (a_n) has *one* convergent subsequence, it must be the case that (a_n) must converge (consider the divergent sequence $a_n = (-1)^n$ and its convergent subsequence $a_{2n} = 1$). If we already knew that (a_n) converged, then any subsequence of (a_n) must converge to the same value (Theorem 3.66). The point here is that we are trying to prove that (a_n) converges, not assuming that it already does. ■

We can summarize Theorems 3.97 and 3.99 into one corollary.

Corollary 3.101. *A sequence of real numbers converges if and only if it is Cauchy.*

Idea of proof. *This corollary follows directly from Theorems 3.97 and 3.99.*

Proof of Corollary 3.101. Let (a_n) be a sequence of real numbers. If (a_n) converges, then it is Cauchy by Theorem 3.97. Conversely, if (a_n) is Cauchy, then it converges by Theorem 3.99. □

Now is a good time to point out why we have subtly used the phrase "real numbers" when referring to sequences in this section. Since we have focused our attention on the analysis of *real numbers*, there is a blanket assumption that our mathematical playground is $\mathbb{R}$, and so it goes without saying that these sequences are sequences of real numbers. So why redundantly reiterate that condition?

It turns out that there are other forms of analysis whose mathematical playground is something other than the real numbers. For example, Complex Analysis studies, among many other things, convergence of sequences of complex numbers. In fact, any set that has some sort of function to measure distance between two elements in the set is a suitable space to study (these are called *metric spaces* and briefly introduced on Page 69). For example, the set of rational numbers $\mathbb{Q}$ is such a set, where the distance function (the absolute value) is inherited from $\mathbb{R}$. Convergence in any such set can be studied (as one of the foundational topics), and as such, the property of being Cauchy can also be studied. It turns out that not every mathematical "playground" enjoys the property that Cauchy sequences converge, although the ones that do are called *complete*, or *Cauchy complete.* In Exercise 8 we ask you to prove that if Cauchy sequences converge, then the Completeness Axiom holds, justifying the use of the word "complete" in "Completeness Axiom". We have shown in this section that $\mathbb{R}$ is complete. For now, we give an example to demonstrate that $\mathbb{Q}$ is not complete: there exists at least one sequence that is Cauchy, but it does not converge to any number in the space.

Example 3.102. Let $A = \{x \in \mathbb{Q} | x^2 < 2\}$. According to Theorem 3.46, there exists a sequence $a_n \in A$ that converges to $\sup(S) = \sqrt{2} \notin \mathbb{Q}$. Since (a_n) converges, it must be Cauchy (Theorem 3.97).

What is interesting about this example is that $a_n \in A \subseteq \mathbb{Q}$ for every n. In addition, (a_n) is Cauchy, but (a_n) does not converge to a number in $\mathbb{Q}$ (it converges to a number *outside* $\mathbb{Q}$). In the sense described above, this example demonstrates that $\mathbb{Q}$ is *not* complete. ■

Recall that one of the main benefits of knowing that a sequence of real numbers is Cauchy is that one knows that the sequence converges, but one does *not* need to know what the limit of the sequence is. We close the section by giving a practice problem designed to exhibit the Cauchy condition (and thus, convergence), but with a sequence where it would be difficult to prove convergence using the methods developed that are previous to this section.

Practice 3.103. Let $a < b$, and set $a_1 = a$, and $a_2 = b$. Recursively define

$$a_n = \frac{1}{2}(a_{n-1} + a_{n-2}) \quad \text{for } n \geq 3.$$

Show that (a_n) is Cauchy, and hence convergent. Then, compute $\lim_{n\to\infty} a_n$.

Methodology. *We will start by using Mathematical Induction to show*

$$a_n - a_{n-1} = \left(-\frac{1}{2}\right)^{n-2} (b-a) \quad \text{for } n \geq 2.$$

From there, we will adapt the methodology of Practice 3.95 to this situation in order to show that (a_n) is Cauchy. In the solution, we will do this quite explicitly by creating a new sequence that will directly relate to Practice 3.95. The end result will be that a constant multiple of a Cauchy sequence is still Cauchy.

To compute $\lim_{n\to\infty} a_n$ one usually uses the limit theorems from Section 3.3 and lets $n \to \infty$ on both sides of the defining equation of (a_n) to produce an equation that characterizes the solution (for an example of this, see Practice 3.75, specifically, the discussion just before and after Equation (3.27) on page 154). In this case, this process gives us, if $a_n \to x$, the equation

$$a_n = \frac{1}{2}(a_{n-1} + a_{n-2}) \longrightarrow x = \frac{1}{2}(x + x), \text{or } x = x.$$

This is both unhelpful and irritating. A tool we ***do*** *have, however, is our alternate expression for $a_n - a_{n-1}$ that we found above. Beginning with $a_n - a_1$, we create the successive differences below to help us determine x:*

$$\begin{aligned} a_n - a_1 &= (a_n - a_{n-1}) + a_{n-1} - a_1 \\ &= (a_n - a_{n-1}) + (a_{n-1} - a_{n-2}) + a_{n-2} - a_1 \\ &= \cdots . \end{aligned}$$

We find a closed form for $a_n - a_1$ based on this idea, and establish it by Mathematical Induction. This formula will help us to determine x.

Solution. The solution will have three phases. The first will be to establish the following by Mathematical Induction:

$$a_n - a_{n-1} = \left(-\frac{1}{2}\right)^{n-2} (b-a) \quad \text{for } n \geq 2. \tag{3.33}$$

The next phase will be a proof that (a_n) is Cauchy (and hence convergent, according to Theorem 3.99) by adapting the method and result of Practice 3.95. The last phase will use Equation (3.33) to help determine $\lim_{n\to\infty} a_n$.

We begin by proving Equation (3.33) by Mathematical Induction. If $n = 2$, then we are left to prove the triviality

$$a_2 - a_1 = \left(\frac{1}{2}\right)^0 (b-a),$$

which is true since we defined $a_2 = b$, and $a_1 = a$. Now suppose Equation (3.33) is true for some $n \geq 2$. Then

$$\begin{aligned} a_{n+1} - a_n &= \tfrac{1}{2}(a_n + a_{n-1}) - a_n \\ &= \left(-\tfrac{1}{2}\right)(a_n - a_{n-1}) \\ &= \left(-\tfrac{1}{2}\right)^{n-1}(b-a), \end{aligned}$$

where, in the last equality above we use the induction hypothesis.

We may now adapt the solution of Practice 3.95 to show (a_n) is Cauchy. Temporarily denote $b_n = \frac{2}{b-a}a_n$. Then

$$\begin{aligned} b_{n+1} - b_n &= \tfrac{2}{b-a}(a_{n+1} - a_n) \\ &= \tfrac{2}{b-a}\left(-\tfrac{1}{2}\right)^{n-1}(b-a) \\ &= \left(-\tfrac{1}{2}\right)^n. \end{aligned}$$

So,

$$|b_{n+1} - b_n| = \left(\frac{1}{2}\right)^n.$$

According to Practice 3.95, (b_n) is Cauchy, and so it converges. Relating this back to (a_n),

$$a_n = \frac{b-a}{2}b_n,$$

and so (a_n) converges by part (3) of Theorem 3.31. Since (a_n) converges, it is Cauchy.

Let $x = \lim_{n\to\infty} a_n$. In order to determine x, we prove the following by Mathematical Induction. For convenience, let $r = \frac{-1}{2}$.

$$a_n - a_1 = (1 + r + \cdots + r^{n-2})(b-a) \quad \text{for } n \geq 2. \tag{3.34}$$

Clearly the trivial case of $n = 2$ holds. Now, using Equation (3.33), we can complete the Mathematical Induction proof of Equation (3.34):

$$\begin{aligned} a_{n+1} - a_1 &= a_{n+1} - a_n + a_n - a_1 \\ &= r^{n-1}(b-a) + (1 + r + \cdots + r^{n-2})(b-a) \\ &= (1 + r + \cdots + r^{n-2} + r^{n-1})(b-a). \end{aligned}$$

Now that we have established Equation (3.34), we use Exercise 21 of Section 3.3 to deduce that, for $n \geq 2$, we have

$$\begin{aligned} a_n - a_1 &= (1 + r + \cdots + r^{n-2})(b-a) \\ &= \tfrac{1-r^{n-1}}{1-r}(b-a), \text{ so since } a_1 = a, \\ a_n &= \tfrac{1-r^{n-1}}{1-r}(b-a) + a \end{aligned}$$

Since $|r| = \left|-\frac{1}{2}\right| = \frac{1}{2} < 1$, we take limits to see that

$$\lim_{n\to\infty} a_n = \lim_{n\to\infty} \frac{1-r^{n-1}}{1-r}(b-a)+a = \frac{1}{1-r}(b-a)+a.$$

Since we set $r = -\frac{1}{2}$, $\frac{1}{1-r} = \frac{2}{3}$, and so

$$x = \lim_{n\to\infty} a_n = \frac{2}{3}(b-a)+a.$$

■

Remark 3.104. In the proof of the above theorem, we demonstrated that $b_n = \frac{2}{b-a}a_n$ was Cauchy, and used the equivalence of convergent and Cauchy sequences (Corollary 3.101) to eventually deduce that (a_n) is Cauchy. Although this is perfectly valid, Exercise 3 asks you to show that basic transformations of Cauchy sequences preserve the property of being Cauchy, and we ask you to do this without using the equivalence of convergent and Cauchy sequences. This would have made that particular part of the solution, perhaps, less awkward. ■

3.7.3 Review of important concepts from Section 3.7

1. The set of convergent sequences and the set of Cauchy sequences are the same. That is, every convergent sequence is Cauchy (Theorem 3.97), and every Cauchy sequence is convergent (Theorem 3.99).
2. Demonstrating that (a_n) is Cauchy is enough to demonstrate that (a_n) is convergent, but with the added benefit that you can do this without specifically knowing $\lim_{n\to\infty} a_n$ in advance.
3. If (a_n) is a recursively defined sequence that you would like to show is Cauchy, the method in Practice 3.95 (and its generalization in Exercise 4) is a very commonly used method. Of course, it is not the only method, but it may be a good place to start.

Exercises for Section 3.7

1. Show that $a_n = \frac{1}{1!} + \frac{1}{2!} + \cdots + \frac{1}{n!}$ is Cauchy.
2. Let $a_n = 1 + \frac{1}{2} + \frac{1}{3} + \cdots + \frac{1}{n}$. Show that $|a_{n+1} - a_n| \to 0$, but that (a_n) is *not* Cauchy. *[Hint: Rather than proceeding as in Practice 3.94, consider Exercise 2 of Section 3.5, and Theorem 3.98.]*
3. Suppose that (a_n) and (b_n) are Cauchy sequences, and that $k \in \mathbb{R}$. Without using the equivalence between convergent and Cauchy sequences (Theorem 3.97, Theorem 3.99, and Corollary 3.101), prove the following.

   (a) Show that $(a_n \pm b_n)$ is Cauchy.

(b) Show that (ka_n) is Cauchy.

4. Suppose that (a_n) is a sequence that eventually satisfies

$$|a_{n+1} - a_n| \leq r|a_n - a_{n-1}|$$

for some $0 \leq r < 1$ (these sequences are sometimes called *contractive*). Show that (a_n) converges. *[Hint: Show (a_n) is Cauchy, and follow along the same lines as in Practice 3.95.]*

5. Let $a_1 > 0$, and set $a_{n+1} = \frac{1}{a_n+2}$ for $n \geq 1$.

   (a) Show that $a_n > 0$ for every n.

   (b) Find an $r \in (0, 1)$ so that $|a_{n+1} - a_n| \leq r|a_n - a_{n-1}|$.

   (c) Use Exercise 4 to conclude (a_n) converges, and find $\lim_{n\to\infty} a_n$.

6. (This exercise is an example of the use of Newton's Method for the function $f(x) = x^2 - \alpha$. The following sequence (x_n) will converge to $\sqrt{\alpha}$, and hence, the numbers x_n give a way to approximate $\sqrt{\alpha}$.) Let $x_1 > 0$, and let $\alpha > 0$. Set $x_{n+1} = \frac{1}{2}\left(x_n + \frac{\alpha}{x_n}\right)$. Use Exercise 4 to show that (x_n) is Cauchy, and hence convergent. Show that $x_n \to \sqrt{\alpha}$.

7. In his construction of the real numbers, Cantor defined two Cauchy sequences (x_n) and (y_n) to be *equivalent* if for every $\epsilon > 0$, there exists a $N \in \mathbb{N}$ so that if $n, m \geq N$, we have $|x_n - y_n| < \epsilon$. Prove that if $x_n \to z$, then $y_n \to z$. (See page 153 of [2] for more details on this.)

8. Prove that if every Cauchy sequence converges, then the Completeness Axiom holds. (This is a justification for using the word "Complete" in "Completeness Axiom.") *[Hint: Follow the same suggestion as that of Exercise 7 of Section 3.5.]*

3.8 Infinite Limits

So far, we have been interested in sequences that converge. The main reason we might be interested in convergent sequences is because there is so much useful structure within the set of all convergent sequences. This entire chapter has been devoted to elucidating this structure and to understanding its usefulness.

In this section, we wish to study those divergent sequences that tend to ∞ or $-\infty$. The set of these sequences has some structure to it that will be very useful. In fact, much of what we have developed for convergent sequences will carry over to those sequences that have infinite limits.

We will begin in Subsection 3.8.1 with a definition of what it means for a sequence to "diverge to ∞", and then go on to describe how our previous results carry over to those sequences that have infinite limits in Subsection 3.8.2.

3.8.1 Infinite limits

Consider for a moment the sequence $a_n = n^2 + 1$ that was in Practice 3.26. The question "Does (a_n) converge?" was answered in Practice 3.26: no, (a_n) diverges because it is unbounded. All the same, the sequence (a_n) is not the least bit bizarre: it is a polynomial in n, and is strictly monotone increasing. In fact, as Figure 3.8.1 shows, as n increases, it could be said that the numbers a_n are jumping in leaps and bounds toward ∞.

n	$a_n = n^2 + 1$		
1	$1^2 + 1$	$=$	2
5	$5^2 + 1$	$=$	26
10	$10^2 + 1$	$=$	101
100	$100^2 + 1$	$=$	$10,001$
1000	$1000^2 + 1$	$=$	$1,000,001$
1 million	$10^{12} + 1$	$=$	$1,000,000,000,001$

FIGURE 3.8.1
The behavior of $a_n = n^2 + 1$ for some large values of n.

It could be said that (a_n) is tending toward ∞ as $n \to \infty$, although we do not yet have any mathematical definition of what exactly this means. **In this section, we define precisely what it means for a sequence to have either ∞ or $-\infty$ as a limit.** We will construct this definition through a process similar to how we defined convergence of sequences in Section 3.2: translate what an infinite limit *should* be into rigorous mathematical statements. For the sake of clarity, we will continue to consider the sequence $a_n = n^2 + 1$ throughout this translation process.

To begin to make the definition of an infinite limit more precise, we observe that the starting statement below seems a little too vague:

"The quantities $n^2 + 1$ tend to ∞ as n gets large."

What seems to be lacking is what it means to "tend to ∞". Back in Section 3.2, we wanted to further clarify what it meant to "tend to" a particular real number, and this task was accomplished by considering eventually minuscule distances between a_n and its proposed real limit a. Since ∞ is not a real number, and quite frankly there is no real number that could be conceivably "close" to ∞, we will have to take a different approach.

Instead, we could rephrase the above as

"The quantities $n^2 + 1$ will exceed any given real number for large enough n."

If we were to give a name (M) to this "given real number", and describe when $n^2 + 1$ exceeds that real number (when $n \geq N$), we will arrive at the actual definition of what it means for $n^2 + 1 \to \infty$:

For every $M \in \mathbb{R}$, there exists an N so that $n^2 + 1 \geq M$ for all $n \geq N$.

Let us indeed verify that this is the case in Practice 3.105.

Practice 3.105. Show that for any $M \in \mathbb{R}$, there exists an N so that $n^2+1 \geq M$ for all $n \geq N$.

Methodology. *If $M \in \mathbb{R}$ is given, then we need to identify when $n^2 + 1$ exceeds it. That is, for which N is $N^2+1 \geq M$? This could be found by solving for N, or by simply pointing out that $n^2 + 1 \geq n$ for all n (we give a separate shorter solution elucidating this line of reasoning). Then we would need to show that $n^2 + 1 \geq M$ for every $n \geq N$. Since $a_n = n^2 + 1$ is increasing, as soon as an N is found with $N^2 + 1 \geq M$, then for every $n \geq N$, we would have*

$$n^2 + 1 \geq N^2 + 1 \geq M.$$

Solution 1. Let $M \in \mathbb{R}$ be given. If it is the case that $M \leq 2$, then $n^2 + 1 \geq M$ for every $n \geq 1$, and so we could choose $N = 1$. If, on the other hand, $M \geq 2$, then

$$\begin{array}{rrcl} & N^2+1 & \geq & M \\ \Leftrightarrow & N & \geq & \sqrt{M-1}. \end{array}$$

So choose N to be any integer larger than $\sqrt{M-1}$. We observe that $a_n = n^2 + 1$ is an increasing sequence. So for every $n \geq N$, we have

$$n^2 + 1 \geq N^2 + 1 \geq M.$$

■

Solution 2. Let $M \in \mathbb{R}$ be given. Let N be any natural number satisfying $N \geq M$. Then for $n \geq N$, we have

$$n^2 + 1 \geq n \geq N \geq M.$$

Using the previous discussion for our inspiration, we may now present the definition of what it means for a sequence to have an infinite limit.

Definition 3.106. Let (a_n) be a sequence.

1. For every $M \in \mathbb{R}$, if there exists an N so that $a_n \geq M$ for all $n \geq N$, then a_n *diverges to* ∞, and we write $a_n \to \infty$ or $\lim_{n\to\infty} a_n = \infty$.

2. For every $M \in \mathbb{R}$, if there exists an N so that $a_n \leq M$ for all $n \geq N$, then a_n *diverges to* $-\infty$, and we write $a_n \to -\infty$ or $\lim_{n\to\infty} a_n = -\infty$.

Henceforth, if $a \in \mathbb{R}^\sharp$, the expression $a_n \to a$ or $\lim_{n\to\infty} a_n = a$ means (a_n) converges to a if $a \in \mathbb{R}$, or (a_n) diverges to a if $a = \infty$ or $-\infty$.

Practice 3.105 demonstrates that (n^2+1) diverges to ∞. Below we give two results: the first is a practice problem involving a specific sequence, while the second is an intuitive result of which we can now give a rigorous mathematical demonstration.

Practice 3.107. Show that $(\sqrt{n})$ diverges to ∞.

Methodology. *We prove this directly by Definition 3.106: if M is our given real number, we need to find an N so that $\sqrt{n} \geq M$ for every $n \geq N$. Since $\sqrt{n}$ is increasing, $\sqrt{n} \geq \sqrt{N}$, so we just need to solve for N in $\sqrt{N} \geq M$. As a result, $N = M^2$ will do.*

Solution. Let $M \in \mathbb{R}$ be given. Our goal is to find an N with $\sqrt{n} \geq M$ for every $n \geq N$. In particular, this must hold for $n = N$, so it would have to be the case $\sqrt{N} \geq M$, or rather, $N \geq M^2$. So let N be any natural number greater than or equal to M^2. Then for $n \geq N$, we have

$$\sqrt{n} \geq \sqrt{N} \geq M.$$

■

The following is a result that is very intuitive, but until now we have not been able to give a formal mathematical proof of it.

Proposition 3.108. *Suppose that $a_n > 0$ for all n. Show that $a_n \to \infty$ if and only if $\frac{1}{a_n} \to 0$.*

Idea of proof. *We prove each implication separately. If $a_n \to \infty$, then it will eventually be larger than whatever number is needed so that $\frac{1}{a_n} < \epsilon$ for some given ϵ. Conversely, if $M \in \mathbb{R}$ is given, then our assumption that $\frac{1}{a_n} \to 0$ implies that $\frac{1}{a_n}$ will eventually be smaller than $\frac{1}{M}$, so that a_n will be larger than M.*

Proof of Proposition 3.108. ($\Rightarrow$) Suppose $a_n \to \infty$. We must show that $\frac{1}{a_n} \to 0$. Let $\epsilon > 0$ be given. If we set $M = \frac{1}{\epsilon}$, then since $a_n \to \infty$, there exists an N so that $a_n > M = \frac{1}{\epsilon}$. In other words, $\epsilon > \frac{1}{a_n}$ for $n \geq N$, and so by definition $\frac{1}{a_n} \to 0$.

($\Leftarrow$) Conversely, suppose now that $\frac{1}{a_n} \to 0$. We must show that $a_n \to \infty$. Let $M \in \mathbb{R}$ be given. We must find an N so that $a_n \geq M$ for every $n \geq N$. If it is the case that $M \leq 0$, then any N will do, since by assumption $a_n > 0 \geq M$. If instead $M > 0$, then since $\frac{1}{a_n} \to 0$, there exists an N so that $\frac{1}{a_n} < \frac{1}{M}$ for every $n \geq N$. In other words, $M < a_n$ for every $n \geq N$, and so $a_n \to \infty$ by definition. □

In Section 3.2, we gave the definition of what it means for a sequence to converge (Definition 3.13), and then gave an alternative (and equivalent) definition of convergence using the language of neighborhoods (Theorem 3.21). Here is the equivalent formulation of infinite limits in terms of neighborhoods. Notice that Theorem 3.109 and Theorem 3.21 are almost identical.

Theorem 3.109. *Let (a_n) be a sequence, and let $a = \pm\infty$. The following are equivalent:*

1. *(a_n) diverges to a.*
2. *For every neighborhood U of a, (a_n) is eventually in U.*

Idea of proof. *We prove (1) $\Rightarrow$ (2), and then (2) $\Rightarrow$ (1). Just like the proof of Theorem 3.21, the idea is to simply unravel what each statement means in terms of the other.*

Proof of Theorem 3.109. (We prove this theorem for the case $a = \infty$, and leave the proof of the theorem for $a = -\infty$ for Exercise 7.)

((1) $\Rightarrow$ (2)) Suppose $a_n \to \infty$. Let $U = (M, \infty)$ be a neighborhood of ∞. Since $a_n \to a$, there exists an N so that $a_n \geq M + 1$. In other words, $a_n \in U$ for every $n \geq N$, and so (a_n) is eventually in U.

((2) $\Rightarrow$ (1)) Conversely, suppose that for every neighborhood U of ∞, the sequence (a_n) is eventually in U. We want to show that $a_n \to \infty$. So, let $M \in \mathbb{R}$ be given. The set $U = (M, \infty)$ is a neighborhood of ∞, and so (a_n) is eventually in U. As such, there exists an N so that $a_n \in U$ for every $n \geq N$. In other words, $a_n > M$ for $n \geq N$, and so by definition, $a_n \to \infty$. □

Before moving on to the infinite analogues of our known results, we pause to prove a lemma.

Lemma 3.110. *If $a_n \to \pm\infty$, then $-a_n \to \mp\infty$.*

Idea of proof. *This is a direct application of Definition 3.106. The N we are looking for is exactly the same N that is provided to us by the hypothesis.*

Proof of Lemma 3.110. Exercise 5. □

3.8.2 Infinite analogues of known results

So far, this entire chapter has been devoted to developing the theory of convergent sequences. Almost all of the important convergence results we have developed have an analogue that holds that concerns divergence to $\pm\infty$. Some of these analogues are simple extensions (see Theorems 3.111, 3.114, 3.116, and 3.118), while others need to be modestly rephrased (Theorems 3.112 and 3.117). In this subsection, we list these analogues, roughly in the same order the analogous result appears in the chapter. The first of these deals with uniqueness of infinite limits, and extends Theorem 3.22.

Theorem 3.111 (Analogue of Theorem 3.22). *If $\lim_{n\to\infty} a_n$ exists, then it is unique. That is, if $a_n \to a$, and $a_n \to b$ for $a, b \in \mathbb{R}^\sharp$, then $a = b$.*

Idea of proof. *If a and b are both real, then the result has already been proven. If one of them is infinite, then we proceed similarly to the proof of Theorem 3.22 and create disjoint neighborhoods separating a and b if $a \neq b$. If $a_n \to a$ and $a_n \to b$, then we would reach a contradiction, as the tail of (a_n) would have to reside in both neighborhoods.*

Proof of Theorem 3.111. If a and b are real, then the result has already been proven as Theorem 3.22, so we only need consider the situation that either a or b (or both) is infinite. Similar to the proof of Theorem 3.22, we claim that there exists a neighborhood U of a, and a neighborhood V of b with $U \cap V = \emptyset$. We will show this for all choices of a and b with $a \neq b$, then we will complete the proof of Theorem 3.111 through a contradiction argument.

Suppose a is infinite, and $b \in \mathbb{R}$. Let $V = (b-1, b+1)$ be a neighborhood of V. If $a = \infty$, choose any M with $M > b+1$, and $U = (M, \infty)$ is a neighborhood of a with $U \cap V = \emptyset$. If $a = -\infty$, choose any M with $M < b-1$, so that $U = (-\infty, M)$ is a neighborhood of a with $U \cap V = \emptyset$. In any case, we have shown that there exist neighborhoods of a and b that are disjoint. If both a and b are infinite, say $a = -\infty$ and $b = \infty$, then the neighborhoods $U = (-\infty, 0)$ of $-\infty$ and $V = (0, \infty)$ of $+\infty$ are disjoint, separating $-\infty$ from $+\infty$.

Now suppose that $a_n \to a$ and $a_n \to b$ where either a or b is infinite and $a \neq b$. We have shown that there exists a neighborhood U of a, and a neighborhood V of b with $U \cap V = \emptyset$. But according to Theorem 3.109, (a_n) is eventually in U because $a_n \to a$, and (a_n) is eventually in V as well since $a_n \to b$. This cannot be the case since $U \cap V = \emptyset$. □

Our next result is an analogy to Theorem 3.31, which was the first of our "supporting results" to help us both demonstrate convergence, but also to help us compute the limits of convergent sequences. Theorem 3.112 will do the same, but considers the possibility that the limits involved are infinite. Keep in mind that there are certain arithmetic operations that are not allowed in $\mathbb{R}^\sharp$, and it is unfortunate that we have to arrange our hypotheses in Theorem 3.112 to avoid these situations, since the expression of these hypotheses appears complicated at first glance. **Generally speaking, however, whenever any conclusion of Theorem 3.31 makes sense for infinite limits, it is true.**

Theorem 3.112 (Analogue of Theorem 3.31). *Suppose $a_n \to a$ and $b_n \to b$, with $a, b \in \mathbb{R}^\sharp$.*

1. *Sums and differences of infinite limits satisfy the following.*

(a) *If $b \in \mathbb{R}$, and $a = \pm\infty$, then $a_n \pm b_n \to a + b$.*
(b) *If $a = b = \pm\infty$, then $a_n + b_n \to a + b$*

2. *If a and b are nonzero and one of them is infinite, then $a_n b_n \to ab$.*

3. *If c is a nonzero real number, then $ca_n \to ca$.*

4. $|a_n| \to |a|$.

5. *Quotients of sequences with infinite limits satisfy the following.*

(a) *If b is infinite and a is finite, then $\frac{a_n}{b_n} \to 0$.*
(b) *If a is infinite and b is finite and nonzero, then $\frac{a_n}{b_n} \to \frac{a}{b}$.*

Idea of proof. *Parts (1) and (2) will be proven by Definition 3.106. Part (3) is a special case of Part (2) in the event that we choose $b_n = c$ to be a constant (nonzero) sequence. Part (4) is left as Exercise 6. The first part of (5) can be proven by using Part (2) of Theorem 3.31 with Proposition 3.108, while the second part of (5) follows from Part (2).*

A general note on the proof: it is a little tedious to repeat essentially the same argument to cover all of the various cases. For example, in Part (2), there are two cases: either a or b is finite, or both a and b are infinite. Then there are subcases of each case, for example: if b is finite but a is infinite, we need to consider all permutations of $b > 0$ and $b < 0$ with $a = \infty$ and $a = -\infty$. We alleviate some of this tedium by using a few tricks and Lemma 3.110.

Proof of Theorem 3.112. (1) We begin with Part (a). Suppose $b \in \mathbb{R}$, and let $M \in \mathbb{R}$ be given. Since (b_n) converges, it is bounded (Theorem 3.28). Suppose that $-P \leq b_n \leq P$ for every n; we may assume $P \geq 0$.

If $a_n \to \infty$ (so that $a + b = \infty$), there exists an N so that $a_n \geq M + P$. Therefore, for $n \geq N$, we have

$$\begin{aligned} a_n + b_n &\geq (M+P) + b_n && (\text{ minimize } a_n) \\ &\geq (M+P) - P && (\text{ minimize } b_n) \\ &= M. \end{aligned}$$

If it is the case that $a = -\infty$ (so that $a + b = -\infty$), then there exists an N so that $a_n \leq M - P$, so that for every $n \geq N$ we have

$$\begin{aligned} a_n + b_n &\leq (M-P) + b_n && (\text{ maximize } a_n) \\ &\leq (M-P) + P && (\text{ maximize } b_n) \\ &= M. \end{aligned}$$

To prove (b), we first assume $a = b = \infty$ (so that $a + b = \infty$). Let $M \in \mathbb{R}$ be given. Since $a_n \to \infty$, there exists an N_1 so that $a_n \geq \frac{M}{2}$ for every $n \geq N_1$. Since $b_n \to \infty$, there exists an N_2 so that $b_n \geq \frac{M}{2}$ for every $n \geq N_2$. Set $N = \max\{N_1, N_2\}$. Then for $n \geq N$, we have

$$a_n + b_n \geq \frac{M}{2} + \frac{M}{2} = M.$$

If it is the case that $a = b = -\infty$, (so that $a + b = -\infty$), then we proceed similarly. Let $M \in \mathbb{R}$ be given. Since $a_n \to -\infty$, there exists an N_1 so that $a_n \leq \frac{M}{2}$ for every $n \geq N_1$. Since $b_n \to -\infty$, there exists an N_2 so that $b_n \leq \frac{M}{2}$ for every $n \geq N_2$. Set $N = \max\{N_1, N_2\}$. Then for $n \geq N$, we have

$$a_n + b_n \leq \frac{M}{2} + \frac{M}{2} = M.$$

(2) Suppose first that b is nonzero and $b \in \mathbb{R}$, and that a is infinite. If $b > 0$, then it is eventually bounded above 0. So there exists an $\epsilon > 0$ and N_1 so that $b_n \geq \epsilon$ for all $n \geq N_1$. Let $M \in \mathbb{R}$ be given.

If $a = \infty$ (so that $ab = \infty$), then since $a_n \to \infty$, there exists an N_2 with $a_n \geq \frac{M}{\epsilon}$ for all $n \geq N_2$. If we choose $N = \max\{N_1, N_2\}$, then for every $n \geq N$, we have

$$a_n b_n \geq \frac{M}{\epsilon}\epsilon = M.$$

If $a = -\infty$ (so that $ab = -\infty$), then since $a_n \to -\infty$, there exists an N_2 with $a_n \leq \frac{M}{\epsilon}$ for all $n \geq N_2$. If we choose $N = \max\{N_1, N_2\}$, then for every $n \geq N$, we have

$$a_n b_n \leq \frac{M}{\epsilon}\epsilon = M.$$

If $b < 0$, then $-b > 0$, and so if a is infinite, by Lemma 3.110, $-a_n \to -a$. Our work in the previous paragraphs then show that

$$a_n b_n = (-a_n)(-b_n) \to (-a)(-b) = ab.$$

Now suppose that a and b are infinite. Suppose first that $a = b = \infty$ (so that $ab = \infty$), and $M \in \mathbb{R}$ is given. Since $a_n \to \infty$, there exists an N_1 so that $a_n \geq \sqrt{|M|}$ for all $n \geq N_1$. Since $b_n \to \infty$, there exists an N_2 so that $b_n \geq \sqrt{|M|}$ for all $n \geq N_2$. If we choose $N = \max\{N_1, N_2\}$, then for every $n \geq N$, we have

$$a_n b_n \geq \sqrt{|M|}\sqrt{|M|} = |M| \geq M.$$

Suppose now that $a = -\infty$, and $b = \infty$ (so that $ab = -\infty$). Then by Lemma 3.110, $-a_n \to \infty$, and so

$$-a_n b_n = (-a_n)(b_n) \to -ab = \infty.$$

Our earlier work then has $(-1)(-a_n b_n) = a_n b_n \to (-1)\infty = -\infty$ as desired.

If $a = b = -\infty$ (so that $ab = +\infty$), then by Lemma 3.110, we have $-a_n \to \infty$, and $-b_n \to \infty$, so by our previous work we have

$$a_n b_n = (-a_n)(-b_n) \to (\infty)(\infty) = \infty.$$

Part (3) is a special case of Part (2): set $b_n = c$, a constant sequence.

The proof of (4) is left as Exercise 6, whereas if $a \in \mathbb{R}$, this fact was already proven as Part (4) of Theorem 3.31.

To prove the first part of Part (5), we may use Part (4) above, Proposition 3.108, Exercise 10 of Section 3.2, and Part (2) of Theorem 3.31. If $a_n \to a \in \mathbb{R}$, and $b_n \to \pm\infty$, then Part (4) above shows $|b_n| \to \infty$.

$$\begin{array}{rcll} \left|\frac{a_n}{b_n}\right| = |a_n|\left(\frac{1}{|b_n|}\right) & \to & |a| \cdot 0 = 0 & \text{(Thm.3.31(2), and Prop. 3.108)} \\ & \Rightarrow & \frac{a_n}{b_n} \to 0. & \text{(Ex. 10 of Section 3.2)} \end{array}$$

Finally, we prove the second part of Part (5) by using Part (2) above. If a is infinite, and $b \in \mathbb{R}$ is nonzero, then we have

$$\frac{a_n}{b_n} = a_n\left(\frac{1}{b_n}\right) \to a\left(\frac{1}{b}\right) = \frac{a}{b}.$$

□

Theorem 3.112 provides an initial framework for evaluating infinite limits. Just like its analogue (Theorem 3.31), the conclusions of Theorem 3.112 not only demonstrate divergence to $\pm\infty$, but also establish the limit of the sequence considered.

Practice 3.113. Compute the limits of the following sequences, if they exist.

1. $a_n = n^2$.
2. $a_n = n^2 - n$.
3. $a_n = \frac{3n^3+n^2+1}{2n-1}$.
4. $a_n = (-1)^n n$.

Methodology. *Each of these parts will need to use Theorem 3.112 in some way, and we will be explicit about its usage (even though, as before, once one has a facility with the results of Theorem 3.112, one rarely needs to specifically mention it. See the discussion just before Subsection 3.3.2 on page 130). These examples have been chosen partly because they also illustrate how to avoid potential pitfalls in evaluating infinite limits. For example, in Part (2) it would be easy to think that the limit is $\infty - \infty$, but this quantity is not defined, and so other methods will need to be used. Similarly in Part (3), it would be tempting to think that the limit is $\frac{\infty}{\infty}$, but this, too, is not defined. The last sequence diverges, and we will show that through a contradiction proof* *(in fact, Theorem 3.112 has no application to Part (4); notice that $(-1)^n$ is bounded and that $n \to \infty$: compare Exercise 15 to this situation.)*

Solution. (1) Notice that the sequence (n) tends to ∞ (Exercise 1). Since $n^2 = n \cdot n$, according to Part (2) of Theorem 3.112, $n^2 = n \cdot n \to \infty \cdot \infty = \infty$.

(2) Notice that $n^2 - n = n(n-1)$. So since $n, n-1 \to \infty$, Part (2) of Theorem 3.112 applies again, and $n(n-1) \to \infty \cdot \infty = \infty$. *(Notice that it would be incorrect to assert that Part (1) of Theorem 3.112 applies, and that the answer is $\infty - \infty$.)*

(3) We separate this sequence along the numerator and rearrange it as follows:

$$\begin{aligned} a_n &= \tfrac{3n^3}{2n-1} + \tfrac{n^2}{2n-1} + \tfrac{1}{2n-1} \\ &= n^2\left(\tfrac{3n}{2n-1}\right) + n\left(\tfrac{n}{2n-1}\right) + \tfrac{1}{2n-1}. \end{aligned}$$

The two expressions in parentheses above have finite and nonzero limits, whereas both n^2 and n have a limit of ∞. So according to Part (2) each of the products

$$n^2\left(\frac{3n}{2n-1}\right) \to \infty, \quad \text{and} \quad n\left(\frac{n}{2n-1}\right) \to \infty.$$

Therefore, according to Part (1) of Theorem 3.112,

$$n^2\left(\frac{3n}{2n-1}\right) + n\left(\frac{n}{2n-1}\right) \to \infty.$$

Finally, the sequence $\frac{1}{2n-1} \to 0$, so again Part (1) of Theorem 3.112 applies, and

$$\left[n^2\left(\frac{3n}{2n-1}\right) + n\left(\frac{n}{2n-1}\right)\right] + \frac{1}{2n-1} \to \infty + 0 = \infty.$$

(4) We claim that $a_n = (-1)^n n$ diverges. Since (a_n) is unbounded, it cannot converge to any real number (Theorem 3.28). So suppose that $a_n \to a$, where a is infinite. If $a = \infty$, then consider the neighborhood $(0, \infty)$. If $a_n \to a$, then (a_n) is eventually in $(0, \infty)$, but if n is any odd number, $a_n \notin (0, \infty)$. Similarly, if $a = -\infty$, then consider the neighborhood $(-\infty, 0)$. Now, if n is even, then $a_n \notin (-\infty, 0)$. So in either case, there exists a neighborhood U of a so that (a_n) is not eventually in U. Since (a_n) does not converge (to a real number), and it does not diverge to ∞ or $-\infty$, it must diverge. ■

Next, we extend the Squeeze Theorem to apply to infinite limits. Notice that we do not need to "squeeze" a sequence with an infinite limit from two sides as we did for sequences with finite limits.

Theorem 3.114 (Analogue of the Squeeze Theorem)**.** *Suppose that $a_n \leq b_n$.*

1. *If $a_n \to \infty$, then $b_n \to \infty$.*
2. *If $b_n \to -\infty$, then $a_n \to -\infty$.*

Idea of proof. *The proof is really quite straightforward for both parts. For the first part, if $a_n \geq M$ eventually, since $b_n \geq a_n$, it follows that $b_n \geq M$ eventually as well. We can prove the second part from the first one by multiplying through by -1 and using Lemma 3.110.*

Proof of Theorem 3.114. We begin by proving Part (1). Suppose $a_n \to \infty$. We wish to show $b_n \to \infty$. Let $M \in \mathbb{R}$ be given. Since $a_n \to \infty$, there exists an N so that $a_n \geq M$ for every $n \geq N$. But since $b_n \geq a_n$, it follows that $b_n \geq a_n \geq M$ for every $n \geq N$.

Now suppose $b_n \to -\infty$ and $a_n \leq b_n$ to prove Part (2). According to Lemma 3.110, $-b_n \to +\infty$. Multiplying through by -1, we have $-b_n \leq -a_n$. By the first part of this theorem, $-a_n \to \infty$. Then by Lemma 3.110, $a_n \to -\infty$. □

Practice 3.115. Show that if $\alpha \geq 1$, then $n^\alpha \to \infty$.

Methodology. *If $\alpha \geq 1$, then $n^\alpha \geq n$, and since $n \to \infty$, the Squeeze Theorem will apply.*

Solution. Let $\alpha \geq 1$. For $n \geq 0$, we then have $n^\alpha \geq n$. Since $n \to \infty$, by the Squeeze Theorem, $n^\alpha \to \infty$ as well. ■

The theory of subsequences can be extended as well. The following result is an extension of Theorem 3.66. See also Exercise 14 for an analogue of Theorem 3.67.

Theorem 3.116 (Analogue of Theorem 3.66)**.** *Suppose $a_n \to a$, where a is infinite. Then every subsequence $a_{n_k} \to a$ as well.*

Idea of proof. *This proof is another straightforward use of Definition 3.106, and it follows in spirit in a way similar to the proof of Theorem 3.66.*

Proof of Theorem 3.116. Exercise 12. □

The Monotone Convergence Theorem paved the way for the Bolzano-Weierstrass Theorem for sequences. It does the same when one considers their analogues for unbounded sequences.

Theorem 3.117 (Analogue of the Monotone Convergence Theorem)**.** *If (a_n) is monotone increasing and not bounded above, then $a_n \to \infty$. If (a_n) is monotone decreasing and not bounded below, then $a_n \to -\infty$.*

Idea of proof. *The proofs of both statements have a similar line of reasoning. If (a_n) is increasing and not bounded above, then there is some N for which $a_N \geq M$ for any given M. Since (a_n) is increasing, all of the rest of $a_n \geq M$ for $n \geq N$ as well.*

Proof of Theorem 3.117. Suppose (a_n) is monotone increasing, and *not* bounded above. Let $M \in \mathbb{R}$ be given. Since (a_n) is not bounded above, there exists an N so that $a_N \geq M$. If $n \geq N$, since (a_n) is increasing, we have $a_n \geq a_N \geq M$, so $a_n \to \infty$.

Suppose now that (a_n) is decreasing and *not* bounded below. Let $M \in \mathbb{R}$ be given. Since (a_n) is not bounded below, there exists an N so that $a_N \leq M$. If $n \geq N$, since (a_n) is decreasing, we have $a_n \leq a_N \leq M$, so $a_n \to -\infty$. □

We can use this to prove our final result of the section: a more general of the Bolzano-Weierstrass Theorem for sequences.

Theorem 3.118 (Analogue of Theorem 3.84). *Every unbounded sequence in $\mathbb{R}$ has a subsequence with an infinite limit.*

Idea of Proof. *If (a_n) is our unbounded sequence, then it is not bounded above or not bounded below. If it is not bounded above, then we create a subsequence (a_{n_k}) by demanding that $a_{n_k} \in [k, \infty)$. The hypothesis that (a_n) is not bounded above will always allow us to find such an a_{n_k} for every k. A similar argument will be used if (a_n) is unbounded below.*

Proof of Theorem 3.118. Suppose first that (a_n) is not bounded above. Then, for every $k \in \mathbb{N}$, there exists an $n_k \in \mathbb{N}$ with $a_{n_k} \geq k$. We claim that $a_{n_k} \to \infty$ by the Squeeze Theorem (3.114): the sequence $c_k = k \to \infty$, and $c_k \leq a_{n_k}$ for every k. So, $a_{n_k} \to \infty$.

If (a_n) is not bounded below, we proceed similarly. For every $k \in \mathbb{N}$, there exists an $n_k \in \mathbb{N}$ with $a_{n_k} \leq -k$. We claim that $a_{n_k} \to -\infty$ by again using the Squeeze Theorem (Theorem 3.114): the sequence $c_k = -k \to -\infty$, and $a_{n_k} \geq c_k$ for every k. So, $a_{n_k} \to -\infty$. □

3.8.3 Review of important concepts from Section 3.8

1. It is possible (and useful) to rigorously define what it means for a sequence to tend toward ∞ or $-\infty$ (Definition 3.106).
2. Theorem 3.109 is a characterization of what it means to diverge to ∞ or $-\infty$ using the language of neighborhoods.
3. Most of the convergence results developed thus far in the chapter have analogues that consider infinite limits.
4. Infinite limits are preserved by many common transformations. See Practice 3.115 and Exercises 8, 9, and 10.

Exercises for Section 3.8

1. Compute the following limits.

   (a) $a_n = n$.

   (b) $a_n = \frac{n^2+1}{2n-1}$.

   (c) $a_n = \sqrt{n^5 - n - 15}, n \geq 2$.

   (d) $a_n = \sqrt[3]{n^2 + n + 1}$.

   (e) $a_n = \frac{n!}{2^n}$.

   (f) $a_n = \frac{n^n}{n!}$.

2. Show that the sequence $a_n = \frac{1}{1} + \frac{1}{2} + \cdots + \frac{1}{n}$ diverges to ∞. *[Hint: See Exercise 2 of Section 3.5.]*

3. (a) Show that if $r > 1$, then $r^n \to \infty$. *[Hint: Consider Exercise 14 in Section 3.2.]*

   (b) Show also that if $r \leq -1$, then (r^n) diverges (but not to ∞ or $-\infty$).

4. (This is an analogue of Exercise 22 of Section 3.3.) Let $a_n = \frac{p(n)}{q(n)}$, where p and q are polynomials. Show that if the degree of p is strictly greater than the degree of q, then (a_n) diverges to either ∞ or $-\infty$, depending on the signs of the leading coefficients of p and q.

5. Prove Lemma 3.110 without using Theorem 3.112. That is, show by the definition that if $a_n \to \infty$, then $-a_n \to -\infty$, and that if $a_n \to -\infty$, then $-a_n \to \infty$.

6. Show that if $a_n \to \pm\infty$, then $|a_n| \to \infty$. Is it true that if $|a_n| \to \infty$, then $a_n \to \infty$ or $a_n \to -\infty$?

7. Complete the proof of Theorem 3.109 by establishing it for the case $a = -\infty$.

8. (a) Show that if $x_n \to \infty$, then $e^{x_n} \to \infty$.

   (b) Show that if $x_n \to -\infty$, then $e^{x_n} \to 0$.

9. Suppose $x_n > 0$ for these problems.

   (a) Show that if $x_n \to \infty$, then $\ln(x_n) \to \infty$.

   (b) Show that if $x_n \to 0$, then $\ln(x_n) \to -\infty$.

10. Show that if $x_n \to \infty$ and $\alpha > 0$, then $x_n^\alpha \to \infty$. *[Hint: As in Exercise 15 of Section 3.3, express $x_n^\alpha = e^{\alpha \ln(x_n)}$, and use Exercises 8 and 9.]*

11. This exercise is designed to reinforce the fact that $\infty - \infty, 0 \cdot \infty$, and $\frac{\infty}{\infty}$ are not outcomes that can be considered in the context of Theorem 3.112.

    (a) Suppose $x \in \mathbb{R}^\sharp$ is given. Construct sequences (a_n) and (b_n) so that $a_n \to \infty$, $b_n \to \infty$, and $a_n - b_n \to x$.

    (b) Suppose $x \in \mathbb{R}^\sharp$ is given. Construct sequences (a_n) and (b_n) so that $a_n \to 0$, $b_n \to \infty$, and $a_n \cdot b_n \to x$.

    (c) Suppose $x \in \mathbb{R}^\sharp$ is given. Construct sequences (a_n) and (b_n) so that $a_n \to \infty$, $b_n \to \infty$, and $\frac{a_n}{b_n} \to x$.

12. Prove Theorem 3.116. That is, prove that if $a_n \to a$, and a is infinite, then every subsequence $a_{n_k} \to a$ as well.

13. Suppose E is any subset of $\mathbb{R}^\sharp$. (This is an extension of the same result when E is bounded, see Exercise 3 of Section 3.5.)

(a) Suppose $\sup(E) = \infty$. Prove that there exists a sequence $x_n \in E$ with $x_n \to \infty$.

(b) Suppose $\inf(E) = -\infty$. Prove that there exists a sequence $x_n \in E$ with $x_n \to -\infty$.

14. This problem is an analogue of Theorem 3.67.

    (a) Show that (x_n) does *not* converge to ∞ if and only if there is a subsequence of (x_n) that is bounded above. (One could rephrase this as: (x_n) does not converge to ∞ if and only if there is a subsequence of (x_n) that is bounded away from ∞.)

    (b) Show that (x_n) does *not* converge to $-\infty$ if and only if there is a subsequence of (x_n) that is bounded below. (One could rephrase this as: (x_n) does not converge to $-\infty$ if and only if there is a subsequence of (x_n) that is bounded away from $-\infty$.)

15. Let $B \in \mathbb{R}$. Show that if $a_n \geq B > 0$, (a_n) is bounded, and $b_n \to \infty$, then $a_n b_n \to \infty$. (This generalizes Part (2) of Theorem 3.112 because it does not assume that (a_n) converges or that it diverges to ∞.)

16. Show that if (a_n) is bounded, and $b_n \to \infty$, then $\frac{a_n}{b_n} \to 0$. (This generalizes Part (5) of Theorem 3.112 because it does not assume that (a_n) converges.)

17. Show there is no sensible unbounded analogue of the Bolzano-Weierstrass Theorem for sets (Theorem 3.91). That is, show that there exists *unbounded* and infinite subsets of $\mathbb{R}$ that have no limit points. Show also, however, that if A is an infinite subset of $\mathbb{R}$ that is unbounded, then there exists a subsequence $a_n \in A$ that has an infinite limit in $\mathbb{R}^\sharp$, so one could consider the notion that ∞ or $-\infty$ is a limit point of such an A. (For the record, however, we do not consider ∞ or $-\infty$ limit points of subsets of $\mathbb{R}$, since we insist that limit points be in $\mathbb{R}$, not $\mathbb{R}^\sharp$.)

18. (This problem is the analogue of Exercise 2 of Section 3.6, in that it does not assume (x_n) is bounded, and it accordingly allows for infinite subsequential limits. It is also expanded upon in Section 3.9.) Let (x_n) be a sequence of real numbers. Let E be the set of *subsequential limits* of (x_n), defined by

$$E = \{x \in \mathbb{R}^\sharp | \text{ there exists a subsequence } x_{n_k} \to x\}.$$

    (a) Show that E is nonempty.

    (b) Show that $x_n \to x$ if and only if $E = \{x\}$ contains only one element.

3.9 The lim sup and lim inf

So far in this chapter, we have studied the convergence (or divergence to $\pm\infty$) of sequences. Generally speaking, if $\lim_{n\to\infty} x_n$ exists, then this quantity is a numerical description of what the sequence tends to as n gets large. Unfortunately, however, we have seen that very often $\lim_{n\to\infty} x_n$ does not exist, and up until this point, we have been without any numerical descriptors of (x_n) in the event that (x_n) diverges. In this section, we solve this problem by introducing and developing the *limit superior of* (x_n) ($\limsup(x_n)$) and the *limit inferior of* (x_n) ($\liminf(x_n)$). These are quantities that *always* exist in $\mathbb{R}^\sharp$ no matter how bizarre or well-behaved (x_n) may be, and generalize the notion of limits of sequences to those sequences that might diverge.

3.9.1 The definition of lim sup and lim inf

Consider for a moment the sequence

$$a_n = \left(\frac{1+(-1)^n}{2}\right)\left(\frac{n+1}{n}\right). \tag{3.35}$$

n	a_n
1	0
2	$\frac{3}{2}$
3	0
4	$\frac{5}{4}$
5	0
6	$\frac{7}{6}$
7	0
8	$\frac{9}{8}$

FIGURE 3.9.1
The first few members of $a_n = \left(\frac{1+(-1)^n}{2}\right)\left(\frac{n+1}{n}\right)$.

We list out the first few members of this sequence in Figure 3.9.1. It is immediately clear that this sequence alternates between the constant sequence (0) and a sequence that has limit 1. These two subsequences are

$$a_{2n+1} = 0 \to 0, \quad \text{and} \quad a_{2n} = \frac{2n+1}{2n} \to 1.$$

According to Theorem 3.66, if (a_n) were to converge, every subsequence would have to converge to the same value. So since these two subsequences above converge to different values, it is the unfortunate case that (a_n) diverges.

All the same, (a_n) is not at all a strange sequence: it is bounded, and it has two monotonic subsequences. The only reason it diverges is because it has (at least) two subsequences that have different limits. **The reason why we have not studied divergent sequences in detail up until this point is because they are simply too bizarre to say anything certain about them.** Since there are divergent sequences (such as (a_n) above) that are not really that bizarre, we want to find a number (or numbers) that describes the sequence, even if that number is not (and could not be) the limit of the sequence. This section describes two such quantities, called the $\limsup$ and $\liminf$ of a sequence.

To this end, let (x_n) be an arbitrary sequence, and consider the following set:

$$E = \{x \in \mathbb{R}^\sharp |\text{ there is a subsequence } x_{n_k} \to x\}.$$

The set E is sometimes called the *set of subsequential limits* of (x_n) (see Exercise 2 of Section 3.6, and Exercise 18 of Section 3.8). It is *always* the case that E is nonempty, since every sequence has at least one subsequence with a limit in $\mathbb{R}^\sharp$ (by the Bolzano-Weierstrass Theorems for sequences: Theorems 3.84 and 3.118). As a result, $\sup(E)$ and $\inf(E)$ exist, even if they could be infinite, and are meaningful descriptors of the set E.

Back to our example defined in Equation (3.35), the sequence (a_n) has $1, 0 \in E$. In fact, E is bounded, $\sup(E) = 1$, and $\inf(E) = 0$. So even though (a_n) diverges, we can still assign two numbers to it ($\sup(E)$ and $\inf(E)$) that relate to the convergence of any of its subsequences. This motivates the following definition.

Definition 3.119. Let (x_n) be any sequence, and let E be the set of its subsequential limits,

$$E = \{x \in \mathbb{R}^\sharp |\text{ there is a subsequence } x_{n_k} \to x\}.$$

Define the *limit superior* of (x_n), denoted $\limsup(x_n)$, as $\sup(E)$. Similarly, define the *limit inferior* of (x_n), denoted $\liminf(x_n)$, as $\inf(E)$.

Remark 3.120. Other authors sometimes denote $\limsup(x_n)$ as $\overline{\lim}(x_n)$, and $\liminf(x_n)$ as $\underline{\lim}(x_n)$. ■

Remark 3.121. Suppose (x_n) is any sequence. It is important to remark that **the set E of its subsequential limits is always nonempty, so that $\inf(E) \leq \sup(E)$, and so $\liminf(x_n) \leq \limsup(x_n)$.**

Let us find $\limsup(a_n)$ and $\liminf(a_n)$, where (a_n) is our example defined in Equation (3.35) above.

Practice 3.122. Let (a_n) be the sequence defined in Equation 3.35. Compute $\limsup(a_n)$ and $\liminf(a_n)$.

Methodology. *The definition of* $\limsup$ *and* $\liminf$ *asks us to find the supremum and infimum of the set E of its subsequential limits. One could accomplish this by computing exactly what E is, and then computing the* sup *and* inf *of E. This turns out to be too much work: we can show directly that* $\sup(E) = 1$, *and* $\inf(E) = 0$ *by using Exercise 18 of Section 3.2. That is, we will show that if $x > 1$, then (a_n) is eventually bounded away from x, and so every subsequence of (a_n) must be eventually bounded from x, and so no subsequence could converge to x. We can demonstrate that $1 \in E$ by producing a subsequence that converges to 1, so 1 is clearly the supremum of E: $1 \in E$, and for every $x > 1$, $x \notin E$. A similar argument will demonstrate* $0 = \inf(E)$.

Solution. Let E be the set of subsequential limits of (a_n). We begin by computing $\limsup(a_n)$. Suppose $x > 1$. Notice that the quantity

$$a_{2n} = \frac{2n+1}{2n} \geq a_k \quad \text{for all} \quad k \geq 2n.$$

Indeed, if k is odd, then $a_k = 0$, and if k is even with $k \geq 2n$, then $a_k \leq a_{2n}$ since the sequence $\frac{2n+1}{2n} = 1 + \frac{1}{2n}$ is decreasing. In other words, a_{2n} is the largest member of the set $\{a_k | k \geq 2n\}$. Since $\frac{2n+1}{2n} \to 1$ and $x > 1$, the sequence $\frac{2n+1}{2n}$ is eventually bounded away from x by Exercise 17. So, every subsequence of (a_n) is also eventually bounded away from x, which means there is no subsequence of (a_n) that has x as its limit. So, if $x > 1$, then $x \notin E$. On the other hand, the subsequence $a_{2n} \to 1$, and $1 \in E$. Therefore, $\sup(E) = \limsup(a_n) = 1$.

We now compute $\liminf(a_n)$. We proceed similarly. Notice that $a_n \geq 0$ for all n. If $a_{n_k} \to a$ is a convergent subsequence of (a_n), then $a \geq 0$ as well (Theorem 3.48), and so if $a \in E$ is any subsequential limit, then $a \geq 0$. Notice also that the subsequence $a_{2n+1} \to 0$, and so $0 \in E$. Therefore, $\inf(E) = \liminf(a_n) = 0$. ■

At first glance, the solution to Practice 3.122 seems needlessly complicated: why can we not simply point out that $a_{2n} \to 1$ and $a_{2n+1} \to 0$ and be done with it? There are two problems with that reasoning, and although one of these problems can be resolved, the other cannot. The first problem is that there is nothing yet that says that there is a subsequence of (x_n) that has $\limsup(x_n)$ or $\liminf(x_n)$ as its limit. After all, if E is an arbitrary nonempty subset of $\mathbb{R}^\sharp$, we cannot expect that E contains its supremum and infimum. We shall show in Theorem 3.125 that if E is the set of subsequential limits of a sequence, then E contains its supremum and infimum, and so $\limsup(x_n)$ and $\liminf(x_n)$ of a sequence can be identified as the limit of some subsequence of (x_n). The second problem with the idea that one can find the lim sup and lim inf by simply considering two obvious subsequential limits is that this is a consideration of only *two* subsequential limits, and not *all* of the subsequential limits. **Generally speaking, one cannot simply identify some subsequential limit of (x_n) and claim it to be $\limsup(x_n)$ or $\liminf(x_n)$ without justification.** One must justify why such a subsequential limit is the "largest" or "smallest" as we did in the solution of Practice 3.122–unlike the first issue, there is no way around this one (there is an equivalent characterization of lim sup and lim inf in Theorem 3.129 that may be of some help in computing these quantities, however).

3.9.2 Results about the $\limsup$ and $\liminf$

There are three results presented in this subsection, Theorems 3.123, 3.125, and 3.127. **One of the main concepts of Section 3.9 is that the $\limsup(x_n)$ and $\liminf(x_n)$ always exist (in $\mathbb{R}^\sharp$), and can numerically describe a sequence even if the sequence diverges.** Theorem 3.123 connects the $\limsup$ and $\liminf$ of a sequence to the limit of a sequence in the event that the limit exists: all three are equal. Theorem 3.125 says that the $\limsup(x_n)$ and $\liminf(x_n)$ are, in fact, subsequential limits of (x_n), providing a method of computing these quantities in addition to providing a new tool to prove theorems about $\limsup(x_n)$ and $\liminf(x_n)$. Theorem 3.127 employs the result in Theorem 3.125 to establish basic properties about the $\limsup$ and $\liminf$.

Theorem 3.123. *Let (x_n) be a sequence, and let E be the set of subsequential limits of (x_n). The following statements are equivalent.*

1. $x_n \to x$.
2. $\limsup(x_n) = \liminf(x_n) = x$.
3. $E = \{x\}$.

Idea of proof. *Parts (2) and (3) are easily equivalent. If a nonempty set has its supremum equal to its infimum, then it must contain only one element. Conversely, if the set of subsequential limits contains only one element, then its supremum and infimum are equal.*

We will then prove (1) and (3) are equivalent. If $x_n \to x$, then by Theorems 3.66 and 3.116, all subsequential limits are the same (i.e., (3) holds). We prove the converse by contradiction: this part is a little more detailed. That is, if $E = \{x\}$ but $\lim_{n\to\infty} x_n \neq x$, then there is a subsequence (y_k) that is bounded away from x by Theorem 3.67 (for finite x) and Exercise 14 of Section 3.8 (for infinite x). This subsequence (y_k) has a subsequence (z_m) that has a limit $z \in \mathbb{R}^\sharp$ by the Bolzano-Weierstrass Theorems for sequences (Theorems 3.84 and 3.118). But, since (y_k) is bounded away from x and (z_m) is a subsequence of (y_k), (z_m) is also bounded away from x, which means again by Theorem 3.67 that $z \neq x$. This supplies the contradiction: (z_m) is a (subsequence of a) subsequence of (x_n) with limit $z \neq x$, and this contradicts the assumption that $E = \{x\}$.

Proof of Theorem 3.123. We first prove (2) and (3) are equivalent, and then show (1) and (3) are equivalent.

((2) $\Leftrightarrow$ (3)) If E is any subset of $\mathbb{R}^\sharp$, then $\sup(E) = \inf(E)$ if and only if E contains only one element.

((1) $\Rightarrow$ (3)) Suppose $x_n \to x$. Then according to Theorem 3.66 (when x is finite) and Theorem 3.116 (when x is infinite), every subsequence of (x_n) must also have x as its limit.

((3) $\Rightarrow$ (1)) Now suppose that $E = \{x\}$, but that $\lim_{n\to\infty} x_n \neq x$. According to Theorem 3.67 (for finite x) and Exercise 14 of Section 3.8 (for infinite x), there exists a subsequence (y_k) of (x_n) that is bounded away from x. By the Bolzano-Weierstrass Theorems for sequences (Theorem 3.84 if (y_k) is bounded, and Theorem 3.118 if (y_k) is not bounded), there is a subsequence (z_m) of (y_k) that has a limit $z \in \mathbb{R}^\sharp$. Since (y_k) is bounded away from x, (z_m) is also bounded away from x, and so $z \neq x$ (by Theorem 3.67 (for finite x) and Exercise 14 of Section 3.8 (for infinite x)). In other words, E contains more than just one element. □

Much like the limit considers a sequence and outputs an element of $\mathbb{R}^\sharp$ (if it exists), so does the lim sup and lim inf. But because the lim sup and lim inf *always* exist in $\mathbb{R}^\sharp$, one would expect that they do not enjoy the same convenient properties that limits do, and the next example illustrates this.

Example 3.124. Let $a_n = (-1)^n$, and let $b_n = (-1)^{n+1}$. Figure 3.9.2 gives the first few values of a_n, b_n, and $a_n + b_n$. Since both $a_n, b_n \in \mathbb{Z}$, so must

n	1	2	3	4	5	6	$\cdots$
a_n	-1	1	-1	1	-1	1	$\cdots$
b_n	1	-1	1	-1	1	-1	$\cdots$
$a_n + b_n$	0	0	0	0	0	0	$\cdots$

FIGURE 3.9.2
The first few values of a_n, b_n, and $a_n + b_n$.

any of their sequences, and so any subsequence of either that has a limit must be eventually constant (Exercise 12 of Section 3.2); these constants must be either 1 or -1 (in other words, the set of subsequential limits E of either (a_n) or (b_n) must satisfy $E \subseteq \{-1, 1\}$) . So since 1 and -1 are both subsequential limits of (a_n) and (b_n), it must be the case that $\{-1, 1\} \subseteq E$. So for both the sequences (a_n) and (b_n), the set E is the same, $E = \{-1, 1\}$. So,

$$\begin{aligned} \limsup(a_n) &= \sup(E) = 1, & \liminf(a_n) &= \inf(E) = -1, \\ \limsup(b_n) &= \sup(E) = 1, & \liminf(b_n) &= \inf(E) = -1. \end{aligned}$$

But now consider the sequence $a_n + b_n = 0$: this is a constant sequence, and hence convergent, $a_n + b_n \to 0$. According to Theorem 3.123, this means that

$$\limsup(a_n + b_n) = \liminf(a_n + b_n) = 0.$$

So we have the following (compare this with part (1) of Theorems 3.31 and 3.112):

$$\begin{aligned} &\limsup(a_n + b_n) = 0 < 2 = \limsup(a_n) + \limsup(b_n), \text{ and} \\ &\liminf(a_n + b_n) = 0 > -2 = \liminf(a_n) + \liminf(b_n). \end{aligned}$$

■

Example 3.124 illustrates that the lim sup and lim inf do not always behave well under sums of sequences. By contrast, if it is the case that $x_n \to x$ and $y_n \to y$, then Theorem 3.31 or Theorem 3.112 (if it applies) says that $x_n + y_n \to x + y$, and Theorem 3.123 says that

$$\begin{aligned}
\limsup(x_n + y_n) &= x + y \\
&= \limsup(x_n) + \limsup(y_n), \text{ and} \\
\liminf(x_n + y_n) &= x + y \\
&= \liminf(x_n) + \liminf(y_n).
\end{aligned}$$

So, in the event that (x_n) and (y_n) have a limit, the lim sup and lim inf *do* behave as we would hope. This makes sense: the lim sup and lim inf generalize the notion of convergence (and divergence to $\pm\infty$), and as such, their properties should be more general as well. Theorem 3.127 illustrates some of these the more general properties of the lim sup and lim inf, but first we pause to give a foundational result that will help us (among other places) in the proof of Theorem 3.127.

Theorem 3.125. *There exists a subsequence of* (x_n) *that has* $\limsup(x_n)$ *as its limit. There also exists a subsequence of* (x_n) *that has* $\liminf(x_n)$ *as its limit. In other words, limsup*(x_n) *and* $\liminf(x_n)$ *are themselves subsequential limits of* (x_n).

Idea of proof. *This proof is a classic one in Real Analysis. We shall prove here that* $\limsup(x_n)$ *is a subsequential limit of* (x_n)*, and leave the proof that* $\liminf(x_n)$ *is a subsequential limit of* (x_n) *to Exercise 10.*

There is one trivial case we can eliminate immediately. Suppose E *is the set of subsequential limits of* (x_n)*. If* $\limsup(x_n) = -\infty$*, then* $\liminf(x_n) = -\infty$ *as well, and according to Theorem 3.123,* $x_n \to \limsup(x_n)$.

That leaves only the possibility that $\limsup(x_n) = \infty$*, or* $\limsup(x_n)$ *is finite. This observation will be crucial:* $\sup(E)$ *is the limit of a sequence in* E *(Exercise 3 of Section 3.5 and Exercise 13 of Section 3.8). We will have to pay special attention to the fact that our* $E \subseteq \mathbb{R}^\sharp$ *so that our sequence that tends to* $\sup(E)$ *consists only of real numbers, but this will not present a problem. We will use this sequence* $z_n \to z = \sup(E)$ *in the following way: each* $z_n \in E$*, and so for each* n *there is a subsequence* $x_{n,k} \overset{k}{\to} z_n$*. For each* n*, we will pick one member out of the subsequence* $(x_{n,k})$*, subject to the condition that the resulting ordered list of members* $(x_{s,k_s})_s$ *is a subsequence of* (x_n)*. We will show that* $x_{s,k_s} \overset{s}{\to} z$ *first in the case that* $z = \infty$ *(using the Squeeze Theorem (Theorem 3.114)), and then in the case* $z \in \mathbb{R}$ *(using the definition of convergence (Definition 3.13)).*

Proof of Theorem 3.125. Let E be the set of subsequential limits of (x_n), and let $z = \sup(E) = \limsup(x_n)$. We want to show that there exists a subsequence of (x_n) that has z as its limit. We will first consider the possibility that $z = -\infty$, and then the possibility that $z = \infty$ or z is finite.

If $z = -\infty$, then

$$\liminf(x_n) \leq \limsup(x_n) = -\infty,$$

and so $\liminf(x_n) = \limsup(x_n) = -\infty$. In that event, Theorem 3.123 asserts that $x_n \to -\infty$, and so the result holds if $z = -\infty$.

Now suppose that $z = \infty$ or that z is finite. We claim that there is only relevant conclusion to consider: there exists a sequence $z_n \to z$ with $z_n \in E \cap \mathbb{R}$ (namely, we need not consider the possibility that such a sequence (z_n) should have to have $z_n = \infty$ for any n). Indeed, if $z = \infty$, then Exercise 3 of Section 3.8 asserts that there exists a sequence $z_n \to z$, with $z_n \in E$. If any of the $z_n = \infty$, then $\infty \in E$, and it would already be the case that z is a subsequential limit. If z is finite, then Exercise 3 of Section 3.5 asserts that there is a sequence $z_n \in E$ with $z_n \to z$; note that if z is finite, then E is bounded above, and so the z_n may be chosen to be finite (i.e., not equal to $-\infty$). So in either case, there exists a sequence (z_n) of real numbers, with $z_n \to z$.

Suppose first that $z = \infty$. We will construct a subsequence of (x_n) that has z as its limit. Suppose $z_n \to z$, with $z_n \in E \cap \mathbb{R}$. In particular, z_1 is a real subsequential limit of (x_n). So, there exists a subsequence $(x_{1,k})$ of (x_n) converging to z_1. As a result, there exists a k_1 with $x_{1,k_1} \geq z_1 - 1$. The element $x_{1,k_1} = x_{n_1}$ for some subscript n_1. Similarly, z_2 is a real subsequential limit of (x_n), and so there exists a subsequence $(x_{2,k})$ that converges to z_2, and so there are infinitely many k with $x_{2,k} \geq z_2 - 1$. Choose $n_2 > n_1$ with $x_{n_2} = x_{2,k}$ for any k satisfying $x_{2,k} \geq z_2 - 1$. Continuing in this manner, we construct a subsequence (x_{n_s}) that satisfies $x_{n_s} \geq z_s - 1$. By the Squeeze Theorem 3.114, $x_{n_s} \to \infty$ since $z_n \to \infty$.

Now suppose that z is finite. We will proceed similarly in finding a subsequence of (x_n) that converges to z. Let (z_n) be a sequence of real numbers in E that converges to z. The number z_1 is a real subsequential limit of (x_n), and so there exists a subsequence $(x_{1,k})$ of (x_n) that converges to z_1. As a result, there exists a k_1 with $|x_{1,k_1} - z_1| < 1$. The element $x_{1,k_1} = x_{n_1}$ for some subscript n_1. Similarly, z_2 is a real subsequential limit of (x_n), and so there exists a subsequence $(x_{2,k})$ that converges to z_2, and so there are infinitely many k satisfying $|x_{2,k} - z_2| < \frac{1}{2}$. Choose $n_2 > n_1$ with $x_{n_2} = x_{2,k}$ for any k satisfying $|x_{2,k} - z_2| < \frac{1}{2}$. Continuing in this manner, we construct a subsequence (x_{n_s}) that satisfies $|x_{n_s} - z_s| < \frac{1}{s}$ for every natural number s. We claim that $x_{n_s} \to z$, and we prove this by the definition of convergence.

Let $\epsilon > 0$ be given. There exists an N_1 with $\frac{1}{N_1} < \frac{\epsilon}{2}$. Since $z_n \to z$, there exists an N_2 with $|z_s - z| < \frac{\epsilon}{2}$ for every $s \geq N_2$. Choose $N = \max\{N_1, N_2\}$, and let $s \geq N$. Then

$$\begin{aligned}
|x_{n_s} - z| &\leq |x_{n_s} - z_s| + |z_s - z| \\
&< \frac{1}{s} + \frac{\epsilon}{2} \\
&\leq \frac{1}{N_1} + \frac{\epsilon}{2} \\
&< \frac{\epsilon}{2} + \frac{\epsilon}{2} \\
&= \epsilon. \qquad \square
\end{aligned}$$

Remark 3.126. The method of proof of Theorem 3.125 can be adapted to show that if E is the set of subsequential limits of (x_n), then every limit point of E is a member of E (Exercise 14).

Using Theorem 3.125, we can establish some of the basic properties of the lim sup and the lim inf. The first property below is sometimes rephrased as "The lim sup is *subadditive.*" The second property below is sometimes rephrased as "The lim inf is *superadditive.*"

Theorem 3.127. *Let (x_n) and (y_n) be sequences. The following statements hold whenever they are defined.*

1. $\limsup(x_n + y_n) \le \limsup(x_n) + \limsup(y_n)$.
2. $\liminf(x_n) + \liminf(y_n) \le \liminf(x_n + y_n)$.
3. *If* $x_n \le y_n$, *then* $\limsup(x_n) \le \limsup(y_n)$, *and* $\liminf(x_n) \le \liminf(y_n)$.
4. $\limsup(-x_n) = -\liminf(x_n)$.

Idea of proof. *First off, the phrase "whenever they are defined" means that we need not consider a situation where, for example,* $\limsup(x_n) = \infty$ *and* $\limsup(y_n) = -\infty$, *since the conclusion would have the undefined* $\infty - \infty$ *in it. Similarly for the other parts. Now to the proof: with exception to the last part, these results can be verified by using Theorem 3.125. For the first part, we find a subsequence* $(x_{n_k} + y_{n_k})$ *of* $(x_n + y_n)$ *that has* $\limsup(x_n + y_n)$ *as its limit. We look at the common subscript* n_k *of* $(x_n + y_n)$ *needed to achieve that, and apply that subscript to* (x_n) *to produce the not necessarily convergent subsequence* (x_{n_k}). *There is a subsequence* $(x_{n_{k_s}})$ *of* (x_{n_k}) *that has a limit* x. *Transferring the subscript of* n_{k_s} *to the sequence* (y_n) *gives us another not necessarily convergent subsequence* $(y_{n_{k_s}})$ *which has its own subsequence* $(y_{n_{k_{s_t}}})$ *with a limit* y. *Using that ridiculous subscript now for both sequences, we have* $(x_{n_{k_{s_t}}} + y_{n_{k_{s_t}}})$ *is a subsequence of* $(x_{n_k} + y_{n_k})$, *and it has limit* $x + y$, *and so we have identified, by the uniqueness of limits,* $\limsup(x_n + y_n) = x + y$. *But* x *and* y *are just subsequential limits of* (x_n) *and* (y_n), *respectively, and so* $x \le \limsup(x_n)$ *and* $y \le \limsup(y_n)$, *and so* $x+y \le \limsup(x_n)+\limsup(y_n)$.

Part (2) is left as Exercise 11, but can be proven in the same manner. Part (3) is also proven using the same idea of extracting subsequences of subsequences that have limits, as described above. For the last part, we use Exercise 2.2.5, which says $\sup(-E) = -\inf(E)$.

Proof of Theorem 3.127. Using Theorem 3.125, there exists a subsequence of $(x_n + y_n)$ that has $\limsup(x_n + y_n)$ as its limit. Suppose $x_{n_k} + y_{n_k} \xrightarrow{k} \limsup(x_n + y_n)$ is that subsequence. The sequence $(x_{n_k})_k$ has a subsequence with a limit in $\mathbb{R}^\sharp$ by the Bolzano-Weierstrass Theorems (Theorems 3.84 and 3.118). Suppose $x_{n_{k_s}} \xrightarrow{s} x \in \mathbb{R}^\sharp$. We apply the same reasoning now to the sequence $(y_{n_{k_s}})_s$: we can find a subsequence $y_{n_{k_{s_t}}} \xrightarrow{t} y \in \mathbb{R}^\sharp$. Then, since $(x_{n_{k_{s_t}}})$

is a subsequence of $(x_{n_{k_s}})$, and $x_{n_{k_s}} \xrightarrow{s} x$, it follows that $x_{n_{k_{s_t}}} \xrightarrow{t} x$ (by Theorems 3.66 and 3.116). Since $y_{n_{k_{s_t}}} \to y$, we have $x_{n_{k_{s_t}}} + y_{n_{k_{s_t}}} \to x+y$ (Part (1) of Theorems 3.31 and 3.112). But $(x_{n_{k_{s_t}}} + y_{n_{k_{s_t}}})$ is a subsequence of $(x_{n_k} + y_{n_k})$ and $x_{n_k} + y_{n_k} \to \limsup(x_n + y_n)$, so $x_{n_{k_{s_t}}} + y_{n_{k_{s_t}}} \to \limsup(x_n + y_n)$ as well (again by Theorems 3.66 and 3.116). By the uniqueness of limits (Theorems 3.22 and 3.111), $x + y = \limsup(x_n + y_n)$. But x is a subsequential limit of (x_n), and so $x \leq \limsup(x_n)$. Similarly, y is a subsequential limit of (y_n), so $y \leq \limsup(y_n)$. So $x+y \leq \limsup(x_n) + \limsup(y_n)$. Part (2) of this theorem follows very similarly to Part (1), and is left as Exercise 11.

We now prove the first part of (3). Suppose (using Theorem 3.125) that $x_{n_k} \to \limsup(x_n)$. The sequence (y_{n_k}) has a subsequence $(y_{n_{k_s}})$ that has limit y by the Bolzano-Weierstrass Theorems (Theorems 3.84 and 3.118). The subsequence $(x_{n_{k_s}})$ of (x_{n_k}) has $\limsup(x_n)$ as its limit (Theorems 3.66 and 3.116). So, we have $x_{n_{k_s}} \to \limsup(x_n)$, $y_{n_{k_s}} \to y$, and $x_{n_{k_s}} \leq y_{n_{k_s}}$. By Theorems 3.48 and 3.114 (in the case $\limsup(x_n) = -\infty$ there is nothing to prove), $\limsup(x_n) \leq y$. But y is a subsequential limit of (y_n), and so $y \leq \limsup(y_n)$. So

$$\limsup(x_n) \leq y \leq \limsup(y_n).$$

We now prove the second part of (3) in the same manner. Suppose now that the subsequence $y_{n_k} \to \liminf(y_n)$. Now, the sequence (x_{n_k}) has a subsequence $x_{n_{k_s}} \to x$. The sequence $(y_{n_{k_s}})$ is a subsequence of (y_{n_k}), and hence has the same limit of $\liminf(y_n)$. So we have $x_{n_{k_s}} \to x$, $y_{n_{k_s}} \to \liminf(y_n)$, and $x_{n_{k_s}} \leq y_{n_{k_s}}$. Therefore $x \leq \liminf(y_n)$. But x is a subsequential limit of (x_n), and so $\liminf(x_n) \leq x$. Now it follows that

$$\liminf(x_n) \leq x \leq \liminf(y_n).$$

We can now establish Part (4). Let E be the set of subsequential limits of (x_n). Since $\sup(-E) = -\inf(E)$ (Exercise 2.2.5), it follows that $\limsup(-x_n) = -\liminf(x_n)$. □

Remark 3.128. The previous proof is a great example of an application of many of the previous results we have developed in this chapter. Although some of these statements have been improved or restated throughout the chapter (for example, above we could have simply referenced Theorem 3.123 instead of referencing Theorems 3.66 and 3.116 above), we have referenced the more contextually specific results in the proof above for the reader's convenience.

3.9.3 Equivalent characterizations of the lim sup and lim inf

We close this section by exhibiting two useful characterizations of the lim sup and lim inf. The first of these characterizations exhibits the lim sup and lim inf as the limit of a certain sequence, and explains why the notation lim sup and lim inf is a good one (see the next paragraph). The second characterization of the lim sup and lim inf points out that each of these numbers is unique: they are the only ones that satisfy certain conditions.

Before stating and proving Theorem 3.129, we pause to define a new sequence (we actually hinted at this construction in the solution of Practice 3.122). Let (x_n) be a sequence, and define

$$\bar{x}_n = \sup\{x_k|k \geq n\}, \quad \text{and} \quad \underline{x}_n = \inf\{x_k|k \geq n\}.$$

In words, $\bar{x}_n$ is the supremum of the set of values (x_n) takes on beyond n. Since

$$\{x_k|k \geq n\} \supseteq \{x_k|k \geq n+1\},$$

the sequence $\bar{x}_n$ is a *decreasing* one. Similarly, $\underline{x}_n$ is the infimum of the set of values (x_n) takes on beyond n, and $\underline{x}_n$ is *increasing*. Below, Theorem 3.129 asserts that

$$\lim_{n\to\infty} \bar{x}_n = \lim_{n\to\infty} \sup\{x_k|k \geq n\} = \limsup(x_n),$$

and

$$\lim_{n\to\infty} \underline{x}_n = \lim_{n\to\infty} \inf\{x_k|k \geq n\} = \liminf(x_n).$$

It is important to point out that it is entirely possible that $\sup\{x_k|k \geq n\} \notin \{x_k|k \geq n\}$, *so that* $(\bar{x}_n)$ *need not be a subsequence of* (x_n). *Similarly for* $(\underline{x}_n)$. We will consider this in the proof below. The usefulness of Theorem 3.129 is that it gives a concrete way to find $\limsup(x_n)$ as the limit of a sequence you can actually determine (as opposed to the limit of *some* subsequence, as in Theorem 3.125).

Theorem 3.129. *If* (x_n) *is any sequence, then*

$$\lim_{n\to\infty} \sup\{x_k|k \geq n\} = \limsup(x_n),$$

and

$$\lim_{n\to\infty} \inf\{x_k|k \geq n\} = \liminf(x_n).$$

Idea of proof. *In our discussion above and in our proof below, we will denote* $\bar{x}_n = \sup\{x_k|k \geq n\}$ *and* $\underline{x}_n = \inf\{x_k|k \geq n\}$. *Both the* lim sup *and* lim inf *statements have similar proofs. The main idea of these proofs is (for the* lim sup *statement) to first exhibit* $\lim_{n\to\infty} \bar{x}_n$ *as being greater than or equal to* $\sup(E)$, *where* E *is the set of subsequential limits* ($\bar{x}_n \to \bar{x}$ *for some* $\bar{x} \in \mathbb{R}^\sharp$ *because* $(\bar{x}_n)$ *is monotone). This will not be hard by the definition of* $\bar{x}_n$ *and by considering a subsequence of* (x_n) *that has* $\limsup(x_n)$ *as its limit. Then, we will show that there is a subsequence of* (x_n) *with* $\bar{x}$ *as its limit using a method similar to the proof of Theorem 3.125. If* $\bar{x} \in E$ *and* $\bar{x} \geq \sup(E)$, *then* $\bar{x} = \sup(E)$.

Proof of Theorem 3.129. For convenience, set

$$\bar{x}_n = \sup\{x_k | k \geq n\}, \quad \text{and} \quad \underline{x}_n = \inf\{x_k | k \geq n\}.$$

We first prove that $\lim_{n\to\infty} \bar{x}_n = \limsup(x_n)$. The sequence, $(\bar{x}_n)$ is monotone decreasing. Therefore, there is a number $\bar{x} \in \mathbb{R}^\sharp$ so that $\bar{x}_n \to \bar{x}$ (Theorems 3.74 and 3.114).

Suppose that $x_{n_k} \to \limsup(x_n)$ (Theorem 3.125). For each k, by the definition of $\bar{x}_k$ and since $n_k \geq k$ (Exercise 3 of Section 3.4), we have $\bar{x}_k \geq x_{n_k}$. Therefore, by Theorems 3.48 and 3.114, $\limsup(x_n) \leq \bar{x}$. If we can show that $\bar{x}$ is a subsequential limit of (x_n), then as the supremum of that set of subsequential limits, $\limsup(x_n) \geq \bar{x}$, and it would follow that $\bar{x} = \limsup(x_n)$ as desired. So all that remains is to prove that $\bar{x}$ is a subsequential limit of (x_n).

We consider three separate possibilities that depend on the $\bar{x}_n$. If it is the case that there are infinitely many n with $\bar{x}_n \in \{x_k | k \geq n\}$, then we may find a subsequence $x_{n_{k_s}} = \bar{x}_{k_s}$, and as such, $x_{n_{k_s}} \to \bar{x}$ since the subsequence $(\bar{x}_{k_s})$ must have the same limit as $(\bar{x}_n)$. If there are only finitely many n with $\bar{x}_n \in \{x_k | k \geq n\}$, then let N be strictly greater than the last of these n, so that $\bar{x}_s \notin \{x_k | k \geq n\}$ for every $s \geq N$.

We break this situation into two cases: either there is an $s \geq N$ with $\bar{x}_s = \infty$, or there is not. If there is such an s, then by Exercise 13 of Section 3.8, there exists a subsequence $x_{n_k} \to \infty$ (with $n_1 \geq s$), and in that case ∞ is a subsequential limit of (x_n), $\limsup(x_n) = \infty$, and since $\limsup(x_n) \leq \bar{x}$, it follows that $\bar{x} = \infty$, and so $x_{n_k} \to \infty$, and we conclude that $\bar{x}$ is a subsequential limit of (x_n) as required. The second case is that there are no $\bar{x}_s$ that are infinite (as supremums of sets that contain real numbers, none of the $\bar{x}_n$ could equal $-\infty$). If all of the $\bar{x}_n$ (for $n \geq N$) are real, and $\bar{x}_n \notin \{x_k | k \geq n\}$, then each of the numbers $\bar{x}_n$ for $n \geq N$ are subsequential limits of (x_n) (Exercise 3 of Section 3.5). An argument identical to the proof of Theorem 3.125 provides a subsequence $x_{n_k} \to \bar{x}$.

The proof that $\lim_{n\to\infty} \inf\{x_k | k \geq n\} = \liminf(x_n)$ is similar and omitted. □

The final results of this section are those that characterize the lim sup and lim inf as the unique numbers which satisfy certain properties. These results are very similar to one another, and they are broken up into Theorems 3.130 and 3.131 (these results can be found in [24], page 56).

Theorem 3.130. *Let E be the set of subsequential limits of (x_n). If $z = \liminf(x_n)$, then z satisfies both of the following:*

1. *$z \in E$.*
2. *If $x < z$, then there exists an N such that $n \geq N$ implies $x_n > x$.*

Moreover, if z satisfies both properties above, then $z = \liminf(x_n)$.

Idea of proof. *The first part of this proof asks us to show that* $\liminf(x_n)$ *satisfies two properties. Both of these properties will follow from results we have already established. The second part of the proof asks us to show that only* $\liminf(x_n)$ *can satisfy the two properties listed above. This, too, will follow easily from previous results.*

Proof of Theorem 3.130. Suppose $z = \liminf(x_n)$. Theorem 3.125 guarantees the existence of a subsequence $x_{n_k} \to z$. Now let $x < z$, and suppose the second property above fails. Then, for every N, there exists an $n \geq N$ with $x_n \leq x$. We create a subsequence out of this hypothesis. If $N = 1$, then there exists $n_1 \geq 1$ with $x_{n_1} \leq x$. Now if $N = n_1 + 1$, then there exists $n_2 \geq n_1 + 1 > n_1$ with $x_{n_2} \leq x$. Continuing in this manner, we have a subsequence $x_{n_k} \leq x$. There exists a subsequence $(x_{n_{k_s}})$ that has a limit y, and by Theorems 3.48 and 3.114, $y \leq x < z$. But y is a subsequential limit of (x_n), and $y < z$ contradicts the definition of $z = \liminf(x_n)$. We conclude that $z = \liminf(x_n)$ satisfies the second property above.

Now suppose that z satisfies the two properties above. Since $z \in E$, and $\liminf(x_n) = \inf(E)$, it must be the case that $\liminf(x_n) \leq z$. Suppose then that $\liminf(x_n) < z$. Then there exists a real number y with

$$\liminf(x_n) < y < z.$$

Since $y < z$, then the second property states that there exists an N such that $n \geq N$ implies $x_n > y$. This means that (x_n) is eventually bounded away from $\liminf(x_n)$, and so no subsequence of (x_n) can have $\liminf(x_n)$ as its limit (Exercise 18 of Section 3.2, and Exercise 14 of Section 3.8). But according to Theorem 3.125, $\liminf(x_n)$ is the limit of some subsequence of (x_n). This contradiction shows that $z = \liminf(x_n)$. □

Its analogue for the lim sup is Theorem 3.131.

Theorem 3.131. *Let E be the set of subsequential limits of (x_n). If $z = \limsup(x_n)$, then z satisfies both of the following:*

1. *There exists a subsequence $x_{n_k} \to z$.*
2. *If $x > z$, then there exists an N such that $n \geq N$ implies $x_n < x$.*

Moreover, if z satisfies both properties above, then $z = \limsup(x_n)$.

Idea of proof. *The proof follows similarly to that of Theorem 3.130, and is left as Exercise 12.*

Proof of Theorem 3.131. Exercise 12. □

3.9.4 Review of important concepts from Section 3.9

1. The lim sup and lim inf are generalizations of limits for every sequence, even those that diverge. The $\limsup(x_n)$, or *limit superior*, is the supremum of the set of subsequential limits of (x_n). The $\liminf(x_n)$, or *limit inferior*, is the infimum of the set of subsequential limits of (x_n) (Definition 3.119).

2. The $\limsup(x_n) = \liminf(x_n)$ if and only if $\lim_{n\to\infty} x_n$ exists and is equal to that common quantity (Theorem 3.123).

3. The lim sup is *subadditive* (property (1) of Theorem 3.127), while the lim inf is *superadditive* (property (2) of Theorem 3.127). Thus, there are properties of limits that do not hold for lim sup and lim inf (compare with Theorems 3.31 and 3.112). This makes sense: if the lim sup and lim inf are generalizations of limits, their properties should be more general than limits.

4. There exists a subsequence of (x_n) that has $\limsup(x_n)$ as its limit, and there exists a (possibly different) subsequence of (x_n) that has $\liminf(x_n)$ as its limit (Theorem 3.125).

5. There are two other ways of characterizing $\limsup(x_n)$ and $\liminf(x_n)$ that are presented here. The first characterizes each of these as the limit of a certain sequence (Theorem 3.129). These sequences are not necessarily subsequences of (x_n). The second characterization demonstrates that the $\limsup(x_n)$ satisfies two properties, and that any quantity that satisfies both properties must be the $\limsup(x_n)$ (Theorem 3.131). A similar result holds for $\liminf(x_n)$ (Theorem 3.130).

Exercises for Section 3.9

1. Compute $\limsup(x_n)$ and $\liminf(x_n)$ and the set of subsequential limits of (x_n) for each part below.

   (a) $x_n = (-1)^n n$.

   (b) $x_n = (-1)^n + \frac{1}{n}$.

   (c) $x_n = \left(1 + \frac{1}{n}\right)^{(-1)^n n}$.

2. Let (p_n) and (q_n) be sequences, and set

$$a_n = \begin{cases} 1 \text{ for even } n \\ p_n \text{ for odd } n \end{cases}, \quad \text{and} \quad b_n = \begin{cases} 2 \text{ for even } n \\ q_n \text{ for odd } n \end{cases}.$$

   (a) If $p_n \geq 4$ and $q_n \geq 7$, then what can you say about $\liminf(a_n)$, $\liminf(b_n)$, and $\liminf(a_n + b_n)$?

   (b) If $p_n \leq 2$ and $q_n \leq 2$, then what can you say about $\limsup(a_n)$, $\limsup(b_n)$, and $\limsup(a_n + b_n)$?

(c) What conditions on (p_n) would make it certain that $\limsup(a_n) = \liminf(a_n)$? What conditions on (p_n) would make it certain that $\limsup(a_n) \neq \liminf(a_n)$?

3. Suppose $x_n \leq y_n$. Must it be true that $\limsup(x_n) \leq \liminf(y_n)$?

4. Suppose $x_n \leq y_n \leq z_n$ for all n, and that $\liminf(x_n) = \limsup(z_n)$. Prove that $\lim_{n\to\infty} y_n$ exists, and compute it.

5. Suppose that (x_n) is a sequence in $\mathbb{R}$.

   (a) Suppose that $\liminf(x_n) = \infty$. What can you say about the sequence (x_n)?

   (b) Suppose that $\limsup(x_n) = -\infty$. What can you say about the sequence (x_n)?

6. Let (q_n) be an enumeration of the rational numbers in $(0, 1)$. Compute the set E of subsequential limits of (q_n), $\limsup(q_n)$ and $\liminf(q_n)$.

7. Suppose that $x_n \to x$, where x is a positive real number. Let (y_n) be any sequence.

   (a) Prove that $\limsup(x_n y_n) = x \limsup(y_n)$. Deduce that if x is any positive real number, then $\limsup(x y_n) = x \limsup(y_n)$. Note that this last statement is true if $x = 0$ as well.

   (b) Prove that $\liminf(x_n y_n) = x \liminf(y_n)$. Deduce that if x is any positive real number, then $\liminf(x y_n) = x \liminf(y_n)$. Note that this is true if $x = 0$ as well.

8. State and prove a result similar to Part (1) of Theorem 3.127 that involves $\limsup(x_n - y_n)$ and $\liminf(x_n - y_n)$.

9. Complete the proof of Theorem 3.129 by proving that

$$\lim_{n\to\infty} \inf\{x_k | k \geq n\} = \liminf(x_n).$$

10. Complete the proof of Theorem 3.125 by proving that $\liminf(x_n)$ is a subsequential limit of (x_n). *[Hint: Model your solution after the proof of Theorem 3.125. Namely, if* $\liminf(x_n) = \infty$*, then there is very little to prove. If* $\liminf(x_n)$ *is finite or* $-\infty$*, then there is a sequence in* E *(the set of subsequential limits of* (x_n)*) of real numbers with that as its limit, and a suitable subsequence of* (x_n) *can be found in a similar manner as in the proof of Theorem 3.125.]*

11. Prove Part (2) of Theorem 3.127. That is, prove that if (x_n) and (y_n) are sequences, then

$$\liminf(x_n) + \liminf(y_n) \leq \liminf(x_n + y_n).$$

12. Prove Theorem 3.131.

13. Assume that Parts (1) and (4) of Theorem 3.127 hold. Establish Part (2) of Theorem 3.127. *[Hint: Replace x_n with $-x_n$ and y_n with $-y_n$.]*

14. Let E be the set of subsequential limits of a sequence (x_n). Show that if z is a limit point of E, then $z \in E$. (It suffices to identify exactly what would change about the proof of Theorem 3.125. Using the terminology of Exercise 14 of Section 3.6, this problem asks you to show that $\bar{E} = E$, or that E is its own closure.)

15. (a) Construct a sequence (x_n) whose set of subsequential limits is $\{-\infty, \infty\}$.

    (b) Construct a sequence (x_n) that has exactly p subsequential limits, where $p \in \mathbb{N}$.

    (c) Construct a sequence (x_n) whose set of subsequential limits is $\mathbb{N} \cup \{\infty\}$. Construct a sequence (x_n) whose set of subsequential limits is $\mathbb{Z} \cup \{-\infty, \infty\}$.

    (d) Is it possible that there is a sequence whose set of subsequential limits is $\mathbb{Q}$? Why or why not? *[Hint: See Exercise 14.]*

    (e) Is it possible that there is a sequence whose set of subsequential limits is $\mathbb{R}$? Why or why not? *[Hint: See Exercise 16.]*

    (f) Is it possible that there is a sequence whose set of subsequential limits is $\mathbb{R}^\sharp$? Why or why not?

16. Let E be the set of subsequential limits of (x_n). Prove that if E is unbounded, then E must contain either ∞ or $-\infty$.

17. (The following has been called L'Hôpital's Theorem for Sequences [11] or the Stolz-Cesàro Theorem.) Suppose $(a_n), (b_n)$ are sequences with $b_n \to \infty$. Suppose also that (b_n) is strictly monotone increasing, and $b_n > 0$. By following the steps below, prove that

$$\frac{a_{n+1} - a_n}{b_{n+1} - b_n} \to L \quad \text{implies} \quad \frac{a_n}{b_n} \to L.$$

    (a) Let $\alpha < \liminf(\frac{a_{n+1}-a_n}{b_{n+1}-b_n})$. Prove that there is an N so that $(b_{n+1} - b_n)\alpha < a_{n+1} - a_n$ for $n \geq N$.

    (b) Sum from $n = N$ to k to conclude that $(b_{k+1} - b_N)\alpha < a_{k+1} - a_N$.

    (c) Divide by b_{k+1}, and show that $\alpha(1 - \frac{b_N}{b_{k+1}}) + \frac{a_N}{b_{k+1}} < \frac{a_{k+1}}{b_{k+1}}$.

    (d) Notice that the left side converges as $k \to \infty$, and so $\alpha \leq \liminf\left(\frac{a_k}{b_k}\right)$. Conclude that $\liminf\left(\frac{a_{n+1}-a_n}{b_{n+1}-b_n}\right) \leq \liminf\left(\frac{a_n}{b_n}\right)$.

    (e) Follow similar steps to show that $\limsup\left(\frac{a_n}{b_n}\right) \leq \limsup\left(\frac{a_{n+1}-a_n}{b_{n+1}-b_n}\right)$.

    (f) Prove that $\frac{a_{n+1}-a_n}{b_{n+1}-b_n} \to L$ implies $\frac{a_n}{b_n} \to L$.

18. Suppose that $a_n \to L \in \mathbb{R}^\sharp$. Prove $\frac{a_1+\cdots+a_n}{n} \to L$. (This result is due to Cauchy.) *[Hint: Use Exercise 17.]*

19. Use the following method to provide a different proof of Exercise 18 in the event that $L \in \mathbb{R}$.

    (a) Let $\sigma_n = \frac{a_1+a_2+\cdots+a_n}{n}$, and suppose $a_n \to L \in \mathbb{R}$. Write

    $$\sigma_n - L = \frac{a_1 + a_2 + \cdots + a_n}{n} - \frac{nL}{n} = \frac{(a_1 - L) + \cdots + (a_n - L)}{n},$$

    so that if $\epsilon > 0$ is given, and a suitable N is chosen, for $n \geq N$ we have

    $$|\sigma_n - L| \leq \frac{|a_1 - L| + \cdots + |a_N - L|}{n} + \left(\frac{n-N}{n}\right)\epsilon.$$

    (b) Prove $\sigma_n \to L$.

20. Refer to Exercise 18. Show that there exist (a_n) that diverge, but that (σ_n) still converges.

21. Use Exercise 17 to prove that

$$\frac{1 + \frac{1}{2} + \frac{1}{3} + \cdots + \frac{1}{n}}{n} \to 0.$$

3.10 Troubleshooting guide for Sequences and Their Limits

Type of difficulty:	*Suggestion for help:*
Section 3.2: Limits and Convergence	
"I don't know how to prove a sequence converges."	Try becoming more familiar with the definition and basic examples. Go back to Section 3.2 and read the material preceding Subsection 3.2.1. Then try reading Definition 3.13, then Example 3.15, and then Practice 3.14, which gives a formal proof of convergence you could try to model your own off of.
"I don't know how to determine what a sequence converges to."	See Section 3.3. Every result in that section is geared toward helping you determine what a sequence converges to. See also the discussion on page 115 where it discusses this problem.
"I don't know which inequalities to use."	See the discussion on page 117.
"I don't know how to find N in a convergence proof."	See the discussion on page 117.
"I don't know how to demonstrate that a sequence *diverges*."	See the discussion on page 117. See also Corollary 3.29. For a treatment of this using subsequential limits, see Exercises 6 and 7 of Section 3.4. See also Exercise 2 of Section 3.6 or Theorem 3.123. For divergence of a sequence to $\pm\infty$, see Section 3.8.
"What does 'prove this sequence converges using the definition' mean?"	Prove convergence of a sequence using only Definition 3.13, and not any supporting results. See Subsection 3.2.2 on page 113.

Type of difficulty:	***Suggestion for help:***
"Must I always try to approximate $\|a_n - L\|$ using $\frac{1}{n}$?"	No, but $\frac{1}{n}$ relates so directly to the Archimedean Principle that it is a desirable approximation. Practice 3.19 gives an estimate using $\frac{1}{n^2}$, and Exercise (3) of Section 3.2 uses an estimate of $\frac{1}{\sqrt{n}}$. The Squeeze Theorem on page 130 says that there are many sequences one can relate $\|a_n - L\|$ to in an effort to prove convergence.
Section 3.3: Limit Theorems	
"I'm having trouble seeing how any of these limit theorems help with showing convergence of a sequence. Don't you always have to use the definition of convergence if you're trying to show a given sequence converges?"	Although Definition 3.13 is the place we first define convergence of sequences, there are many supporting results that will demonstrate convergence so that you do not always have to refer back to Definition 3.13. See Practice 3.35 for a careful explanation of the use of some basic limit theorems. All of Section 3.3 is devoted to these supporting results as well.
"Why do I keep seeing things like '$\frac{\epsilon}{2}$' and/or 'choose $N = \max\{N_1, N_2\}$'? What purpose would something like these serve?"	See separate entries under the General Troubleshooting for Chapter 3 heading below.
"I don't know how to find the right sort of estimate to help me employ the Squeeze Theorem."	Every situation is different, and generally one has to be quite creative. Read over Practices 3.39 to 3.43 , and also read Remarks 3.41 and 3.44. For a general comment about inequalities, see the discussion on page 117.
Section 3.4: Subsequences	
"Why is there a different index variable for subsequences? That is, why do we write (a_{n_k})?"	n_k is the way that we traverse through the subsequence $b_k = a_{n_k}$. See Example 3.58 on page 143 for a more detailed explanation.

Type of difficulty:	*Suggestion for help:*
Section 3.5: Monotone Sequences	
"I don't know how to show a sequence is monotone."	There are several common methods, although generally such a task requires creativity, and the more practice you have the better. Read Definition 3.68, the paragraph after Definition 3.68, and then Practices 3.69–3.73 for guidance.
Section 3.6: The Bolzano-Weierstrass Theorems	
"What's the deal with limit points and isolated points?"	Accumulation points are part of The Bolzano-Weierstrass Theorem for sets (Theorem 3.91). Isolated points are a sort of opposite of accumulation points, and these two sorts of points form the basis for one's intuition going forward into the field of Topology.
"I find it difficult to find the limit points and/or the isolated points of a given set."	To narrow your search, remember that an isolated point must be already within the set (see Definition 3.87 and Theorem 3.90), while limit points of a set cannot be "too far away" from the set (see Definition 3.86, and Theorem 3.89, and Exercise 7 of Section 3.6). See also Practice 3.88 for practice with how to locate these points.
Section 3.7: Cauchy Sequences	
"I don't know how I'm supposed to show (a_n) is Cauchy."	The following two methods are by far the most common: One can use the definition (see Definition 3.92 on page 171). See also Practices 3.93, 3.94, and 3.95 that follow for examples of the use of Definition 3.92. One could also prove (a_n) is convergent, and use Theorem 3.97.
"I'm looking at a weird recursively defined sequence. How do I show it is Cauchy?"	A very common method (but not the only one–see above) is to show that the consecutive differences $\|a_{n+1} - a_n\|$ are bounded above by an appropriate geometric sequence. See the discussion just before Practice 3.95, and Exercise 4 of Section 3.7. For an example of the use of such a method, see Practice 3.103 and then Exercise 5 of Section 3.7.

Type of difficulty:	***Suggestion for help:***
"I don't know how to show (a_n) is *not* Cauchy."	The most common method would be to negate the conclusion of a theorem that begins with "If (a_n) is Cauchy, then ...." As such, one could show (a_n) is not bounded (Theorem 3.98), or not convergent (Theorem 3.99).
"I don't understand the difference between Cauchy sequences and convergent sequences."	Within the set of real numbers, there is no difference. See Corollary 3.101. Read also, however, the discussion on page 177 about "completeness" for an interesting perspective.
"I don't know what is so special about Cauchy sequences."	Demonstrating that a sequence is Cauchy will be enough to show that it is convergent (Theorem 3.99), but with the added benefit that you need not know what its limit is (contrast this with the definition of convergence (Definition 3.13)).
Section 3.8: Infinite Limits	
"I don't understand exactly what it means to 'diverge to ∞'."	Definition 3.106 is the actual definition of this, although for a discussion that helps put it in perspective, see the discussion that begins with Subsection 3.8.1 on page 182. After that, try reading Practice 3.105.
"If $a_n \to \infty$, why don't we just say that it 'converges'? Why the different terminology of 'diverges to ∞'?"	If we did do that, then we would have to rewrite all of our results in Section 3.3 to avoid certain considerations. It is both frustration and source of fundamental intrigue in this subject that "converges," "diverges to ∞," and "diverges to $-\infty$" must be treated differently. See the discussion at the beginning of Subsection 3.8.2, and the subsequent discussion in that subsection.
"I don't understand how to prove that $a_n \to \infty$ using only the definition."	To prove $a_n \to \infty$ "using only the definition" requires you to use only Definition 3.106 and not any supporting results. For examples of how to do this, see Practices 3.105 and 3.107. For a more general theoretical example, see Proposition 3.108 or the proof of Theorem 3.112.

Type of difficulty:	***Suggestion for help:***
"Why do the supporting results for infinite limits appear so complicated? Aren't they just analogues of what we already know for finite limits?"	They appear more complicated because, among other things, we have to make sure to avoid the sort of absurdities one could encounter with working with $\pm\infty$, and arithmetic expressions that are not allowed in $\mathbb{R}^\sharp$, such as $\infty - \infty, 0 \cdot \infty$, and $\frac{\infty}{\infty}$. Generally speaking, most conclusions about finite limits that make sense to include infinite limits are true, after one makes appropriate changes. See Subsection 3.8.2 for an account of these results: between each result is a short description of how it relates to a previously known result involving finite limits.
"Why do I keep seeing 'choose $N = \max\{N_1, N_2\}$'? What purpose would something like this serve?"	See below under the General Troubleshooting for Chapter 3 heading below.
Section 3.9: The lim sup **and** lim inf	
"I don't really understand the difference between the limit of a sequence and the lim sup of a sequence."	The limit of a sequence exists, by definition, when the sequence converges (or diverges to something infinite). The lim sup of a sequence *always* exists. See the discussion in Subsection 3.9.1. See also the relationship between sequences that do have limits and their lim sup and lim inf in Theorem 3.123.
"Finding $\limsup(x_n)$ seems needlessly complicated. Why can't I just find a subsequence that seems to have the largest limit, and call its limit the lim sup?"	As it turns out, you can do that, especially since $\limsup(x_n)$ is a subsequential limit (Theorem 3.125). It is crucial, however, that you justify that the subsequential limit you find is in fact the "largest". See the discussion immediately after Practice 3.122.
"Given an actual sequence, how do I actually compute the lim inf and lim sup?"	Often these subsequential limits are easy to see. See Practice 3.122 for a basic example, and see Theorem 3.129 for a way to compute it directly. An example where we compute the set of subsequential limits explicitly is in Example 3.124.

Type of difficulty:	*Suggestion for help:*
"I don't understand the proof of Theorem 3.125 at all."	The proof is admittedly a bit complicated, but try to see it one case at a time. The theorem is established for $\sup(E) = -\infty$, then for $\sup(E) = \infty$ or $\sup(E) \in \mathbb{R}$. Many people are confused by the notation $(x_{n,k})$: this is meant to be a subsequence that converges to the element z_n, where k is the index variable. The main idea is to create a sequence of subsequences, each subsequence converges to a z_n. One can then create a new subsequence of (x_n) that will have $\sup(E)$ as its limit by choosing the first member from the subsequence that converges to z_1, the second member from the subsequence that converges to z_2, etc. Care must be taken to make sure that the ordered list of elements you construct is indeed a subsequence. Consider rereading the idea of the proof that precedes the actual proof of Theorem 3.125.
General Troubleshooting for Chapter 3	
"How do I find the limit of a convergent sequence that is recursively defined?"	There are many methods, although adapting the methodology of the following examples may help: Practice 3.75 (specifically, the discussion just before and after Equation (3.27) on page 154), and Practice 3.103. See also Exercises 5 and 6 of Section 3.7. More elaborate methods that are specific to the situation may need to be employed, see Practice 3.103.
"Why do I keep seeing '$\frac{\epsilon}{2}$,' and things of that sort?"	There are a variety of reasons, although usually such an expression comes up when one is trying to show a particular distance is less than ϵ, and does so by comparing that distance to two (or more) others. See, for example, the proof of Theorems 3.31, 3.97, 3.99, and 3.125. In all cases, we are estimating a desired distance by other distances that we hypothesize can be made to be as small as we like.

Type of difficulty:	***Suggestion for help:***
"I don't understand why something like 'choose $N = \max\{N_1, N_2\}$' is needed."	Such a choice will require any $n \geq N$ to satisfy $n \geq N_1$ and $n \geq N_2$. Usually, this will allow you to simultaneously use multiple inequalities previously developed. For examples of this with convergence (i.e., finite limits), see Practice 3.23 on page 115, and then Remark 3.24 on page 117. A theoretical example can be found in the proof of Theorem 3.31 on page 124. For examples of this involving infinite limits, see the proof of Theorem 3.112.

4

Series of Real Numbers

Chapter Summary:

The theory of infinite series is a natural next topic, since convergence of these series is defined by convergence of their associated sequence of partial sums. We introduce infinite series in Section 4.1, present various convergence tests in Sections 4.2 (for nonnegative terms) and 4.3. We close the chapter by presenting information concerning rearrangements of series, including Riemann's Rearrangement Theorem in Section 4.4.

Section 4.1: Infinite Series. Infinite series are defined here. Telescoping and geometric series are discussed, along with the Test for Divergence (Theorem 4.6), a result concerning sums, differences, and constant multiples of series (Theorem 4.9), and the Cauchy Criterion (Theorem 4.11).

Section 4.2: Convergence Tests for Series with Nonnegative Terms. We present convergence tests for series with nonnegative terms. We use the Cauchy Condensation Test (Theorem 4.14) to determine the convergence or divergence of $p-$series in Theorem 4.15, and then present the Direct and Limit Comparison Tests (Theorems 4.17 and 4.20).

Section 4.3: Other Convergence Tests for Series. We discuss convergence of series whose terms need not be nonnegative. The Alternating Series Test (Theorem 4.24) follows immediately from Dirichlet's Test (Theorem 4.23). Absolute and conditional convergence is discussed, and the Ratio and Root Tests (Theorems 4.31 and 4.32) are presented.

Section 4.4: Rearrangements. Riemann's Rearrangement Theorem (Theorem 4.40) is presented. In addition, absolutely convergent series remain absolutely convergent under rearrangement and converge to the same value (Theorem 4.38).

Section 4.5: Troubleshooting Guide for Series of Real Numbers. A list of frequently asked questions and their answers, arranged by section. Come here if you are struggling with any material from this chapter.

4.1 Infinite Series

The theory of infinite series is an extremely important one throughout mathematics. The first task will be to understand just what

$$\sum_{n=1}^{\infty} a_n = a_1 + a_2 + a_3 + \cdots$$

means. Indeed, it seems one cannot simply "add infinitely many numbers together." Since we can add finitely many real numbers together, the natural suggestion from an analyst's point of view would be to try to approximate this infinite series by a sequence of finite sums, and take a limit. It turns out this is exactly how we define the meaning of $\sum_{n=1}^{\infty} a_n$. This sequence of finite sums is the new "partial sum" sequence $s_n = \sum_{k=1}^{n} a_k$, and it seems that the larger n is, the closer s_n should be to $\sum_{n=1}^{\infty} a_n$. Thus, the convergence of (s_n) is the only factor that determines whether or not $\sum_{n=1}^{\infty} a_n$ means anything at all.

We start the section by continuing this discussion more formally, and defining the convergence of $\sum_{n=1}^{\infty} a_n$ in terms of the convergence of these partial sums in Definition 4.1. After introducing telescoping and geometric series in Examples 4.4 and 4.5, we prove three important basic facts concerning series: the Test for Divergence (Theorem 4.6), a result concerning how the convergence of series is preserved through addition, subtraction, and scalar multiplication (Theorem 4.9), and a characterization of convergence of series known as the Cauchy Criterion (Theorem 4.11).

4.1.1 Infinite series and convergence

Given a sequence (a_n), we define (s_n) as the *partial sum sequence of* (a_n) as

$$s_n = a_1 + a_2 + \cdots + a_n = \sum_{k=1}^{n} a_k.$$

An infinite series is the sum of a given sequence, and it exists (or converges), by definition, if its sequence of partial sums converges.

Definition 4.1. Let (a_n) be a sequence, and define $s_n = \sum_{k=1}^{n} a_k$ to be its associated partial sum sequence. The infinite series $\sum_{k=1}^{\infty} a_k$ *converges* if the sequence (s_n) converges, in which case we write

$$\sum_{k=1}^{\infty} a_k = \lim_{n\to\infty} s_n.$$

If the partial sum sequence diverges, then we say the series is *divergent.*

Here are a few basic remarks about the definition, before we give two concrete examples of convergent infinite series.

Remark 4.2. **The convergence of a series is not affected by where the series begins.** Exercise 2 illustrates this precisely. It is for this reason that we generally will simplify our notation and write $\sum a_n$ instead of $\sum_{n=1}^{\infty} a_n$ when the index of the sum is unimportant. An example of a situation where this index would be important to write down would be if we were to actually compute what $\sum_{n=1}^{\infty} a_n$ converges to. This would generally be a different number than $\sum_{n=2}^{\infty} a_n = (\sum_{n=1}^{\infty} a_n) - a_1$.

Remark 4.3. **In many instances, we care only about whether or not $\sum a_n$ converges, as opposed to computing what it converges to.** Most of this chapter presents "tests for convergence", which will conclude that a series converges (or, sometimes that it diverges) under certain circumstances.

There are two common types of infinite series whose sequence of partial sums can be determined explicitly. As a result, we cannot only determine that the series converges, but we will be able to determine what it converges to. We go over each of these in the next two examples.

Example 4.4 (Telescoping Series). This example gives an example of a series whose partial sum sequence is particularly easy to find a closed form for due to successive additive cancellations. Such series are called *telescoping series.*

Perhaps the most straightforward example of a telescoping series is $\sum_{n=1}^{\infty} \frac{1}{n(n+1)}$. Here, $a_n = \frac{1}{n(n+1)}$. Through uninspired computation, we see that the first few partial sum sequence terms proceed as $s_1 = a_1 = \frac{1}{2}$, $s_2 = a_1 + a_2 = \frac{1}{2} + \frac{1}{6} = \frac{2}{3}$, and $s_3 = \frac{1}{2} + \frac{1}{6} + \frac{1}{12} = \frac{3}{4}$. One could continue this process in hopes of finding a pattern; however, if we consider[1] that

$$a_n = \frac{1}{n(n+1)} = \frac{1}{n} - \frac{1}{n+1},$$

we could compute the partial sums as follows. Note the cancellations that make it particularly easy to sum these terms:

$$\begin{array}{rcllcl} s_1 & = & a_1 & & & \\ & = & \left(1 - \frac{1}{2}\right) & & = & 1 - \frac{1}{2}, \\ s_2 & = & a_1 + a_2 & & & \\ & = & \left(1 - \frac{1}{2}\right) + \left(\frac{1}{2} - \frac{1}{3}\right) & & = & 1 - \frac{1}{3} \\ s_3 & = & a_1 + a_2 + a_3 & & & \\ & = & \left(1 - \frac{1}{2}\right) + \left(\frac{1}{2} - \frac{1}{3}\right) + \left(\frac{1}{3} - \frac{1}{4}\right) & & = & 1 - \frac{1}{4} \end{array}$$

More generally, since[2] $s_n = s_{n-1} + a_n$, one could inductively prove that $s_n = 1 - \frac{1}{n+1}$. We can now easily appeal to Definition 4.1 to not only demonstrate

[1]One could arrive at this through the method of partial fractions.

[2]Summing up the first n terms of (a_n) could be achieved by summing the first $n - 1$ terms of (a_n), and then adding the final term a_n. See also the proof of Theorem 4.6.

that this series converges, but to compute what it converges to: the partial sum sequence $s_n = 1 - \frac{1}{n+1}$ converges, and so by definition, $\sum_{n=1}^{\infty} \frac{1}{n(n+1)}$ converges. Since we have determined $s_n \to 1$, we conclude that $\sum_{n=1}^{\infty} \frac{1}{n(n+1)} = 1$. ■

Example 4.5 (Geometric Series)**.** This example illustrates exactly for which r the *geometric series* $\sum_{n=0}^{\infty} r^n$ converges, and what it converges to. We base our discussion off of the result of Exercise 21 of Section 3.3. It states that for $r \neq 1$ and any $n \in \mathbb{N}$, we have

$$1 + r + r^2 + \cdots + r^n = \frac{1 - r^{n+1}}{1 - r}.$$

So if $a_n = r^n$, the partial sum sequence

$$\begin{aligned} s_n &= a_1 + a_2 + \cdots + a_n \\ &= 1 + r + r^2 + \cdots + r^n = \tfrac{1-r^{n+1}}{1-r}. \end{aligned}$$

Now if $|r| < 1$, we see that $s_n \to \frac{1}{1-r}$ since $r^{n+1} = r \cdot r^n \to 0$ (see Theorem 3.55). By Exercise 3 of Section 3.8, if $|r| > 1$, the partial sum sequence (s_n) diverges. If $r = -1$, the partial sum sequence becomes $s_n = \frac{1}{2}(1 - (-1)^{n+1})$, which diverges because the subsequences $s_{2n} \to 1$ and $s_{2n+1} \to 0$ (see Exercise 6 of Section 3.4). Finally, if $r = 1$, then $s_n = 1 + 1 + \cdots + 1 = n$, which clearly diverges.

To summarize, we have shown that

$$\boxed{\sum_{n=0}^{\infty} r^n \text{ converges to } \tfrac{1}{1-r} \text{ if and only if } |r| < 1.}$$

For all other values of r, this series diverges. ■

4.1.2 Basic convergence facts

There are three basic convergence facts to discuss: the first will provide a useful first requirement that a series converge, and the last two will follow directly from our work with sequences.

Theorem 4.6 (Test for Divergence)**.** *If $\sum a_n$ converges, then $a_n \to 0$. Consequently, if (a_n) does not converge to 0, then $\sum a_n$ diverges.*

Idea of proof. *The idea is to notice that the partial sum sequence satisfies $s_n = s_{n-1} + a_n$, so that $s_n - s_{n-1} = a_n$. If $\sum a_n$ converges, then by Definition 4.1, its partial sum sequence $s_n \to s$. So $s_{n-1} \to s$ and so $a_n \to 0$.*

Proof of Theorem 4.6. Let $s_n = \sum_{k=1}^{n} a_n$ be the partial sum sequence of the series $\sum a_n$. Then

$$s_n = \sum_{k=1}^{n} a_n = \sum_{k=1}^{n-1} a_k + a_n = s_{n-1} + a_n,$$

so that $s_n - s_{n-1} = a_n$. If $\sum a_n$ converges, then by Definition 4.1, (s_n) converges to some number, say $s_n \to s$. Then $s_{n-1} \to s$ (Exercise 6 of Section 3.2), and

$$a_n = s_n - s_{n-1} \to s - s = 0.$$

The second statement in the theorem is just the contrapositive of the first. □

Practice 4.7. Show that $\sum \frac{n-1}{n+1}$ and $\sum(-1)^n$ diverge.

Methodology. *We apply the Test for Divergence (Theorem 4.6) to both series.*

Solution. Since the sequence $\frac{n-1}{n+1} \to 1 \neq 0$, the Test for Divergence (Theorem 4.6) applies, and $\sum \frac{n-1}{n+1}$ diverges. The Test for Divergence applies also to $\sum(-1)^n$: the sequence $(-1)^n$ diverges, and so it does not converge to 0. ■

The following example illustrates that the converse of the Test for Divergence does *not* hold: there are instances where $a_n \to 0$ but that $\sum a_n$ diverges.

Example 4.8. This example illustrates that $\sum \frac{1}{n}$ diverges, even though the sequence $\frac{1}{n} \to 0$. We consider the partial sum sequence

$$s_n = \sum_{k=1}^{n} \frac{1}{k} = 1 + \frac{1}{2} + \frac{1}{3} + \cdots \frac{1}{n}.$$

Exercise 2 of Section 3.8 shows that this sequence diverges to ∞. ■

The next two results will follow from easy applications of some of our previous work in Chapter 3.

Theorem 4.9. *Suppose $\sum a_n$ and $\sum b_n$ converge, and $\alpha \in \mathbb{R}$.*

1. *$\sum(a_n \pm b_n)$ converges to $(\sum a_n) \pm (\sum b_n)$.*
2. *$\sum(\alpha a_n)$ converges to $\alpha\left(\sum a_n\right)$.*

Idea of proof. *The proofs of both of these facts relies on Theorem 3.31: the partial sum sequence for $\sum(a_n \pm b_n)$ is the sum or difference of the partial sum sequence for each of $\sum a_n$ and $\sum b_n$, and thus converges to the appropriate number. The second assertion has a similar line of reasoning.*

Proof of Theorem 4.9. We compute the partial sum sequence for the series $\sum(a_n + b_n)$, with the computation being similar for the series $\sum(a_n - b_n)$ (see Exercise 3). Define

$$s_n = \sum_{k=1}^{n} a_k, t_n = \sum_{k=1}^{n} b_k, \text{and } u_n = \sum_{k=1}^{n} (a_k + b_k).$$

We rearrange the following finite[3] partial sums to find:

$$\begin{aligned} u_n &= (a_1 + b_1) + (a_2 + b_2) + \cdots + (a_n + b_n) \\ &= (a_1 + a_2 + \cdots + a_n) + (b_1 + b_2 + \cdots + b_n) \\ &= s_n + t_n. \end{aligned}$$

Since $\sum a_n$ and $\sum b_n$ converge, by Definition 4.1, (s_n) and (t_n) converge. Suppose $s_n \to s = \sum a_n$ and $t_n \to t = \sum b_n$. Then by Theorem 3.31, (u_n) converges to $s + t = \sum a_n + \sum b_n$.

We prove the next assertion similarly. If (s_n) is the partial sum sequence of $\sum a_n$ as above, and $v_n = \sum_{k=1}^{n} (\alpha a_n)$ is the partial sum sequence for $\sum(\alpha a_n)$, then

$$v_n = \alpha a_1 + \alpha a_2 + \cdots + \alpha a_n = \alpha(a_1 + a_2 + \cdots + a_n) = \alpha s_n.$$

Again, since $\sum a_n$, by definition $s_n \to s = \sum a_n$ is convergent. By Theorem 3.31, $v_n = \alpha s_n \to \alpha s = \alpha \sum a_n$ converges as well. □

Remark 4.10. It should be noted that by correct interpretations of the contrapositive, certain series can be shown to diverge. For example, it is shown in Theorem 4.15 that $\sum \frac{1}{n}$ diverges. One can show that $\sum \frac{2}{n}$ diverges as well: if $\sum \frac{2}{n}$ converges, then so must $\frac{1}{2} \sum \frac{2}{n}$, which by Assertion 2 we would erroneously conclude that $\frac{1}{2} \sum \frac{2}{n} = \sum \frac{1}{n}$ converges as well. Similar arguments can be made, but with care: see Exercises 4 and 5.

Our last result of this section is called the *Cauchy criterion* for convergence, which simply equates convergence of the partial sum sequence in terms of it being a Cauchy sequence.

Theorem 4.11 (Cauchy Criterion for Convergence). *The series $\sum a_n$ converges if and only if for every $\epsilon > 0$ there exists an N so that for every $n > m \geq N$ we have $|\sum_{k=m+1}^{n} a_k| < \epsilon$.*

Idea of proof. *There is nothing to do except notice that by Corollary 3.101, the partial sum sequence (s_n) converges if and only if it is Cauchy. The latter condition is simply written in terms of the original sum as opposed to leaving it as "(s_n) is Cauchy."*

Proof of Theorem 4.11. If $\sum a_n$ converges, then by Definition 4.1, its sequence of partial sums (s_n) is Cauchy. So, for any given $\epsilon > 0$, there exists a N so that if $n, m \geq N$, then $|s_n - s_m| < \epsilon$. If $n = m \geq N$ then this holds trivially, and in the event that $n > m \geq N$, we have

$$\begin{aligned} |s_n - s_m| &= |\sum_{k=1}^{n} a_k - \sum_{k=1}^{m} a_k| \\ &= |\sum_{k=m+1}^{n} a_k| < \epsilon. \end{aligned}$$

[3]Rearrangements of infinite sums behave rather unusually. We discuss this in Section 4.4.

Conversely, if the Cauchy Criterion is satisfied (that is, for every $\epsilon > 0$ there exists a N so that for every $n > m \geq N$ we have $|\sum_{k=m+1}^{n} a_k| < \epsilon$), then the partial sum sequence (s_n) is Cauchy, and hence converges by Corollary 3.101. So, $\sum a_n$ converges by Definition 4.1. □

4.1.3 Review of important concepts from Section 4.1

1. The convergence of an infinite series is determined by the convergence of its sequence of partial sums (see Definition 4.1). The addition or removal of finitely many terms does not affect this convergence, and so $\sum_{n=1}^{\infty} a_n$ is sometimes written as $\sum a_n$ where there is no possibility for confusion.
2. Telescoping (Example 4.4) and Geometric series (Example 4.5) are common types of series where one can easily find a closed form expression for the partial sum sequence, allowing one to explicitly compute the values of these series (i.e., what they converge to).
3. The Test for Divergence (Theorem 4.6) gives an easy way to check if a series diverges. It does not always apply: there are series $\sum a_n$ that diverge even though $a_n \to 0$, as in Example 4.8.
4. Sums and constant multiples of convergent series behave nicely, as in Theorem 4.9.
5. The Cauchy Criterion (Theorem 4.11) gives an equivalent characterization of convergence of infinite series.

Exercises for Section 4.1

1. Determine whether or not the following series converge. In the event a series converges, compute its sum.

   (a) $\sum_{n=2}^{\infty} \left(\frac{1}{5}\right)^n$
   (b) $\sum_{n=1}^{\infty} \frac{2}{4n^2+8n+3}$
   (c) $\sum_{n=2}^{\infty} \left[\frac{1}{\sqrt{n-1}} - \frac{1}{\sqrt{n+1}}\right]$
   (d) $.999\ldots = \sum_{n=1}^{\infty} 9\left(\frac{1}{10}\right)^n$
   (e) $\sum_{n=1}^{\infty} r^n$, if $|r| < 1$
   (f) $\sum_{n=N}^{\infty} r^n$, if $N \geq 1, |r| < 1$
   (g) $\sum_{n=0}^{\infty} \frac{2^n+3^n}{6^n}$
   (h) $\sum_{n=0}^{\infty} \frac{p^n+q^n}{(pq)^n}$, if $p, q > 0$

2. Prove that for any $N \in \mathbb{N}$, that $\sum_{n=1}^{\infty} a_n$ converges if and only if $\sum_{n=N}^{\infty} a_n$ converges. Thus, the convergence of a series is not dependent on any finite number of terms we skip, and only the sum of the tail of (a_n).
3. Compute the partial sums of $\sum(a_n - b_n)$, and prove that this series converges to $(\sum a_n) - (\sum b_n)$ if both $\sum a_n$ and $\sum b_n$ converge.
4. Prove that if $\sum a_n$ converges and $\sum b_n$ diverges, then $\sum(a_n + b_n)$ diverges.

5. Give examples to show that if both $\sum a_n$ and $\sum b_n$ diverge, then in some cases $\sum(a_n + b_n)$ diverges and in other cases $\sum(a_n + b_n)$ converges.
6. Give an example to show that $(\sum a_n)(\sum b_n)$ is generally not equal to $\sum(a_n b_n)$.
7. If $\sum a_n$ converges and $a_n > 0$, prove that $\sum \sqrt[n]{a_n + 1}$ diverges.
8. Find the flaw[4] in the following argument that $0 = 1$: Compute the sum $\sum_{n=0}^{\infty}(-1)^n$ in two different ways. On the one hand,

$$\begin{aligned} \textstyle\sum_{n=0}^{\infty}(-1)^n &= 1 - 1 + 1 - 1 + 1 - 1 + \cdots \\ &= (1-1) + (1-1) + (1-1) + \cdots \\ &= 0 + 0 + 0 \cdots \\ &= 0. \end{aligned}$$

On the other hand,

$$\begin{aligned} \textstyle\sum_{n=0}^{\infty}(-1)^n &= 1 - 1 + 1 - 1 + 1 - 1 + \cdots \\ &= 1 + (-1+1) + (-1+1) + (-1+1) + \cdots \\ &= 1 + 0 + 0 \cdots \\ &= 1. \end{aligned}$$

[4]It could be argued that there are actually at least two flaws here. See Page 157 of [8].

4.2 Convergence Tests for Series with Nonnegative Terms

In this section and the next, we explore a variety of convergence tests. These will tell you whether a given series converges or diverges in the event they meet certain criteria that are easy to check. This section focuses on series with nonnegative terms, since such series possesses a desirable quality in its sequence of partial sums, as illustrated in the following straightforward[5] Lemma:

Lemma 4.12. *If (a_n) is a sequence of nonnegative numbers, then the partial sum sequence $s_n = \sum_{k=1}^{n} a_k$ is monotone increasing. Thus, by the Monotone Convergence Theorem (Theorems 3.74 and 3.117), $\sum a_n$ converges if (s_n) is bounded, and $\sum a_n$ diverges if $s_n \to \infty$.*

In this section, we prove that $p-$series converge for $p > 1$ and diverge for all over values of p in Theorem 4.15. Instead of using the integral test, we establish this theorem by using the Cauchy Condensation Test (Theorem 4.14), which is a useful convergence test on its own. We then establish the Direct and Limit Comparison Tests in Theorems 4.17 and 4.20, and give examples of its use.

4.2.1 The Cauchy Condensation Test and $p-$series

The goal of this subsection is to provide a multitude of series for which their convergence or divergence is already known. We will then compare these series to others whose convergence is unknown. We first introduce these series, called $p-$series, and then give the means to demonstrate when they converge in the Cauchy Condensation Test (Theorem 4.14).

Definition 4.13. Let $p \in \mathbb{R}$. A *$p-$series* is an infinite series of the form

$$\sum_{n=1}^{\infty} \frac{1}{n^p}.$$

We prove in Theorem 4.15 that $p-$series converge if and only if $p > 1$. The proof of that result uses the Cauchy Condensation Test.

Theorem 4.14 (Cauchy Condensation Test). *Suppose (a_n) is a nonnegative sequence that is monotone decreasing. Then*

$$\sum a_n \text{ converges if and only if } \sum 2^n a_{2^n} \text{ converges.}$$

Idea of proof. *We prove that if $\sum a_n$ converges then $\sum 2^n a_{2^n}$ converges, and conversely, if $\sum a_n$ diverges then $\sum 2^n a_{2^n}$ diverges. The first of these*

[5] Since $s_{n+1} = s_n + a_{n+1}$, if $a_{n+1} \geq 0$, then $s_{n+1} \geq s_n$.

statements will follow from Lemma 4.12 and a careful overestimate of a partial sum sequence. The latter will follow from Lemma 4.12 after a careful overestimate. Both estimates involve either reducing or enlarging the partial sum sequence of $\sum 2^n a_{2^n}$ *by considering groupings of 1, 2, 4, 8, ... terms of* $\sum a_n$ *and using the hypotheses.*

Proof of Theorem 4.14. For the entire proof, let $s_n = \sum_{k=1}^{n} a_n$ and $t_n = \sum_{k=1}^{n} 2^{n-1} a_{2^n}$ be the partial sum sequences of $\sum a_n$ and $\sum 2^{n-1} a_{2^n}$, respectively. Since both $\sum a_n$ and $\sum 2^{n-1} a_{2^n}$ have nonnegative terms, we have (s_n) and (t_n) are monotone. Because of this, by Lemma 4.12, if $\sum a_n = s$ converges then $s_n \leq s$ for all n, and if $\sum a_n$ diverges then $s_n \to \infty$.

We begin by proving that if $\sum a_n$ converges, then $\sum 2^{n-1} a_{2^n}$ converges. It will then follow that $\sum 2^n a_{2^n}$ converges by Part 2 of Theorem 4.9 (see Remark 4.10), since $2 \sum 2^{n-1} a_{2^n} = \sum 2^n a_{2^n}$. We now underestimate (s_{2^n}) by noticing that (a_n) is decreasing, and so for every n,

$$a_{2^n+1}, a_{2^n+2}, \ldots, a_{2^{n+1}} \leq a_{2^{n+1}}. \tag{4.1}$$

With this observation, we prove that $s_{2^n} \geq t_n$ for every n by Mathematical Induction. The base case holds since

$$s_{2^1} = s_2 = a_1 + a_2 \geq a_2 = t_1.$$

Now suppose that $s_{2^n} \geq t_n$ for some n. We establish the induction step by using the induction hypothesis and the estimate[6] in Equation (4.1):

$$\begin{aligned} s_{2^{n+1}} &= s_{2^n} + a_{2^n+1} + a_{2^n+2} + \cdots + a_{2^{n+1}} \\ &\geq t_n + a_{2^n+1} + a_{2^n+2} + \cdots + a_{2^{n+1}} \\ &\geq t_n + a_{2^{n+1}} + a_{2^{n+1}} + \cdots + a_{2^{n+1}} \\ &= t_n + 2^n a_{2^{n+1}} \\ &= t_{n+1}. \end{aligned}$$

Since $\sum a_n = s$ converges, by definition, $s_n \to s$. As a subsequence, by Theorem 3.66 $s_{2^n} \to s$ as well. With our above remarks, we have

$$t_n \leq s_{2^n} \leq s,$$

so (t_n) is bounded. Hence $\sum 2^{n-1} a_{2^n}$ converges by Lemma 4.12.

We now show that if $\sum a_n$ diverges, then $\sum 2^{n-1} a_{2^n}$ diverges. It will follow from Theorem 4.9 that $\sum 2^n a_{2^n}$ diverges, since if it converged so would $\frac{1}{2} \sum 2^n a_{2^n} = \sum 2^{n-1} a_{2^n}$ (see Remark 4.10).

Before we begin, we establish a helpful estimate similar to Equation (4.1). Since (a_n) is decreasing:

$$a_{2^{n+1}}, a_{2^{n+1}+1}, \ldots, a_{2^{n+2}-1} \leq a_{2^{n+1}}. \tag{4.2}$$

[6] There are 2^n terms in Equation (4.1), and 2^{n+1} terms in Equation (4.2).

We prove that $s_{2^{n+1}-1} \leq a_1 + 2t_n$ by Mathematical Induction. The base case holds since

$$s_{2^{1+1}-1} = s_3 = a_1 + a_2 + a_3 \leq a_1 + 2a_2 = a_1 + 2t_1.$$

Now suppose that $s_{2^{n+1}-1} \leq a_1 + 2t_n$ holds for some n. Then by the induction hypothesis and the estimate in Equation (4.2), we have

$$\begin{array}{rcl} s_{2^{n+2}-1} & = & s_{2^{n+1}-1} + a_{2^{n+1}} + a_{2^{n+1}+1} + \cdots + a_{2^{n+2}-1} \\ & \leq & a_1 + 2t_n + a_{2^{n+1}} + a_{2^{n+1}+1} + \cdots + a_{2^{n+2}-1} \\ & \leq & a_1 + 2t_n + a_{2^{n+1}} + a_{2^{n+1}} + \cdots + a_{2^{n+1}} \\ & = & a_1 + 2t_n + 2^{n+1}a_{2^{n+1}} \\ & = & a_1 + 2(t_n + 2^n a_{2^{n+1}}) \\ & = & a_1 + 2t_{n+1}. \end{array}$$

We have proven that $s_{2^{n+1}-1} \leq a_1 + 2t_n$ for all n, and therefore

$$\frac{1}{2}(s_{2^{n+1}-1} - a_1) \leq t_n.$$

Since $\sum a_n$ diverges, its partial sum sequence $s_n \to \infty$ by Lemma 4.12. As a subsequence, $s_{2^{n+1}} \to \infty$ as well by Theorem 3.116, and so by the Squeeze Theorem (Theorem 3.114), $t_n \to \infty$. By Lemma 4.12, $\sum 2^{n-1}a_{2^n}$ diverges, completing the proof. □

We can now easily demonstrate that $p-$series converge if $p > 1$, and diverge otherwise[7].

Theorem 4.15. *The $p-$series $\sum \frac{1}{n^p}$ converges if $p > 1$, and diverges for all other values of p.*

Idea of proof. *This will follow from an easy application of the Cauchy Condensation Test (Theorem 4.14).*

Proof of Theorem 4.15. If $a_n = \frac{1}{n^p}$, then

$$2^n a_{2^n} = 2^n \frac{1}{(2^n)^p} = \left(\frac{1}{2^{p-1}}\right)^n.$$

This is a geometric series with $r = \frac{1}{2^{p-1}}$. By Example 4.5, this converges if $|r| < 1$, and diverges otherwise. If $p > 1$, then $r < 1$. If $p \leq 1$, then $|r| \geq 1$. Thus the series $\sum 2^n a_{2^n}$ converges when $p > 1$ and diverges if $p \leq 1$. By the Cauchy Condensation Test (Theorem 4.14), the $p-$series will converge if $p > 1$ and diverge for all other values of p. □

[7] In most calculus books, the Integral Test is used to prove this fact. We proceed without the need for that machinery.

For the intriguing value of $p = 1$ the series $\sum \frac{1}{n}$ diverges (see also Exercise 2 of Section 3.5). This series is called the *harmonic series*, a distant reference to a harmonic progression in music and their corresponding wavelengths.

Before moving on to discuss comparison tests in the next subsection, we give a short practice problem briefly illustrating a use of Theorem 4.15.

Practice 4.16. Determine whether or not $\sum \frac{1}{n \ln n}$ converges.

Methodology. *We will first apply the Cauchy Condensation Test to see that the convergence of the given series is linked to the convergence of* $\sum \frac{1}{\ln 2} \frac{1}{n}$*, a multiple of the divergent p–series* $\sum \frac{1}{n}$ *(here,* $p = 1$*). Thus,* $\sum \frac{1}{n \ln n}$ *diverges.*

Solution. If $a_n = \frac{1}{n \ln n}$, then

$$2^n a_{2^n} = 2^n \frac{1}{2^n \ln(2^n)} = \frac{1}{\ln 2} \cdot \frac{1}{n}.$$

The series $\sum \frac{1}{n}$ diverges, and so $\sum \frac{1}{\ln 2} \frac{1}{n}$ diverges by Theorem 4.9 (see Remark 4.10). By the Cauchy Condensation Test (Theorem 4.14), $\sum \frac{1}{n \ln n}$ diverges as well. ■

4.2.2 Comparison Tests

There are two common comparison tests for series with nonnegative terms: the Direct Comparison Test (Theorem 4.17), and the Limit Comparison Test (Theorem 4.20). The Direct Comparison Test intuitively asserts that a series "smaller" than a convergent series must also converge, and a series that is "larger" than a divergent series must also diverge. In a more subtle way, the Limit Comparison Test links the convergence or divergence of a series to another that behaves in roughly the same way. Generally, p–series are commonly (but not exclusively) used to make these comparisons.

Theorem 4.17 (Direct Comparison Test). *Let* $\sum a_n$ *and* $\sum b_n$ *be series with nonnegative terms. Suppose that* $a_n \leq b_n$.

1. *If* $\sum b_n$ *converges, then* $\sum a_n$ *converges, and* $\sum a_n \leq \sum b_n$.
2. *If* $\sum a_n$ *diverges, then* $\sum b_n$ *diverges.*

Idea of proof. *We need only prove the first assertion, as the second is the contrapositive of the first. If* $\sum b_n$ *converges and* $b_n \geq 0$*, then the partial sum sequence for* $\sum b_n$ *is increasing and bounded above by* $\sum b_n$*. Since* $a_n \leq b_n$*, the partial sum sequence for* $\sum a_n$ *must also be bounded (and increasing), so* $\sum a_n$ *converges by Lemma 4.12.*

Proof of Theorem 4.17. Let $s_n = \sum_{k=1}^n a_k$ and $t_n = \sum_{k=1}^n b_k$ be the partial sum sequences for $\sum a_n$ and $\sum b_n$, respectively. Since $a_n \leq b_n$, we have $s_n \leq t_n$. Since $\sum b_n$ converges and $b_n \geq 0$, the sequence $t_n \to t$ is convergent and increasing by Lemma 4.12. Thus, $t_n \leq t$ for all n, and so $s_n \leq t_n \leq t$ for all

n. So $\sum a_n$ converges by Lemma 4.12. In addition, since $s_n \leq t$, and $s_n \to s$ converges, we have $s \leq t$. Or, in other words, $\sum a_n \leq \sum b_n$. □

Remark 4.18. The Direct Comparison Test is also true if $a_n \leq b_n$ *eventually*, see Exercise 2.

We give an example illustrating both a feature and a shortcoming of the Direct Comparison Test.

Example 4.19. This example shows that $\sum \frac{1}{n^2+2}$ can easily be shown to converge using the Direct Comparison Test, but that at first glance this test does not seem to apply to the similar series $\sum \frac{1}{n^2-2}$. Since $\frac{1}{n^2-2} \geq 0$ for $n \geq 2$, all infinite sums in this example are from $n = 2$ to ∞ to allow for the use of the Direct Comparison Test. Our results will be the same should the sums begin at $n = 1$, since this differs by only a finite number of terms (see Exercise 2 of Section 4.1).

The $p-$series $\sum \frac{1}{n^2}$ converges (Theorem 4.15), and $\frac{1}{n^2+2} \leq \frac{1}{n^2}$. So by direct comparison (Theorem 4.17), $\sum \frac{1}{n^2+2}$ converges.

The problem with applying this test to compare $\sum \frac{1}{n^2}$ and $\sum \frac{1}{n^2-2}$ is that $\frac{1}{n^2} \leq \frac{1}{n^2-2}$, and so the hypotheses of the Direct Comparison Test are not met. It seems reasonable that since $\frac{1}{n^2+2}$ and $\frac{1}{n^2-2}$ behave so similarly, and that $\sum \frac{1}{n^2+2}$ converges, that $\sum \frac{1}{n^2-2}$ should converge as well. We simply need a different estimate, which we provide below.

It can be easily checked that for $n \geq 2$, we have $\frac{1}{n^2-2} \leq \frac{2}{n^2}$. Since $\sum \frac{1}{n^2}$ converges (Theorem 4.15, for $p = 2$), $\sum \frac{2}{n^2}$ converges as well by Theorem 4.9, and so by direct comparison, $\sum \frac{1}{n^2-2}$ converges as well. ■

In the last example, we saw that sometimes sequences that behave similarly should form series that either both converge or both diverge. However, we also saw that a direct comparison may not be so obvious. The Limit Comparison Test helps to alleviate this issue by comparing the long-term behavior of the sequences involved.

Theorem 4.20 (Limit Comparison Test). *Let $\sum a_n$ and $\sum b_n$ be series with nonnegative terms. Suppose also that $b_n > 0$, and that $\frac{a_n}{b_n} \to L$.*

1. *If L is positive and finite, then both $\sum a_n$ and $\sum b_n$ converge or both diverge.*
2. *If $L = 0$ and $\sum b_n$ converges, then $\sum a_n$ converges.*
3. *If $L = \infty$ and $\sum b_n$ diverges, then $\sum a_n$ diverges.*

Idea of proof. *In Part 1, since $\frac{a_n}{b_n} \to L \in (0, \infty)$, we know that $\frac{a_n}{b_n}$ is eventually between $L/2$ and $3L/2$, and so a_n is trapped between multiples of b_n for these n. Applying a direct comparison to the result will complete the proof. The proofs of the other two assertions are similar.*

Proof of Theorem 4.20. To prove Part 1, suppose $\frac{a_n}{b_n} \to L \in (0, \infty)$. Eventually (choosing $\epsilon = \frac{L}{2} > 0$),

$$\frac{L}{2} < \frac{a_n}{b_n} < \frac{3L}{2}, \text{ so } \frac{L}{2} b_n < a_n < \frac{3L}{2} b_n.$$

If $\sum a_n$ converges, then so must $\sum \frac{L}{2} b_n$ by the Direct Comparison Test (see Exercise 2), and so must $\sum b_n$ by Theorem 4.9. Similarly, if $\sum a_n$ diverges, then so must $\sum \frac{3L}{2} b_n$, and so must $\sum b_n$.

Now suppose $L = 0$, and that $\sum b_n$ converges. Eventually, $\frac{a_n}{b_n} < 1$, and so $a_n < b_n$ for these n. Since $\sum b_n$ converges, $\sum a_n$ converges by the Direct Comparison Test (Theorem 4.17 and Exercise 2).

The proof of the last assertion is given as Exercise 5. □

We close the section with a practice problem that illustrates the use of the Limit Comparison Test. Generally, one wants to choose a series $\sum b_n$ to compare the given series in a way where the Limit Comparison Test would be useful. This requires that you know whether or not $\sum b_n$ converges or diverges, and that such a choice would allow the Limit Comparison Test to apply.

Practice 4.21. Determine the convergence or divergence of the following series.

(a) $\sum \frac{n}{n^5-3}$

(b) $\sum \frac{\ln n}{n^3+1}$

(c) $\sum \frac{\sin(n)+5}{\sqrt{n}}$

Methodology. *For Part (a), we choose $b_n = \frac{1}{n^4}$. We arrived at this choice by noting that polynomials tend to behave most like their leading term, so $n^5 - 3$ is very similar to n^5 for large n. As a result, $\frac{n}{n^5-3}$ should be very similar to $\frac{n}{n^5} = \frac{1}{n^4}$ for large n. For Part (b), we find that $b_n = \frac{1}{n^3}$ is not a suitable comparison, since $\frac{a_n}{b_n} \to \infty$ but $\sum b_n$ converges. Choosing $b_n = \frac{1}{n^2}$ will work instead, giving us $L = 0$. In Part (c), we find that comparing this to $b_n = \frac{1}{\sqrt{n}}$ is not suitable, since it appears that $\frac{a_n}{b_n}$ does not converge. Instead, we try $b_n = \frac{1}{n}$.*

Solution. (a) Let $b_n = \frac{1}{n^4}$. Then

$$\frac{a_n}{b_n} = \frac{\frac{n}{n^5-3}}{\frac{1}{n^4}} = \frac{n^5}{n^5 - 3} \to 1.$$

Since $\sum \frac{1}{n^4}$ is convergent by Theorem 4.15, the series $\sum \frac{n}{n^5-3}$ converges by the Limit Comparison Test (Theorem 4.20).

(b) Let $b_n = \frac{1}{n^2}$. Then

$$\frac{a_n}{b_n} = \frac{\frac{\ln n}{n^3+1}}{\frac{1}{n^2}} = \frac{\ln n}{n} \frac{n^3}{n^3 + 1} \to 0.$$

Since $\sum \frac{1}{n^2}$ converges by Theorem 4.15, and $\frac{a_n}{b_n} \to 0$, the series $\sum \frac{\ln n}{n^3+1}$ converges.

(c) Let $b_n = \frac{1}{n}$. Then

$$\frac{a_n}{b_n} = \frac{\frac{\sin(n)+5}{\sqrt{n}}}{\frac{1}{n}} = (\sin(n) + 5)\sqrt{n}.$$

Notice that $3 < \sin(n) + 5 < 7$ is bounded, and $\sqrt{n} \to \infty$. By Exercise 15 of Section 3.8, $(\sin(n)+5)\sqrt{n} \to \infty$. So, since $\sum \frac{1}{n}$ diverges (Theorem 4.15), the Limit Comparison Test says that $\sum \frac{\sin(n)+5}{\sqrt{n}}$ diverges as well. ■

4.2.3 Review of important concepts from Section 4.2

1. Series with nonnegative terms are particularly nice since their sequence of partial sums is increasing. Thus, such a series will converge if this partial sum sequence is bounded, and diverge if it is unbounded (Lemma 4.12). This observation is critical in proving the convergence tests found in this section.
2. The Cauchy Condensation Test (Theorem 4.14) provides a characterization for the convergence of series with nonnegative terms. We use this to determine the convergence of $p-$series.
3. The $p-$series $\sum \frac{1}{n^p}$ converges if $p > 1$ and diverges for all other p. We use these series often in comparisons since we have such a clear picture as to whether or not any of them converge.
4. The Direct Comparison Test (Theorem 4.17) and Limit Comparison Test (Theorem 4.20) provide a means to check for convergence or divergence if you are able to adequately compare your series to one that is known to either converge or diverge.

Exercises for Section 4.2

1. Determine if the following series converge.

   (a) $\sum \frac{1}{\sqrt{2n-1}}$
   (b) $\sum \frac{n}{\sqrt{n^5+1}}$
   (c) $\sum \frac{\sqrt{n+1}-\sqrt{n}}{n}$
   (d) $\sum(\sqrt{n+1} - \sqrt{n})$
   (e) $\sum \frac{2^n}{3^n+1}$
   (f) $\sum \frac{3^n}{2^n+1}$
   (g) $\sum \frac{\ln n}{n^2+1}$
   (h) $\sum(\sqrt[n]{n} - 1)^n$
   (i) $\sum(\sqrt[n]{n} - 1)$

2. Prove that if $0 \leq a_n \leq b_n$ eventually, and $\sum b_n$ converges, then $\sum a_n$ converges. (This is a slightly more general version of the Direct Comparison Test, Theorem 4.17.)

3. Suppose $a_n, b_n \geq 0$. If $\sum a_n$ converges and (b_n) is bounded, prove that $\sum a_n b_n$ converges.
4. Prove that $\sum \frac{\sin^2(n)}{n^2}$ converges.
5. Suppose that $a_n \geq 0$ and $b_n > 0$. If $\frac{a_n}{b_n} \to \infty$ and $\sum b_n$ diverges, prove that $\sum a_n$ diverges.
6. If $a_n > 0$ and $\sum a_n$ converges, prove that $\sum a_n^2$ converges.
7. If $a_n > 0$ and $\sum a_n$ converges, provide examples that show that $\sum \sqrt{a_n}$ could converge or diverge.
8. This exercise demonstrates that $\gamma_n = (1 + \frac{1}{2} + \cdots + \frac{1}{n}) - \ln(n)$ converges. Its limit is known as the Euler-Mascheroni constant, and is usually denoted γ. For this problem, we shall use the result of Exercise 5 of Section 6.2, namely, we assume that for $x \geq 1$, we have
$$\frac{1}{x+1} \leq \ln(x+1) - \ln(x) = \ln\left(1 + \frac{1}{x}\right) \leq \frac{1}{x}.$$
   (a) Show that $1 + \frac{1}{2} + \cdots + \frac{1}{n} - \ln(n) = \frac{1}{n} + \sum_{k=1}^{n-1}[\frac{1}{k} - \ln(1 + \frac{1}{k})]$.
   (b) Observe that $\frac{1}{k} - \ln(1+\frac{1}{k}) \geq 0$, so the terms in $\sum_{k=1}^{\infty}[\frac{1}{k} - \ln(1+\frac{1}{k})]$ are positive.
   (c) Prove that $\sum_{k=1}^{\infty}[\frac{1}{k} - \ln(1 + \frac{1}{k})]$ converges by comparing $\frac{1}{k} - \ln(1+\frac{1}{k}) \leq \frac{1}{k^2}$. Deduce that $\frac{1}{n} + \sum_{k=1}^{n-1}[\frac{1}{k} - \ln(1+\frac{1}{k})]$ converges.
9. This exercise is one of the main results of [14], and provides a different proof that $p-$series converges for $p > 1$. More partial sum estimates can be found there as well. Let $p > 1$ and let $s_n = \sum_{k=1}^{n} \frac{1}{n^p}$ be the partial sum sequence of the $p-$series $\sum_{n=1}^{\infty} \frac{1}{n^p}$.
   (a) Prove that $s_n \leq s_{2n}$.
   (b) Prove that
$$\left[\frac{1}{2^p} + \frac{1}{4^p} + \cdots + \frac{1}{(2n)^p}\right] + \left[\frac{1}{3^p} + \frac{1}{5^p} + \cdots + \frac{1}{(2n-1)^p}\right] < \frac{2}{2^p}s_n,$$
   and that as a result,
$$s_{2n} < 1 + \frac{2}{2^p}s_n.$$
   (c) Use the previous estimates to show that $s_n < \frac{2^p}{2^p - 2}$, hence it is bounded, and that (s_n) converges[8].
   (d) Use this estimate to find an upper bound for $\sum_{n=1}^{\infty} \frac{1}{n^2}$ and $\sum_{n=1}^{\infty} \frac{1}{n^3}$.

[8]Where do you use the hypothesis that $p > 1$?

4.3 Other Convergence Tests for Series

The last section outlined a number of tests for convergence, although these only applied to series with nonnegative terms. The question of convergence becomes more difficult without this assumption. This section introduces the Alternating Series Test (Theorem 4.24) as derived from Dirichlet's Test (Theorem 4.23) as the first convergence tests we have that apply to series with terms that could be positive or negative. We also introduce the notions of absolute and conditional convergence, and introduce the Ratio and Root Tests (Theorems 4.31 and 4.32).

4.3.1 Dirichlet's Test and the Alternating Series Test

Our first test for convergence of series whose terms may be negative is Dirichlet's Test (Theorem 4.23). This will follow from Abel's Lemma (Lemma 4.22), and will imply the familiar Alternating Series Test (Theorem 4.24).

We begin by presenting Abel's Lemma, sometimes also called *summation by parts*.

Lemma 4.22 (Abel's Lemma). *Let (a_n) and (b_n) be sequences, and let $s_n = \sum_{k=1}^n a_k$ be the sequence of partial sums of (a_n). Then*

$$\sum_{k=1}^{n} a_k b_k = \sum_{k=1}^{n} s_k(b_k - b_{k+1}) + s_n b_{n+1}.$$

Idea of proof. *Notice again that $s_k - s_{k-1} = a_k$ for every n. Using this and a reindexing of the sum will yield the result.*

Proof of Theorem 4.22. For the sake of ease, define $s_0 = 0$. Then

$$\begin{array}{rcl} \sum_{k=1}^n a_k b_k & = & \sum_{k=1}^n (s_k - s_{k-1}) b_k \\ & = & \sum_{k=1}^n s_k b_k - \sum_{k=1}^n s_{k-1} b_k \\ & = & \sum_{k=1}^n s_k b_k - \sum_{k=1}^n s_k b_{k+1} + s_n b_{n+1} \\ & = & \sum_{k=1}^n s_k (b_k - b_{k+1}) + s_n b_{n+1}. \end{array}$$

□

We can now prove Dirichlet's Test, which gives conditions on when $\sum a_n b_n$ converges.

Theorem 4.23 (Dirichlet's Test). *Suppose $\sum a_n$ is a series whose partial sum sequence $s_n = \sum_{k=1}^n a_k$ is bounded. If $b_n \to 0$ and is monotone decreasing, then $\sum a_n b_n$ converges.*

Idea of proof. *We express the partial sum sequence of $\sum a_n b_n$ in terms of Abel's Lemma (Lemma 4.22). By Exercise 23 of Section 3.3, since $b_n \to 0$*

and (s_n) *is bounded,* $s_n b_{n+1} \to 0$*. Thus, the only concern is to show that* $\sum_{k=1}^{n} s_k(b_k - b_{k+1})$ *converges. Using the assumption that* (s_n) *is bounded, we can apply the Cauchy Criterion (Theorem 4.11) to show it converges.*

Proof of Theorem 4.23. We show that the partial sum sequence $\sum_{k=1}^{n} a_k b_k$ converges. By Abel's Lemma (Lemma 4.22), we have

$$\sum_{k=1}^{n} a_k b_k = \sum_{k=1}^{n} s_k(b_k - b_{k+1}) + s_n b_{n+1}.$$

Since $b_n \to 0$ and (s_n) is bounded, by Exercise 23 of Section 3.3, $s_n b_{n+1} \to 0$. Thus, if we can show that $\sum_{k=1}^{n} s_k(b_k - b_{k+1})$ converges, we will be done.

This sequence is, however, the partial sum sequence of the series $\sum_{k=1}^{\infty} s_k(b_k - b_{k+1})$, and so the sequence $\sum_{k=1}^{n} s_k(b_k - b_{k+1})$ converges if and only if $\sum_{k=1}^{\infty} s_k(b_k - b_{k+1})$ converges, by Definition 4.1. We show that this series converges using the Cauchy Criterion for Convergence (Theorem 4.11). Let $\epsilon > 0$ be given. Since (s_n) is bounded, there exists an M so that $|s_n| \leq M$ for all n. This condition is vacuous unless $M \geq 0$. Thus for every $n > m$, we have

$$\begin{aligned} |\textstyle\sum_{k=m+1}^{n} s_k(b_k - b_{k+1})| &\leq \textstyle\sum_{k=m+1}^{n} |s_k(b_k - b_{k+1})| \\ &\leq \textstyle\sum_{k=m+1}^{n} M|b_k - b_{k+1}|. \end{aligned} \tag{4.3}$$

Since (b_n) is decreasing, we have $b_k \geq b_{k+1}$, so $|b_k - b_{k+1}| = b_k - b_{k+1}$. Moreover, the following sum telescopes:

$$\begin{aligned} \textstyle\sum_{k=m+1}^{n}(b_k - b_{k+1}) &= (b_{m+1} - b_{m+2}) + (b_{m+2} - b_{m+3}) + \cdots \\ &\qquad \cdots + (b_n - b_{n+1}) \\ &= b_{m+1} - b_{n+1}. \end{aligned}$$

Applying this information to Equation (4.3), we have

$$\sum_{k=m+1}^{n} M|b_k - b_{k+1}| = M(b_{m+1} - b_{n+1}).$$

Since $b_n \to 0$, it is Cauchy (Theorem 3.97). So there exists a N so that for every $n, m \geq N$, we have $|b_m - b_n| < \frac{\epsilon}{M+1}$. Thus for $n, m \geq N$, $n+1, m+1 \geq N$ as well, and

$$M(b_{m+1} - b_{n+1}) < M\frac{\epsilon}{M+1} < \epsilon.$$

□

Dirichlet's Test may be used to easily prove the familiar Alternating Series Test. We note that series of the form $\sum(-1)^n b_n$ are called *alternating* if $b_n > 0$ because the sign of successive terms in this sum alternates from positive to negative[9].

[9]Since $(-1)\sum(-1)^n b_n = \sum(-1)^{n+1} b_n$ when it converges, these sums are also called alternating if $b_n > 0$.

Theorem 4.24 (Alternating Series Test). *If $b_n \to 0$ and is decreasing, then $\sum(-1)^n b_n$ converges.*

Idea of proof. *Since the partial sums of the sum $\sum(-1)^n$ are bounded, this is a direct consequence of Dirichlet's Test (Theorem 4.23).*

Proof of Theorem 4.24. Let $a_n = (-1)^n$. The partial sums of $\sum a_n$ proceed as $-1, 0, -1, 0, \ldots$, and are therefore bounded. Thus $\sum a_n b_n = \sum(-1)^n b_n$ converges by Dirichlet's Test (Theorem 4.23). □

Practice 4.25. Show that $\sum \frac{(-1)^n}{n}$ and $\sum \frac{(-2)^n}{3^n}$ converge.

Methodology. *Both of these are alternating series, so we apply the Alternating Series Test (Theorem 4.24).*

Solution. For the first series, we notice that $\frac{1}{n} \to 0$, and that $\frac{1}{n} \geq \frac{1}{n+1}$ is decreasing. So, the Alternating Series Test applies, and $\sum \frac{(-1)^n}{n}$ converges.

In the second series, we have $(-2)^n = (-1)^n 2^n$, and so $\frac{(-2)^n}{3^n} = (-1)^n \left(\frac{2}{3}\right)^n$. Since $\left(\frac{2}{3}\right)^n \to 0$ and is decreasing, the Alternating Series Test again applies, and this series converges. ■

Another powerful test known as Abel's Test can be derived from Dirichlet's Test, see Exercise 5. We give an example here of how Dirichlet's Test can be helpful in the study of trigonometric series.

Example 4.26. This example illustrates how to use Dirichlet's Test (Theorem 4.23) to illustrate that if $b_n \to 0$ and is monotonic, then $\sum b_n \sin(nx)$ converges for all $x \in \mathbb{R}$. We use the following trigonometric identity to help us, which is valid for every $x \neq 2\pi m$ for $m \in \mathbb{Z}$:

$$\sum_{k=1}^{n} \sin(kx) = \frac{\cos(\frac{1}{2}x) - \cos(n + \frac{1}{2})x}{2\sin(\frac{1}{2}x)}.$$

So that

$$\left|\sum_{k=1}^{n} \sin(kx)\right| = \left|\frac{\cos(\frac{1}{2}x) - \cos((n + \frac{1}{2})x)}{2\sin(\frac{1}{2}x)}\right| \leq \frac{1}{|\sin(\frac{1}{2}x)|}.$$

If $x = 2\pi m$ for some integer m, then $\sum_{k=1}^{n} \sin(kx) = 0$. In either case, the partial sums of $a_n = \sin(kx)$ are bounded for every $x \in \mathbb{R}$, and so for any sequence $b_n \to 0$ that is decreasing, we conclude $\sum b_n \sin(nx)$ converges by Dirichlet's Test. For example, $b_n = \frac{1}{n}$ is such a sequence, and so when $x = 1$ we have $\sum \frac{\sin(n)}{n}$ converges. ■

4.3.2 Absolute and conditional convergence

It seems that series with nonnegative terms behave better, since there are a number of straightforward tests for convergence of these that we saw in the last section. One might wonder if the theory of convergence of series could be reduced to only considering series with nonnegative terms in a hope that it would be easier. Of course, this is not the case. Consider for a moment both of the alternating series in Practice 4.25. We saw that both of these series converge, but we note that if we were to sum the absolute value of each term in each series, we would find

$$\sum \left| \frac{(-1)^n}{n} \right| = \sum \frac{1}{n} \text{ diverges, while} \sum \left| \frac{(-2)^n}{3^n} \right| = \sum \left(\frac{2}{3} \right)^n \text{ converges.}$$

Thus there is sometimes a significant difference if we were to simply replace each term in our sum with its positive counterpart. There is, however, an advantage of doing this in terms of testing convergence, as we will see in Theorem 4.28. We distinguish series that behave a certain way in the event we take the absolute value of each term as follows.

Definition 4.27. If $\sum a_n$ is a series, and

1. $\sum |a_n|$ converges, then we say that $\sum a_n$ is *absolutely convergent.*

2. $\sum |a_n|$ diverges while $\sum a_n$ converges, then we say that $\sum a_n$ is *conditionally convergent.*

Thus, our discussion above illustrates that $\sum \frac{(-1)^n}{n}$ conditionally converges, while $\sum \frac{(-2)^n}{3^n}$ absolutely converges. The following theorem demonstrates that if $\sum a_n$ absolutely converges, then it converges.

Theorem 4.28. *If $\sum a_n$ absolutely converges, then $\sum a_n$ converges.*

Idea of proof. *We use the Cauchy Criterion for Convergence (Theorem 4.11) to show $\sum a_n$ converges. This will follow by a use of the Triangle Inequality.*

Proof of Theorem 4.28. We show $\sum a_n$ converges using the Cauchy Criterion for Convergence (Theorem 4.11). Let $\epsilon > 0$ be given. Since $\sum |a_n|$ converges, there exists an N so that if $n > m \geq N$, $\sum_{k=m+1}^{n} |a_k| < \epsilon$. For the same $n > m \geq N$, by the Triangle Inequality we have

$$\left| \sum_{k=m+1}^{n} a_k \right| \leq \sum_{k=m+1}^{n} |a_k| < \epsilon.$$

□

Remark 4.29. If $a_n \geq 0$, then there is no difference between absolute convergence and convergence of $\sum a_n$. Thus, the tests we encountered in Section 4.2 concerning series with nonnegative terms are all tests for absolute convergence.

Practice 4.30. Show that $\sum \frac{(-1)^n}{n^p}$ converges conditionally for $0 < p \leq 1$, and absolutely converges for $p > 1$.

Methodology. *This will follow from a knowledge of p−series and the Alternating Series Test (Theorem 4.24).*

Solution. Notice that $\left|\frac{(-1)^n}{n^p}\right| = \frac{1}{n^p}$. As a p−series, $\sum \frac{1}{n^p}$ converges for $p > 1$ (Theorem 4.15), so $\sum \frac{(-1)^n}{n^p}$ absolutely converges for $p > 1$.

If $p > 0$, then $\frac{1}{n^p} \to 0$ and is decreasing, so $\sum \frac{(-1)^n}{n^p}$ converges by the Alternating Series Test (Theorem 4.24). However, if $0 < p \leq 1$, then $\sum \frac{1}{n^p}$ diverges. Thus, $\sum \frac{(-1)^n}{n^p}$ conditionally converges if $0 < p \leq 1$. ■

4.3.3 The Ratio and Root Tests

There are two common tests for absolute convergence that should be familiar: The Ratio Test (Theorem 4.31) and The Root Test (Theorem 4.32). Both of these tests seek to find a relationship between a given series and a geometric one; the following discussion is a brief motivation for the Ratio Test (The Root Test has a similar motivation).

Notice that one of the defining features of the sequence $a_n = r^n$ is that $\frac{a_{n+1}}{a_n} = r$. Clearly, the sequence $b_n = nr^n$ is not geometric, although

$$\frac{|b_{n+1}|}{|b_n|} = \frac{(n+1)|r|^{n+1}}{n|r|^n} = |r|\frac{n+1}{n}.$$

We find that

$$\frac{|a_{n+1}|}{|a_n|} \to |r|, \text{ and } \frac{|b_{n+1}|}{|b_n|} \to |r| \text{ as well.}$$

Thus it seems that $b_n = nr^n$ behaves somewhat similarly to a geometric sequence, and that through comparison we should be able to determine if $\sum nb_n$ as a result. Since comparisons of series are only valid for series with nonnegative terms, we could only really hope to conclude that the series either absolutely converges or diverges. After presenting the Ratio and Root Tests, Practice 4.34 continues this thought in a concrete example.

Theorem 4.31 (Ratio Test). *Suppose $\sum a_n$ is a series, and*

$$L = \lim_{n\to\infty} \frac{|a_{n+1}|}{|a_n|}.$$

If $L < 1$, then $\sum a_n$ absolutely converges. If $L > 1$ then $\sum a_n$ diverges.

Idea of proof. *The hypotheses allow us to make a direct comparison of $|a_n|$ to a geometric sequence. From there, we use facts about geometric series to reach our conclusions.*

Proof of Theorem 4.31. First note that since $\frac{|a_{n+1}|}{|a_n|} \geq 0$, we must have $L \geq 0$

(Theorem 3.48). Suppose $L < 1$. Choose r with $L < r < 1$. Since $\frac{|a_{n+1}|}{|a_n|} \to L$, there exists an N so that for every $n \geq N$, $\frac{|a_{n+1}|}{|a_n|} < r$. For these n, we must have $|a_{n+1}| < r|a_n|$. Thus for $n > N$,

$$|a_n| \leq r|a_{n-1}| \leq r^2|a_{n-2}| \leq \cdots \leq r^{n-N}|a_N| = \frac{|a_N|}{r^N}r^n.$$

In other words, eventually $|a_n| \leq \frac{|a_N|}{r^N}r^n$. By the Direct Comparison Test (Theorem 4.17) and the fact that $\sum \frac{|a_N|}{r^N}r^n$ converges since $0 < r < 1$, we conclude that $\sum |a_n|$ converges. Thus, if $L < 1$ then $\sum a_n$ absolutely converges.

Now suppose $L > 1$, and choose r with $1 < r < L$. Similar to above, since $\frac{|a_{n+1}|}{|a_n|} \to L$, there exists an N so that for every $n \geq N$, $\frac{|a_{n+1}|}{|a_n|} > r$. For these n, we must have $|a_{n+1}| > r|a_n|$. Thus for $n > N$,

$$|a_n| \geq r|a_{n-1}| \geq r^2|a_{n-2}| \geq \cdots \geq r^{n-N}|a_N| = \frac{|a_N|}{r^N}r^n.$$

In other words, eventually $|a_n| \geq \frac{|a_N|}{r^N}r^n$. Since $r > 1$, $r^n \to \infty$, and so $|a_n| \to \infty$. Thus (a_n) does not converge to 0. So, by the Test for Divergence (Theorem 4.6), $\sum a_n$ diverges. □

The Root Test makes use of another sort of comparison to a geometric series in its proof.

Theorem 4.32 (Root Test). *Suppose $\sum a_n$ is a series, and*

$$L = \lim_{n\to\infty} \sqrt[n]{|a_n|}.$$

If $L < 1$, then $\sum a_n$ absolutely converges. If $L > 1$ then $\sum a_n$ diverges.

Idea of proof. *Similar to the proof of the Ratio Test, our hypotheses allow us to make a direct comparison of $|a_n|$ to a geometric sequence. From there, we use facts about geometric series to reach our conclusions.*

Proof of Theorem 4.32. First note that since $\sqrt[n]{|a_n|} \geq 0$, we must have $L \geq 0$ (Theorem 3.48). Suppose $L < 1$. Choose r with $L < r < 1$. Since $\sqrt[n]{|a_n|} \to L$, there exists an N so that for every $n \geq N$, $\sqrt[n]{|a_n|} < r$, or that $|a_n| < r^n$. By the Direct Comparison Test (Theorem 4.17) and the fact that $\sum r^n$ converges since $0 < r < 1$, we conclude that $\sum |a_n|$ converges. Thus, if $L < 1$ then $\sum a_n$ absolutely converges.

Now suppose $L > 1$, and choose r with $1 < r < L$. Similar to above, since $\sqrt[n]{|a_n|} \to L$, there exists an N so that for every $n \geq N$, $\sqrt[n]{|a_n|} > r$, or that $|a_n| > r^n$. Since $r > 1$, $r^n \to \infty$, and so $|a_n| \to \infty$. Thus (a_n) does not converge to 0. So, by the Test for Divergence (Theorem 4.6), $\sum a_n$ diverges. □

Although the Root Test is a powerful tool in determining the convergence of series, it is unfortunately the case that $\lim_{n\to\infty} \sqrt[n]{|a_n|}$ does not always exist. The lim sup of a sequence always exists, and the following is a generalization of the Root Test (also referred to as "the Root Test") that we shall need in Section 7.4.

Theorem 4.33 (Root Test). *Suppose $\sum a_n$ is a series, and*

$$L = \limsup(\sqrt[n]{|a_n|}).$$

If $L < 1$, then $\sum a_n$ absolutely converges. If $L > 1$, then $\sum a_n$ diverges.

Idea of proof. *Similar to the proof of the Root and Ratio Tests, our hypotheses allow us to make a direct comparison of $|a_n|$ to a geometric sequence. The difference in replacing $\lim_{n\to\infty}$ with $\limsup$ will only be that a certain comparison may only be made after a certain point for a different reason. Other than this, everything is the same as the proof of Theorem 4.32.*

Proof of Theorem 4.33. First note that since $\sqrt[n]{|a_n|} \geq 0$, we must have $L \geq 0$ (Theorem 3.48). Suppose $L < 1$. Choose r with $L < r < 1$. There must be only finitely many of the $\sqrt[n]{|a_n|}$ that are greater than or equal to r: if there were infinitely many, we could treat these as a subsequence of $\sqrt[n]{|a_n|}$, and this sequence would have a subsequential limit greater than or equal to r. This would be a subsequential limit of $\sqrt[n]{|a_n|}$ larger than L, which contradicts $L = \limsup(\sqrt[n]{|a_n|})$. So, there exists an N so that for every $n \geq N$, $\sqrt[n]{|a_n|} < r$, or that $|a_n| < r^n$. By the Direct Comparison Test (Theorem 4.17) and the fact that $\sum r^n$ converges since $0 < r < 1$, we conclude that $\sum |a_n|$ converges. Thus, if $L < 1$ then $\sum a_n$ absolutely converges.

Now suppose $L > 1$, and choose r with $1 < r < L$. Similar to above, only finitely many of the $\sqrt[n]{|a_n|}$ are less than or equal to r, so there exists an N so that for every $n \geq N$, $\sqrt[n]{|a_n|} > r$, or that $|a_n| > r^n$. Since $r > 1$, $r^n \to \infty$, and so $|a_n| \to \infty$. Thus (a_n) does not converge to 0. So, by the Test for Divergence (Theorem 4.6), $\sum a_n$ diverges. □

The following practice problem is an example of our previous discussion motivating these theorems.

Practice 4.34. Determine whether or not $\sum \frac{(-1)^n n}{2^n}$ diverges, converges conditionally, or converges absolutely.

Methodology. *We apply the Ratio Test to see that $L = \frac{1}{2}$. Thus, this series absolutely converges.*

Solution. If $a_n = \frac{(-1)^n n}{2^n}$, then

$$\frac{|a_{n+1}|}{|a_n|} = \frac{\left|\frac{(-1)^{n+1}(n+1)}{2^{n+1}}\right|}{\left|\frac{(-1)^n n}{2^n}\right|} = \frac{1}{2}\frac{n+1}{n} \to \frac{1}{2} < 1.$$

This series absolutely converges by the Ratio Test (Theorem 4.31). ■

The following example points out that if $L = 1$ in either the Ratio or Root Tests, no information can be garnered and something else must be tried.

Example 4.35. This example is designed to illustrate that if one encounters $L = 1$ in either the Ratio or the Root Tests, there is nothing certain that can be said about the convergence of the series. We consider the series $\sum \frac{1}{n}, \sum \frac{(-1)^n}{n}$, and $\sum \frac{1}{n^2}$. Our results concerning $p-$series (Theorem 4.15) and the Alternating Series Test (Theorem 4.24) conclude that $\sum \frac{1}{n}$ diverges, $\sum \frac{(-1)^n}{n}$ conditionally converges, and $\sum \frac{1}{n^2}$ absolutely converges.

We attempt to apply the Ratio Test (Theorem 4.31) to these series. If $a_n = \frac{1}{n}$ or $\frac{(-1)^n}{n}$, then

$$\frac{|a_{n+1}|}{|a_n|} = \frac{n}{n+1} \to 1.$$

If $a_n = \frac{1}{n^2}$, then

$$\frac{|a_{n+1}|}{|a_n|} = \frac{n^2}{(n+1)^2} \to 1 \text{ as well.}$$

Now we attempt to apply the Root Test (Theorem 4.32) to these series. Recall from Theorem 3.55 that $\sqrt[n]{n} \to 1$. If $a_n = \frac{1}{n}$ or $\frac{(-1)^n}{n}$, then

$$\sqrt[n]{|a_n|} = \frac{1}{\sqrt[n]{n}} \to 1.$$

If $a_n = \frac{1}{n^2}$, then

$$\sqrt[n]{|a_n|} = \frac{1}{\sqrt[n]{n^2}} = \frac{1}{\sqrt[n]{n}^2} \to 1 \text{ as well.}$$

Thus if $L = 1$ in either of these tests, there is nothing certain that can be said about $\sum a_n$. ■

Since each of the Ratio and Root Tests utilizes a similar comparison in its proofs, it would seem that the two are equivalent in the sense that both will be conclusive or both will be inconclusive. We close this section with one final example from [8] that is designed to illustrate that this is not the case: the Root Test is a stronger result. The interested reader will find the Ratio and Root Tests stated more generally and a result indicating the dominance of the Root Test in Theorem 3.37 on page 68 in [24].

Example 4.36. This example will illustrate an instance where the Ratio Test fails, but the Root Test does not. Let $0 < r < 1$, and consider the series $\sum a_n = 2r + r^2 + 2r^3 + r^4 + \cdots$. In an attempt to apply the Ratio Test (Theorem 4.31),

$$\frac{|a_{n+1}|}{|a_n|} = \begin{cases} 2r & \text{if } n \text{ is even,} \\ \frac{1}{2}r & \text{if } n \text{ is odd.} \end{cases}$$

Here, $L = \lim_{n\to\infty} \frac{|a_{n+1}|}{|a_n|}$ does not exist. However, our attempt to use the Root Test will be more useful:

$$\sqrt[n]{|a_n|} = \begin{cases} r & \text{if } n \text{ is even,} \\ 2^{1/n} r & \text{if } n \text{ is odd.} \end{cases}$$

Here, we find that $\lim_{n\to\infty} \sqrt[n]{|a_n|} = r \in (0, 1)$. Thus, this series absolutely converges according to the Root Test, even though the Ratio Test is inconclusive. ■

4.3.4 Review of important concepts from Section 4.3

1. Dirichlet's Test (Theorem 4.23) is our first convergence test for series whose terms need not be nonnegative. The Alternating Series Test (Theorem 4.24) is an easy corollary of Dirichlet's Test.
2. The notion of Absolute and Conditional Convergence is defined in Definition 4.27. Absolute convergence implies convergence (Theorem 4.28).
3. The Ratio and Root Tests (Theorems 4.31 and 4.32) are common ways of testing for absolute convergence.

Exercises for Section 4.3

1. Determine if the following converge conditionally, converge absolutely, or diverge.

   (a) $\sum \frac{(-1)^n \ln n}{n}$

   (b) $\sum (-1)^n \frac{\sin(n\frac{\pi}{2})}{n^2}$

   (c) $\sum \frac{r^n}{n!}$ for any $r \in \mathbb{R}$

   (d) $\sum (-n) 2^n 3^{-n}$

   (e) $\sum \frac{(-1)^n n^n}{(n+1)^n}$

   (f) $\sum \frac{n!}{n^n}$

2. Give an example showing that the assumption that (b_n) is decreasing cannot be removed from the Alternating Series Test (Theorem 4.24).
3. Suppose that $b_n \to 0$, and that (b_n) is decreasing. Show that the following series converges:

$$b_1 + b_2 - b_3 - b_4 + b_5 + b_6 - b_7 - b_8 + \cdots .$$

4. Is it always the case that if $\sum a_n$ converges, then $\sum a_n^2$ converges? (Compare this with Exercise 6 from the last section.)
5. The following is known as **Abel's Test**: Suppose that $\sum a_n$ converges and (b_n) is monotone and bounded. Prove $\sum a_n b_n$ converges. *[Hint: Use Dirichlet's Test (Theorem 4.23).]*

6. Suppose $a_n > 0$, and that $\sum a_n$ converges. If (s_n) is the sequence of partial sums of $\sum a_n$, show that $\sum a_n s_n$ and $\sum \frac{a_n}{s_n}$ converge. *[Hint: Use Abel's Test in Exercise 5.]*

7. The following will prove the Alternating Series Test (Theorem 4.24) without relying on Dirichlet's Test (Theorem 4.23). It will also provide a way of estimating the sum of a convergent alternating series. Let $a_n \to 0$ and suppose (a_n) is decreasing. Let $s_n = \sum_{k=1}^{n}(-1)^{k+1}a_k$ be the partial sum sequence of $\sum_{n=1}^{\infty}(-1)^{n+1}a_n$.
   (a) Prove that (s_{2n}) is monotone increasing.
   (b) Prove that $s_{2n} \leq a_1$.
   (c) Prove that (s_{2n}) converges to some number s.
   (d) Prove that $s_{2n+1} \to s$ as well.
   (e) Prove that $s_n \to s$.
   (f) Prove that $|s_n - s| \leq a_{n+1}$.

8. Given that $\sum_{n=1}^{\infty} \frac{(-1)^{n+1}}{n} = \ln 2$ (see Exercise 9), use the estimate in Exercise 7 to determine N so that $\sum_{n=1}^{N} \frac{(-1)^{n+1}}{n}$ is within .0001 of $\ln 2$.

9. This exercise demonstrates that the alternating harmonic series $\sum_{n=1}^{\infty} \frac{(-1)^{n+1}}{n} = \ln(2)$.
   (a) Recall from Exercise 8 in Section 4.2 that

$$\gamma_n = 1 + \frac{1}{2} + \cdots + \frac{1}{n} - \ln(n) \to \gamma.$$

   Prove that $1 + \frac{1}{2} + \cdots + \frac{1}{2n-1} + \frac{1}{2n} - \ln(2n) \to \gamma$.
   (b) Prove that $t_n = 1 - \frac{1}{2} + \frac{1}{3} - \cdots + \frac{1}{2n-1} - \frac{1}{2n}$ converges, and that its limit, $s = \sum_{k=1}^{\infty} \frac{(-1)^{k+1}}{k}$.
   (c) Prove that $s = \sum_{k=1}^{\infty} \frac{(-1)^{k+1}}{k}$ is equal to $\ln(2)$ by showing that $t_n = \gamma_{2n} - \gamma_n + \ln(2)$, and taking the limit as $n \to \infty$.

10. Prove that if $\sum a_n$ absolutely converges, then the triangle inequality holds for infinite sums: $|\sum_{n=1}^{\infty} a_n| \leq \sum_{n=1}^{\infty} |a_n|$.

4.4 Rearrangements

Having discussed various convergence tests, we point out that not much has been done by way of actually determining what various series converge to. There are a variety of methods in a multitude of areas of mathematics to determine the value of certain series. With the exception of our brief discussion of Power Series in Chapter 7, we do not discuss these methods here. Rather, we point out something that Dirichlet noticed in 1827: certain infinite series would converge to different values if their terms were rearranged! Riemann eventually came to understand the specifics of this phenomenon in 1853, and a paper was published posthumously in 1866 containing what is now referred to as Riemann's Rearrangement Theorem (Theorem 4.40). The interested reader may find more about this history in [12].

This section presents Riemann's result after proving that rearrangements of absolutely convergent series remain absolutely convergent and converge to the same value (Theorem 4.38).

4.4.1 A rearrangement

We begin by considering a rearrangement of the conditionally convergent alternating harmonic series: $\sum_{n=1}^{\infty} \frac{(-1)^{n+1}}{n} = \ln(2)$ (we saw that this series converges to $\ln(2)$ in Exercise 9 in the last section). We rearrange this series by writing one (positive) odd term followed by two (negative) even terms[10]. We shall find something rather strange when we compare it to the original series:

$$\begin{aligned} & 1 - \tfrac{1}{2} - \tfrac{1}{4} + \tfrac{1}{3} - \tfrac{1}{6} - \tfrac{1}{8} + \tfrac{1}{5} - \tfrac{1}{10} - \tfrac{1}{12} + \cdots \\ = & \left(1 - \tfrac{1}{2}\right) - \tfrac{1}{4} + \left(\tfrac{1}{3} - \tfrac{1}{6}\right) - \tfrac{1}{8} + \left(\tfrac{1}{5} - \tfrac{1}{10}\right) - \tfrac{1}{12} + \cdots \\ = & \tfrac{1}{2} - \tfrac{1}{4} + \tfrac{1}{6} - \tfrac{1}{8} + \tfrac{1}{10} - \cdots \\ = & \tfrac{1}{2} \sum_{n=1}^{\infty} \tfrac{(-1)^{n+1}}{n} \\ = & \tfrac{1}{2} \ln 2. \end{aligned}$$

There are only two explanations for the apparent outcome that we can rearrange $\sum_{n=1}^{\infty} \frac{(-1)^{n+1}}{n} = \ln(2)$ into something that equals $\frac{1}{2}\ln 2$: either inserting parentheses is not allowed, or rearranging a sum may change its value. Inserting parentheses has the effect of passing to a subsequence of partial sums, and so if the series converges there must be no change in its value (see page 157 of [8] for a more formal illustration of this). Therefore, we have to accept the somewhat surprising conclusion that a rearrangement of a series can alter its value.

[10] It is suggested in [21] that Laurent is responsible for this rearrangement. In fact, the interested reader will find many examples of rearrangements in [21], published in 1906.

4.4.2 Riemann's Rearrangement Theorem

Upon deeper reflection, it should actually come as no surprise that a rearranged sum might actually differ in value from the original sum: by definition, the value of a series is the limit of its partial sum sequence. Changing the order of summation very likely changes this sequence[11], and so it very likely changes what its limit is. We shall find out that absolutely convergent series will converge to the same value under any rearrangement, while conditionally convergent series may not. We begin by formally defining what a rearrangement is, and then go on to prove these results.

Definition 4.37. Let $\sigma : \mathbb{N} \to \mathbb{N}$ be a bijection. The series $\sum b_n$ is a *rearrangement* of $\sum a_n$ if $b_n = a_{\sigma(n)}$.

Absolutely convergent series behave very well under rearrangement, in that the rearranged series must still be absolutely convergent, and the rearranged series must converge to the same value.

Theorem 4.38. *Suppose $\sum a_n$ is absolutely convergent, and $\sum_{n=1}^{\infty} a_n = s$. If $\sum b_n$ is a rearrangement of $\sum a_n$, then $\sum b_n$ is absolutely convergent and $\sum_{n=1}^{\infty} b_n = s$.*

Idea of proof. *We use the Cauchy Criterion (Theorem 4.11) to prove that $\sum b_n$ is absolutely convergent. To prove that $\sum b_n$ has the same value, we show that the difference of their partial sums tends to 0.*

Proof of Theorem 4.38. Let $\sigma : \mathbb{N} \to \mathbb{N}$ be a bijection, and define $b_{\sigma(n)}$ so that $\sum b_n$ is a rearrangement of $\sum a_n$. Assume $\sum a_n$ is absolutely convergent, and let $\epsilon > 0$ be given. Since $\sum |a_n|$ converges, the Cauchy Criterion (Theorem 4.11) guarantees the existence of an $N \in \mathbb{N}$ so that if $n > m \geq N$, we have $\sum_{k=m+1}^{n} |a_k| < \frac{\epsilon}{2}$. In particular, choosing $m = N$, for every n we must have $\sum_{k=N+1}^{n} |a_k| < \frac{\epsilon}{2}$. The series $\sum_{k=N+1}^{\infty} |a_k|$ converges, and each of its partial sums is bounded by $\frac{\epsilon}{2}$, so

$$\sum_{k=N+1}^{\infty} |a_k| \leq \infty.$$

Since σ is a bijection, σ^{-1} exists. Set

$$N_1 = \max\{\sigma^{-1}(1), \sigma^{-1}(2), \ldots, \sigma^{-1}(N)\}.$$

If $k > N_1$, it must be the case that $\sigma(k) > N$. If $\sigma(k) = j \leq N$, then $\sigma^{-1}(j) = k \leq N_1$, a contradiction. Now for $n > m \geq N_1$,

$$\sum_{k=m+1}^{n} |b_k| = \sum_{k=m+1}^{n} |a_{\sigma(k)}| \leq \sum_{k=N+1}^{\infty} |a_k| \leq \frac{\epsilon}{2} < \epsilon.$$

[11]This change is very likely not a mere rearrangement of the partial sum sequence. See Exercise 6.

We have shown that $\sum |b_n|$ converges by the Cauchy Criterion (Theorem 4.11), and so by definition $\sum b_n$ is absolutely convergent.

We now show that $\sum_{n=1}^{\infty} a_n = \sum_{n=1}^{\infty} b_n$. Let $s_n = \sum_{k=1}^{n} a_n$ and $t_n = \sum_{k=1}^{n} b_n$ be the partial sum sequences of $\sum_{n=1}^{\infty} a_n$ and $\sum_{n=1}^{\infty} b_n$. Since $\sum_{n=1}^{\infty} a_n = s$, we have $s_n \to s$. We complete the proof by showing that $t_n - s_n \to 0$, since this would imply that $t_n = (t_n - s_n) + s_n \to 0 + s = s$.

Notice that for any $n > N_1$, the partial sums (s_n) and (t_n) each contains exactly one summand of $a_1, a_2, \ldots, a_N$. This is obvious in (s_n), since $N_1 \geq N$. For each $k \leq N$, $\sigma^{-1}(k) = j \leq N_1$, and so $k = \sigma(j)$ so that $a_k = a_{\sigma(j)}$ is that summand in (t_n). Thus the difference $t_n - s_n$ contains at most one instance of a_k for $k > N$. As a result, if $N_2 = \max\{\sigma(i) | i = 1, 2, \ldots, n\} > N$,

$$|t_n - s_n| \leq \left| \sum_{k=N+1}^{N_2} a_k \right| \leq \sum_{k=N+1}^{N_2} |a_k| \leq \sum_{k=N+1}^{\infty} |a_k| \leq \frac{\epsilon}{2} = \epsilon.$$

□

The proof of Riemann's Rearrangement Theorem proceeds by alternately considering the positive and negative terms of a given conditionally convergent series. Note that if $x \in \mathbb{R}$, then

$$P = \frac{|x| + x}{2} = \begin{cases} x \text{ if } x \geq 0, \text{ and} \\ 0 \text{ if } x < 0. \end{cases}$$

Furthermore,

$$Q = \frac{|x| - x}{2} = \begin{cases} 0 \text{ if } x \geq 0, \text{ and} \\ x \text{ if } x < 0. \end{cases}$$

We call P and Q the positive and negative parts of x, respectively (even though $Q > 0$). It is easy to see that $x = P - Q$, and $|x| = P + Q$. We use this idea in the next lemma.

Lemma 4.39. *Suppose $\sum a_n$ is a conditionally convergent series. Let $P_n = \frac{|x|+x}{2}$, and $Q_n = \frac{|x|-x}{2}$. Both $\sum P_n$ and $\sum Q_n$ diverge.*

Idea of proof. *The idea is to use the relationships between P_n, Q_n to a_n, as discussed above. If at least one of $\sum P_n$ or $\sum Q_n$ converges, those relationships can be exploited to find a contradiction.*

Proof of Lemma 4.39. Define $P_n = \frac{|x|+x}{2}$ and $Q_n = \frac{|x|-x}{2}$. Note that $a_n = P_n - Q_n$, and $|a_n| = P_n + Q_n$. If $\sum P_n$ and $\sum Q_n$ both converge, then so does $\sum P_n + \sum Q_n = \sum (P_n + Q_n) = \sum |a_n|$, a contradiction to the assumption that $\sum a_n$ conditionally converges. If $\sum P_n$ converges and $\sum Q_n$ diverges, then $\sum P_n - \sum a_n = \sum (P_n - a_n) = \sum Q_n$ converges, another contradiction. If $\sum P_n$ diverges and $\sum Q_n$ converges, then $\sum a_n + \sum Q_n = \sum (a_n + Q_n) = \sum P_n$ converges, a final contradiction. □

We are now ready to state and prove Riemann's Rearrangement Theorem, which says that every conditionally convergent series can be rearranged to converge to *any* real number.

Theorem 4.40 (Riemann's Rearrangement Theorem). *If $\sum a_n$ is a conditionally convergent series and $x \in \mathbb{R}$, there exists a rearrangement of $\sum a_n$ that converges to x.*

Idea of proof. *The idea is actually quite intuitive: rearrange the series to start by adding positive terms until you exceed x. Then subtract negative members until you fall below x. Repeating this process yields the desired rearrangement. That this rearrangement converges to x is also straightforward: If we stop adding or subtracting as soon as we exceed or fall below x (not anytime after), then the difference between the partial sums and x is less than one of these positive or negative terms, depending on what n is. Since $\sum a_n$ converges, the general term $a_n \to 0$, so these differences between the partial sums of the rearranged series and x also tend to 0.*

Proof of Theorem 4.40. Define $P_n = \frac{|x|+x}{2}$ and $Q_n = \frac{|x|-x}{2}$ as above. Let p_k and q_k be the kth nonzero term in (P_n) and (Q_n), respectively. The series $\sum P_n$ and $\sum Q_n$ diverge by Lemma 4.39, and since (p_n) and (q_n) differ from (P_n) and (Q_n) only by the deletion of terms which are 0, $\sum p_n$ and $\sum q_n$ also diverge. Since $p_n, q_n > 0$, it must be the case that $\sum p_n = \sum q_n = \infty$.

Let $x \in \mathbb{R}$ be given. We construct the desired rearrangement as follows. Let m_1 be the first natural number for which

$$p_1 + p_2 + \cdots + p_{m_1} \geq x.$$

Such an m_1 exists since $\sum p_n = \infty$. In the event that $x \leq 0$, we allow for $m_1 = 0$. Then let n_1 be the first natural number for which

$$(p_1 + \cdots + p_{m_1}) - (q_1 + \cdots + q_{n_1}) \leq x.$$

Again, such an n_1 exists since $\sum q_n = \infty$. Let m_2 be the first positive integer beyond m_1 so that

$$(p_1 + \cdots + p_{m_1}) - (q_1 + \cdots + q_{n_1}) + (p_1 + \cdots + p_{m_2}) \geq x,$$

and so on (see Figure 4.4.1 for an illustration of this construction).

We now show that this rearranged series converges to x. Let (s_n) be the partial sum sequence of this rearranged series. We consider the following cases. If the last term in the partial sum is a p_{m_j}, then $|s_n - s| \leq p_{m_j}$. If the last term in the partial sum is an n_j, then $|s_n - x| < q_{n_j}$. If n is an index between these two, then $|s_n - x| < p_{m_j}$, and finally if the last term of the partial sum is between a q_{n_j} and a $p_{m_{j+1}}$, then $|s_n - x| < q_{n_j}$. In every case, $|s_n - x|$ is less than or equal to either p_{m_j} or q_{n_j} for some j. Since $\sum a_n$ converges, $a_n \to 0$ (Theorem 4.6). Since these p_{m_j} and q_{n_j} form a subsequence of (a_n), it must be the case that $p_n \to 0$ and $q_n \to 0$ (Theorem 3.66). Hence $|s_n - x| \to 0$, and so $s_n \to x$. □

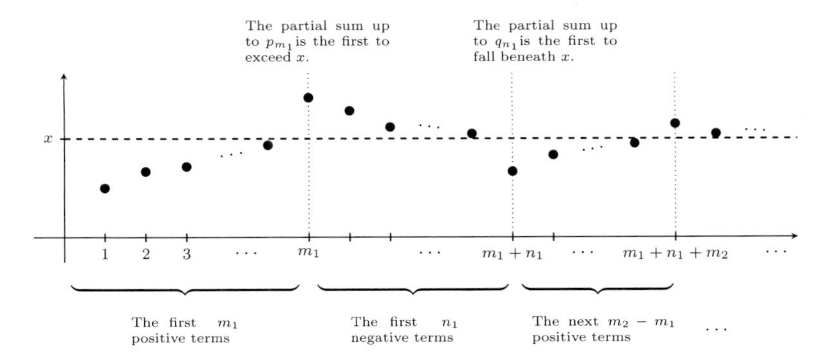

FIGURE 4.4.1
The partial sums of the rearranged series are approaching x.

It is not difficult to use this method to show that a conditionally convergent series can be rearranged so that its sum is ∞ or $-\infty$, see Exercise 4.

4.4.3 Review of important concepts from Section 4.4

1. Rearranging the terms in an infinite series will alter its sequence of partial sums, thus it is at least reasonable to suggest that such a rearrangement might alter what the series converges to (if it converges).
2. Any rearrangement of an absolutely convergent series will be absolutely convergent and converge to the same value (Theorem 4.38).
3. Riemann's Rearrangement Theorem (Theorem 4.40) states that a conditionally convergent series can be rearranged to converge to any real number. In addition, one could rearrange such a series to have ∞ or $-\infty$ as its limit as well (Exercise 4).

Exercises for Section 4.4

1. Prove that if $\sum b_n$ is a finite rearrangement of $\sum a_n$, then $\sum a_n$ converges if and only if $\sum b_n$ does, and in that case $\sum a_n = \sum b_n$. That is, the only rearrangements that could affect the value of a series are infinite rearrangements. (A rearrangement is *finite* if $a_n = b_n$ eventually.)
2. The following rearrangement can be found in [22][12]. Consider the

[12]It is shown in [22] that by considering alternately M positive terms followed by N negative terms, the rearrangement converges to $\ln\left(\frac{M}{N}\right)\big)$. This problem uses $M = 2, N = 1$.

following series:

$$\frac{1}{1} - \frac{1}{1} + \frac{1}{2} - \frac{1}{2} + \frac{1}{3} - \frac{1}{3} + \cdots.$$

(a) Prove that this series conditionally converges and find its sum.

(b) Rearrange the series to have two positive terms followed by one negative term as follows:

$$\frac{1}{1} + \left(\frac{1}{2} - \frac{1}{1}\right) + \frac{1}{3} + \left(\frac{1}{4} - \frac{1}{2}\right) + \cdots.$$

What does this rearranged series equal?

3. Can a divergent series be rearranged to form an absolutely convergent series? Why or why not?

4. If $\sum a_n$ is a conditionally convergent series, prove that there exists a rearrangement of $\sum a_n$ that converges to ∞, and another rearrangement that converges to $-\infty$.

5. Suppose $a_n > 0$ and $\sum_{n=1}^{\infty} a_n = \infty$. Prove that if $b_n = a_{\sigma(n)}$ is a rearrangement of (a_n), then $\sum b_n = \infty$ as well. (This is a modest generalization of Theorem 4.38.)

6. If $a_n \to a$ is a convergent *sequence*, and $b_n = a_{\sigma(n)}$ is a rearrangement of (a_n), must it be the case that $b_n \to a$ as well?

4.5 Troubleshooting Guide for Series of Real Numbers

Type of difficulty:	***Suggestion for help:***
Section 4.1: Infinite Series	
"Why doesn't $\sum \frac{n-1}{n+1}$ converge to 1?"	The limit of the sequence (a_n) that forms the series is hardly ever what $\sum a_n$ converges to. Do not confuse the limit of (a_n) with the limit of the partial sum sequence (s_n). See Definition 4.1.
"Why do I see most series starting at 1, while I see others starting at 0. Does this matter?"	It only matters in what the series actually *equals*, if it converges. For example, we see the geometric series (starting at $n = 0$) $\sum_{n=0}^{\infty} r^n$ of Example 4.5 converges to $\frac{1}{1-r}$. By contrast, the same sum beginning at $n = 1$ is $\sum_{n=1}^{\infty} r^n = \frac{1}{1-r} - 1$. We simply wanted to study the former series in this example, although perhaps at some other time we could be interested in the latter as well.
"I thought that if $a_n \to 0$, then $\sum a_n$ converges. Why is that wrong?"	This is the *converse* of the test for divergence (Theorem 4.6), and it is not generally true. Example 4.8 shows that $\sum \frac{1}{n}$ diverges even though $\frac{1}{n} \to 0$.
"Why do I see $\sum a_k$ sometimes and $\sum a_n$ some other times? Is there a difference in using either k or n?"	The k or the n are the index of summation and make no difference in the sum. Generally one is used along with the other in the presence of a partial sum, where the index of summation cannot be reused as the index of the partial sum.
Section 4.2: Tests for Series with Nonnegative Terms	
"I can't figure out what is in the subscript and the exponents of the subscript in the proof of the Cauchy Condensation Test (Theorem 4.14)."	It is admittedly difficult to see what is intended without extra typesetting emphasis. This could be a learning opportunity: try to determine for yourself what *must* these be for the argument to make sense.

Type of difficulty:	***Suggestion for help:***
"How did they come up with the estimate $\frac{1}{n^2-2} \leq \frac{2}{n^2}$ in Example 4.19? It seems rather convenient and I'm not sure I could come up with that on my own."	Generally, for large inputs, all degree d polynomials behave like multiples of other degree d polynomials, it is just a matter of finding that multiple to serve your purposes. In this case, a comparison to $\frac{1}{n^2}$ is desirable, which led us to search for how these are related.
"What does it mean to 'test the following series for convergence'?"	It means that you are to determine if the given series converges or diverges.
Section 4.3: Other Convergence Tests for Series	
"I've never heard of Dirichlet's Test (Theorem 4.23), but I have heard of the Alternating Series Test (Theorem 4.24). Why?"	Dirichlet's Test is not usually taught in basic Calculus, but the Alternating Series Test is. Dirichlet's test is a more general test, which makes it a more robust tool. However, it is more complicated than the Alternating Series Test, which is why it is often left out of the Calculus curriculum. Exercise 7 gives an independent proof of the Alternating Series Test.
"Why do the three series in Example 4.35 all have $L = 1$? Is there some pattern about these series that makes that happen?"	The proofs of the Ratio and Root Tests make use of a comparison to geometric series. If there is no such comparison, then generally $L = 1$ and these tests will be inconclusive.
Section 4.4: Rearrangements	
"I'm still not sure I believe that rearranging terms in a series could make it converge to something else."	It's true! See the example in the first subsection to this section, the discussion before Definition 4.37 about how this is at least plausible, and then see Theorem 4.40.
"Are there other examples of rearranged series, like the one given at the beginning of this section?"	Yes, there are many, see [21]. See also Exercise 2.

5

Limits and Continuity

Chapter Summary:

Limits and continuity are the natural next topic in our study of introductory analysis, leading to the next chapter on differentiation. The material on sequences is heavily used throughout this chapter.

Section 5.1: Limits of Functions. Limits of functions are introduced, and are formulated in terms of neighborhoods. The uniqueness of limits is also shown, and one-sided limits are briefly discussed.

Section 5.2: Properties of Limits. Limits of functions are characterized in terms of convergent sequences, providing a link to Section 3.3. Analogous supporting results about limits are proven using this result and results in Section 3.3, such as basic limit properties, the Squeeze Theorem for limits, and the fact that inequalities are preserved in limits.

Section 5.3: Infinite Limits. Infinite limits are introduced, and basic results concerning infinite limits are established as analogs of their finite counterparts.

Section 5.4: Continuity and Discontinuity. Continuity of functions is defined and related to limits of functions at limit points of the domain. Basic properties of functions are given, along with a short discussion of discontinuities.

Section 5.5: Consequences of Continuity. We state and prove important consequences of continuous functions defined on intervals of the form $[a, b]$. This includes the Boundedness Theorem (Theorem 5.41), the Extreme Value Theorem (Theorem 5.44), the Intermediate Value Theorem (Theorem 5.45), and several other results.

Section 5.6: Uniform Continuity. We define uniform continuity in this section, and introduce some important consequences: a continuous function on $[a, b]$ is uniformly continuous (Theorem 5.51), the interaction of Cauchy sequences with uniformly continuous functions (Theorems 5.52 and 5.53), and continuous extensions.

Section 5.7: Troubleshooting Guide for Limits and Continuity. A list of frequently asked questions and their answers, arranged by section. Come here if you are struggling with any material from this chapter.

5.1 Limits of Functions

We begin this chapter with the definition of a limit of a real-valued function, one of the most important tools in all of real analysis. We begin with presenting a formulation of its definition, followed by some examples. We close the section with a convenient reformulation of limits in terms of neighborhoods, and a proof that limits which exist are unique. We close the section with a short discussion of one-sided limits, and a result that relates one-sided limits to "regular" limits.

5.1.1 Formulating the limit of a function

We wish to formulate a precise definition of the *limit of $f(x)$ as x approaches* c, and will proceed similarly to the beginning of last chapter when we began with an imprecise guiding statement, and slowly transformed it into a useable mathematical definition. **The general idea of a limit of a function at a point c is to determine what value–if any–the function's outputs are tending toward as the inputs get closer to c, but without considering what the function is actually doing at c.** For now, we will only consider the situation that the proposed limit L and c itself are real numbers, and leave the infinite cases for later in this chapter, in Section 5.3.

And so suppose $f : D \to \mathbb{R}$ is a function, let c be a real number, and suppose L is the limit of $f(x)$ as x approaches c. For convenience, we will write $\lim_{x\to c} f(x)$ to be this limit. Using the motivation in boldface in the previous paragraph, it seems reasonable to say

"$\lim_{x\to c} f(x) = L$ if and only if $f(x)$ is near L whenever x is near c."

There are two major problems with the imprecise statement above that require improvement. First, x needs to be in the domain D of f for us to even discuss the object "$f(x)$", so we will include "$x \in D$" in our formulation. Second, the statement has no reference to the fact that we want to specifically forbid x from equaling c, so we will include "$x \neq c$" as well. With this in mind, consider the more precise statement:

"$\lim_{x\to c} f(x) = L$ if and only if the distance between $f(x)$ and L is negligible whenever the distance between x and c is negligible, where $x \in D$ and $x \neq c$."

This is better, but a bit cumbersome and still not precise insofar as what the word "negligible" means in each of the two different contexts it is used in above. There seems to be a conditional nature between how close $f(x)$ is to L and how close x is to c: for any given $\epsilon > 0$, one is challenged to find how close x must be from c (call this δ) to ensure that $f(x)$ is within ϵ from L:

"$\lim_{x \to c} f(x) = L$ if and only if for every $\epsilon > 0$, there exists a δ so that if $x \in D, x \neq c$, and within δ from c, then $f(x)$ is within ϵ from L."

We put this wording into more formal mathematical symbolism in Definition 5.1 below as our official definition of a limit after one more issue that has not been discussed: what if there are no inputs of f that are "close" (within δ, for some δ) to c? If this were the case, then we could choose *any* number L to be our limit since the hypothesis "if x is within δ of c and $x \neq c$" would not be satisfied for certain δ (and so logically, *any* conclusion as to what the limit would be is valid). In order to avoid this quagmire, we only define limits at points c so that each neighborhood about c contains elements of the domain D of f other than c. One may recall that this is *precisely* the situation that c is a limit point of D (see Definition 3.86 on Page 163 for a review of limit points if necessary).

Definition 5.1. Let $f : D \to \mathbb{R}$, and let c be a limit point of D. The *limit of* $f(x)$ *as* x *approaches* c is $L \in \mathbb{R}$ if, for every $\epsilon > 0$ there exists a δ so that if $x \in D$ and $0 < |x - c| < \delta$, then $|f(x) - L| < \epsilon$. In the event this limit exists, we write $\lim_{x \to c} f(x) = L$, or $f(x) \to L$ as $x \to c$.

Remark 5.2. **The number $f(c)$, if it is defined, is utterly irrelevant when computing** $\lim_{x \to c} f(x)$. In our discussion of continuity in Section 5.4 we will discuss the relationship between the limit of a function at c and the number $f(c)$, but besides this very important exception, it is *never* justifiable to evaluate a limit at c by simply computing $f(c)$. So until it is further discussed, one should not assume that $\lim_{x \to c} f(x)$ is equal to $f(c)$.

Remark 5.3. In Definition 5.1, we have the phrase "$\ldots$if $x \in D$ and $0 < |x - c| < \delta \ldots$". Since we insist that $0 < |x - c|$, we never consider the case that $x = c$, which fulfills one of the major requirements of what we would like a limit to be, as stated in bold at the beginning of this subsection.

Remark 5.4. In the definition above, one insists that c is a limit point of the domain D of f, whereas most of the time one is asked to simply compute $\lim_{x \to c} f(x)$ without any reference to what D is. In this case, it may be unclear whether or not c is a limit point of D, preventing us from proceeding. Generally speaking, the domain of f is assumed to be the largest domain for which f makes sense, unless an alternate domain is specified.

The following practice problems illustrate the use of Definition 5.1. In these problems, the domain of the functions involved is all of $\mathbb{R}$, so we omit any reference to the "$x \in D$" portion of Definition 5.1.

Practice 5.5. Show that $\lim_{x \to 4} 3x - 1 = 11$.

Methodology. *We aim to apply Definition 5.1 directly, with* $f(x) = 3x - 1$, $c = 4$, *and* $L = 11$. *We begin with an arbitrary* $\epsilon > 0$, *and search for a* δ *so that if* x *is within* δ *from 4, then* $f(x)$ *is within* ϵ *from* 11. *We work backward* *to achieve this goal, starting with the required conclusion* $|f(x) - 11| < \epsilon$, *and determining* δ *in terms of the given* ϵ.

Solution. Let $\epsilon > 0$ be given. We must find a δ so that if $0 < |x - 4| < \delta$, then $|(3x - 1) - 11| < \epsilon$. But

$$\begin{aligned} |(3x-1) - 11| &= |3x - 12| \\ &= 3|x - 4|. \end{aligned}$$

Set $\delta = \frac{\epsilon}{3}$. If x satisfies $0 < |x - 4| < \delta$, then

$$|(3x - 1) - 11| = 3|x - 4| < 3\delta = 3\frac{\epsilon}{3} = \epsilon.$$

■

Generally speaking, this "working backward" approach is a useful one in practice, although sometimes we may need to resort to more detailed methods in order to find δ. It would be wise to carefully read the methodology of Practice 5.6, as it introduces a common tool that will be of use.

Practice 5.6. Determine $\lim_{x\to 2} x^2 + 1$.

Methodology. *Just like the last practice problem, we use Definition 5.1, but this time we need to decide for ourselves what the limit is supposed to be. According to Definition 5.1, we begin with an arbitrary* $\epsilon > 0$, *and attempt to find a* δ *so that whenever* x *is within* δ *of* 2 *(but not equal to* 2*), then the number* $x^2 + 1$ *is within* ϵ *of* 5 *(see Figure 5.1.1 for a graph of this situation). In the last practice problem there was only some simple factoring to uncover the relationship between* ϵ *and* δ, *but in this case as we work backward,*

$$|f(x) - L| = |(x^2 + 1) - 5| = |x - 2||x + 2|.$$

We only consider the situation that $0 < |x - 2| < \delta$, *which leaves us to wonder how big or small* $|x+2|$ *will be: after all, we are trying to make* $|x-2||x+2| < \epsilon$. *There are a variety of ways to bound* $|x + 2|$ *in this situation, consider the following reasoning. Suppose, for just a moment, that* $|x - 2| < 1$, *that is, that a* δ *that we are looking for has been found, and that it is no bigger than 1. We can extrapolate just how big* $|x + 2|$ *must be in this event:*

$$\begin{aligned} |x - 2| < 1 &\iff -1 < x - 2 < 1 \\ \text{(adding 4)} &\iff 3 < x + 2 < 5, \\ & \quad\ \text{so } |x + 2| < 5. \end{aligned}$$

In other words, if the distance from x *to* 2 *is smaller than 1 (*$|x - 2| < 1$*), then the distance from* x *to* -2 *is smaller than 5 (*$|x + 2| < 5$*). We can now impose this condition in our measurement of* $|f(x) - L|$ *to see that*

$$|(x^2 + 1) - 5| = |x - 2||x + 2| < 5|x - 2|,$$

and if $0 < |x - 2| < \delta$ *as required, we find* $|f(x) - L| < 5\delta$. *So provided both of the following are met, the problem will be finished:*

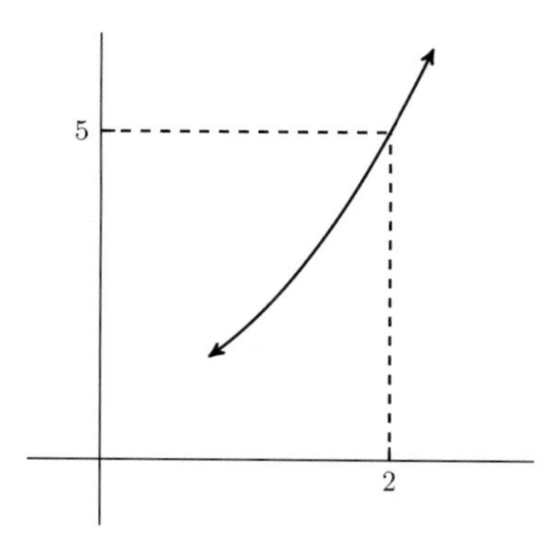

FIGURE 5.1.1
A graph of $f(x) = x^2 + 1$ near $x = 2$.

1. *We must have* δ *no bigger than* 1 *to enforce* $|x + 2| < 5$ *when* $0 < |x - 2| < \delta$, *and*
2. *We must have* $\delta \leq \frac{\epsilon}{5}$ *so that if* $0 < |x - 2| < \delta$, *then*

$$5|x - 2| < 5\delta \leq 5\frac{\epsilon}{5} = \epsilon.$$

The best way to make sure that $\delta \leq 1$ *and* $\delta \leq \frac{\epsilon}{5}$ *is to set* δ *to be the minimum of these two numbers:* $\delta := \min\{1, \frac{\epsilon}{5}\}$.

Solution. We claim that $\lim_{x \to 2} x^2 + 1 = 5$, and we shall prove this using Definition 5.1. Let $\epsilon > 0$ be arbitrary. Notice that if $0 < |x - 2| < 1$, then $|x + 2| < 5$. So set $\delta = \min\{1, \frac{\epsilon}{5}\}$, and if $0 < |x - 2| < \delta$, we have

$$|f(x) - L| = |(x^2 + 1) - 5| = |x - 2||x + 2| < 5\delta \leq \epsilon.$$

■

It is certainly the case that a limit might not exist, and to establish this, we will need to negate Definition 5.1, which we may do as follows:

$\lim_{x \to c} f(x)$ is *not* equal to L if and only if there exists an $\epsilon > 0$ so that for every $\delta > 0$ there exists an $x \in D$ with $0 < |x - c| < \delta$ but $|f(x) - L| \geq \epsilon$.

We illustrate the inexistence of a limit in the following practice problem.

Practice 5.7. Show that $\lim_{x \to 0} \frac{x}{|x|}$ does not exist.

Methodology. *The method is to assume that some real* L *is this limit, and to use the negation of Definition 5.1 above. As such, we find an* $\epsilon > 0$ *so that no matter what* $\delta > 0$ *we consider with* $0 < |x - 0| < \delta$, *the distance between* $f(x)$ *and this* L *is at least as large as our* ϵ. *Since this function's outputs are* *all a distance of 2 apart (which we see from its graph), a choice of* $\epsilon = 1$ *should work.*

Solution. Suppose that the limit did exist, and that $\frac{x}{|x|} \to L$ as $x \to 0$. Set $\epsilon = 1$, and let $\delta > 0$ be arbitrary. Suppose that $0 < |x - 0| < \delta$, or rather, $0 < |x| < \delta$. There are two possibilities: either $0 < x < \delta$, or $-\delta < x < 0$. If $0 < x < \delta$, then $f(x) = 1$, and if $-\delta < x < 0$, then $f(x) = -1$. Thus, if $L \geq 0$, we choose a negative x to have $|f(x) - L| \geq 1$, whereas if $L < 0$, we choose a positive x to have $|f(x) - L| \geq 1$. ■

5.1.2 Basic properties of limits

There are two useful and basic facts about limits that we shall find useful in subsequent sections, and we close the section by presenting these results. The first of these results is a useful alternative phrasing of Definition 5.1 in Theorem 5.9. In this theorem we shall find our first use for what is called a *deleted neighborhood*, defined below. The second basic result is that limits of functions are unique, provided of course that they exist. Our proof will use the characterization of limits of functions in Theorem 5.9.

Definition 5.8. Let $c \in \mathbb{R}^\sharp$. A *deleted neighborhood* U^- of c is a neighborhood of c in $\mathbb{R}$, but with c deleted. The *radius* of the deleted neighborhood U^- is the same as the radius of the neighborhood U from which it is derived.

Thus, if $c \in \mathbb{R}$, and $U = (c - \delta, c + \delta)$ is a neighborhood of (or about) c of radius δ (for some $\delta > 0$), then the deleted neighborhood

$$U^- = U - \{c\} = (c - \delta, c) \cup (c, c + \delta).$$

Theorem 5.9. *Suppose* $f : D \to \mathbb{R}$, *and that* c *is a limit point of* D. *The* $\lim_{x\to c} f(x) = L$ *if and only if for every neighborhood* V *of* L, *there exists a deleted neighborhood* U^- *of* c *so that if* $x \in U^- \cap D$ *then* $f(x) \in V$.

Idea of proof. *The proof of this Theorem will simply unravel Definition 5.1, and put it into language of neighborhoods, very similar in nature to Theorem 3.21.*

Proof of Theorem 5.9. ($\Leftarrow$) Suppose that $\lim_{x\to c} f(x) = L$, and let V be any neighborhood of L. Therefore, there exists an ϵ so that $V = (L - \epsilon, L + \epsilon)$. Since $\lim_{x\to c} f(x) = L$, there exists a δ so that if $0 < |x - c| < \delta$ and $x \in D$, then $|f(x) - L| < \epsilon$. The last statement may be rephrased as saying that there is a deleted neighborhood U^- (of radius δ) about c so that if $x \in U^- \cap D$, then $f(x) \in V$.

($\Rightarrow$) Suppose now that for every neighborhood V of L, there exists a deleted neighborhood U^- of c so that if $x \in U^- \cap D$, then $f(x) \in V$. We prove that $\lim_{x\to c} f(x) = L$ according to Definition 5.1. Let $\epsilon > 0$ be given, and let V be the $\epsilon-$neighborhood about L. By our hypothesis, there exists a deleted neighborhood U^- of c so that if $x \in U^- \cap D$, then $f(x) \in V$. Suppose the radius of this deleted neighborhood is δ. We may rephrase our hypothesis as the following: if $x \in D$ and $0 < |x - c| < \delta$, then $|f(x) - L| < \epsilon$, proving that $\lim_{x \to c} f(x) = L$. □

We close this section with a proof that limits are unique, provided they exist.

Theorem 5.10. *Suppose $f : D \to \mathbb{R}$, and let c be a limit point of D. If $\lim_{x \to c} f(x)$ exists, then it is unique. That is, if $\lim_{x \to c} f(x) = L_1$ and $\lim_{x \to c} f(x) = L_2$, then $L_1 = L_2$.*

Idea of proof. *We will use Theorem 5.9 to provide a contradiction if $L_1 \neq L_2$, since if this were the case we could construct neighborhoods V_1 about L_1 and V_2 about L_2 that do not have any points in common. Then, using Theorem 5.9, we could find a deleted neighborhood U^- about c with f mapping every element of U^- into both V_1 and V_2, which would be a contradiction.*

Proof of Theorem 5.10. Suppose that $L_1 \neq L_2$. Since the Real numbers are Hausdorff (Theorem 2.14), there exists a neighborhood V_1 of L_1 and V_2 of L_2 with $V_1 \cap V_2 = \emptyset$. By hypothesis $f(x) \to L_1$ as $x \to c$, so there exists a deleted neighborhood U_1^- of radius δ_1 about c so that if $x \in U_1^-$ then $f(x) \in V_1$ (by Theorem 5.9). By hypothesis $f(x) \to L_2$ as $x \to c$ as well, so there exists a deleted neighborhood U_2^- of radius δ_2 so that if $x \in U_2^-$ then $f(x) \in V_2$ (again by Theorem 5.9). Set $\delta = \min\{\delta_1, \delta_2\}$, so that the deleted neighborhood U^- of radius δ is contained in both U_1^- and U_2^-, and as a result, if $x \in U^-$ then $f(x) \in V_1$ and $f(x) \in V_2$. This is a contradiction to V_1 and V_2 being disjoint. □

5.1.3 One-sided limits

The limit of a function is designed to extract information about a function as its inputs approach a certain quantity, c. It is sometimes the case that such a complete picture of this behavior is both unnecessary and unwanted. For example, if $f(t)$ is the amount of your student debt t years after you graduate, you may be interested in the behavior of $f(t)$ *only* as t increases (that is, only as time goes forward), or put differently, as t approaches some time using quantities below (or less than) that time. Proposition 5.30 is another example illustrating that there is sometimes meaningful information to be extracted by only approaching c from one side. These are *one-sided limits*.

Definition 5.11. Let $f : D \to \mathbb{R}$.

1. If c is a limit point of $D \cap (c, \infty)$, then the *limit of $f(x)$ from above is L*, written $\lim_{x \to c^+} f(x) = L$, if for every $\epsilon > 0$, there exists a δ so that if $0 < x - c < \delta$ then $|f(x) - L| < \epsilon$.
2. If c is a limit point of $D \cap (-\infty, c)$, then the *limit of $f(x)$ from below is L*, written $\lim_{x \to c^-} f(x) = L$, if for every $\epsilon > 0$, there exists a δ so that if $0 < c - x < \delta$ then $|f(x) - L| < \epsilon$.

First, notice that Definition 5.11 is almost identical to Definition 5.1 except in two places. First, a one-sided limit requires that c be a limit point of $D \cap (c, \infty)$ (for limits from above) or $D \cap (-\infty, c)$ (for limits from below). Second, one encounters "$0 < x - c < \delta$" (for limits from above) or "$0 < c - x < \delta$" (for limits from below), as opposed to $0 < |x - c| < \delta$ in Definition 5.1. Both of these differences emphasize the condition in these one-sided limits that we are measuring the behavior of f for only $x > c$ (for limits from above) or $x < c$ for limits from below, as opposed to any value of $x \neq c$ that is near c as in Definition 5.1.

There is a strong relationship between these one-sided limits and the limits defined in Definition 5.1 which we describe using the following example. Let $f : [0, \infty) \to \mathbb{R}$ defined as $f(x) = \sqrt{x}$. Notice that 0 is a limit point of $[0, \infty)$, and so we may consider $\lim_{x \to 0} \sqrt{x}$ as in Definition 5.1. Notice also that 0 is a limit point of $[0, \infty) \cap (0, \infty) = (0, \infty)$, and so we may also consider the one-sided limit $\lim_{x \to 0^+} \sqrt{x}$ as in Definition 5.11. However, 0 is not a limit point of $[0, \infty) \cap (-\infty, 0) = \emptyset$, and so we cannot consider[1] $\lim_{x \to 0^-} \sqrt{x}$. This example illustrates that **sometimes** $\lim_{x \to c} f(x)$ **is already a one-sided limit**. As a result, one only usually considers one-sided limits in the event that, for whatever reason, $\lim_{x \to c} f(x)$ is not the desired quantity. This was the case in our school loan example above: although it makes sense mathematically to consider the full limit as t approaches some number, one might only really be interested in what happens as you approach that number from below, so a one-sided limit would be appropriate.

One-sided limits and "regular" limits (in the sense of Definition 5.1) are related to one another, as we can see below in Theorem 5.12.

Theorem 5.12. *Let $f : D \to \mathbb{R}$ and let c be a limit point of D. Then $\lim_{x \to c} f(x) = L$ if and only if both one-sided limits exist and equal L.*

Idea of proof. *If the limit exists, then the hypothesis $0 < |x - c| < \delta$ means that both $0 < x - c < \delta$ and $0 < c - x < \delta$ hold, which is how we will prove that both one-sided limits exist. Conversely, if both limits exist and are equal, then this will give us two deltas: one each that tells us how far above and how far below c we may be to ensure that the function is within ϵ from L. Taking the minimum of these will be a distance x can be from c on either side that ensures $f(x)$ is within ϵ from L, proving that the limit exists.*

Proof of Theorem 5.12. Suppose $\lim_{x \to c} f(x) = L$, and let $\epsilon > 0$ be given. According to Definition 5.1, there exists a δ so that if $0 < |x - c| < \delta$ then $|f(x) - L| < \epsilon$. The hypothesis that $0 < |x - c| < \delta$ means that for $x > c$ we have $0 < x - c < \delta$, and for $x < c$ we have[2] $0 < c - x < \delta$. Since, for these x in both situations, $|f(x) - L| < \epsilon$, we conclude that both one-sided limits exist and equal L, according to Definition 5.11.

---

[1]This is different from saying "$\lim_{x \to 0^-} \sqrt{x}$ does not exist," since the latter assumes the limit could be considered in the first place, i.e., that 0 is a limit point of $[0, \infty) \cap (-\infty, 0)$.

[2]If $x < c$, then $x - c < 0$ so that $|x - c| = c - x$.

Conversely, suppose that both one-sided limits exist and equal L. Let $\epsilon > 0$ be given. Since $\lim_{x\to c^+} f(x) = L$, there exists a δ_+ so that if $0 < x - c < \delta_+$, then $|f(x) - L| < \epsilon$. Similarly, since $\lim_{x\to c^-} f(x) = L$, there exists a δ_- so that if $0 < c - x < \delta_-$, then $|f(x) - L| < \epsilon$. Choose $\delta = \min\{\delta_+, \delta_-\}$. Then, for any $0 < |x - c| < \delta$, we have

$$\begin{aligned} 0 < x - c < \delta \leq \delta_+ \quad &\text{for} \quad x > c, \text{ and} \\ 0 < c - x < \delta \leq \delta_- \quad &\text{for} \quad x < c. \end{aligned}$$

Thus, whether $x > c$ or $x < c$, we still have $|f(x) - L| < \epsilon$, so $\lim_{x\to c} f(x) = L$. □

The next few sections exhibit some important facts about limits, and all of these facts will apply to one-sided limits as well. We shall generally not offer separate formal proofs of these facts, since, unless otherwise noted, these proofs will be identical for one-sided limits except for exchanging the phrase "if $0 < |x - c| < \delta$" with "if $0 < x - c < \delta$" or "if $0 < c - x < \delta$".

5.1.4 Review of important concepts from Section 5.1

1. The limit of $f(x)$ as x approaches c is designed to measure to what number (if any) the outputs of $f(x)$ tend to as the inputs approach c. Definition 5.1 is a formal definition of this.
2. There is an equivalent formulation of Definition 5.1 in Theorem 5.9 that is put in terms of neighborhoods. Using this characterization of limits can be useful and convenient.
3. If the limit of a function exists, then it is unique (see Theorem 5.10).
4. One-sided limits measure the behavior of $f(x)$ as x approaches c from below or from above. Sometimes "regular" limits (as in Definition 5.1) are already one-sided limits depending on the domain of f, and sometimes a one-sided limit is the desired quantity even if a regular limit is available.
5. Theorem 5.12 characterizes limits in terms of both one-sided limits existing and being equal.

Exercises for Section 5.1

1. Determine the following limits. Be sure to include a proof of your claim.

   (a) $\lim_{x\to 2} 4x - 7$

   (b) $\lim_{x\to -3} 5x + 1$

   (c) $\lim_{x\to 2} 3x^2 + 1$

2. Find a function $f(x)$ defined on $\mathbb{R}$, and an input c for which $\lim_{x\to c} f(x)$ is not equal to $f(c)$.

3. Suppose $f(x)$ is the piecewise defined function

$$f(x) = \begin{cases} 1 & \text{for } x \leq 4 \\ 3x - 10 & \text{for } x > 4. \end{cases}$$

   Show that $\lim_{x\to 4} f(x)$ does not exist. (This is one way of demonstrating that one should not simply plug 4 into this function to compute its limit.)

4. Prove that if c is any real number, then $\lim_{x\to c} k = k$. That is, the limit of a constant function is always equal to that constant.

5. Suppose that a and b are real numbers. Prove that $\lim_{x\to c} ax + b = ac + b$. That is to compute the limit of a linear function, one need only input c into it. *[Hint: Consider the case where $a = 0$ and then $a \neq 0$.]*

5.2 Properties of Limits

In Section 3.3, we were able to prove several facts about the convergence of sequences that were of great help to us. Before we encountered these results, we could only show that a sequence converged by the definition. The present section functions in a similar way: at this point, we can only demonstrate that a limit exists by using the definition, and it would be very helpful for us to have a collection of facts to aid us. This section provides these "supporting results".

The main result of this section is Theorem 5.13: it asserts, more or less, that the limit of a function f as $x \to c$ exists if and only if for every sequence of inputs $x_n \to c$ has the sequence of outputs $f(x_n) \to L$. After proving this theorem in the first part of this section, we go on to use it to prove the same sort of "supporting results" found in Section 3.3. This includes Theorem 5.16 which tells us that common combinations of functions are preserved in their limits, and the Squeeze Theorem for limits which is an analogy to the Squeeze Theorem for sequences (Theorem 3.38). Theorem 5.22 asserts that inequalities are preserved by limits, and Corollary 5.23 tells us that the limit of a function must be in an interval if the outputs of the function are in that interval. Note that all of the results in this section apply to one-sided limits as well.

5.2.1 Connecting limits and convergence of sequences

There is a fundamental result that will be of great use to us that links limits of functions to convergence of sequences. With this result, we may easily prove certain statements about limits by using our previously developed language of sequences and convergence. We prove this theorem first before illustrating some of its uses.

Theorem 5.13. *Let $f : D \to \mathbb{R}$, and let c be a limit point of D. Then $\lim_{x\to c} f(x) = L$ if and only if for every sequence $x_n \in D$ with $x_n \to c$, and $x_n \neq c$, we have $f(x_n) \to L$.*

Idea of proof. *There are two statements to prove: if the limit exists, and $x_n \to c$ is a convergent sequence in D that is never equal to c, then $f(x_n) \to L$. Then: If every sequence $x_n \to c$ in D with $x_n \neq c$ has $f(x_n) \to L$, then the limit exists.*

To prove the first statement, we suppose there is a sequence $x_n \to c$ with $x_n \in D$, and $x_n \neq c$. We use the hypothesis that $\lim_{x\to c} f(x) = L$ to prove that $f(x_n) \to L$ using the language of neighborhoods from Theorem 5.9. We prove the converse by contradiction: we suppose that for every $x_n \to c$ in D and $x_n \neq c$ we have $f(x_n) \to L$, but that the limit does not equal L. Using the language on Page 257 about what it means for a limit to not *equal L, we can construct a sequence $x_n \to c$ with $x_n \in D$ and $x_n \neq c$, but with $f(x_n)$ not converging to L: a contradiction.*

Proof of Theorem 5.13. Suppose that $\lim_{x\to c} f(x) = L$, and that $x_n \to c$ is a sequence in D. We prove that $f(x_n) \to L$ using Theorem 5.9. Let V be a neighborhood of L. Since $\lim_{x\to c} f(x) = L$, according to Theorem 5.9, there exists a deleted neighborhood U^- of c so that if $x \in U^- \cap D$ then $f(x) \in V$. Since $x_n \in D$, $x_n \to c$, and U^- is a deleted neighborhood of c, there exists an N so that $x_n \in U^- \cap D$ for all $n \geq N$. Since $x_n \neq c$, we are guaranteed that x_n is a member of the *deleted* neighborhood U^-. Since $x_n \in U^- \cap D$ for $n \geq N$, $f(x_n) \in V$ for $n \geq N$. To summarize, given a neighborhood V of L, we have shown that $f(x_n)$ is eventually in V. According to Theorem 3.21, we have shown that $f(x_n) \to L$.

To prove the converse, suppose that every sequence $x_n \in D$ with $x_n \to c$ and $x_n \neq c$ has $f(x_n) \to L$, but that $\lim_{x\to c} f(x)$ does not equal L. Using the statement on Page 257, we conclude that there exists an $\epsilon > 0$ so that for every $\delta > 0$ there exists an $x \in D$ with $0 < |x - c| < \delta$, but with $|f(x) - L| \geq \epsilon$. For each $n \in \mathbb{N}$, choose $\delta = \frac{1}{n}$. Then there exists an $x_n \in D$ for which $0 < |x_n - c| < \frac{1}{n}$, but with $|f(x_n) - L| \geq \epsilon$. We note the following about the sequences (x_n) and $(f(x_n))$:

1. By construction, $x_n \in D$.
2. Since $0 < |x_n - c|$, we have $x_n \neq c$.
3. $x_n \to c$. By the Squeeze Theorem (Theorem 3.38), since $0 < |x_n - c| < \frac{1}{n}$ and since $\frac{1}{n} \to 0$, we have $|x_n - c| \to 0$. By Exercise 10 of Section 3.2, we have $x_n - c \to 0$, or, that $x_n \to c$.
4. Since $|f(x_n) - L| \geq \epsilon$, the sequence $(f(x_n))$ is bounded away from L. Thus, according to Exercise 18 of Section 3.2, $(f(x_n))$ does not converge to L.

Thus, we have constructed a sequence (x_n), where $x_n \in D$, $x_n \neq c$, and $x_n \to c$, but that $(f(x_n))$ does not converge to L, a contradiction. □

With Theorem 5.13 and previously known results about sequences, we can quickly establish a number of useful facts, which we do in the next subsection. Before that, we offer a pair of practice problems demonstrating how one might use this theorem.

Practice 5.14. Determine $\lim_{x\to 2} x^2 + 1$. (This redoes Practice 5.6 using Theorem 5.13.)

Methodology. *We use Theorem 5.13 to show that this limit is 5. If (x_n) is an arbitrary sequence in $\mathbb{R}$ (here, $\mathbb{R} = D$ is the domain of this function) converging to 2, but never equaling 2, then we wish to show that $(x_n)^2 + 1$ is a sequence converging to 5. Using Theorem 3.31, this is indeed the case.*

Solution. Let $x_n \to 2$, where $x_n \in \mathbb{R}$ and $x_n \neq 2$. According to Theorem 3.31, the sequence $(x_n)^2 + 1 \to 5$. Thus, since (x_n) was arbitrary, $\lim_{x\to 2} x^2 + 1 = 5$ by Theorem 5.13. ■

Practice 5.15. Show that $\lim_{x\to 0} \frac{x}{|x|}$ does not exist. (This redoes Practice 5.7 using Theorem 5.13.)

Methodology. *If this limit were to exist and equal L, then Theorem 5.13 asserts that every sequence of inputs $x_n \to 0$, with $x_n \neq 0$, we would have to have $\frac{x_n}{|x_n|} \to L$. Thus, if we were to find two sequences (x_n) and (y_n) converging to 0 in $\mathbb{R} - \{0\}$ for which $\frac{x_n}{|x_n|}$ and $\frac{y_n}{|y_n|}$ converged to two different values, the limit $\lim_{x\to 0} \frac{x}{|x|}$ must not exist. Here, we have to specifically define the sequences (x_n) and (y_n), whereas in the previous practice problem, we considered an arbitrary sequence.*

Solution. Let $x_n = \frac{1}{n}$. Then $x_n \to 0$ and $x_n \neq 0$ for every n. We have

$$\frac{x_n}{|x_n|} = \frac{1/n}{|1/n|} = 1 \to 1.$$

Thus, according to Theorem 5.13, if $\lim_{x\to 0} \frac{x}{|x|}$ exists, it must equal 1.

Now let $y_n = \frac{-1}{n}$. Then $y_n \to 0$ and $y_n \neq 0$ for every n. We have

$$\frac{y_n}{|y_n|} = \frac{-1/n}{|-1/n|} = \frac{-1/n}{1/n} = -1 \to -1.$$

Thus, according to Theorem 5.13, if $\lim_{x\to 0} \frac{x}{|x|}$ exists, it must equal -1. But the limit of a sequence is unique if it exists (Theorem 5.10), and since we have shown that if this limit exists it must equal both 1 and -1. Therefore, it must be the case that this limit does not exist. ■

5.2.2 Basic limit properties

The remainder of this section mostly mirrors Section 3.3: for each applicable result there about sequences, we can use Theorem 5.13 to produce a corresponding result about limits. The general scheme of proofs of this sort is as follows:

- Consider an arbitrary sequence in D converging to c as in Theorem 5.13.
- Any hypotheses of the limits of certain functions existing will force the images of this sequence under those functions to converge. This uses one of the statements ($\Rightarrow$) in Theorem 5.13.
- Previous results from Section 3.3 will usually apply giving you a desired result.
- Since the sequence you considered above was arbitrary, the other statement ($\Leftarrow$) in Theorem 5.13 establishes the result.

Before we begin, it is important to point out that this scheme is not *required* for proving the following results. That is, one can prove all of what follows directly, using only the definition of a limit (Definition 5.1).

The first collection of these results tells us how to compute limits of functions that are constructed from other functions. This is similar to Theorem 3.31 where we considered the same sort of constructions of sequences.

Theorem 5.16. *Suppose that $f, g : D \to \mathbb{R}$, and that c is a limit point of D. Suppose also that*[3] $\lim_{x\to c} f(x) = L$ *and* $\lim_{x\to c} g(x) = M$.

1. $\lim_{x\to c}[f(x) \pm g(x)] = L \pm M$.
2. $\lim_{x\to c}[f(x)g(x)] = LM$.
3. *If k is a real number, then* $\lim_{x\to c}[kf(x)] = kL$.
4. $\lim_{x\to c} |f(x)| = |L|$.
5. *If $M \neq 0$, then* $\lim_{x\to c} \frac{f(x)}{g(x)} = \frac{L}{M}$.

Idea of proof. *For each of these assertions, we consider an arbitrary sequence $x_n \to c$, where $x_n \in D$ and $x_n \neq c$; this is the first step in the scheme we have just outlined. Since we have assumed $\lim_{x\to c} f(x) = L$ and $\lim_{x\to c} g(x) = M$, Theorem 5.13 asserts that $f(x_n) \to L$ and $g(x_n) \to M$; this is the second step in the scheme we have just outlined. The corresponding assertion in Theorem 3.31 will be what is needed to use Theorem 5.13 to establish the result; these applications are the third step in the scheme we have just outlined. For example, to prove the first assertion $\lim_{x\to c}[f(x) \pm g(x)] = L \pm M$, we use the first assertion of Theorem 3.31 to establish that $f(x_n) \pm g(x_n) \to L \pm M$. Thus, in this situation we have shown that given the arbitrary sequence $x_n \to c$, $x_n \in D$ and $x_n \neq c$, we have $f(x_n) \pm g(x_n) \to L \pm M$. So, by Theorem 5.13, $\lim_{x\to c}[f(x) \pm g(x)] = L \pm M$; this is the final step in the scheme we have just outlined. The proofs of the other assertions follow in the same way. There is a straightforward way to prove these assertions without the use of Theorem 5.13, be sure to see the remark after the proof.*

Proof of Theorem 5.16. Let $x_n \to c$ be an arbitrary sequence in D, with $x_n \neq c$. Since $\lim_{x\to c} f(x) = L$ and $\lim_{x\to c} g(x) = M$, we must have $f(x_n) \to L$ and $g(x_n) \to M$ by Theorem 5.13.

Assertion (1) of Theorem 3.31 has $f(x_n) \pm g(x_n) \to L \pm M$. Thus, by Theorem 5.13, $\lim_{x\to c}[f(x) \pm g(x)] = L \pm M$.

Assertion (2) of Theorem 3.31 has $f(x_n)g(x_n) \to LM$. Thus, by Theorem 5.13, $\lim_{x\to c}[f(x)g(x)] = LM$.

Assertion (3) of Theorem 3.31 has $kf(x_n) \to kL$. Thus, by Theorem 5.13, $\lim_{x\to c}[kf(x)] = kL$.

The remainder of this proof is left as Exercises 4 and 5. □

As was mentioned earlier, each of the assertions in Theorem 5.16 can be proven directly and without the use of Theorem 5.13. Such a direct proof will use the $\epsilon - \delta$ definition of a limit (Definition 5.1), and understandably employs an estimate similar to that found in the corresponding assertion of

[3]This statement implicitly assumes that both of these limits exist.

Theorem 3.31. As an example, we prove the first assertion of Theorem 5.16 in this manner below (it may be helpful to review the proof of Assertion (1) of Theorem 3.31 to better see the similarity in the estimate used). See also Exercise 7, which asks the reader to prove the other Assertions of Theorem 5.16 in this manner.

Direct proof of Assertion (1) of Theorem 5.16. Let $\epsilon > 0$ be given. Since $\lim_{x\to c} f(x) = L$, there exists a δ_1 so that if $0 < |x-c| < \delta_1$, then $|f(x)-L| < \frac{\epsilon}{2}$. Since $\lim_{x\to c} g(x) = M$, there exists a δ_2 so that if $0 < |x-c| < \delta_2$, then $|g(x) - M| < \frac{\epsilon}{2}$. Choose $\delta = \min\{\delta_1, \delta_2\}$, and choose any x so that $0 < |x-c| < \delta$. Then, using the Triangle Inequality, we have

$$\begin{aligned} |[f(x) \pm g(x)] - [L \pm M]| &= |(f(x) - L) \pm (g(x) - M)| \\ &\leq |f(x) - L| + |g(x) - M| \\ &< \tfrac{\epsilon}{2} + \tfrac{\epsilon}{2} \\ &= \epsilon. \end{aligned}$$

□

We can illustrate the usefulness of Theorem 5.16 in the practice problems below.

Practice 5.17. Prove that if k is a constant, then $\lim_{x\to c} k = k$ and that $\lim_{x\to c} x = c$ for any c. Then use this fact to prove that if $p(x)$ is a polynomial, then $\lim_{x\to c} p(x) = p(c)$.

Methodology. *Although we could prove the first two assertions using Theorem 5.13, we will prove it directly by choosing*[4] $\epsilon = \delta$ *for both assertions. We will use Assertion (3) of Theorem 5.16, followed by repeated uses of Assertions (1) and (2) of that theorem to complete the proof.*

Solution. We first prove that $\lim_{x\to c} k = k$ and $\lim_{x\to c} x = c$ by using Definition 5.1. Let $\epsilon > 0$ be given, and choose $\delta = \epsilon$. If $0 < |x-c| < \delta$, then $|k-k| = 0 < \delta$ (proving that $\lim_{x\to c} k = k$), while $|x-c| < \delta = \epsilon$ (proving that $\lim_{x\to c} x = c$).

Now, let $p(x) = a_n x^n + \cdots + a_1 x + a_0$ be a polynomial. Notice that for each $k = 1, \ldots, n$, $x^k \to c^k$ as $x \to c$ by repeatedly using Assertion (2) of Theorem 5.16 and the previously proven fact that $\lim_{x\to c} x = c$. Thus, $\lim_{x\to c} a_k x^k = a_k c^k$ as well for each of these k by Assertion (3) of Theorem 5.16. So, by using these facts and repeatedly using Assertion (1) of Theorem 5.16, we have

$$\begin{aligned} \lim_{x\to c}[a_n x^n + \cdots + a_1 x + a_0] &= \lim_{x\to c}[a_n x^n] + \cdots \\ &\qquad + \lim_{x\to c}[a_1 x] + \lim_{x\to c}[a_0] \\ &= a_n c^n + \cdots + a_1 c + a_0 \\ &= p(c). \end{aligned}$$

■

[4]For the first assertion, one could choose δ to be literally any positive number and this proof would still work. This is not the case in almost any other limit.

Practice 5.18. Suppose $f, g : D \to \mathbb{R}$, and that c is a limit point of D. Suppose also that $\lim_{x\to c} f(x) = L$ exists, but that $\lim_{x\to c} g(x)$ does not exist. Prove that $\lim_{x\to c}[f(x) + g(x)]$ does not exist.

Methodology. *We proceed by contradiction under the assumption that* $\lim_{x\to c}[f(x) + g(x)] = M$ *does exist. Then, since both limits exist, we can prove that* $\lim_{x\to c} g(x)$ *exists as the difference of two limits by using the first assertion of Theorem 5.16. The existence of this limit is a contradiction to the hypotheses, so our assumption that* $\lim_{x\to c}[f(x) + g(x)]$ *exists must be false.*

Solution. Suppose that $\lim_{x\to c}[f(x) + g(x)] = M$ exists. Then according to Assertion (1) of Theorem 5.16:

$$\begin{aligned} L - M &= \lim_{x\to c}[f(x) + g(x)] - \lim_{x\to c} f(x) \\ &= \lim_{x\to c}[(f(x) + g(x)) - f(x)] \\ &= \lim_{x\to c} g(x). \end{aligned}$$

Thus, $\lim_{x\to c} g(x)$ exists, which contradicts our assumption that it does not. ■

Sometimes it is the case that Theorem 5.16 does not apply. For example, consider

$$\lim_{x\to 1}\left(\frac{x^2 - 1}{x - 1}\right).$$

This is a quotient of functions, and so at first glance it seems as though Assertion (5) of Theorem 5.16 would apply. However, we see that $\lim_{x\to 1}[x - 1] = 0$, violating one of the hypotheses. As it was pointed out in Remark 5.2, *the value of a limit at c has nothing to do with the value of the function at c.* Notice that

$$\frac{x^2 - 1}{x - 1} = x + 1 \text{ when } x \neq 1.$$

Since the limit of $\frac{x^2-1}{x-1}$ at 1 has nothing to do with the value of the function at 1, we should be able to use either function when determining the limit. Thus,

$$\lim_{x\to 1}\left(\frac{x^2 - 1}{x - 1}\right) = \lim_{x\to 1}(x + 1) = 2.$$

In the above equation, we have implicitly used a result that we will formally state and prove below in Theorem 5.19, that if two functions differ only at one point, then their limits are the same.

Theorem 5.19. *Suppose* $f, g : D \to \mathbb{R}$, *and let* c *be a limit point of* D. *Suppose that* $f(x) = g(x)$ *for all* $x \in D$, *except possibly at* $x = c$. *If* $\lim_{x\to c} g(x) = L$, *then* $\lim_{x\to c} f(x) = L$ *as well.*

The proof of this result below will reveal that in the event that $c \notin D$, the domain of g can include c. In the example above this is the case: $x = 1$ is not in the domain of $f(x) = \frac{x^2-1}{x-1}$, whereas it is in the domain of $g(x) = x + 1$.

Idea of proof. *Although one can use Theorem 5.13, we prove it directly. The main idea is to notice that in Definition 5.1, $x = c$ is not considered. Here, since $f(x) = g(x)$ for every x except possibly when $x = c$, everything relating to $\lim_{x\to c} g(x)$ applies equally to $\lim_{x\to c} f(x)$.*

Proof of Theorem 5.19. Let $\epsilon > 0$ be given. Since $\lim_{x\to c} g(x) = L$, there exists a δ so that if $0 < |x - c| < \delta$ and $x \in D$, then $|g(x) - L| < \epsilon$. Since $f(x) = g(x)$ for each $x \neq c$, we conclude for that $|f(x) - L| < \epsilon$ as well. Thus, $\lim_{x\to c} f(x) = L$. □

The next practice problem illustrates what a formal proof using Theorem 5.19 would look like.

Practice 5.20. Determine, with proof, $\lim_{x\to 3} \left(\frac{x^2-2x-3}{x-3}\right)$.

Methodology. *Notice first that $x^2 - 2x - 3 = (x-3)(x+1)$. So, $\frac{x^2-2x-3}{x-3} = x+1$ for every x except at $x = 3$. Thus, according to Theorem 5.19, their limits as $x \to 3$ must be the same.*

Solution. Using Theorem 5.19 in the second line and Practice 5.17 in the third line, we have

$$\begin{aligned} \lim_{x\to 3} \left(\frac{x^2-2x-3}{x-3}\right) &= \lim_{x\to 3} \left(\frac{(x-3)(x+1)}{x-3}\right) \\ &= \lim_{x\to 3}(x+1) \\ &= 4. \end{aligned}$$

■

5.2.3 Limit Inequalities

We can derive several results about inequalities of limits by using Theorem 5.13 and Theorems 3.38, and 3.48, and Corollary 3.49 from Section 3.3.

The first result asserts that the squeeze theorem holds for limits.

Theorem 5.21 (Squeeze Theorem for Limits). *Suppose $f, g, h : D \to \mathbb{R}$, and that c is a limit point of D. Suppose also that $f(x) \leq g(x) \leq h(x)$ for all $x \in D$, and that $\lim_{x\to c} f(x) = L = \lim_{x\to c} h(x)$. Then $\lim_{x\to c} g(x) = L$ as well.*

Idea of proof. *We use Theorem 5.13: let $x_n \in D$ be a sequence with $x_n \neq c$ and $x_n \to c$; this is the first step in the scheme outlined at the beginning of this subsection. The hypothesis that $\lim_{x\to c} f(x) = L = \lim_{x\to c} h(x)$ (in particular, that these limits exist and are equal) imply that $f(x_n) \to L$ and $h(x_n) \to L$ as well; this is the second step in this scheme. Since $f(x) \leq g(x) \leq h(x)$ for every $x \in D$ means that $f(x_n) \leq g(x_n) \leq h(x_n)$, and so the Squeeze Theorem for limits (Theorem 3.38) tells us that $g(x_n) \to L$ as well; this is the third step in this scheme. Theorem 5.13 now concludes that $\lim_{x\to c} g(x) = L$; this is the final step in this scheme.*

Proof of Theorem 5.21. Let $x_n \in D$, with $x_n \to c$ and $x_n \neq c$. Since $\lim_{x\to c} f(x) = L = \lim_{x\to c} h(x)$, we have $f(x_n) \to L$ and $h(x_n) \to L$ by Theorem 5.13. Since $f(x_n) \leq g(x_n) \leq h(x_n)$, the Squeeze Theorem for Sequences (Theorem 3.38) asserts that $g(x_n) \to L$ as well. Now by Theorem 5.13, $\lim_{x\to c} g(x) = L$. □

It was remarked earlier that one does not need to use Theorem 5.13 to prove these results. Below, we offer an alternative proof demonstrating this. As before, the estimates used are very similar to those in Theorem 3.38, and it may be helpful to review that proof before proceeding.

Direct proof of Theorem 5.21. Let $\epsilon > 0$ be given. Since $\lim_{x\to c} f(x) = L$, there exists a δ_1 so that if $x \in D$ and $0 < |x - c| < \delta_1$, then $|f(x) - L| < \epsilon$. That is, for these x,

$$-\epsilon < f(x) - L < \epsilon. \tag{5.1}$$

Since $\lim_{x\to c} h(x) = L$, there exists a δ_2 so that if $x \in D$ and $0 < |x-c| < \delta_2$, then $|h(x) - L| < \epsilon$. That is, for these x,

$$-\epsilon < h(x) - L < \epsilon. \tag{5.2}$$

Now since $f(x) \leq g(x) \leq h(x)$ for every $x \in D$, it follows that

$$f(x) - L \leq g(x) - L \leq h(x) - L.$$

So, using the left inequality in Equation (5.1) and the right inequality in Equation (5.2), we conclude that for $\delta = \min\{\delta_1, \delta_2\}$, and $x \in D$ satisfying $0 < |x - c| < \delta$,

$$-\epsilon < f(x) - L \leq g(x) - L \leq h(x) - L < \epsilon,$$

or, that $|g(x) - L| < \epsilon$. □

We continue to mirror the results in Section 3.3 by demonstrating that limits of functions preserve inequalities, as in Theorem 3.48.

Theorem 5.22. *Suppose $f, g : D \to \mathbb{R}$, and c is a limit point of D. If $\lim_{x\to c} f(x) = L, \lim_{x\to c} g(x) = M$, and $f(x) \leq g(x)$ for every $x \in D$, then $L \leq M$.*

Idea of proof. *As in the first step of our scheme of proof outlined at the beginning of the section, we consider an arbitrary sequence in D with $x_n \to c$ and $x_n \neq c$. Since $f(x) \leq g(x)$ and these limits exist, by Theorem 5.13 we have $f(x_n) \to L$ and $g(x_n) \to M$ as in the second step of our scheme. Theorem 3.48 asserts that $L \leq M$ as in the last part of our scheme.*

Proof of Theorem 5.22. Let $x_n \to c$ be a sequence in D with $x_n \neq c$. By Theorem 5.13, $f(x_n) \to L$ and $g(x_n) \to M$. By Theorem 3.48, since $f(x) \leq g(x)$ for all $x \in D$, we have $L \leq M$. □

As before, there is a direct proof of this result that does not use Theorem 5.13; see Exercise 9.

Our final result of the section mirrors that of Corollary 3.49.

Corollary 5.23. *Suppose $f : D \to \mathbb{R}$, and c is a limit point of D. If $f(x) \in [a, b]$ for all $x \in D$, and $\lim_{x\to c} f(x) = L$ exists, then $L \in [a, b]$.*

Idea of proof. *In contrast to the approach above, we prove this result directly and leave a possible proof using Theorem 5.13 as Exercise 10. This can be proven by using Theorem 5.22 twice: since $f(x) \le b$, we have $L \le b$. Since $a \le f(x)$, we have $a \le L$, and so $L \in [a, b]$.*

Proof of Corollary 5.23. Since $f(x) \in [a, b]$ for all $x \in D$, we have $f(x) \le b$. Treating $g(x) = b$ as a constant function, $\lim_{x\to c} b = b$ (Practice 5.17). So, Theorem 5.22 asserts that $L \le b$. Similarly, since $a \le f(x)$, we have $a \le L$. So, $L \in [a, b]$. □

The careful reader will notice one difference between this section and Section 3.3: some of the results in Section 3.3 involves certain sequence satisfying a hypothesis *eventually*. The appropriate translation to limits would be to hypothesize that a certain property holds on some neighborhood of c. For example, Theorems 5.21 and 5.22, and Corollary 5.23 can be generalized as follows. See Exercises 11, 12, and 13 for their proofs:

Theorem 5.24. *Suppose $f, g, h : D \to \mathbb{R}$, and that c is a limit point of D. Suppose also that $f(x) \le g(x) \le h(x)$ for all $x \in U \cap D$ where U is some neighborhood of c, and that $\lim_{x\to c} f(x) = L = \lim_{x\to c} h(x)$. Then $\lim_{x\to c} g(x) = L$ as well.*

Theorem 5.25. *Suppose $f, g : D \to \mathbb{R}$, and c is a limit point of D. If $\lim_{x\to c} f(x) = L, \lim_{x\to c} g(x) = M$, and for some neighborhood U of c we have $f(x) \le g(x)$ for every $x \in U \cap D$, then $L \le M$.*

Corollary 5.26. *Suppose $f : D \to \mathbb{R}$, and c is a limit point of D. If for some neighborhood U of c, $f(x) \in [a, b]$ for all $x \in U \cap D$, and $\lim_{x\to c} f(x) = L$ exists, then $L \in [a, b]$.*

5.2.4 Review of important concepts from Section 5.2

1. Theorem 5.13 provides a link between establishing limits of functions and convergence of sequences. Many new results about limits of functions can be proven using this theorem.
2. Theorem 5.16 asserts that basic combinations of functions (for instance, sums and differences) translate into basic combinations of limits.
3. Functions that agree at all but possibly one point can be used interchangeably when computing limits (see Theorem 5.19). This result proves useful mainly in examples.

4. Theorems 5.21 (the Squeeze Theorem for limits) and 5.22, and Corollary 5.23 all relate inequalities of functions to corresponding inequalities of their limits.

Exercises for Section 5.2

1. Determine the following limits. Be sure to include a proof.

   (a) $\lim_{x\to 0} \frac{x^2+1}{2x+1}$.

   (b) $\lim_{x\to 1} \left(\frac{x^2+x-2}{x-1}\right)$.

   (c) $\lim_{x\to 0} \frac{8}{x+2}$.

   (d) $\lim_{x\to 0} x\sin\left(\frac{1}{x}\right)$. *[Hint: Try the Squeeze Theorem (Theorem 5.21). Or, use Exercise 6 and recall that the* sin *function is bounded. One cannot use Assertion (2) of Theorem 5.16 since* $\lim_{x\to 0}\sin(\frac{1}{x})$ *does not exist, see Exercise 2.]*

2. Prove that $\lim_{x\to 0}\sin(\frac{1}{x})$ does not exist.

3. Let $F : D \to \mathbb{R}$, and let c be a limit point of D. If it exists, prove that

$$\lim_{x\to c} F(x) = \lim_{h\to 0} F(c+h).$$

4. Prove Assertion (4) of Theorem 5.16. That is, prove that if $\lim_{x\to c} f(x) = L$, then $\lim_{x\to c} |f(x)| = |L|$. *[Hint: The method of proof of the other assertions of Theorem 5.16 can be used here.]*

5. Prove Assertion (5) of Theorem 5.16. That is, prove that if $\lim_{x\to c} f(x) = L$ and $\lim_{x\to c} g(x) = M \neq 0$, then $\lim_{x\to c}\frac{f(x)}{g(x)} = \frac{L}{M}$. *[Hint: The method of proof of the other assertions of Theorem 5.16 can be used here.]*

6. A function $g : D \to \mathbb{R}$ is *bounded* if $g(D)$ is bounded. Let c be a limit point of D. Prove that if $f, g : D \to \mathbb{R}$, $\lim_{x\to c} f(x) = 0$ and g is bounded, then $\lim_{x\to c} f(x)g(x) = 0$. Is the same result true if $\lim_{x\to c} f(x)$ is equal to something other than 0? *[Notice that Assertion (2) of Theorem 5.16 does not apply here since we specifically do not assume that* $\lim_{x\to c} g(x)$ *exists. Compare this with Exercise 23 of Section 3.3.]*

7. Prove directly and without using Theorem 5.13 any of Assertions (2) – (5) of Theorem 5.16. *[Hint: See the remark on Page 266 for a discussion of this manner of proof regarding Assertion (1).]*

8. If $\lim_{x\to c} f(x) = L$ exists, but that $\lim_{x\to c} g(x)$ does not exist, what can you say about the existence of $\lim_{x\to c}(f(x)+g(x))$? What about $\lim_{x\to c}(f(x)g(x))$?

9. Prove Theorem 5.22 directly (without using Theorem 5.13). That is, let $f, g : D \to \mathbb{R}$, and c is a limit point of D. If $\lim_{x \to c} f(x) = L$, $\lim_{x \to c} g(x) = M$, and $f(x) \leq g(x)$ for all $x \in D$, then $L \leq M$.

10. Prove Corollary 5.23 using Theorem 5.13. That is, prove the following using Theorem 5.13: suppose $f : D \to \mathbb{R}$, and c is a limit point of D. If $f(x) \in [a, b]$ for all $x \in D$, and $\lim_{x \to c} f(x) = L$ exists, then $L \in [a, b]$.

11. Prove Theorem 5.24. That is: suppose $f, g, h : D \to \mathbb{R}$, and that c is a limit point of D. Suppose also that $f(x) \leq g(x) \leq h(x)$ for all $x \in U \cap D$ where U is some neighborhood of c, and that $\lim_{x \to c} f(x) = L = \lim_{x \to c} h(x)$. Then $\lim_{x \to c} g(x) = L$ as well.

12. Prove Theorem 5.25. That is: suppose $f, g : D \to \mathbb{R}$, and c is a limit point of D. If $\lim_{x \to c} f(x) = L, \lim_{x \to c} g(x) = M$, and for some neighborhood U of c we have $f(x) \leq g(x)$ for every $x \in U \cap D$, then $L \leq M$.

13. Prove Corollary 5.26. That is: suppose $f : D \to \mathbb{R}$, and c is a limit point of D. If for some neighborhood U of c, $f(x) \in [a, b]$ for all $x \in U \cap D$, and $\lim_{x \to c} f(x) = L$ exists, then $L \in [a, b]$.

5.3 Infinite Limits

In the previous sections, we have insisted that both c and L are real in our discussion of $\lim_{x \to c} f(x) = L$. It is possible–and quite reasonable–to discuss this with either c or L are infinite, and these are referred to as *infinite limits.* We begin this section with a formal definition of these infinite limits (Definition 5.27). In the same way that Section 3.8 about sequences with infinite limits mirrored that of Section 3.3 concerning finite limits, this section presents infinite analogs of finite limit theorems found in Section 5.2. The first such result extends Theorem 5.13 to this setting, which states that infinite limits may be approximated by sequences, see Theorem 5.29. Using this result, it is an easy matter to establish uniqueness of infinite limits (Theorem 5.32), other basic results concerning infinite limits (Theorem 5.16), and the Squeeze Theorem for infinite limits (Theorem 3.114).

5.3.1 Infinite limits

We shall have a need for *infinite limits*: those where either L or c (or both) are infinite. We may conveniently combine all of the various cases ($L = \pm\infty$, and/or $c = \pm\infty$, and even Definition 5.1) into one definition using the language of neighborhoods. Care must be taken, however, because the blanket assumption "c is a limit point of D" does not apply in the event c is infinite, since we do not consider ∞ or $-\infty$ to be a limit point of any subset of $\mathbb{R}$. To take the appropriate care, we enumerate the various possibilities for c to ensure the limit makes sense.

Definition 5.27. Suppose $f : D \to \mathbb{R}$, and suppose $c, L \in \mathbb{R}^\sharp$. Suppose c satisfies one of the following conditions:

1. If $c \in \mathbb{R}$, then c is a limit point of D.
2. If $c = \infty$, then D is unbounded above.
3. If $c = -\infty$, then D is unbounded below.

We say the *limit of $f(x)$ as x approaches c* is L if, for every neighborhood V of L, there exists a deleted neighborhood U^- of c so that if $x \in U^-$, then $f(x) \in V$. In this case, we write $\lim_{x \to c} f(x) = L$, or, $f(x) \to L$ as $x \to c$.

We remark that **if c and L are finite, then Theorem 5.9 equates Definition 5.27 with Definition 5.1**, so that Definition 5.27 is an honest generalization of Definition 5.1 with the possibility that one (or both) of c and L are infinite. We next point out that a deleted neighborhood of $\pm\infty$ is the same as a neighborhood of $\pm\infty$ as defined in Definition 2.13. Since it will be cumbersome to continue to list the hypotheses on c and L in Definition 5.27, we shall omit them henceforth, unless there is some possibility for confusion or if extra clarity is needed[5]. Finally, Definition 5.27 extends to define infinite one-sided limits in the obvious way: replace the reference to the deleted neighborhood U^- of c with $0 < x - c < \delta$ or $0 < c - x < \delta$ as appropriate if c is finite. One does not define $\lim_{x\to\infty^+} f(x)$ or $\lim_{x\to-\infty^-} f(x)$, and otherwise if $c = \pm\infty$, $\lim_{x\to c} f(x)$ is already a one-sided limit (you can only approach ∞ from below, and $-\infty$ from above). All of the results in this section apply to one-sided limits as well, although see Theorem 5.30 for a result that requires a one-sided limit.

Before proceeding, it will also be useful to obtain the negation of the above definition in terms of neighborhoods, in the event we wish to show that a limit does not exist. Provided c meets one of the three criteria given in Definition 5.27 (so that one may even speak of a limit in the first place), the boxed statement below is valid even if c and/or L are infinite.

> $\lim_{x\to c} f(x)$ is *not* equal to L if and only if there exists a neighborhood V of L so that for every deleted neighborhood U^- of c, there exists an $x \in U^- \cap D$ with $f(x) \notin V$.

Practice 5.28. Show that $\lim_{x\to\infty} \frac{1}{x} = 0$, and that $\lim_{x\to 0} \frac{1}{x}$ does not exist.

Methodology. *We use Definition 5.27 to establish the first statement. We will consider an $\epsilon-$neighborhood V of 0, and show that there is a deleted neighborhood of ∞ that maps its elements to V (in this case we do not need a "deleted" neighborhood of ∞, since neighborhoods of ∞ already do not contain ∞.)*

We use the negation of Definition 5.27 to establish the second statement: if $L \in \mathbb{R}^\sharp$ is a proposed limit of $\frac{1}{x}$ at 0, then any deleted neighborhood of 0 will map outside any chosen neighborhood V of L.

Solution. Let $\epsilon > 0$ be given, and set V as the $\epsilon-$neighborhood $V = (-\epsilon, \epsilon)$ of 0. Let $M = \frac{1}{\epsilon}$, and set $U^- = (M, \infty)$ as a deleted neighborhood of ∞. If $x \in M$ ($x \neq 0$, since $M > 0$), then $x > M = \frac{1}{\epsilon}$, so $0 < \frac{1}{x} < \epsilon$, and so $f(x) \in V$.

To prove that $\lim_{x\to 0} \frac{1}{x}$ does not exist, let $L \in \mathbb{R}^\sharp$ be given as a possible limit. If $L \in \mathbb{R}$, then define the neighborhood $V = (L-1, L+1)$, and notice that if U^- is an arbitrary deleted $\delta-$neighborhood of 0, then there exists an $x \in U^-$ with $\frac{1}{x} \geq L+1$ (namely, if $L \leq 0$, we can choose $x = \min\{1, \frac{\delta}{2}\}$, and if $L > 0$, then we may choose $x = \min\{\frac{1}{L+1}, \frac{\delta}{2}\}$). In either event, f maps such an element of U^- outside V.

If $L = \infty$, we may set $V = (1, \infty)$ as a neighborhood of ∞. Now if U^- is an arbitrary deleted $\delta-$neighborhood of 0, then $x = -\frac{\delta}{2} \in U^-$, but $f(x) = -\frac{2}{\delta} \notin V$. Similarly, if $L = -\infty$, we may set $V = (-\infty, -1)$ as a neighborhood of $-\infty$. Now, if U^- is an arbitrary deleted $\delta-$neighborhood of 0, then $x = \frac{\delta}{2} \in U^-$, but $f(x) = \frac{2}{\delta} \notin V$. ■

[5]As an example of when one might need extra clarity, we shall typically need $c \in D$ in Section 5.4, an extra hypothesis that we will be sure to list when it is needed.

5.3.2 Infinite limits and sequences

In the previous section, we found that Theorem 5.13 was extremely valuable: it characterized limits in terms of convergence of sequences. So, since we already have a firm grasp on convergence of sequences, we were able to adapt the results of Section 3.3 to results about limits. Similar results will hold about infinite limits, and we shall go about it the same way: prove a result that links infinite limits (that is, those limits $\lim_{x\to c} f(x) = L$ where either c or L is infinite) to the limits of sequences, and then adapt the results of Section 3.8. The link between infinite limits of functions and limits of sequences is almost exactly the same as Theorem 5.13.

Theorem 5.29. *Let $f : D \to \mathbb{R}$, and let $c, L \in \mathbb{R}^\sharp$. Then $\lim_{x\to c} f(x) = L$ if and only if for every sequence $x_n \in D$ with $x_n \to c$ and $x_n \neq c$, we have $f(x_n) \to L$.*

Idea of proof. *If the limit exists, then to prove $f(x_n) \to L$ it is a simple matter of unraveling Definition 5.27 and applying it in the same manner as in the proof of Theorem 5.13. Conversely, we suppose that the limit does not exist and use the negation of Definition 5.27 on Page 275. We consider different cases as to when c or L is infinite, and in each case, we fill find a contradiction similar in style to that found in the proof of Theorem 5.13.*

Proof of Theorem 5.29. Suppose first that $\lim_{x\to c} f(x) = L$, and $x_n \in D$ is a sequence with $x_n \to c$ and $x_n \neq c$. Let V be a neighborhood of L. Since $\lim_{x\to c} f(x) = L$, according to Definition 5.27 there exists a deleted neighborhood U^- of c so that if $x \in U^- \cap D$ then $f(x) \in V$. Since $x_n \to c$ and $x_n \neq c$, the sequence (x_n) is eventually in U^-, so that $f(x_n)$ is eventually in V. Thus, $f(x_n) \to L$ by either Theorem 3.21 ($L \in \mathbb{R}$) or Theorem 3.109 ($L = \pm\infty$).

To prove the converse, suppose that every sequence $x_n \in D$ with $x_n \to c$ and $x_n \neq c$ had $f(x_n) \to L$, but that $\lim_{x\to c} f(x) \neq L$. Thus, there exists a neighborhood V of L so that for every deleted neighborhood U^- of c, there exists an $x \in U^- \cap D$ with $f(x) \notin V$ (see the statement on Page 275). We complete the proof by finding a contradiction depending on whether or not c is infinite.

If both c and L are finite, then we have already proven this in Theorem 5.13. If c is finite and L is infinite, then for every $n \in \mathbb{N}$ there exists an $x_n \in D$ with $0 < |x_n - c| < \frac{1}{n}$ but that $f(x_n) \notin V$ (here, U^- is the deleted neighborhood about c of radius $\frac{1}{n}$). As in the proof of Theorem 5.13, $x_n \to c$ and $x_n \neq c$, but $f(x_n) \notin V$ so it is bounded away from L. Thus, it is not possible that $f(x_n) \to L$, a contradiction.

If $c = \infty$, then for every $n \in \mathbb{N}$, there exists an $x_n \in D$ with $x_n \in (n, \infty) = U^-$, but with $f(x_n) \notin V$. Then since $n \to \infty$, by the infinite analog to the Squeeze Theorem (Theorem 3.114), $x_n \to \infty = c$. Since $f(x_n) \notin V$, it is bounded away from L, and so $f(x_n)$ must not have L as its limit, a contradiction. If $c = -\infty$, then one replaces U^- with $(-\infty, -n)$ to produce a similar contradiction. □

As an example of how this theorem can be used in theory and in practice, we present a useful way of relating infinite limits to finite ones in the following Proposition, an analog to Proposition 3.108. Afterward, we give an example of its use.

Proposition 5.30. *Suppose $f : D \to \mathbb{R}$, and $L \in \mathbb{R}^\sharp$.*

1. *If $f(x) > 0$ for all $x \in D$, then for any $c \in \mathbb{R}^\sharp$, $\lim_{x\to c} f(x) = \infty$ if and only if $\lim_{x\to c} \frac{1}{f(x)} = 0$.*

2. *Suppose $D \supseteq (0, \infty)$. $\lim_{x\to\infty} f(x) = L$ if and only if $\lim_{x\to 0^+} f(\frac{1}{x}) = L$.*

Idea of proof. *The first assertion is a straightforward application of Proposition 3.108 with Theorem 5.29: if $x_n \in D$ with $x_n \to c$ and $x_n \neq c$, then Proposition 3.108 asserts that $f(x_n) \to \infty$ if and only if $\frac{1}{f(x_n)} \to 0$. So if $\lim_{x\to c} f(x) = \infty$ then $\lim_{x\to c} \frac{1}{f(x)} = 0$ and vice versa. We use the same two results to prove the next assertion: if every sequence $x_n \in (0, \infty)$ with $x_n \to \infty$ had $f(x_n) \to L$, then we consider any other sequence $x_n \to 0^+$, and find that $\frac{1}{x_n} \to \infty$ by Proposition 3.108, so $f(\frac{1}{x_n}) \to L$ as well. The reasoning in proving the converse is similar. We have assumed that $D \supseteq (0, \infty)$ in the second assertion since we need to ensure that if $x \in D$ and $x > 0$ then $\frac{1}{x} \in D$ as well as part of the proof. Domains containing other than $(0, \infty)$ will do, but this seems to be the most reasonable.*

Proof of Proposition 5.30. To prove the first assertion, let $x_n \in D$ be any sequence with $x_n \to c$ and $x_n \neq c$. If $\lim_{x\to c} f(x) = \infty$ then $f(x_n) \to \infty$, and if $\lim_{x\to c} \frac{1}{f(x)} = 0$ then $\frac{1}{f(x_n)} \to 0$ by Theorem 5.29. So, $\lim_{x\to c} f(x) = \infty$ if and only if for this arbitrary sequence (x_n) we have $f(x_n) \to \infty$ (Theorem 5.29) if and only if for this same arbitrary sequence (x_n) we have $\frac{1}{f(x_n)} \to 0$ (Proposition 3.108) if and only if $\lim_{x\to c} \frac{1}{f(x)} = 0$.

We now prove Assertion (2). Assume $\lim_{x\to\infty} f(x) = L$, and let $x_n \in D$ with $x_n \to 0^+$. By Proposition 3.108, $y_n = \frac{1}{x_n} \to \infty$. Since $\lim_{x\to\infty} f(x) = L$ and $y_n \to \infty$, $f(y_n) = f(\frac{1}{x_n}) \to L$ as well. Conversely, if the latter limit holds and $x_n \to \infty$, then $\frac{1}{x_n} \to 0^+$ (Proposition 3.108), so $f(\frac{1}{1/x_n}) = f(x_n) \to L$ by assumption. Thus, by Theorem 5.29, $\lim_{x\to\infty} f(x) = L$. □

Here is an example that illustrates the use of Proposition 5.30 by translating an "old" (finite) limit into a "new" (infinite) limit.

Example 5.31. This example illustrates how one can use Proposition 5.30 to easily get new information about certain limits. This Example will do Practice 5.28 in a different way, although Proposition 5.30 provides for more flexibility as we shall see.

Let $f(x) = \frac{1}{x}$. Then $\lim_{x\to 0^+} f(\frac{1}{x}) = \lim_{x\to 0^+} x = 0$. By Proposition 5.30, $\lim_{x\to\infty} f(x) = \lim_{x\to\infty} \frac{1}{x} = 0$ as well. In fact, if $p(x)$ is any polynomial, we have $\lim_{x\to\infty} p(\frac{1}{x}) = p(0)$ by the same reasoning: By Practice 5.17 (and Theorem 5.12), $\lim_{x\to 0^+} p(x) = p(0)$, so by Proposition 5.30 we have $\lim_{x\to\infty} p(\frac{1}{x}) = p(0)$ as well. ■

5.3.3 Infinite analogs of limit theorems

In Section 5.2 we proved Theorem 5.13 which related (finite) limits of functions to sequences. Subsequently, we were able to easily prove a number of useful limit theorems that mirrored the results found in Section 3.3. In this section, we proved Theorem 5.29 which extended the relationship in Theorem 5.13 to infinite limits. In the same way as before, we may now prove a number of useful limit theorems concerning infinite limits by mirroring the results about infinite limits of sequences in Section 3.8. There are three main infinite analogs to be proven from Section 5.2: if infinite limits exist, then they are unique (Theorem 5.32), basic combinations of functions will have infinite limits that are combined in the same way (Theorem 5.33), and an infinite analog of the Squeeze Theorem for Limits (Theorem 5.34). Exercises 7 and 8 are other useful analogs as well.

Theorem 5.32. *If a limit exists, then it is unique. That is, for $c, L, M \in \mathbb{R}^\sharp$, if $f : D \to \mathbb{R}$ and $\lim_{x\to c} f(x) = L$ and $\lim_{x\to c} f(x) = M$ then $L = M$.*

Idea of proof. *We use Theorem 5.29 and Theorem 3.111. If $x_n \in D$ is a sequence with $x_n \to c$ and $x_n \neq c$, then the assumptions and Theorem 5.29 tell us that $f(x_n) \to L$ and $f(x_n) \to M$. Theorem 3.111 now tells us that $L = M$.*

Proof of Theorem 5.32. Let $x_n \in D$ with $x_n \to c$ and $x_n \neq c$. Our assumptions that the limit equals both L and M with Theorem 5.29 tell us that $f(x_n) \to L$ and $f(x_n) \to M$. Theorem 3.111 concludes that $L = M$. □

The next result is an analog of Theorem 5.16: basic combinations of functions will lead to basic combinations of limits, even if those limits are infinite.

Theorem 5.33. *Suppose that $f, g : D \to \mathbb{R}$. Suppose also that $\lim_{x\to c} f(x) = L$ and $\lim_{x\to c} g(x) = M$ for $c, L, M \in \mathbb{R}^\sharp$. Whenever they are defined:*

1. $\lim_{x\to c}[f(x) \pm g(x)] = L \pm M$.
2. $\lim_{x\to c}[f(x)g(x)] = LM$.
3. *If k is a real number, then* $\lim_{x\to c}[kf(x)] = kL$.
4. $\lim_{x\to c} |f(x)| = |L|$.
5. $\lim_{x\to c} \frac{f(x)}{g(x)} = \frac{L}{M}$.

Idea of proof. *All of these facts follow directly from Theorem 5.29 and Theorem 3.112. We start with a sequence $x_n \in D$ with $x_n \to c$ and $x_n \neq c$. Each result in the theorem corresponds to a result about sequences already proven in Theorem 3.112.*

Proof of Theorem 5.33. Let $x_n \in D$ with $x_n \to c$ and $x_n \neq c$. By Theorem 5.29, $f(x_n) \to L$ and $g(x_n) \to M$. By Theorem 3.112, whenever they are defined, we have

$$\begin{array}{ll} f(x_n) \pm g(x_n) \to L \pm M & \text{(Assertion (1))}, \\ f(x_n)g(x_n) \to LM & \text{(Assertion (2))}, \\ kf(x_n) \to kL & \text{(Assertion (3))}, \\ |f(x_n)| \to |L| & \text{(Assertion (4))}, \\ \frac{f(x_n)}{g(x_n)} \to \frac{L}{M} & \text{(Assertion (5))}. \end{array}$$

□

Theorem 5.34 (Squeeze Theorem for Infinite Limits). *Suppose $f, g : D \to \mathbb{R}$, and $f(x) \leq g(x)$ for all $x \in D$.*

1. *If $\lim_{x\to c} f(x) = \infty$, then $\lim_{x\to c} g(x) = \infty$.*
2. *If $\lim_{x\to c} g(x) = -\infty$, then $\lim_{x\to c} f(x) = -\infty$.*

Idea of proof. *This theorem is an easy application of Theorems 5.29 and 3.114. Let $x_n \to c$ be a sequence in D with $x_n \neq c$. In Assertion (1), by Theorem 3.114 we have $f(x_n) \to \infty$. Since $g(x_n) \geq f(x_n)$, by Theorem 3.114 we have $g(x_n) \to \infty$ as well. So, by Theorem 5.29, $\lim_{x\to c} g(x) = \infty$. The next assertion is proved similarly.*

Proof of Theorem 5.34. Let $x_n \to c$ be a sequence in D with $x_n \neq c$. Assume $\lim_{x\to c} f(x) = \infty$ to prove Assertion (1). Then $f(x_n) \to \infty$ by Theorem 5.29, and since $f(x_n) \leq g(x_n)$, $g(x_n) \to \infty$ as well by Theorem 3.114. So $\lim_{x\to c} g(x) = \infty$ by Theorem 5.29.

To prove Assertion (2), if $\lim_{x\to c} g(x) = -\infty$, then $g(x_n) \to -\infty$ and so must $f(x_n)$ by Theorem 3.114. Thus $\lim_{x\to c} f(x) = -\infty$ by Theorem 5.29. □

Note that with minor changes to the proof, the Squeeze Theorem presented in the last section (Theorem 5.21) also holds for limits which are finite, but as $x \to \pm\infty$.

5.3.4 Review of important concepts from Section 5.3

1. Infinite limits are defined in a reasonable way (Definition 5.27) in terms of neighborhoods. There are a number of infinite analogs to limit theorems presented in this section.

2. Theorem 5.29 is an analog to Theorem 5.13: the limit of $f(x)$ at c exists and equals L if and only if for every sequence $x_n \to c$ in D with $x_n \neq D$, $f(x_n) \to L$. This holds if c or L are infinite. We use this theorem quite a bit to prove other results concerning infinite limits.

3. A collection of infinite analogs of limits are proved here: infinite limits are unique (Theorem 5.32), basic combinations of functions give rise to basic combinations of limits (Theorem 5.33), and The Squeeze Theorem holds (Theorem 3.114). Other analogs can be found in Exercises 7 and 8.

Exercises for Section 5.3

1. Give an example of a function $f : (0, 1] \to \mathbb{R}$ and a Cauchy sequence $x_n \in (0, 1]$ but with $f(x_n) \to \infty$.

2. Determine, with proof, the following limits:

   (a) $\lim_{x\to\infty} \frac{x^2+x+1}{x^2+5}$.

   (b) $\lim_{x\to\infty} \frac{x^3}{x^2+1}$.

   (c) $\lim_{x\to\infty} 2^{\frac{1}{x}}$. (Careful, concluding that this is equal to $2^{\lim_{x\to\infty} \frac{1}{x}}$ needs justification.)

   (d) $\lim_{x\to\infty} \frac{\sin x}{x}$.

   (e) $\lim_{x\to\infty} r^x$, if $|r| < 1$.

3. Suppose $f, g : (0, \infty) \to (0, \infty)$, and that

$$\lim_{x\to\infty} \frac{f(x)}{g(x)} = L > 0 \quad \text{for some } L \in \mathbb{R}.$$

   Prove that $\lim_{x\to\infty} f(x) = \infty$ if and only if $\lim_{x\to\infty} g(x) = \infty$.

4. Prove that if $p(x)$ is a polynomial of even degree, and its leading coefficient is positive. Prove that $\lim_{x\to\pm\infty} p(x) = \infty$. Prove that if p has odd degree, then $\lim_{x\to\pm\infty} p(x) = \pm\infty$. *[Hint: Consider Exercise 3 where $g(x) = x^n$, and n is the degree of p.]*

5. Suppose $f : (0, \infty) \to \mathbb{R}$, and that for some $L \in \mathbb{R}$ we have $\lim_{x\to\infty} xf(x) = L$. Prove that $\lim_{x\to\infty} f(x) = 0$.

6. Suppose $f : \mathbb{R} \to \mathbb{R}$. Prove that $\lim_{x\to-\infty} f(x) = \lim_{x\to\infty} f(-x)$ in the event either limit exists.

7. Suppose that $f, g : D \to \mathbb{R}$ and $f(x) \leq g(x)$ for all $x \in D$. Assume also that D is not bounded above. Prove that if $\lim_{x\to\infty} f(x) = L$ and $\lim_{x\to\infty} g(x) = M$, then $L \leq M$. *[This is an analog of Theorem 5.22.]*

8. Suppose $f : D \to [a, b]$. Prove that if $\lim_{x\to\infty} f(x) = L$, then $L \in [a, b]$. *[This is an analog of Corollary 5.23.]*

5.4 Continuity and Discontinuity

Continuous functions are a foundational object in Analysis, and this section introduces some of their basic properties. We begin by giving a definition of continuity, followed by some very important remarks that relate continuity to limits of functions. After that, we establish some basic facts about continuity and give a list of continuous functions that we have encountered. We close the section with a short discussion of types of discontinuities.

5.4.1 Definition and remarks

In basic calculus one is sometimes taught to think of a continuous function as one whose graph could be drawn without picking up your pen. Although this is not technical enough, it does provide an insight as to what a continuous function should be defined as. Given such a function, and a point $(c, f(c))$ on its graph, the ability to draw the graph with one stroke of the pen means that we could approach the input c and find the corresponding outputs approaching $f(c)$. This is not unlike the definition of a limit, as we shall see.

Described more specifically, if we choose some distance $\epsilon > 0$ around $f(c)$, we should be able to find some distance δ so that if an input x is within δ from c, then $f(x)$ is withing ϵ from $f(c)$. We give a formal definition of continuity with this in mind, followed by some very important remarks.

Definition 5.35. Let $f : D \to \mathbb{R}$, and let $c \in D$. We say f *is continuous at* c if for every $\epsilon > 0$ there exists a $\delta > 0$ so that if $|x - c| < \delta$ then $|f(x) - f(c)| < \epsilon$. If f is continuous at every point of D, then we say f *is continuous on* D. If f is not continuous at c, we say f is *discontinuous at* c, or simply *discontinuous*.

This definition looks very similar to that of limits in Definition 5.1. We take a moment to discover exactly what these similarities and differences are.

There are two main differences between Definitions 5.35 and 5.1: "$|f(x) - f(c)| < \epsilon$" instead of "$|f(x) - L| < \epsilon$," and "$|x - c| < \delta$" instead of "$0 < |x - c| < \delta$." Both of these stem from the idea that the outputs of a continuous function are to approach one specific output, $f(c)$. In addition, since $c \in D$ we are specifically considering what $f(c)$ is in the context of continuity, whereas we specifically do not consider this when talking about limits.

If c is a limit point of D, then f is continuous at c if and only if $\lim_{x\to c} f(x) = f(c)$. That c is a limit point of D is an important requirement to equate these two notions, since limits can only be considered at limit points of the domain, D, of f. But with this assumption, Definitions 5.35 and 5.1 are identical if $L = f(c)$. The following is now an immediate corollary of Theorem 5.13 and will be of great use:

Theorem 5.36. *Let $f : D \to \mathbb{R}$. If c is a limit point of D, then f is continuous at c if and only if for every sequence $x_n \to c$ we have $f(x_n) \to f(c)$.*

Every function is continuous at an isolated point of D. If $c \in D$ and is not a limit point, it must be an isolated point (Definition 3.87). For each isolated point of D there is a neighborhood around it that contains only that point from D (Theorem 3.90). Thus, for a sufficiently small δ, in Definition 5.35 the hypothesis "... if $|x - c| < \delta$" is met only when $x = c$, for which the conclusion "$|f(x) - f(c)| < \epsilon$" surely holds, since $|f(x) - f(c)| = |f(c) - f(c)| = 0 < \epsilon$. It is for this reason that we generally only need to consider functions that are continuous at a limit point of D.

5.4.2 Basic properties of continuous functions

Using Theorems 5.36 and 5.16 and the fact that all functions are continuous at isolated points, the following results about limits from Section 5.2 are now immediate:

Theorem 5.37. *Suppose $f, g : D \to \mathbb{R}$ and $c \in D$. Suppose also that f and g are continuous at c.*

1. *$f \pm g$ is continuous at c.*
2. *fg is continuous at c.*
3. *If k is a real number, then kf is continuous at c.*
4. *$|f|$ is continuous at c.*
5. *If $g(c) \neq 0$, then $\frac{f}{g}$ is continuous at c.*

Furthermore, if f and g are continuous on D, then $f \pm g, fg, kf$, and $|f|$ are continuous on D, and $\frac{f}{g}$ is continuous on D if $g(x) \neq 0$ for all $x \in D$.

There is one fact about continuous functions that is distinct from our previous work on limits: the composition of continuous functions is continuous.

Theorem 5.38. *The composition of continuous functions is continuous. More specifically, let $D, \tilde{D} \subseteq \mathbb{R}$, $f : D \to \tilde{D}$ and $g : \tilde{D} \to \mathbb{R}$. If f is continuous at $c \in D$ and g is continuous at $f(c) \in \tilde{D}$, then $g \circ f$ is continuous at c.*

Idea of proof. *There is a proof using sequences that we reserve for Exercise 8. We prove this by using only Definition 5.35. We let $\epsilon > 0$ be given and try to find a δ so that if x is within δ of c, then $g(f(x))$ is within ϵ of $g(f(c))$ (see Figure 5.4.1). Since $g : \tilde{D} \to \mathbb{R}$ is continuous at $f(c)$, we can find a $\tilde{\delta}$ so that if $|y - f(c)| < \tilde{\delta}$ then $|g(y) - g(f(c))| < \epsilon$. We now use the hypothesis that f is continuous at c (and treat $\tilde{\delta}$ as our new ϵ), and find a δ so that if $|x - c| < \delta$ then $|f(x) - f(c)| < \tilde{\delta}$. Thus, for these x, we have $g(f(x))$ within ϵ of $g(f(c))$.*

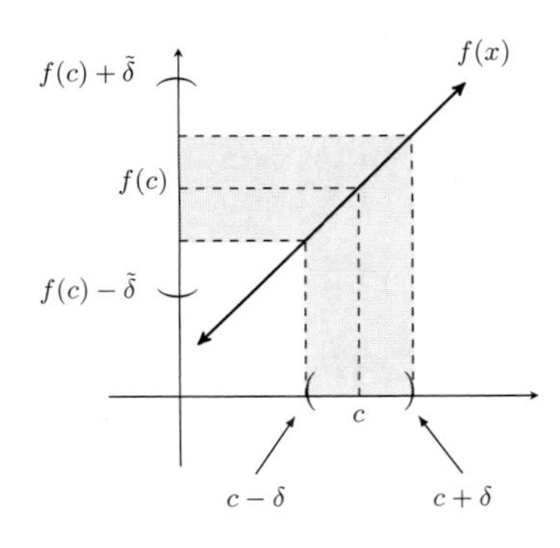

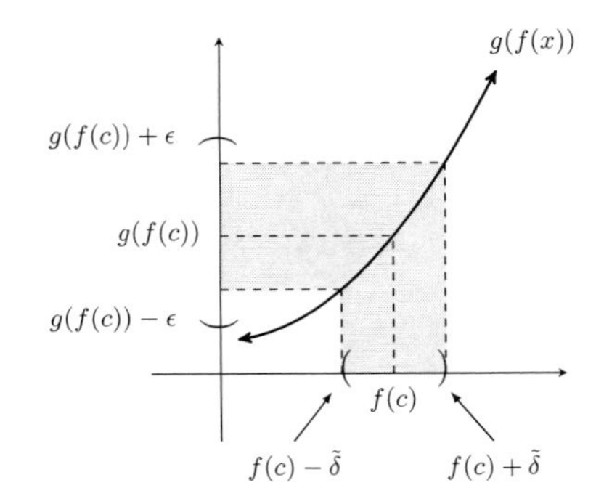

FIGURE 5.4.1
If x is within δ of c, then $f(x)$ is within $\tilde{\delta}$ of $f(c)$. Subsequently, if $f(x)$ is within $\tilde{\delta}$ of $f(c)$, then $g(f(x))$ is within ϵ of $g(f(c))$.

Proof of Theorem 5.38. Let $\epsilon > 0$ be given. Since g is continuous at $f(c)$, there exists a $\tilde{\delta}$ so that if $|y-f(c)| < \tilde{\delta}$, then $|g(y)-g(f(c))| < \epsilon$. Since f is continuous at c, there exists a δ so that if $|x-c| < \delta$, then $|f(x)-f(c)| < \tilde{\delta}$. Now, for $|x-c| < \delta$, $y = f(x)$ satisfies $|y-f(c)| < \tilde{\delta}$, and so $|g(f(x))-g(f(c))| < \epsilon$. □

In fact, a slightly more general result is true, which we give as a practice problem. Nontechnically put, it says that the limit of a continuous function may be "moved inside" of the function.

Practice 5.39. Suppose $f : D \to \mathbb{R}$ is continuous at a limit point c of D. Suppose also that $g : \tilde{D} \to D$ has $\lim_{x\to\tilde{c}} g(x) = c$. Then

$$\lim_{x\to\tilde{c}} f(g(x)) = f(\lim_{x\to\tilde{c}} g(x)) = f(c).$$

Methodology. *We use the sequential criterion for limits and continuity found in Theorems 5.13 and 5.36. If $\tilde{x}_n \to \tilde{c}$ is a sequence in $\tilde{D}$ with $\tilde{x}_n \neq \tilde{c}$, then $x_n = g(\tilde{x}_n) \to c$. Now since f is continuous at c and $x_n \to c$, a limit point of D, $f(x_n) \to f(c)$.*

Solution. Let $\tilde{x}_n \to \tilde{c}$ be a sequence in $\tilde{D}$, with $\tilde{x}_n \neq \tilde{c}$. Since $g(x) \to c$ as $x \to \tilde{c}$, $x_n = g(\tilde{x}_n) \to c$ by Theorem 5.13. Since c is a limit point of D, f is continuous at the limit point c, and $x_n \to c$, we have $f(x_n) \to f(c)$ by Theorem 5.36. The result now follows by Theorem 5.13. ■

5.4.3 A small library of continuous functions

Having properly set the stage with previous results, there can now be a list of number of familiar functions that are continuous on their natural domain:

Function:	Justification:
$\ln(x)$	(Exercise 11 in Section 3.3)
$\log_b(x)$, for $b > 0$	(Exercise 12 in Section 3.3)
e^x	(Exercise 13 in Section 3.3)
a^x, for $a > 0$	(Exercise 14 in Section 3.3)
For any α, x^α	(Exercise 15 in Section 3.3)
All polynomials	(Practice 5.17)
All rational functions	(Assertion 5 of Theorem 5.37)

The list above is not meant to be exhaustive: these are of course not the only continuous functions that exist, for instance, one could endlessly build new continuous functions just using Theorem 5.37. Additionally, there are other common functions we have not discussed yet that are also continuous (for instance, basic trig functions).

5.4.4 Types of discontinuities

There are a variety of ways to draw a graph of a function that is discontinuous at a point c. Perhaps the easiest way is to place two smooth curves together on the same graph creating a "break" in the curve. Alternatively, one could patch these pieces together on either side of c but artificially place $f(c)$ somewhere else. This behavior, pictured in Figure 5.4.2, is not altogether difficult to study, provided the one sided limits $\lim_{x\to c^+} f(x)$ and $\lim_{x\to c^-} f(x)$ exist. Such a discontinuity is called a *discontinuity of the first kind.*

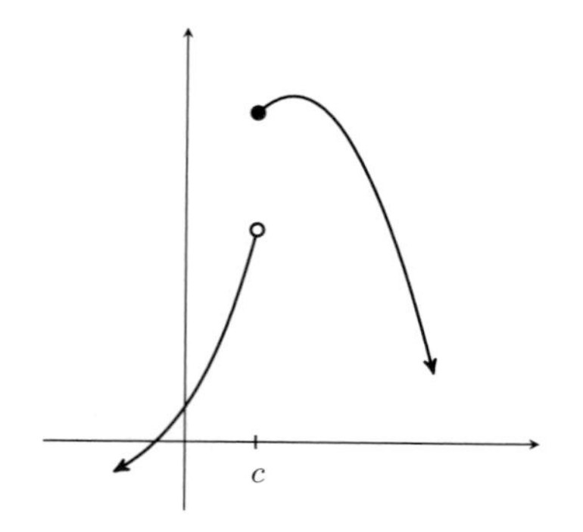

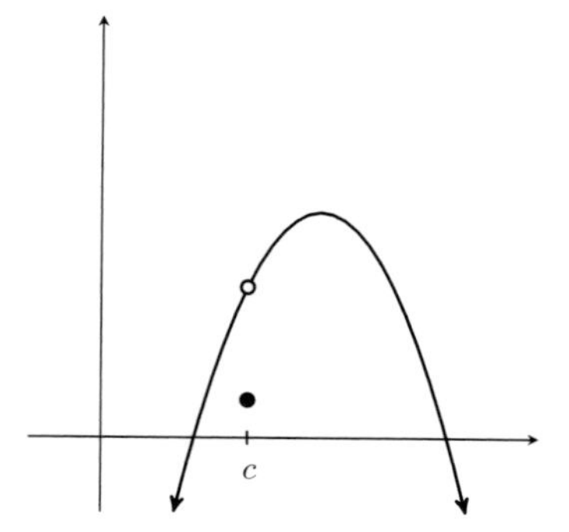

FIGURE 5.4.2
Discontinuity of the first kind.

Other discontinuities, pictured in Figure 5.4.3, are not so nice. A discontinuity not of the first kind is called a *discontinuity of the second kind.*

Definition 5.40. Suppose $f : D \to \mathbb{R}$ is discontinuous at $c \in D$.

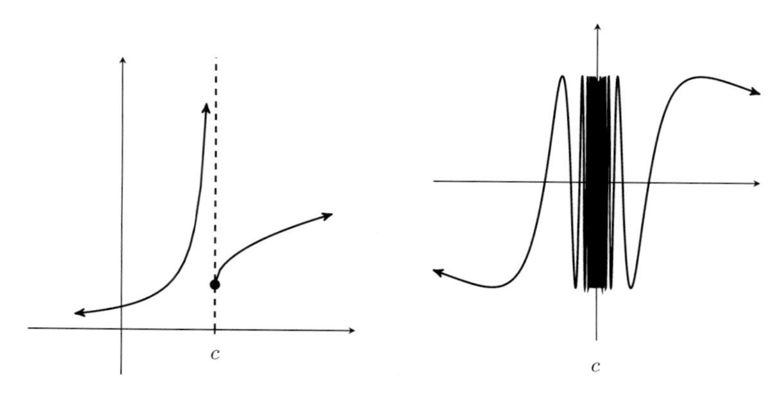

FIGURE 5.4.3
Discontinuities of the second kind.

1. A *discontinuity of the first kind* is one where $\lim_{x \to c^+} f(x)$ and $\lim_{x \to c^-} f(x)$ exist.

2. A *removable discontinuity* is a discontinuity of the first kind where $\lim_{x \to c^+} f(x) = \lim_{x \to c^-} f(x)$.

3. A *discontinuity of the second kind* is one where either $\lim_{x \to c^+} f(x)$ or $\lim_{x \to c^-} f(x)$ does not exist (in $\mathbb{R}$).

The term "removable discontinuity" makes sense: if f has a removable discontinuity at c, then there is a (unique) way to assign a value of $f(c)$ so that the resulting function is continuous. We have seen this already in the discussion (and subsequent practice problem) on Page 268. If $f(x) = \frac{x^2-1}{x-1}$, then f is continuous on $\mathbb{R} - \{1\}$.

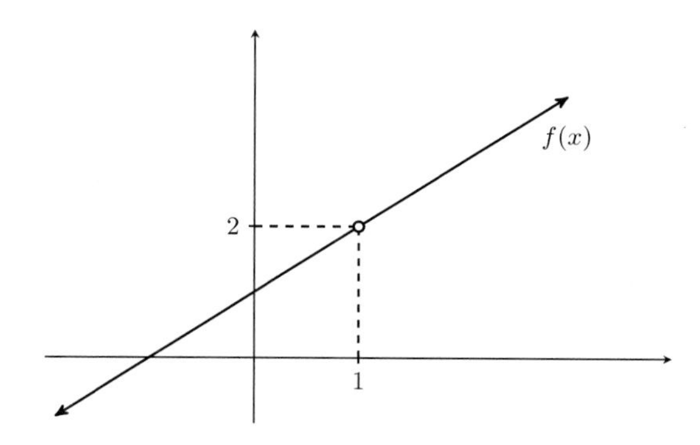

FIGURE 5.4.4
The graph of $f(x) = \frac{x^2-1}{x-1}$.

Since

$$2 = \lim_{x\to 1} \frac{x^2-1}{x-1} = \lim_{x\to 1^+} \frac{x^2-1}{x-1} = \lim_{x\to 1^-} \frac{x^2-1}{x-1},$$

if we were to define $f(1)$ to be anything other than 2, this function would have a removable discontinuity at $x = 1$ (see Figure 5.4.4). But defining $f(1) = 2$ would "remove" this discontinuity, and produce a continuous function on all of $\mathbb{R}$:

$$f(x) \text{ extends to } x+1 = \begin{cases} \frac{x^2-1}{x-1} & \text{for } x \neq 1, \\ 2 & \text{for } x = 1. \end{cases}$$

The study of discontinuities is a very interesting one throughout analysis. One basic realm is in the study of discontinuities of monotone functions. Not only can it be shown that every discontinuity of a monotone function is of the first kind, but that every monotone function has only a countable number of discontinuities! A more advanced example are Baire functions: these are (continuous and some discontinuous) functions which can be approximated by continuous functions. The interested reader can find more on these topics in [8] or [26].

5.4.5 Review of important concepts from Section 5.4

1. Continuity of functions is defined in Definition 5.35. If c is a limit point of D, Theorem 5.36 asserts that f is continuous at c if and only if $\lim_{x\to c} f(x) = f(c)$. Every function is continuous at an isolated point of its domain.
2. Since there is a close link between continuity and limits, it is easy to show that reasonable combinations–in particular, composition–of continuous functions are continuous (see Theorems 5.37 and 5.38). Note that these facts could be proven directly and without the use of Theorem 5.36.
3. Using our previous work, a number of familiar functions can be shown to be continuous. See the table on Page 284.
4. We briefly discuss types of discontinuities at the end of the section.

Exercises for Section 5.4

1. Let $f : D \to \mathbb{R}$. Prove that f is continuous on D if and only if the following is true for every $c \in D$: for every neighborhood V of $f(c)$, there exists a neighborhood U of c so that $f(U \cap D) \subseteq V$. (This fact is a very important one in understanding a more general notion of continuity of functions in the broader field of *topology*.)
2. Give an example of a function that has a discontinuity of the first kind, and give another example of a function with a discontinuity of the second kind.

3. Give an example of a function that is discontinuous at every real number. For your function, identify the type of discontinuity as being of the first or second kind.

4. Suppose $f : D \to \mathbb{R}$ and f is continuous at $c \in D$. If $f(c) > 0$, prove that there exists a neighborhood U of c so that $f(x) > 0$ for every $x \in U \cap D$.

5. Suppose $f : \mathbb{R} \to \mathbb{R}$ is continuous. Suppose that if D is a dense subset of $\mathbb{R}$ and $f(x) = 0$ for all $x \in D$, then $f(x) = 0$ for all $x \in \mathbb{R}$.

6. Suppose $f : D \to \mathbb{R}$ and f is continuous at $c \in D$. Prove there exists a neighborhood U of c so that the set $f(U \cap D)$ is bounded.

7. Suppose $f, g : D \to \mathbb{R}$, and f and g are continuous at $c \in D$. Define

$$\begin{aligned} \max\{f, g\}(x) &= \max\{f(x), g(x)\}, \text{ and} \\ \min\{f, g\}(x) &= \min\{f(x), g(x)\}. \end{aligned}$$

   Prove $\max\{f, g\}$ and $\min\{f, g\}$ are continuous at c. *[Hint: The result of Exercise 4 of Section 1.4 and Theorem 5.37 will be helpful.]*

8. Prove Theorem 5.38 by using Theorem 5.36 when c is a limit point of D and $\tilde{c}$ is a limit point of $\tilde{D}$. More specifically, let $f : D \to \tilde{D}$ and $g : \tilde{D} \to \mathbb{R}$, and suppose f is continuous at the limit point $c \in D$ and g is continuous at the limit point $f(c) = \tilde{c} \in \tilde{D}$. If $x_n \to c$ is an arbitrary sequence in D, prove that $g(f(x_n)) \to g(f(c))$.

5.5 Consequences of Continuity

One of the reasons we are concerned with continuous functions is because they behave so nicely, something we shall see in this section. These are the Boundedness Theorem, the Extreme Value Theorem, and the Intermediate Value Theorem. We also present two other important consequences of continuity concerning the continuous image of an interval and the existence of fixed points.

Before we begin, it is important to point out that the results in this section are not the only important results concerning continuous functions, and the the results in this section more or less involve a continuous function defined on an interval of the form $[a, b]$. There are more general versions of these theorems in more general settings, although we content ourselves with the versions of these results one might recall from basic calculus.

5.5.1 The Boundedness Theorem

Recall from Exercise 6 of Section 5.2 that a function $f : D \to \mathbb{R}$ is *bounded* if its set of outputs $f(D)$ is bounded. Our first important result is the *Boundedness Theorem*, which states that a continuous function on $[a, b]$ must be bounded. The proof of this theorem relies on several previous results from this chapter and Chapter 3.

Theorem 5.41 (Boundedness Theorem). *If $f : [a, b] \to \mathbb{R}$ is continuous, then it is bounded.*

Idea of proof. *We prove this by contradiction and pay special attention to where the hypothesis of the domain being $[a, b]$ is used. If the set $f([a, b])$ were unbounded above, then by Exercise 13 of Section 3.8, there would exist a sequence $f(x_n) \to \infty$. We would like to claim that the sequence (x_n) converges to assist us in finding a contradiction, but it does not have to: here is where we use the hypothesis that the domain of f is $[a, b]$. Since $x_n \in [a, b]$, it is a bounded sequence. The Bolzano-Weierstrass Theorem (Theorem 3.84) asserts that there exists a convergent subsequence $x_{n_k} \to c$. Since $x_{n_k} \in [a, b]$, by Corollary 3.49, $c \in [a, b]$ as well. We now have a convergent sequence $x_{n_k} \in [a, b]$ and since f is continuous on $[a, b]$, namely at c, it must be the case that $f(x_{n_k}) \to f(c)$ by Theorem 5.36. But $(f(x_{n_k}))$ is a subsequence of $(f(x_n))$ and since $f(x_n) \to \infty$, Theorem 3.116 asserts that $f(x_{n_k}) \to \infty$ as well. This is a contradiction, since $f(x_{n_k}) \to f(c) \neq \infty$. If $f([a, b])$ is unbounded below, a similar argument is used.*

Proof of Theorem 5.41. Suppose $f([a, b])$ were unbounded above. By Exercise 13 of Section 3.8, there exists a sequence $f(x_n) \to \infty$. Since $x_n \in [a, b]$ is bounded, there exists a convergent subsequence $x_{n_k} \to c$ by the Bolzano-Weierstrass Theorem (Theorem 3.84). By Corollary 3.49, $c \in [a, b]$ as well.

But f is continuous on $[a, b]$, and so by Theorem 5.36, $f(x_{n_k}) \to f(c)$. But $(f(x_{n_k}))$ is a subsequence of $(f(x_n))$, and since $f(x_n) \to \infty$, by Theorem 3.116, $f(x_{n_k}) \to \infty$ as well. This is a contradiction, since $f(x_{n_k}) \to f(c) \neq \infty$.

In the event that $f([a, b])$ is unbounded below, we replace ∞ with $-\infty$ in the argument above to obtain a similar contradiction. □

This theorem does not hold if we remove any of the hypotheses, as the following example illustrates.

Example 5.42. This example illustrates how the Boundedness Theorem fails if the hypothesis of continuity is removed, see also Exercise 1. We define the function $f : [0, 1] \to \mathbb{R}$ as

$$f(x) = \begin{cases} \frac{1}{x} & \text{for } x \in (0, 1] \\ 0 & \text{for } x = 0. \end{cases}$$

This function is clearly not continuous at $x = 0$: if we define the sequence $x_n = \frac{1}{n}$, then $x_n \in [0, 1]$ and $x_n \to 0$, although $f(x_n) = \frac{1}{1/n} = n \to \infty \neq f(0)$, as it should if f were continuous by Theorem 5.36. This sequence of outputs also illustrates that this function is not bounded on $[0, 1]$. ■

5.5.2 The Extreme Value Theorem

The Extreme Value Theorem goes a step further than the Boundedness Theorem and asserts that there exist a global maximum and global minimum of a continuous function defined on $[a, b]$. It goes without saying that this theorem has a number of very important applications. To properly state the theorem, we need a short definition.

Definition 5.43. Let $f : D \to \mathbb{R}$. If there exists a $c \in D$ for which $f(c) \geq f(x)$ for every $x \in D$, then $f(c)$ is a *global maximum* of f. If there exists a $c \in D$ for which $f(c) \leq f(x)$ for every $x \in D$, then $f(c)$ is a *global minimum* of f.

The Extreme Value Theorem essentially states that every continuous function defined on $[a, b]$ has a global maximum and global minimum.

Theorem 5.44 (Extreme Value Theorem). *Let $f : [a, b] \to \mathbb{R}$. If f is continuous, then f has a global maximum and global minimum. That is, there exists $m, M \in [a, b]$ so that*

$$f(m) \leq f(x) \leq f(M) \text{ for all } x \in [a, b].$$

Idea of proof. *We prove that a global maximum exists, and leave the existence of a global minimum to Exercise 5. By the Boundedness Theorem above, the set $f([a, b])$ is bounded. The Completeness Axiom (Axiom 2.28) guarantees that $\alpha = \sup(f([a, b]))$ is real, and this will be our global max; it remains to show that $\alpha = f(M)$ for some $M \in [a, b]$. By Theorem 3.46, there is a sequence of outputs $f(x_n) \to \alpha$. As in the proof of the Boundedness*

Theorem, we would like this sequence of inputs (x_n) to converge, and use the continuity of f and Theorem 5.36 to complete the proof. But once again, (x_n) may not converge. Here is where we use the hypothesis that the domain of f is $[a,b]$: $x_n \in [a,b]$ is bounded, and so the Bolzano-Weierstrass Theorem (Theorem 3.84) says there is a convergent subsequence $x_{n_k} \to M$. As before, since $x_{n_k} \in [a,b]$, $M \in [a,b]$ by Corollary 3.49. As a subsequence, $f(x_{n_k}) \to \alpha$ since $f(x_n) \to \alpha$ (Theorem 3.66). Now, since f is continuous, by Theorem 5.36 we have $f(x_{n_k}) \to f(M)$. So, $f(x_{n_k}) \to \alpha$ and $f(x_{n_k}) \to f(M)$. Since limits are unique (Theorem 3.22), $f(M) = \alpha$. This proves that $f(M)$ is a global maximum of f: $\alpha = f(M)$ is the supremum of $f([a,b])$, so for every $f(x) \in f([a,b])$, we have $f(x) \leq f(M)$.

Proof of Theorem 5.44. Since f is a continuous function on $[a,b]$, the Boundedness Theorem (Theorem 5.41) tells us that $f([a,b])$ is bounded. Since it is also nonempty, the Completeness Axiom (Axiom 2.28) tells us that $\alpha = \sup(f([a,b])) \in \mathbb{R}$. Thus, for all $x \in [a,b]$, $f(x) \leq \alpha$. We aim to prove that $\alpha = f(M)$ for some $M \in [a,b]$ so that this is the global maximum of f. The existence of a global minimum is left as Exercise 5.

By Theorem 3.46, there exists a sequence $f(x_n) \to \alpha$. Since $x_n \in [a,b]$, this sequence is bounded, and by the Bolzano-Weierstrass Theorem (Theorem 3.84) there exists a convergent subsequence $x_{n_k} \to M$. Since $x_{n_k} \in [a,b]$, $M \in [a,b]$ by Corollary 3.49. Since $(f(x_{n_k}))$ is a subsequence of $(f(x_n))$, and $f(x_n) \to \alpha$, we have $f(x_{n_k}) \to \alpha$ as well by Theorem 3.66. But since f is continuous and $x_{n_k} \to M$, we have $f(x_{n_k}) \to f(M)$ by Theorem 5.36. Now we have $f(x_{n_k}) \to \alpha$ and $f(x_{n_k}) \to f(M)$, and since limits of sequences are unique (Theorem 3.22), $\alpha = f(M)$, completing the proof. □

5.5.3 The Intermediate Value Theorem

The Intermediate Value Theorem is another fundamental result concerning continuous functions defined on intervals. In the statement of the theorem below, we assume that the domain of our function is the interval $[a,b]$, and consider the endpoints a and b. Usually, this theorem is applied to functions defined on some larger domain D, with $[a,b] \subseteq D$; this amounts to applying this theorem to a function restricted to $[a,b]$. See Exercise 8 for an example of this.

Theorem 5.45 (Intermediate Value Theorem). *Let $f : [a,b] \to \mathbb{R}$ be continuous. For each k between $f(a)$ and $f(b)$, there exists a $c \in [a,b]$ so that $f(c) = k$.*

Idea of proof. *We construct a set that will help us:*

$$S = \{x \in [a,b] | f(x) < k\}.$$

This set measures how much of $[a,b]$ is mapped beneath k. In the event that $f(a) < f(b)$ (the other case handled similarly), $a \in S$ so $S \neq \emptyset$. It is an easy *matter to show that $c = \sup(S) \in [a, b]$ (Exercise 3 of Section 2.2), but what requires more care is to show that $f(c) = k$ so that this c is the input we are looking for. If $f(c) < k$, then the continuity of f would guarantee there is an input x slightly larger than c for which $f(x) < k$ as well. In this event, c is not an upper bound of S since $c < x$ and $x \in S$. If $f(c) > k$ then by the same reasoning, we could find any input x which is slightly smaller than c so that $f(x) > k$ as well. As a result, there is a smaller upper bound for S, contradicting the assumption that c is the least upper bound of S.*

Proof of Theorem 5.45. If k is equal to either $f(a)$ or $f(b)$, then we are done as $c = a$ or b will do. For k to not equal either of these values, we must have $f(a) \neq f(b)$. Assume that $f(a) < k < f(b)$, the case that $f(b) < f(a)$ is handled by essentially replacing a with b in what follows. Let

$$S = \{x \in [a, b] | f(x) < k\}.$$

Clearly $a \in S$ since by assumption $f(a) < k$. Also S is bounded, so $c = \sup(S)$ is real, and $c \in [a, b]$ by Exercise 3 in Section 2.2. We claim $f(c) = k$, and prove this by eliminating the possibility that $f(c) < k$ or that $f(c) > k$.

Suppose $f(c) < k$, and let $\epsilon = \frac{1}{2}(k - f(c)) > 0$. Since f is continuous, there exists a δ so that if $|x - c| < \delta$, then $|f(x) - f(c)| < \epsilon$. It is not possible that $c = b$ since we have assumed $f(c) < k$ while at the same time we have assumed $f(b) > k$. Choose any element $x \in (c, c + \delta) \cap [a, b] \neq \emptyset$. Then $|x - c| < \delta$ and so

$$|f(x) - f(c)| < \epsilon = \frac{1}{2}(k - f(c)).$$

And so by considering $f(x) - f(c) < \frac{1}{2}(k - f(c))$ and adding $f(c)$, we have

$$\begin{array}{rcll} f(x) & < & \frac{1}{2}k + \frac{1}{2}f(c) & \\ & < & \frac{1}{2}k + \frac{1}{2}k & \text{(since } f(c) < k) \\ & = & k. & \end{array}$$

Thus, $x \in S$ and $x > c$, which contradicts c being an upper bound for S.

Now suppose $f(c) > k$, and reselect $\epsilon = \frac{1}{2}(f(c) - k) > 0$. Since f is continuous, there exists a δ so that if $|x - c| < \delta$, then $|f(x) - f(c)| < \epsilon$. It is not possible that $c = a$ since we have assumed $f(c) > k$ while at the same time we originally assumed $f(a) < k$. Choose any $x \in (c - \delta, c) \cap [a, b] \neq \emptyset$. Then

$$|f(x) - f(c)| < \epsilon = \frac{1}{2}(f(c) - k).$$

Considering that $-\frac{1}{2}(f(c) - k) < f(x) - f(c)$ and again adding $f(c)$, we have

$$\begin{array}{rcll} f(x) & > & \frac{1}{2}k + \frac{1}{2}f(c) & \\ & > & \frac{1}{2}k + \frac{1}{2}k & \text{(since } f(c) > k) \\ & = & k. & \end{array}$$

We have shown that under the assumption that $f(c) > k$, every element

$x \in (c-\delta, c) \cap [a,b]$ has $f(x) > c$. Thus, $(c-\delta, c) \cap S = \emptyset$ and as a result $c-\delta$ is another upper bound for S that is smaller than c, contradicting c being the *least* upper bound. Note that $c-\delta$ is not less than a, since $a \in S$ and we have eliminated the possibility that any element from S is in $(c-\delta, c)$.

In conclusion, we have eliminated all possibilities except $f(c) = k$, and the proof is complete. □

Here is an example of a possible (and somewhat common) use of the Intermediate Value Theorem.

Practice 5.46. Prove that there exists a root of the polynomial $f(x) = x^5 - x - 1$ between $x = 0$ and $x = 2$.

Methodology. *We use the Intermediate Value Theorem (Theorem 5.45): f is a polynomial and hence continuous (Practice 5.17). Considering this as a function $f : [0,2] \to \mathbb{R}$, we have $f(0) = -1$ and $f(2) = 29$. By the Intermediate Value Theorem (Theorem 5.45), since 0 is between -1 and 29, there must be some input c so that $f(c) = 0$.*

Solution. Since f is a polynomial, it is continuous (Practice 5.17). Thus, the Intermediate Value Theorem (Theorem 5.45) applies: $f(0) = -1$ and $f(2) = 29$, and 0 is between -1 and 29. So there exists a $c \in [0,2]$ so that $f(c) = 0$. ■

5.5.4 Other important results

There are two other main results we present that relate to continuous functions defined on intervals. The first essentially states that continuous functions send intervals of the form $[a,b]$ to another interval of the same sort (see Exercise 11 for what could happen if the intervals are not of that same form). The second result asserts the existence of fixed points for some continuous functions.

Theorem 5.47. *Let $f : [a,b] \to \mathbb{R}$. If f is continuous, then $f([a,b]) = [c,d]$ for some real numbers $c \leq d$.*

Idea of proof. *This is an easy consequence of the Extreme Value Theorem (Theorem 5.44) and the Intermediate Value Theorem (Theorem 5.45). By the Extreme Value Theorem, the image of f is bounded between two numbers which are already outputs and thus also in the image of f. The Intermediate Value Theorem says that every number between those outputs is also targeted by f from some input.*

Proof of Theorem 5.47. By the Extreme Value Theorem (Theorem 5.44), there exists $m, M \in [a,b]$ so that

$$f(m) \leq f(x) \leq f(M),$$

so that if $c = f(m)$ and $d = f(M)$, then $f([a,b]) \subseteq [c,d]$. For each k between c and d, the Intermediate Value Theorem (Theorem 5.45) gives us a $c \in [a,b]$ so that $f(c) = k$. Thus, $[c,d] \subseteq f([a,b])$, and the proof is complete. □

"Fixed Point Theory" is an important field of mathematics devoted to the existence of *fixed points*: an input of a function that is sent to itself. The following result is a basic fact in this subject.

Theorem 5.48. *If $f : [0,1] \to [0,1]$ is continuous, then there exists a fixed point for f. That is, there exists a $c \in [0,1]$ so that $f(c) = c$.*

Idea of proof. *If $f(0) = 0$ or $f(1) = 1$, then our fixed point will be one of those. If not, then $f(0) > 0$ and $f(1) < 1$. As the combination of continuous functions, $g(x) = f(x) - x$ is continuous. Moreover, according to our original assumptions about $f(0)$ and $f(1)$, $g(0) = f(0) - 0 = f(0) > 0$ and $g(1) = f(1) - 1 < 0$. Thus, by the Intermediate Value Theorem (Theorem 5.45), there is a $c \in [0,1]$ so that $g(c) = 0$. But $0 = g(c) = f(c) - c$, so $f(c) = c$.*

Proof of Theorem 5.48. If $f(0) = 0$ or $f(1) = 1$, then we are done. If not, since the image of f is $[0,1]$, we assume $f(0) > 0$ and $f(1) < 1$. Define $g(x) = f(x) - x$, which is continuous by Assertion 1 of Theorem 5.37 and since the function $h(x) = x$ is continuous (Practice 5.17). By assumption, $g(0) = f(0) - 0 > 0$, and $g(1) = f(1) - 1 < 0$. By the Intermediate Value Theorem (Theorem 5.45), there exists a $c \in [0,1]$ so that $0 = g(c) = f(c) - c$. That is, $f(c) = c$ is our fixed point. □

5.5.5 Review of important concepts from Section 5.5

1. The Boundedness Theorem (Theorem 5.41) asserts that any continuous function defined on an interval of the form $[a, b]$ is bounded. The Extreme Value Theorem (Theorem 5.44) goes further to say that there is a largest and smallest output in the range of f.
2. The Intermediate Value Theorem (Theorem 5.45) asserts that all intermediate values between two outputs are attained through some input.
3. The continuous image of the interval $[a, b]$ is another interval of the same type (Theorem 5.47).
4. If $f : [0,1] \to [0,1]$ and is continuous, there exists a fixed point of f (Theorem 5.48).

Exercises for Section 5.5

1. Find an example of a continuous function defined on (a, b) that is not bounded. (This illustrates that the Boundedness Theorem can fail if one removes a hypothesis.)
2. Give an example of how the Intermediate Value Theorem (Theorem 5.45) fails if f is allowed to be discontinuous.

3. In the proof of the Boundedness Theorem (Theorem 5.41), we encounter a sequence of outputs $f(x_n) \to \infty$, and go on to claim that the corresponding sequence of inputs (x_n) need not converge. Give an example of a function $f : [a, b] \to \mathbb{R}$ that is continuous, and a sequence $(f(x_n))$ converging to $\sup(f([a, b]))$, but for which (x_n) does not converge.

4. Suppose $f : D \to \mathbb{R}$, and that f is continuous on D. Suppose also that D contains all of its limit points and that D is bounded[6]. Prove that f is bounded. *[Hint: One could use the same argument in the Boundedness Theorem (Theorem 5.41).]*

5. Complete the proof of the Extreme Value Theorem by proving that if $f : [a, b] \to \mathbb{R}$ is continuous, then there exists a global minimum of f.

6. Suppose $f : [a, b] \to \mathbb{R}$ is continuous. Show that if $f(a)f(b) < 0$, then there exists a $c \in [a, b]$ with $f(c) = 0$.

7. Suppose $f : \mathbb{R} \to \mathbb{Z}$ is continuous. Prove f is constant.

8. Prove that every polynomial of odd degree has a real root (a *root* of a polynomial is a number c that is mapped to 0). *[Hint: You may find Exercise 4 of Section 5.3 helpful.]*

9. Show that for any $n \in \mathbb{N}$, the nth root of a positive number exists. That is, for every $y > 0$, there exists an $c \in \mathbb{R}$ so that $c^n = y$. *[Hint: Let $f(x) = x^n$. Prove that y is between $f(0)$ and $f(y+1)$, and use the Intermediate Value Theorem (Theorem 5.45).]*

10. Prove that there are always two antipodal points on the equator that have the same temperature. *[Hint: This is a special case of the Borsak-Ulam Theorem. Assume that the temperature T on the earth varies continuously, and that the equator is a circle. Then consider the temperature function $T : [0, 2\pi] \to \mathbb{R}$ as being measured from an angle $\theta \in [0, 2\pi]$ from where the International Date Line intersects the equator. Then $f(\theta) = T(\theta) - T(\theta + \pi)$ is continuous, and we may use the Intermediate Value Theorem.]*

11. Give an example of a continuous function $f : (a, b) \to \mathbb{R}$ for which $f((a, b)) \neq (c, d)$ for any $c < d$. (This demonstrates that only intervals of the form $[a, b]$ are necessarily preserved by continuous functions, see Theorem 5.47.)

12. On every Sunday in November, college football and men's basketball teams are each ranked. During one weekend in November, the Oregon football team was ranked lower than the Oregon men's basketball team. Later in the month, the football team was ranked higher than the basketball team, and yet there was no one week

[6] Such a subset of $\mathbb{R}$ is called *compact*, according to the Heine-Borel Theorem.

in which their rankings were equal. Why does this not violate the Intermediate Value Theorem (Theorem 5.45)?

5.6 Uniform Continuity

In this section we introduce a stronger form of continuity: uniform continuity. This stronger type of continuity has a number of very important properties, and this section is subsequently devoted to presenting these.

5.6.1 Formulating uniform continuity

We introduce the notion of Uniform Continuity through the use of an example that, at this point, is elementary. Suppose we were to try to show that $f(x) = x^2$ is continuous at $x = 1$ using only the definition of continuity (Definition 5.35). We let $\epsilon > 0$ be given, and since $f(1) = 1$, we attempt to find a δ so that if $|x-1| < \delta$ then $|x^2 - 1| < \epsilon$. Since $|x^2-1| = |x-1||x+1|$, and $|x-1| < \delta$ will be "small", the real issue is trying to determine just how large $|x+1|$ will be. We use reasoning similar to Practice 5.6: If $\delta \leq 1$ and $|x-1| < \delta$, then

$$-1 \leq -\delta < x - 1 < \delta \leq 1, \tag{5.3}$$

and so by adding 2 everywhere, we conclude that

$$1 < x + 1 < 3, \text{so that } |x+1| < 3.$$

In conclusion, we have shown that if $\delta \leq 1$, then $|x+1| < 3$. We can now complete the proof that f is continuous at 1 by defining $\delta = \min\{1, \epsilon/3\}$, and so when $|x-1| < \delta$ we have

$$|x^2 - 1| = |x-1||x+1| < \delta \cdot 3 \leq \frac{\epsilon}{3} \cdot 3 = \epsilon.$$

Now let us repeat the same example, except this time showing $f(x) = x^2$ is continuous at $x = 100$. Referring to Equation (5.3), when $\delta \leq 1$, we have

$$-1 \leq -\delta < x - 100 < \delta \leq 1,$$

and so by adding 200 everywhere, we conclude that

$$199 < x + 100 < 201, \text{so that } |x+100| < 201.$$

Now, $\delta = \min\{1, \epsilon/201\}$, since when $|x-100| < \delta$, we have

$$|x^2 - 100^2| = |x-100||x+100| < \delta \cdot 201 \leq \frac{\epsilon}{201} \cdot 201 = \epsilon.$$

The point to be made here is that the δ we search for in these examples depends on ϵ, but also at the point in question. It would be useful if the δ we were searching for depended only on ϵ: this is the difference between "regular" continuity and uniform continuity, as shown in the following definition. The term "uniform" refers to the choice of δ being made uniformly across D, without any reference to any point in D.

Definition 5.49. Let $f : D \to \mathbb{R}$. We say f is *uniformly continuous* if for every $\epsilon > 0$, there exists a $\delta > 0$ so that whenever $x, y \in D$ and $|x - y| < \delta$ then $|f(x) - f(y)| < \epsilon$.

There are several remarks to be made before proceeding. First, **if a function is uniformly continuous on D, then it is continuous on D.** This is easily seen in the definition. Second and perhaps more importantly, **the notion of uniform continuity depends on the domain D, not on any one of its points.** It turns out that the same function could be uniformly continuous on one domain, and not uniformly continuous on another, as the next example illustrates.

Example 5.50. Let $f(x) = x^2$. This example illustrates that f may or may not be uniformly continuous, depending on its domain.

First, suppose the domain D of f is $D = [0, 1]$. We show that f is uniformly continuous on this D. Let $\epsilon > 0$ be given. We must find a δ so that whenever $x, y \in [0, 1]$ and $|x - y| < \delta$, then $|x^2 - y^2| < \epsilon$. The discussion above drives our choice: since $|x^2 - y^2| = |x - y||x + y|$, and $|x - y|$ can be made to be "small", we investigate how large $|x + y|$ could be. Here is where the domain of f becomes important: since $x, y \in [0, 1]$, it must be that $|x + y| \leq 2$! So, we let $\delta = \epsilon/2$, and when $x, y \in [0, 1]$ and $|x - y| < \delta$, we have

$$|x^2 - y^2| = |x - y||x + y| < 2\delta = \epsilon.$$

Next, suppose that the domain D of f is $D = \mathbb{R}$. We show that f is *not* uniformly continuous on this D. Suppose that it were, and choose $\epsilon = 1$. There exists a δ so that whenever $x, y \in \mathbb{R}$ and $|x - y| < \delta$, we have $|x^2 - y^2| < \epsilon = 1$. Consider any x and choosing $y = x + \frac{\delta}{2}$. Clearly $|x - y| = |x - (x + \frac{\delta}{2})| = \frac{\delta}{2} < \delta$, but when $x \leq 0$, we have

$$|x^2 - y^2| = \left|x^2 - \left(x + \frac{\delta}{2}\right)^2\right| = \left|-x\delta - \frac{\delta^2}{4}\right| = \delta\left(x + \frac{\delta}{4}\right).$$

As $x \to \infty$, $\delta\left(x + \frac{\delta}{4}\right) \to \infty$, and so there exists some x for which $\delta\left(x + \frac{\delta}{4}\right) \geq 1$. This violates our assumption that this quantity be less than $\epsilon = 1$. As a result, $f(x) = x^2$ is not uniformly continuous on $\mathbb{R}$. ■

5.6.2 Consequences of uniform continuity

Before we present important results concerning uniform continuity, we pause to negate Definition 5.49: A function is *not* uniformly continuous if there exists an $\epsilon > 0$ so that for every δ, there exists $x, y \in D$ with $|x - y| < \delta$, but with $|f(x) - f(y)| \geq \epsilon$. This will be very helpful in what follows.

Although uniform continuity implies continuity, it is not always the case that the converse is true (this can be seen in Example 5.50). If the domain of the function is $[a, b]$, then this converse *is* in fact true.

Theorem 5.51. *Let $f : [a,b] \to \mathbb{R}$. If f is continuous on $[a,b]$, then it is uniformly continuous on $[a,b]$.*

Idea of proof. *We prove this by contradiction. If f were continuous but not uniformly continuous, there would be an ϵ so that for every δ there are $x, y \in [0,1]$ with $|x-y| < \delta$ but $|f(x) - f(y)| \geq \epsilon$. We choose $\delta = \frac{1}{n}$ and find sequences $x_n, y_n \in [a,b]$ with $|x_n - y_n| < \frac{1}{n}$ but $|f(x_n) - f(y_n)| \geq \epsilon$. Here is where the domain of f being $[a,b]$ is important: since $x_n \in [a,b]$ there exists a convergent subsequence $x_{n_k} \to c \in [a,b]$ by the Bolzano-Weierstrass Theorem (Theorem 3.84). We show that $y_{n_k} \to c$ as well, but that $f(x_{n_k})$ and $f(y_{n_k})$ cannot both converge to $f(c)$, as they must since f is assumed to be continuous (Theorem 5.36).*

Proof of Theorem 5.51. Suppose f is continuous on $[a,b]$ but not uniformly continuous on $[a,b]$. There exists an $\epsilon > 0$ so that for every δ there are $x, y \in [a,b]$ with $|x-y| < \delta$ but $|f(x) - f(y)| \geq \epsilon$. We apply this fact for $\delta = \frac{1}{n}$ for any $n \in \mathbb{N}$. Thus there are sequences $x_n, y_n \in [a,b]$ with $|x_n - y_n| < \frac{1}{n}$ but $|f(x_n) - f(y_n)| \geq \epsilon$. Since $x_n \in [a,b]$ there is a convergent subsequence $x_{n_k} \to c \in [a,b]$ by the Bolzano-Weierstrass Theorem (Theorem 3.84).

We claim that the subsequence $y_{n_k} \to c$ as well. To see this, we note that for every k, $n_k \geq k$ (Exercise 3 of Section 3.4), and so $|x_{n_k} - y_{n_k}| < \frac{1}{n_k} \leq \frac{1}{k}$. Since $\frac{1}{k} \to 0$, it follows that $y_{n_k} - x_{n_k} \to 0$ as well. Then

$$y_{n_k} = x_{n_k} - (x_{n_k} - y_{n_k}) \to c - 0 = c.$$

Since f is continuous and $x_{n_k}, y_{n_k} \to c$, we must have $f(x_{n_k}), f(y_{n_k}) \to f(c)$ by Theorem 5.36. If this were the case, then eventually $|f(x_{n_k}) - f(y_{n_k})| < \epsilon$, but for every k we have $|f(x_{n_k}) - f(y_{n_k})| \geq \epsilon$, a contradiction. □

Uniformly continuous functions interact with Cauchy sequences in a way that other functions do not, as the next two results illustrate. Theorem 5.52 says that uniformly continuous functions will map Cauchy sequences to Cauchy sequences. The converse of this is not true unless the domain of the function is bounded. Theorem 5.53 proves this, and Example 5.54 illustrates that this extra boundedness hypothesis is required.

Theorem 5.52. *Every uniformly continuous function maps Cauchy sequences to Cauchy sequences. That is, if $f : D \to \mathbb{R}$ is uniformly continuous on D and (x_n) is any Cauchy sequence in D, then $(f(x_n))$ is also a Cauchy sequence in D.*

Idea of proof. *This proof is really just a straightforward unraveling of what it means to be Cauchy and uniformly continuous. If (x_n) is Cauchy and $\epsilon > 0$ is given, there exists a δ so that whenever $x, y \in D$ and $|x-y| < \delta$, we have $|f(x) - f(y)| < \epsilon$. Since (x_n) is Cauchy, eventually $|x_n - x_m| < \delta$ and so eventually $|f(x_n) - f(x_m)| < \epsilon$.*

Proof of Theorem 5.52. Let $x_n \in D$ be Cauchy and suppose f is uniformly continuous. Let $\epsilon > 0$ be given. There exists a δ so that if $|x - y| < \delta$, then $|f(x) - f(y)| < \epsilon$. Since (x_n) is Cauchy, there exists an $N \in \mathbb{N}$ so that for every $n, m \geq N$ we have $|x_n - x_m| < \delta$. So, for these same $n, m \geq N$, we have $|f(x_n) - f(x_m)| < \epsilon$, and so $(f(x_n))$ is Cauchy. $\square$

Here is the partial converse to Theorem 5.52. Note again the extra hypothesis here concerning the boundedness of the domain.

Theorem 5.53. *Suppose $f : D \to \mathbb{R}$ sends Cauchy sequences to Cauchy sequences and D is bounded. Then f is uniformly continuous.*

Idea of proof. *We prove this by contradiction: if f is not uniformly continuous, we can create a situation similar to that in the proof of Theorem 5.51. There is some ϵ for which every $\delta = \frac{1}{n}$ has elements $x_n, y_n \in D$ and $|x_n - y_n| < \frac{1}{n}$ while $|f(x_n) - f(y_n)| \geq \epsilon$. Also similarly to that proof, since the domain D is bounded, there is a convergent subsequence $x_{n_k} \to c$ by the Bolzano-Weierstrass Theorem (Theorem 3.84), and $y_{n_k} \to c$ as well is convergent. Dissimilarly to that proof, however, c may or may not be in D, which is why the hypothesis about Cauchy sequences is important. We can form the convergent sequence $z_k \to c$ by interlacing the (sub)sequences (x_{n_k}) and (y_{n_k}) as in Exercise 9 from Section 3.4. Since (z_k) converges, it is Cauchy by Corollary 3.101 of Section 3.7. But f sends Cauchy sequences to Cauchy sequences, so $(f(z_k))$ should be Cauchy, although we see that for every k that $|f(z_{2k+1}) - f(z_{2k})| \geq \epsilon$: a contradiction.*

Proof of Theorem 5.53. We suppose that f is not uniformly continuous and initially proceed as in the proof of Theorem 5.51. There exists an $\epsilon > 0$ so that for every δ there are $x, y \in D$ with $|x - y| < \delta$ but $|f(x) - f(y)| \geq \epsilon$. We apply this fact for $\delta = \frac{1}{n}$ for any $n \in \mathbb{N}$. Thus there are sequences $x_n, y_n \in D$ with $|x_n - y_n| < \frac{1}{n}$ but $|f(x_n) - f(y_n)| \geq \epsilon$. Since D is bounded, there exists a convergent subsequence $x_{n_k} \to c$ by the Bolzano-Weierstrass Theorem[7] (Theorem 3.84). By the same reasoning in the proof of Theorem 5.51, $y_{n_k} \to c$ as well.

Define the sequence (z_k) by interlacing (x_{n_k}) and (y_{n_k}) by proceeding as $x_{n_1}, y_{n_1}, x_{n_2}, y_{n_2}, x_{n_3}, y_{n_3}, \ldots$. By Exercise 9 from Section 3.4, (z_k) converges, and so by Corollary 3.101, (z_k) is Cauchy. By assumption $(f(z_k))$ is Cauchy as well. But for every k, $|f(z_{2k+1}) - f(z_{2k})| \geq \epsilon$, and so $(f(z_k))$ is not Cauchy, a contradiction. $\square$

The next example illustrates that Theorem 5.53 need not hold if the domain of the function is not bounded.

Example 5.54. Let $f : \mathbb{R} \to \mathbb{R}$ be defined as $f(x) = x^2$. We show here that f sends Cauchy sequences to Cauchy sequences. In Example 5.50, we proved

[7]In contrast to the proof of Theorem 5.51, c may or may not be in D.

that f is not uniformly continuous on $\mathbb{R}$, so that this example illustrates that the extra boundedness hypothesis in Theorem 5.53 is necessary.

Suppose (x_n) is Cauchy. By Theorem 3.98 of Section 3.7, (x_n) is bounded. So, $x_n \in [a, b]$ for some $a, b \in \mathbb{R}$. Since (x_n) is Cauchy, it converges (Corollary 3.101), so $x_n \to c \in [a, b]$. Since f is continuous, $f(x_n) \to f(c)$, and again by Corollary 3.101, $(f(x_n))$ is Cauchy.

5.6.3 Continuous extensions

One of the major benefits of having a uniformly continuous function is that there exists a unique continuous extension of it to any limit point of its domain (Theorem 5.56). This is a property not necessarily shared by those functions which are not uniformly continuous, and thus provides an easy way to distinguish between these types of functions; see Practice 5.58 below.

To prove this theorem, we begin with a lemma whose result is of interest on its own.

Lemma 5.55. *Let $f : D \to \mathbb{R}$ and c be a limit point of D. If f is uniformly continuous on D, then $\lim_{x \to c} f(x)$ exists.*

Idea of proof. *We use Theorem 5.13 and Corollary 3.101 often. We first determine what $\lim_{x \to c} f(x)$ would have to be by letting $x_n \to c$ with $x_n \neq c$ be a sequence in D converging to c (which we may find since c is a limit point of D). Since (x_n) converges, it is Cauchy. Since f is uniformly continuous, $(f(x_n))$ is Cauchy by Theorem 5.52, and so it converges to some number L. We have not shown $\lim_{x \to c} f(x) = L$, since we have considered only one sequence $x_n \to c$, not an arbitrary one as Theorem 5.13 requires. We consider an arbitrary sequence $y_n \to c$ and try to show $f(y_n) \to L$. To do this, we interlace (y_n) with (x_n) to form a new convergent (hence Cauchy) sequence $z_k \to c$, and so $(f(z_k))$ is Cauchy (hence convergent). The subsequence $f(x_n) \to L$, and since every subsequence of the convergent $(f(z_k))$ must converge to the same thing, so must $f(y_n) \to L$. This completes the proof using Theorem 5.13: (y_n) is an arbitrary sequence converging to c, and we have shown $f(y_n) \to L$.*

Proof of Lemma 5.55. Since c is a limit point of D, there exists a sequence $x_n \to c$ with $x_n \in D$ and $x_n \neq c$. This sequence is Cauchy by Corollary 3.101. Since f is uniformly continuous, the sequence $(f(x_n))$ is Cauchy (Theorem 5.52), and hence converges (Corollary 3.101) to some number L. We claim $L = \lim_{x \to c} f(x)$, which we prove using Theorem 5.13.

Suppose $y_n \to c$ is an arbitrary sequence in D with $y_n \neq c$. Since $x_n \to c$ as well, we form the sequence (z_k) by proceeding as $x_1, y_1, x_2, y_2, \ldots$ as in the proof of Theorem 5.53. By Exercise 9 from Section 3.4, $z_k \to c$. As above, (z_k) is Cauchy, and again since f is uniformly continuous, $(f(z_k))$ is Cauchy. So, $f(z_k) \to \tilde{L}$ for some $\tilde{L}$. But $(f(x_n))$ is a subsequence of $(f(z_k))$, and so since $f(z_k) \to \tilde{L}$, we must have $f(x_n) \to \tilde{L}$ (Theorem 3.66). But $f(x_n) \to L$, and since limits are unique, $L = \tilde{L}$. Finally, $(f(y_n))$ is also a subsequence of

$(f(z_k))$, and so $f(y_n) \to \tilde{L} = L$ as well. We have shown that the arbitrary sequence $y_n \to c$ in D with $y_n \neq c$ has $f(y_n) \to L$, completing the proof. □

We use Lemma 5.55 to prove Theorem 5.56.

Theorem 5.56. *Let $f : D \to \mathbb{R}$. If f is uniformly continuous and c is a limit point of D, then there exists a continuous $\tilde{f} : D \cup \{c\} \to \mathbb{R}$ extending f. This extension is unique, and uniformly continuous on $D \cup \{c\}$.*

Idea of proof. *We are asked to do two things: show a continuous extension exists and then show that such an extension is unique. For the extension to be a continuous one, we have no choice but to define $\tilde{f}(c) = \lim_{x \to c} f(x)$, and this limit exists by Lemma 5.55. Thus, a continuous extension exists, and is unique. To show it is uniformly continuous, we consider a δ afforded to us by the uniform continuity of f on D, and another one based on the fact that $\tilde{f}$ is uniformly continuous on $D \cup \{c\}$.*

Proof of Theorem 5.56. By Lemma 5.55, $L = \lim_{x \to c} f(x)$ exists. For this to be a continuous extension, we must define $\tilde{f}(c) = \lim_{x \to c} f(x) = L$. Thus, an extension exists, and it is uniquely determined.

To prove $\tilde{f}$ is uniformly continuous on $D \cup \{c\}$, let $\epsilon > 0$ be given. Since f is uniformly continuous on D, there exists a δ_1 so that if $x, y, \in D$ and $|x - y| < \delta_1$, then $|f(x) - f(y)| < \epsilon$. Since $\tilde{f}$ is continuous at c, there exists a δ_2 so that if $x \in D \cup \{c\}$ and $|x - c| < \delta_2$, then $|\tilde{f}(x) - \tilde{f}(y)| < \epsilon$. Choose $\delta = \min\{\delta_1, \delta_2\}$, and suppose $x, y \in D \cup \{c\}$ with $|x - y| < \delta$. If neither x nor y is equal to c, then since $\delta \leq \delta_1$, we have $|f(x) - f(y)| < \epsilon$. If both are equal to c, then we clearly have the result (the difference is 0). If $y = c$ and x is not, then since $\delta \leq \delta_2$, we have $|\tilde{f}(x) - \tilde{f}(c)| < \epsilon$ by our choice of δ_2. So, $\tilde{f}$ is uniformly continuous on $D \cup \{c\}$. □

Typically, one applies the above theorem to the domain $D = (a, b)$, and continuously extends the function $f : (a, b) \to \mathbb{R}$ to $[a, b]$, as listed in the following corollary.

Corollary 5.57. *If $f : (a, b) \to \mathbb{R}$ is uniformly continuous, then there exists a unique uniformly continuous extension of f to $[a, b]$.*

This has certain appealing consequences: first, continuous functions on $[a, b]$ are the subject of most of the previous section, and a number of useful facts may then be said about either f or its continuous extension. Second, since we extend f to be continuous on $[a, b]$, our extension will be uniformly continuous on $[a, b]$ by Theorem 5.51.

We close this section with some practice problems concerning uniform continuity that utilize some of the results from this section.

Practice 5.58. Determine, with proof, which of the following functions are uniformly continuous on the given domain.

(a) $f(x) = \frac{1}{x}$ on $[1, \infty)$.

(b) $f(x) = \frac{1}{x^2}$ on $(0, 1]$.

Methodology. *(a) That this is uniformly continuous can be demonstrated by using Definition 5.49. (b) This is not uniformly continuous. One could use Theorem 5.52 and the Cauchy sequence* $x_n = \frac{1}{n}$ *and see that* $f(x_n) = n^2$ *diverges and is hence not Cauchy. Or, one could point out that* 0 *is a limit point of* $(0, 1]$ *and observe that* $\lim_{x \to 0^+} \frac{1}{x^2}$ *does not exist, as in Lemma 5.55.*

Solution. (a) We use Definition 5.49. Let $\epsilon > 0$ be given. Notice that for $x, y \in [1, \infty)$,

$$\left| \frac{1}{x} - \frac{1}{y} \right| = \frac{|x - y|}{xy} \leq |x - y|.$$

So we set $\delta = \epsilon$, and now whenever $|x - y| < \delta$, we have $|f(x) - f(y)| < \epsilon$.

(b) Let $x_n = \frac{1}{n}$. Notice that $x_n \to 0$, and so it is Cauchy (Corollary 3.101). So, by Theorem 5.52, $f(x_n) = n^2$ would have to be Cauchy as well. But it is not because it does not converge. So, $f(x) = \frac{1}{x^2}$ is not uniformly continuous on $(0, 1]$. ■

5.6.4 Review of important concepts from Section 5.6

1. Uniform continuity (Definition 5.49) is a stronger form of continuity. It is something a function satisfies on its domain (a set) instead of at some point.
2. Continuous functions defined on $[a, b]$ are uniformly continuous (Theorem 5.51).
3. Uniformly continuous functions send Cauchy sequences to Cauchy sequences. The converse is true if the domain is bounded (Theorems 5.52 and 5.53). See also Exercise 6.
4. Uniformly continuous functions can be uniquely extended to continuous functions at limit points of their domain (Theorem 5.56). This is usually used when the domain is (a, b) and the function is continuously extended to $[a, b]$, as in Corollary 5.57.

Exercises for Section 5.6

1. Determine which of the following functions are uniformly continuous on the given domain.

   (a) $f(x) = \frac{1}{x^2+1}$ on $D = [0, \infty)$.
   (b) $f(x) = \frac{\sqrt{x-1}}{x}$ on $D = [1, 5]$.
   (c) $f(x) = \frac{1}{x-7}$ on $D = (7, \infty)$.
   (d) $f(x) = \frac{x}{|x|}$ on $D = [-1, 0) \cup (0, 1]$.
   (e) $f(x) = \sqrt{x}$ on $D = [1, \infty)$.
   (f) $f(x) = \sin\left(\frac{1}{x}\right)$ on $D = (0, \infty)$.

2. Show that if D is bounded and $f : D \to \mathbb{R}$ is uniformly continuous on D, then it is bounded on D. Give an example of a function $f : (0, 1) \to \mathbb{R}$ that is continuous, but not bounded. Also give an example of a uniformly continuous function $f : \mathbb{R} \to \mathbb{R}$ that is not bounded on $\mathbb{R}$. (These examples illustrate that both hypotheses are necessary in the original problem.)

3. Suppose $f : [a, b] \cup I \to \mathbb{R}$, where $I = [b, c), [b, c]$, or $[b, \infty)$. Show that if f is uniformly continuous on $[a, b]$ and I, then it is uniformly continuous on $[a, b] \cup I$.

4. Show that $f(x) = \sqrt{x}$ is uniformly continuous on $[0, \infty)$. *[Hint: Consider* $[0, \infty) = [0, 1] \cup [1, \infty)$, *and use Theorem 5.51 and Exercises 1 and 3.]*

5. Show that the function $f(x) = \frac{1}{x}$ is uniformly continuous on $[1, \infty)$, but not uniformly continuous on $(0, 1]$.

6. Let $f(x) = x^2$ on $\mathbb{R}$. Show that f sends Cauchy sequences to Cauchy sequences, but is not uniformly continuous on $\mathbb{R}$.

7. The number $c = 1$ is a limit point of $D = [0, 1) \cup (1, 2]$. Find a function $f : D \to \mathbb{R}$ that is continuous on D but has no continuous extension to $[0, 2]$ (and is therefore not uniformly continuous by Theorem 5.56).

8. Suppose $f : (a, b) \to \mathbb{R}$ is continuous, and that $\lim_{x \to a^+} f(x)$ and $\lim_{x \to b^-} f(x)$ exist and are real. Prove that f is uniformly continuous on (a, b).

9. Suppose f is continuous on $\mathbb{R}$. Prove that f is uniformly continuous on every bounded subset of $\mathbb{R}$.

10. A function $f : D \to \mathbb{R}$ is *Lipschitz* if there exists a positive constant M so that $|f(x) - f(y)| \leq M|x - y|$. Prove that if f is Lipschitz, then it is uniformly continuous on D.

5.7 Troubleshooting Guide for Limits and Continuity

Type of difficulty:	***Suggestion for help:***
Section 5.1: Limits of Functions	
"Why can't I plug c into f to compute the limit as $\lim_{x\to c} f(x) = f(c)$? It seems as though it's always that way."	The definition of a limit *specifically forbids you* from solely evaluating limits in this manner, and generally the value $f(c)$ (if it exists) is irrelevant in computing $\lim_{x\to c} f(x)$. See Definition 5.1, and the discussion at the beginning of Section 5.1. There is one situation where $\lim_{x\to c} f(x) = f(c)$, see the discussion following Definition 5.35 in Section 5.4.
"I always thought that you can just plug c into $f(x)$ when you compute $\lim_{x\to c} f(x)$. What am I missing in all this extra stuff?"	It is definitely not the case that $\lim_{x\to c} f(x)$ can be computed in this way. Certainly there are instances where this is true, but the "extra stuff" you are missing describes this difference. See Remark 5.2 on page 255 for information about this, and Exercises 2 and 3 for examples.
"I don't understand why we would have to deal with any situation that δ is the minimum of two numbers."	Generally speaking, in establishing a limit by using Definition 5.1 only, one needs to use whatever reasoning they can, and sometimes choosing δ to be the minimum of several numbers will ensure that several properties or estimates hold simultaneously. See Practice 5.6 for a detailed discussion of this, and the proof of Theorem 5.10 for a theoretical use. (This concept is not unlike finding N to be the maximum of several numbers as in Practice 3.23.)
"How do I show that the limit of a function does not exist?"	This can be found on page 257 as the negation of Definition 5.1. See also Practice 5.7, and page 275 concerning infinite limits not existing.
Section 5.2: Properties of Limits	
"Can I prove these results without using Theorem 5.13?"	Yes, all of the results in this section (and chapter) can be proven directly and without the use of this theorem. See such a proof on page 267, and Exercises 7 and 9 as well.

Type of difficulty:	***Suggestion for help:***
"I remember from calculus that to compute limits there was always something you had to factor out before you could just plug something in. How does that relate to what's going on here?"	That is certainly one way to compute limits, see the discussion on page 268. Theorem 5.19 illustrates the theory behind this approach, although be advised that sometimes the situation will not let you compute the limit so easily: try Exercise 1 and try to identify exactly the methods that you use.
"It looks as though you're just using previous results about sequences to prove all of the theorems here. It doesn't look at all as if you're proving results about limits."	This is one of the main points of the section. Theorem 5.13 relates limits of functions to convergence of sequences. This theorem is what allows us to use this reasoning in our proofs concerning limits. See the introduction to the use of this method on page 265 (for finite limits) and page 276 (for infinite limits).
Section 5.3: Infinite Limits	
"Why isn't the definition of infinite limits stated similarly to Definition 5.1 (for finite limits)? Why state the intricate Definition 5.1 separately if we just end up generalizing (and simplifying) it anyway?"	If we listed out each definition required for the various infinite limit cases (L infinite or not, c infinite or not, etc.), we would need 6 additional definitions that essentially say the same thing with minor modifications. Combining them into one is an efficient way to remember all 6 at once. We listed the $\epsilon - \delta$ version of a limit in Definition 5.1 because of its instructional value as we formulated the definition. One can consider, however, Definition 5.27 the official definition of a limit in all of the various cases (L and/or c infinite or finite).
"It seems as if we're not really proving anything in some of these results. For instance, in Theorem 5.33 all that's done is reference some other theorem about sequences."	Almost all of these results (including Theorem 5.33) are proven by using previous results. This is one of the reasons we developed the previous results in the first place, so that we could use the results later and make certain other results easier to prove. See also the troubleshooting questions from the previous section.
Section 5.4: Continuity	

Type of difficulty:	***Suggestion for help:***
"It seems as if you don't prove hardly anything in this section, and just make passing references to previous results, as in Theorem 5.37. This doesn't seem rigorous."	It seems this way because there is no formal proof appearing directly after the statements of these theorems. However, the work that goes into proving these results has been done previously and is noted before these "proof-less" theorems are stated. Try reviewing the referred to results to see why a new proof would be redundant–the only reason a proof may be omitted is if it is fairly obvious or derivable from previous results.
"Why does it seem as if you just plug in c in the limits in this section, whereas a big point was made in Section 5.1 not to do that?"	Section 5.1 dealt with limits, whereas this section deals with the stronger concept of continuity. This stronger concept equates $\lim_{x\to c} f(x)$ with $f(c)$ in the event that c is a limit point of D. See the discussion after Definition 5.35.
Section 5.5: Consequences of Continuity	
"I'm not particularly impressed with any of the so-called important results from this section. What's so important about these facts?"	One reason why these results are so important is that they are used quite a lot in a number of subsequent topics. Another reason is that these results highlight exactly what a continuous function can do, and how a discontinuous function might not be able to do the same thing. If anything, the proofs of these results beautifully showcase a number of previous results we have developed (such as the Bolzano-Weierstrass Theorem (Theorem 3.84).
"Is there any way you could narrow down what c is in the Intermediate Value Theorem (Theorem 5.45)?"	Without any further information, no. Remember, there may be several values of c that "work." One suggestion: with the method of "successive bisection", you could consider bisecting the interval $[a, b]$ into two subintervals (similar to the second proof of the Bolzano-Weierstrass Theorem (Theorem 3.84) on Page 162) and testing the hypotheses on the Intermediate Value Theorem on each subinterval.
Section 5.6: Uniform Continuity	

Type of difficulty:	***Suggestion for help:***
"I don't really understand what's going on at the beginning of this section that introduces uniform continuity."	This discussion is meant to illustrate that the δ you find when studying (regular) continuity depends on ϵ but also the point $x = c$, while in the context of uniform continuity this δ is only dependent on ϵ. See also Practice 5.6 for a review of this process.
"I don't really see a difference between uniform continuity and regular continuity."	Comparing the definitions (Definitions 5.35 and 5.49), we see that the definition of continuity asks that you specify one point, $x = c$, and proceed from there. In the definition of Uniform Continuity, you consider the $\epsilon - \delta$ relationship without regard to any one point. Rather, you consider what happens whenever *any* two points are close to each other, not just when some point is close to c.
"How can I show a function is uniformly continuous? How would I show it is not?"	There are several ways. The first, of course, is by Definition 5.49. Otherwise, you could use Theorems 5.51 or 5.53 if appropriate. You could show that something is *not* uniformly continuous by using Theorem 5.52 (find a Cauchy sequence that is not sent to another Cauchy sequence) or Lemma 5.55 (show that a certain limit does not exist).
"Why, all of a sudden, are we back to talking about Cauchy sequences?"	Cauchy sequences appear in the proofs of Theorems 5.52 and 5.53 because we specifically allow for the domain D to not contain a limit point–this is important later in the section when discussing continuous extensions. If such a sequence converged to a point not in D, our arguments might not work if we had to discuss this limit point as member of D. Cauchy sequences are the natural next best thing, since they are convergent (Corollary 3.101), but we do not have to speak of their limit.

Type of difficulty:	***Suggestion for help:***
"Why in the proof of Theorem 5.56 must we define this continuous extension at c by $\tilde{f}(c) = \lim_{x\to c} f(x)$?"	Since c is a limit point of D, continuity is characterized by this property. See Definition 5.35 and the remarks afterward.
"I don't understand why we can't just consider the sequence $x_n \to c$ in the proof of Theorem 5.56. It's an arbitrary sequence, why do we have to consider another one?"	We used this sequence to uncover what L is as $\lim_{n\to\infty} f(x_n)$. How do we know that if we choose another sequence $y_n \to c$ we would not get a different answer for what $(f(y_n))$ converges to?

6

Differentiation

Chapter Summary:

The derivative of a function can reveal much about its behavior. This chapter aims to introduce the derivative and to present several very important results, including the Mean Value Theorem, Taylor's Theorem, and L'Hôpital's Rule.

Section 6.1: The Derivative. The derivative of a function is defined, and basic results concerning differentiation are established, such as the Product, Quotient, and Chain Rules. Afterward, extrema, Rolle's Theorem, and an intermediate value property are discussed.

Section 6.2: The Mean Value Theorem. The Mean Value Theorem is proven here, and several applications are presented.

Section 6.3: Taylor Polynomials. Taylor Polynomials are introduced and motivated. We prove Taylor's Theorem (Theorem 6.25), and illustrate how one uses it to obtain useful approximations for functions and their values.

Section 6.4: L'Hôpital's Rule. L'Hôpital's Rule (Theorem 6.29) is stated and proved as it applies to the $\frac{0}{0}$ and $\frac{\infty}{\infty}$ indeterminate forms. A description of applications of this to other indeterminate forms is given.

Section 6.5: Troubleshooting Guide for Differentiation. A list of frequently asked questions and their answers, arranged by section. Come here if you are struggling with any material from this chapter.

6.1 The Derivative

The derivative of a function is intended to be a measurement of the rate at which a function's outputs change. Beyond this usefulness, however, the entire theory of differentiation is of interest because of its vast reach in many other fields of mathematics.

We begin this subject with a definition (Definition 6.1), and go on to establish some elementary facts about the derivative which are very likely familiar from basic calculus. The discussion then turns to finding extrema of a function (Theorem 6.9), the intuitive and fundamental Rolle's Theorem (Theorem 6.11), and an intermediate value property.

6.1.1 Derivatives

The derivative is a fundamental concept in the subject of analysis. The motivation for this object is that it is to measure how a function changes at a given point. Graphically, this is seen as the slope of a secant line approaching the slope of the tangent line at a given point as shown in Figure 6.1.1.

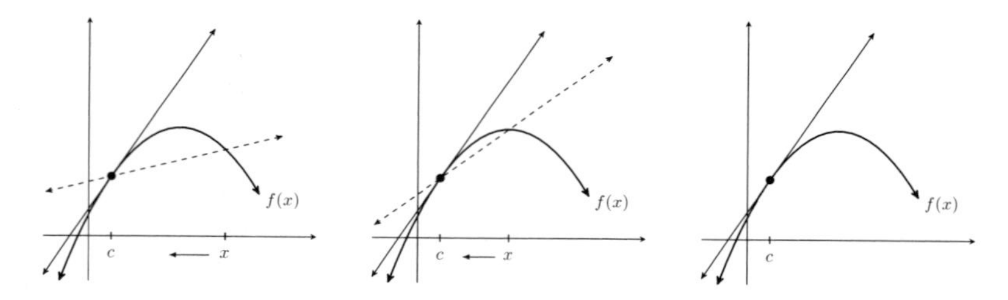

FIGURE 6.1.1
The slope of the secant line through $(c, f(c))$ and $(x, f(x))$ approaches the slope of the line tangent to $(c, f(c))$.

Motivated by this idea, this derivative is defined as follows.

Definition 6.1. Let $f : D \to \mathbb{R}$, and let c be a limit point of D. The *derivative of f at c*, written $f'(c)$, is

$$\lim_{x \to c} \frac{f(x) - f(c)}{x - c}.$$

If $f'(c)$ exists, then we say f is *differentiable at c*, and if f is differentiable at every point in D, then we say f is *differentiable on D*. The quotient in the limit above is referred to as the *difference quotient*.

We recall from basic calculus that as c varies, the function $c \mapsto f'(c)$ is a function we denote $f'(x)$. Thus, we may treat the derivative of a function as another function. One could then ask if f' is differentiable, and if it is, its derivative is $f''(x)$ and is said to be the *second derivative of* f. Provided these derivatives exist, one can iterate this differentiation process to create any number of derivatives; the nth derivative of f is denoted $f^{(n)}(x)$, with the parentheses intended to remove any confusion with the nth power of f. By convention, we define $f^{(0)}(x) = f(x)$. These are the *higher order derivatives* of f.

The notation $f'(c)$ (Lagrange's notation) is not the only notation one uses to express the derivative of f at c. The Leibniz notation for the derivative of f is $\frac{df}{dx}$, or $\frac{d}{dx}(f)$, $\frac{d}{dx}[f(x)]$, or $\frac{dy}{dx}$ when expressing $y = f(x)$, with $|_c$ or (c) appended in the event that we wish to evaluate the derivative at $x = c$. This notation correctly suggests that the derivative of a function is an operation of sorts, and that repeated operation yielding the second derivative (and higher order derivatives) should be expressed as

$$\frac{d}{dx}\left(\frac{d}{dx}(f)\right) = \left(\frac{d}{dx}\right)^2 (f) = \frac{d^2}{dx^2}(f) = \frac{d^2 f}{dx^2}.$$

Euler's notation for the derivative of f is Df, or $D_x f$ when it is important to emphasize the input variable, and $D^2 f, D^3 f, \ldots$ for the higher order derivatives of f. Newton's notation of $\dot{y}$ for y' is also used (similarly for higher order derivatives).

Before moving on to discuss the basic properties of derivatives, we point out that by replacing x with $x+h$ in Definition 6.1 (see Exercise 3 in Section 5.2), we obtain an equivalent definition for $f'(c)$ as

$$f'(c) = \lim_{h\to 0} \frac{f(c+h) - f(c)}{h}.$$

6.1.2 Computing derivatives

In what one usually encounters, there is a theoretical use for derivatives where we tend to use Definition 6.1 more often, and a practical use for derivatives requiring us to actually differentiate functions. To assist us in the latter, we use the theorems below that should be familiar from basic calculus.

For the remainder of the chapter, we will assume that the domain of our function is an interval I of the form $(a, b), [a, b), (a, b]$, or $[a, b]$, with $a, b \in \mathbb{R}$ or equal to $\pm\infty$ where appropriate (so the interval could be infinite in width, but of course not actually include $\pm\infty$). We begin with a result that will be helpful, but also is of interest in its own right.

Theorem 6.2. *If $f : I \to \mathbb{R}$ is differentiable at c, then it is continuous at c.*

Idea of proof. *We prove that $f(x) - f(c) \to 0$ instead of $f(x) \to f(c)$, which will suffice since every point of an interval is a limit point. This limit will follow from a straightforward manipulation of $f(x) - f(c)$ and from the* *assumption that $f'(c)$ exists, in particular, that the limit in Definition 6.1 exists.*

Proof of Theorem 6.2. We prove that $\lim_{x \to c}[f(x) - f(c)] = 0$ at the limit point c, since this will imply $\lim_{x \to c} f(x) = f(c)$. Since $f'(c)$ is assumed to exist, by Assertion (2) of Theorem 5.16 we have

$$\begin{aligned} \lim_{x \to c}[f(x) - f(c)] &= \lim_{x \to c}\left[\left(\tfrac{f(x)-f(c)}{x-c}\right)(x-c)\right] \\ &= \lim_{x \to c}\left[\tfrac{f(x)-f(c)}{x-c}\right] \cdot \lim_{x \to c}[x-c] \\ &= f'(c) \cdot 0 \\ &= 0. \end{aligned}$$

□

It is easy to construct a function that illustrates that the converse of the above theorem is false.

Example 6.3. This example illustrates that the function $f(x) = |x|$ is not differentiable at $x = 0$, even though it is continuous there, see Figure 6.1.2. Indeed, the difference quotient for this function is

$$\frac{f(x) - f(0)}{x - 0} = \frac{|x|}{x},$$

and we claim that $\lim_{x \to 0} \frac{|x|}{x}$ does not exist. If it did and equaled L, then by Theorem 5.13, every sequence $x_n \to 0$ must have $f(x_n) \to L$. Choosing $x_n = \frac{1}{n}$ has $\frac{|x|}{x} = \frac{1/n}{1/n} = 1 \to 1$. However, choosing $x_n = \frac{-1}{n}$ has $f(x_n) = \frac{-1/n}{1/n} = -1 \to -1$.

Generally speaking, it is relatively straightforward to use various combinations of the absolute value function to create a function that is continuous but not differentiable at a point. In 1872, Karl Weierstrass gave an example of a function that is continuous everywhere but differentiable nowhere. An account of the history and relevance of this discovery can be found in [17]. ■

We now introduce some of the basic facts of derivatives that should be familiar.

Theorem 6.4. *Let $f, g : I \to \mathbb{R}$, and let $k \in \mathbb{R}$. If f and g are differentiable at c, then*

1. $\frac{d}{dx}[f(x) \pm g(x)]|_c = f'(c) \pm g'(c)$, *and*
2. $\frac{d}{dx}[kf(x)]|_c = kf'(c)$.

Idea of proof. *Both of these facts follow from a manipulation of the difference quotient, and by basic limit facts.*

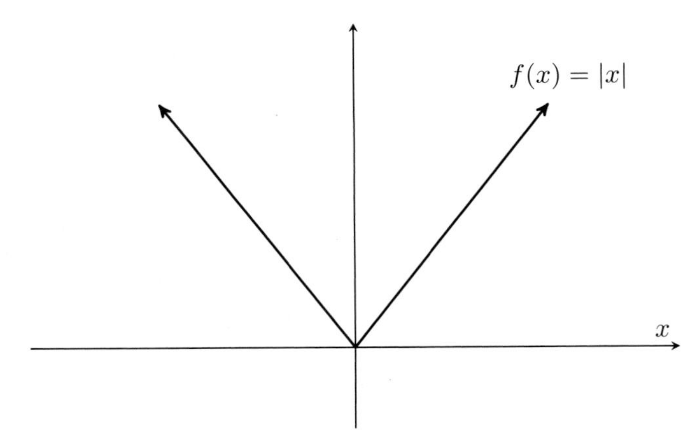

FIGURE 6.1.2
A graph of $f(x) = |x|$. It is not differentiable at $x = 0$.

Proof of Theorem 6.4. To prove the first assertion, rewrite

$$\begin{aligned}\tfrac{(f(x)\pm g(x))-(f(c)\pm g(c))}{x-c} &= \tfrac{f(x)-f(c)}{x-c} \pm \tfrac{g(x)-g(c)}{x-c} \\ &\to f'(c) \pm g'(c).\end{aligned}$$

For the second assertion, factor the difference quotient as

$$\begin{aligned}\tfrac{kf(x)-kf(c)}{x-c} &= k\tfrac{f(x)-f(c)}{x-c} \\ &\to kf'(c).\end{aligned}$$

□

The following are better known as the Product Rule and the Quotient Rule, whose proofs also require a manipulation of the relevant difference quotient, but will also require Theorem 6.2.

Theorem 6.5. *Let $f, g : I \to \mathbb{R}$. If f and g are differentiable at c, then*

1. *(Product Rule)* $\frac{d}{dx}[f(x)g(x)]|_c = f(c)g'(c) + f'(c)g(c)$, *and*
2. *(Quotient Rule)* $\frac{d}{dx}\left[\frac{f(x)}{g(x)}\right]|_c = \frac{g(c)f'(c)-f(c)g'(c)}{(g(c))^2}$ *if* $g(c) \neq 0$.

Idea of proof. *Like the last theorem, these assertions can be proven by manipulating the difference quotient to exploit the hypothesis of differentiability, although we shall also need to use Theorem 6.2. In the proof of the Product Rule, we subtract and add the quantity $f(x)g(c)$, and in the proof of the Quotient Rule we add and subtract $f(c)g(c)$. In taking the subsequent limit, we will need to use Theorem 6.2.*

Proof of Theorem 6.5. To prove Assertion (1), we manipulate the following difference quotient by adding and subtracting $f(x)g(c)$:

$$\begin{aligned}\frac{f(x)g(x)-f(c)g(c)}{x-c} &= \frac{f(x)g(x)-f(x)g(c)+f(x)g(c)-f(c)g(c)}{x-c} \\ &= f(x)\frac{g(x)-g(c)}{x-c} + \frac{f(x)-f(c)}{x-c}g(c).\end{aligned}$$

In taking the limit as $x \to c$, the second of the two summands above tends to $f'(c)g(c)$ by Assertion (3) of Theorem 5.16 and since f is differentiable at c. The first of the two summands above tends to $f(c)g'(c)$ by the following reasoning: $f(x) \to f(c)$ since f being differentiable at c implies it is continuous at c (Theorem 6.2), and so

$$f(x)\frac{g(x)-g(c)}{x-c} \to f(c)g'(c)$$

by Assertion (2) of Theorem 5.16 and since g is differentiable at c. So, by Assertion (1) of Theorem 5.16,

$$f(x)\frac{g(x)-g(c)}{x-c} + \frac{f(x)-f(c)}{x-c}g(c) \to f(c)g'(c) + f'(c)g(c).$$

We proceed similarly to prove Assertion (2). By subtracting and adding $f(c)g(c)$ to the difference quotient, we have

$$\begin{aligned}\frac{\frac{f(x)}{g(x)}-\frac{f(c)}{g(c)}}{x-c} &= \frac{f(x)g(c)-f(c)g(x)}{g(x)g(c)(x-c)} \\ &= \frac{f(x)g(c)-f(c)g(c)+f(c)g(c)-f(c)g(x)}{g(x)g(c)(x-c)} \\ &= \frac{f(x)-f(c)}{x-c}\cdot g(c)\cdot\frac{1}{g(x)g(c)} - f(c)\cdot\frac{g(x)-g(c)}{x-c}\cdot\frac{1}{g(x)g(c)}\end{aligned} \tag{6.1}$$

We separately compute the limits of each of the above terms in the difference above. The first term has

$$\lim_{x\to c}\left[\frac{f(x)-f(c)}{x-c}\cdot g(c)\cdot\frac{1}{g(x)g(c)}\right] = f'(c)\cdot g(c)\cdot\frac{1}{(g(c))^2},$$

since f is differentiable at c, $g(c)$ is a constant, and g is continuous at c (since it is differentiable, by Theorem 6.2) and $g(c) \neq 0$ by assumption. Using Theorem 5.16 completes our reasoning.

The second term in the difference above is

$$\lim_{x\to c}\left[f(c)\cdot\frac{g(x)-g(c)}{x-c}\cdot\frac{1}{g(x)g(c)}\right] = f(c)\cdot g'(c)\cdot\frac{1}{(g(c))^2}$$

by the same reasoning, except here we use the assumption that g is differentiable at c. Thus our difference quotient from Equation (6.1) has, as $x \to c$,

$$\begin{aligned}\frac{\frac{f(x)}{g(x)}-\frac{f(c)}{g(c)}}{x-c} &= \frac{f(x)-f(c)}{x-c}\cdot g(c)\cdot\frac{1}{g(x)g(c)} - f(c)\cdot\frac{g(x)-g(c)}{x-c}\cdot\frac{1}{g(x)g(c)} \\ &\to f'(c)\cdot g(c)\cdot\frac{1}{(g(c))^2} - f(c)\cdot g'(c)\cdot\frac{1}{(g(c))^2} \\ &= \frac{g(c)f'(c)-f(c)g'(c)}{(g(c))^2}.\end{aligned}$$

□

Our final result that assists us in actually computing derivatives in practice is the Chain Rule. An intuitive approach to the proof would be to manipulate the difference quotient in the following way:

$$\frac{f(g(x)) - f(g(c))}{x - c} = \frac{f(g(x)) - f(g(c))}{g(x) - g(c)} \cdot \frac{g(x) - g(c)}{x - c}, \tag{6.2}$$

and to take the limit as $x \to c$. The problem here is that there may be quite a few points[1] at which $g(x) = g(c)$ so that the expression on the right is not defined at those values of x. Thus, the proof of the Chain Rule is more delicate. Although it appears to be a common approach, we use the proof in [19].

Theorem 6.6 (The Chain Rule). *Suppose $g : \tilde{I} \to I$, and $f : I \to \mathbb{R}$. If g is differentiable at $\tilde{c} \in \tilde{I}$ and f is differentiable at $g(\tilde{c})$, then*

$$\frac{d}{dx}[f(g(x))]|_{\tilde{c}} = f'(g(\tilde{c}))g'(\tilde{c}).$$

Idea of proof. *To get around the difficulty described above, we define a function h that is similar to the difference quotient for f except at those points that would make the denominator equal to 0. From there, it is a simple although somewhat uncomfortable manipulation to arrive at the result. To understand this proof, pay special attention to quantifying statements, such as those related to continuity or for what values of x or y certain equations hold.*

Proof of Theorem 6.6. For simplicity, let $c = g(\tilde{c})$. Define the function $h : I \to \mathbb{R}$ as

$$h(y) = \begin{cases} \frac{f(y)-f(c)}{y-c} & \text{for } y \neq c \\ f'(c) & \text{for } y = c \end{cases}.$$

Since f is differentiable at c, remembering that in the limit we have $y \neq c$, we have

$$\lim_{y \to c} h(y) = \lim_{y \to c} \frac{f(y) - f(c)}{y - c} = f'(c) = h(c).$$

Thus h is continuous at c. Since g is differentiable at $\tilde{c}$, it is continuous at $\tilde{c}$ (Theorem 6.2), and so $h \circ g$ is continuous at $\tilde{c}$ by Theorem 5.38. As a result,

$$\lim_{x \to \tilde{c}} h(g(x)) = h(g(\tilde{c})) = h(c) = f'(c) = f'(g(\tilde{c})).$$

By manipulating our formula for h and remembering that $c = g(\tilde{c})$, we note that

$$f(y) - f(g(\tilde{c})) = h(y)(y - g(\tilde{c}))$$

for *every* $y \in I$, *even when* $y = g(\tilde{c})$. Now any $x \in \tilde{I}$ will give us a $g(x) = y \in I$, and so

$$f(g(x)) - f(g(\tilde{c})) = h(g(x))(g(x) - g(\tilde{c})).$$

[1]Indeed, consider the possibility that $g(x) = g(c)$ for all x!

For any $x \neq \tilde{c}$, we then have

$$\frac{f(g(x)) - f(g(\tilde{c}))}{x - \tilde{c}} = (h \circ g)(x) \frac{g(x) - g(\tilde{c})}{x - \tilde{c}}.$$

The left side of this equation is the difference quotient for $f \circ g$. We evaluate the limit as $x \to \tilde{c}$ by inspecting each factor on the right side. Since $h \circ g$ is continuous at $\tilde{c}$,

$$\lim_{x \to \tilde{c}} (h \circ g)(x) = (h \circ g)(\tilde{c}) = h(g(\tilde{c})) = f'(g(\tilde{c})).$$

Since g is differentiable at $\tilde{c}$,

$$\lim_{x \to \tilde{c}} \frac{g(x) - g(\tilde{c})}{x - \tilde{c}} = g'(\tilde{c}).$$

Thus by Assertion (2) of Theorem 5.16, we have

$$\begin{aligned} \lim_{x \to \tilde{c}} \left[\tfrac{f(g(x)) - f(g(\tilde{c}))}{x - \tilde{c}} \right] &= \lim_{x \to \tilde{c}} \left[(h \circ g)(x) \tfrac{g(x) - g(\tilde{c})}{x - \tilde{c}} \right] \\ &= \lim_{x \to \tilde{c}} \left[(h \circ g)(x) \right] \lim_{x \to \tilde{c}} \left[\tfrac{g(x) - g(\tilde{c})}{x - \tilde{c}} \right] \\ &= f'(g(\tilde{c}))g'(c). \end{aligned}$$

□

With these results and a couple of known derivative formulas, we can differentiate a number of functions we encounter in practice. The following practice problem is one such example.

Practice 6.7. Compute the derivative of the constant function $f(x) = k$, and $g(x) = x^n$ for $n \in \mathbb{Z}$.

Methodology. *The derivative of the constant function is easily seen to be 0 by Definition 6.1. We show that $g'(x) = nx^{n-1}$ by Mathematical Induction for $n \in \mathbb{N}$. The base case of $\frac{d}{dx}[x] = 1$ is easy to establish using Definition 6.1, and to prove the inductive step we express $x^n = x \cdot x^{n-1}$ and use the Product Rule (Theorem 6.5). Finally we consider the case where $-n$ is an integer less than zero by noting that at any $x \neq 0$, $x^{-n} = \frac{1}{x^n}$, $n > 0$ and an integer, so the quotient rule could be used along with what we have already discovered.*

Solution. Let $c \in \mathbb{R}$ be arbitrary. We first compute the derivative of $f(x) = k$ by noting that

$$\lim_{x \to c} \frac{f(x) - f(c)}{x - c} = \lim_{x \to c} \frac{k - k}{x - c} = \lim_{x \to c} 0 = 0.$$

Now let n be a positive integer. We prove that if $g(x) = x^n$ then $g'(x) = nx^{n-1}$ by Mathematical Induction. When $n = 1$ we have $g(x) = x$, and we establish the base case by the following:

$$\lim_{x \to c} \frac{g(x) - g(c)}{x - c} = \lim_{x \to c} \frac{x - c}{x - c} = \lim_{x \to c} 1 = 1.$$

Now let $n \geq 2$. The induction hypothesis tells us that $\frac{d}{dx}[x^{n-1}] = (n-1)x^{n-2}$. By the Product Rule (Theorem 6.5) and the fact that $\frac{d}{dx}[x] = 1$,

$$\begin{aligned} \tfrac{d}{dx}[x^n] &= \tfrac{d}{dx}[x \cdot x^{n-1}] \\ &= x \cdot (n-1)x^{n-2} + (1)x^{n-1} \\ &= nx^{n-1}. \end{aligned}$$

All that remains is to consider the negative integers. Let $n \in \mathbb{N}$ so that $-n \in \mathbb{Z}$ and $-n < 0$. By using the Quotient Rule (Theorem 6.5), and the facts we have already proven here,

$$\begin{aligned} \tfrac{d}{dx}[x^{-n}] &= \tfrac{d}{dx}\left[\tfrac{1}{x^n}\right] \\ &= \tfrac{x^n \cdot \frac{d}{dx}[1] - 1 \cdot \frac{d}{dx}[x^n]}{(x^n)^2} \\ &= \tfrac{-nx^{n-1}}{x^{2n}} \\ &= -nx^{-n-1}. \end{aligned}$$

■

Unless specifically noted, you may use all of the familiar differentiation formulas from basic calculus. For example, $\frac{d}{dx}[\sin(x)] = \cos(x)$, $\frac{d}{dx}[e^x] = e^x$, $\frac{d}{dx}[\ln(x)] = \frac{1}{x}$ and so on.

6.1.3 Basic facts involving the derivative

There are a number of immediate theoretical uses for the derivative. These include the ability to locate possible extrema[2] of a function, Rollie's Theorem, and an intermediate value property. We begin with a discussion of extrema.

In Definition 5.43, we defined what a global maximum and minimum of a function is. A *local maximum* or *minimum* is similar in concept, and should be familiar from basic calculus.

Definition 6.8. Suppose $f : D \to \mathbb{R}$. If there exists a $c \in D$ and a δ−neighborhood U of c so that $f(x) \leq f(c)$ for all $x \in D \cap U$, then $f(c)$ is a *local maximum* of f. If there exists a $c \in D$ and a δ−neighborhood U of c so that $f(x) \geq f(c)$ for all $x \in D \cap U$, then $f(c)$ is a *local minimum* of f. A local or global maximum or minimum of f is referred to as an *extremum* of f.

Recall from page 71 that a point $x \in D$ is an *interior point* if there is a neighborhood about x that is a subset of D. If $D = I$ is an interval, the only points that are not interior to I are its endpoints (and only if those endpoints are in I). The following theorem locates the possible extrema of a differentiable function provided they are interior points of the interval. There are at most two other endpoints one could therefore find extrema.

Theorem 6.9. *Let $f : I \to \mathbb{R}$. If $f(c)$ is a local extremum and f is differentiable at the interior point c, then $f'(c) = 0$.*

[2] "Extrema" is the plural of the singular "extremum."

Idea of proof. *We consider what $f'(c)$ might actually have to be by considering the difference quotient. In the event that $f(c)$ is a local (or global) maximum, then $f(x) \leq f(c)$ for all x nearby c. Since c is an interior point, we may consider each one-sided limit of the difference quotient. This limit as $x \to c^-$ must be positive, since both the numerator and the denominator are negative, hence the quotient is positive. On the other hand, this same difference quotient as $x \to c^+$ must be negative, since now the denominator is positive while the numerator is still negative. Since the limit as $x \to c$ exists (we assumed f is differentiable at c) and must equal each of the one-sided limits (Theorem 5.12), the only possibility for $f'(c)$ is 0. The situation where $f(c)$ is a local (or global) minimum is Exercise 3.*

Proof of Theorem 6.9. Since f is differentiable at c and c is an interior point (so that each one-sided limit can be considered), we consider what $f'(c)$ could be by computing the one-sided limits of the difference quotient, which are equal by Theorem 5.12:

$$f'(c) = \lim_{x \to c} \frac{f(x) - f(c)}{x - c} = \lim_{x \to c^-} \frac{f(x) - f(c)}{x - c} = \lim_{x \to c^+} \frac{f(x) - f(c)}{x - c}.$$

Suppose $f(c)$ is a local maximum, so that for any x in a neighborhood of c, $f(x) \leq f(c)$. Written differently, for these x we have $f(x) - f(c) \leq 0$. If $x < c$, then $x - c < 0$ and so the difference quotient in the one-sided limit as $x \to c^-$ is the quotient of two nonpositive numbers and hence nonnegative (the numerator could be 0). Thus by Theorem 5.22,

$$f'(c) = \lim_{x \to c} \frac{f(x) - f(c)}{x - c} = \lim_{x \to c^-} \frac{f(x) - f(c)}{x - c} \geq 0.$$

If $x > c$, then $x - c > 0$ and so the difference quotient in the one-sided limit as $x \to c^+$ is the quotient of a nonpositive and positive number (the numerator could again be 0). Thus by Theorem 5.22,

$$f'(c) = \lim_{x \to c} \frac{f(x) - f(c)}{x - c} = \lim_{x \to c^+} \frac{f(x) - f(c)}{x - c} \geq 0.$$

We have shown that $f'(c) \geq 0$ and $f'(c) \leq 0$, so $f'(c) = 0$. □

Theorem 6.9 is a fundamental and useful result, as it reduces the task of locating possible extrema of a differentiable function on an interval to only those interior points where the derivative vanishes or to the endpoints. If there are any places where f fails to be differentiable, then Theorem 6.9 does not apply at those points, and so it would be possible to find extrema there as well. With this in mind, if $f : I \to \mathbb{R}$, the set of points c where $f'(c) = 0$ or $f'(c)$ does not exist are called *critical points*. The following excellent practice problem is found in [18] and illustrates the importance of considering the endpoints of the interval and the points where the derivative does not exist when computing extrema.

Practice 6.10. Find the critical points and global extrema of the function $f(x) = 2x - 3x^{2/3}$ on the interval $[-1, 3]$. This function is graphed in Figure 6.1.3.

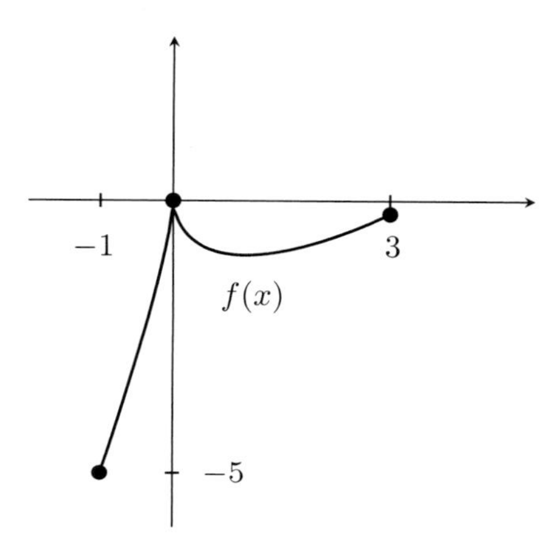

FIGURE 6.1.3
$f(x) = 2x - 3x^{2/3}$ on the interval $[-1, 3]$.

Methodology. *To find the critical points, we look for those values which make the derivative vanish or not exist. Differentiating the function and setting it equal to 0 will find the places where the derivative vanishes, although we must also notice that the derivative does not exist at $x = 0$. Thus, the global extrema can be found by comparing the outputs of the function at each critical point and at the endpoints, $x = -1$ and $x = 3$.*

Solution. Computing the derivative, we have

$$f'(x) = 2 - 2x^{-1/3} = 2\left(\frac{x^{1/3} - 1}{x^{1/3}}\right).$$

It is apparent that $f'(0)$ does not exist, and otherwise, $f'(x) = 0$ only when $x = 1$. So the critical points of this function on the given interval are $x = 0, 1$.

The global extrema may therefore only be the outputs of the critical points, $f(0) = 0$ and $f(1) = -1$, or the outputs of the endpoints, $f(-1) = -5$ or $f(3) = 6 - 3\sqrt[3]{9} \approx -.240$. Thus the global minimum is $f(-1) = -5$ and the global maximum is $f(0) = 0$. ■

Using Theorem 6.9, we may establish another fundamental result, named after Michel Rolle.

Theorem 6.11 (Rolle's Theorem). *Let $f : [a, b] \to \mathbb{R}$ be continuous on $[a, b]$ and differentiable on (a, b). If $f(a) = f(b)$, then there exists a point $c \in (a, b)$ with $f'(c) = 0$.*

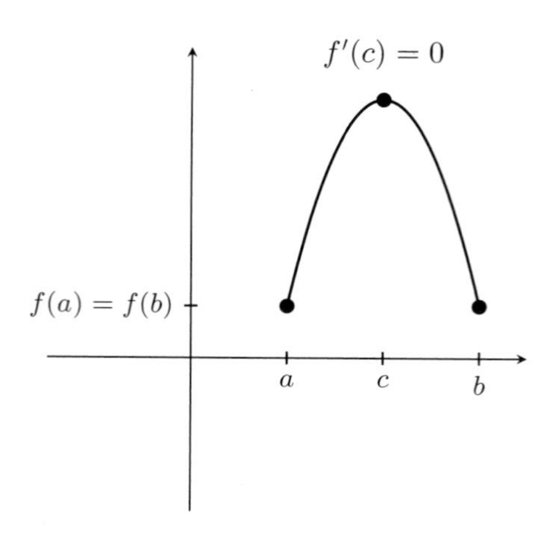

FIGURE 6.1.4
Rolle's Theorem guarantees the derivative vanishes at some interior point.

Idea of proof. *Since f is continuous on $[a, b]$, the Extreme Value Theorem (Theorem 5.44) asserts the existence of both a global maximum and a global minimum. Either of these could only be at an endpoint (at a or b), or at an interior point. If the function is not constant (which we consider first), then this maximum and minimum cannot both be at the endpoints, since $f(a) = f(b)$, and so there must be an extrema at an interior point. We may then use Theorem 6.9 to find the derivative at that point to vanish. See Figure 6.1.4 for a possible graph of this situation.*

Proof of Theorem 6.11. If f is the constant function, then $f'(c) = 0$ for *every* point $c \in (a, b)$ (see Practice 6.7). So, assume f is nonconstant. Since f is continuous on $[a, b]$, there exists a global maximum $f(M)$ and a global minimum $f(m)$ by the Extreme Value Theorem (Theorem 5.44). Since f is now assumed to be nonconstant, $f(m) \neq f(M)$ and so $m \neq M$. The points $m, M \in [a, b]$, and so either both m and M are the endpoints a and b, or at least one of m or M is not an endpoint and is therefore an interior point. If m and M are both endpoints (that is, $m = a$ and $M = b$, or $M = a$ and $m = b$), then

$$\begin{array}{ll} f(m) = f(a) = f(b) = f(M), & \text{if } m = a \text{ and } M = b, \text{ or} \\ f(M) = f(a) = f(b) = f(m), & \text{if } M = a \text{ and } m = b. \end{array}$$

In either case, we would have $f(m) = f(M)$ which is not the case since f is assumed to be nonconstant. Therefore, one of m or M is an interior point c. Since $f(c)$ is an extremum and f is differentiable at every point $c \in (a, b)$, we have $f'(c) = 0$ by Theorem 6.9. □

We close this section by presenting an interesting property that derivatives must satisfy: although derivatives of differentiable functions need not be continuous (see Exercise 8), they must satisfy what is called the *intermediate value property*: every number between two outputs is another output. This is similar to the conclusion of the Intermediate Value Theorem (Theorem 5.45), but without the hypothesis of continuity of the function involved. Our proof is essentially reproduced from [8].

Theorem 6.12. *If $f : I \to \mathbb{R}$ is differentiable, then f' has the intermediate value property. That is, for every $a, b \in I$ and number k between $f'(a)$ and $f'(b)$, there exists a $c \in I$ between a and b with $f'(c) = k$.*

Idea of proof. *We define the differentiable function $F(x) = f(x) - kx$, whose derivative is $F'(x) = f'(x) - k$. We prove that F has an extrema at an interior point c, so that $F'(c) = 0$, or rather, $f'(c) = k$. An analysis similar to the proof of Rolle's Theorem (Theorem 6.11, F is continuous on $[a, b]$ (if $a < b$), and hence there exist extrema of F by the Extreme Value Theorem (Theorem 5.44). We simply eliminate the possibility that both extrema occur at the endpoints of $[a, b]$, so that one extremum c occurs at an interior point. Theorem 6.9 now finds $F'(c) = 0$ as desired.*

Proof of Theorem 6.12. For simplicity, assume $a, b \in I$ with $a < b$ and $f'(a) \leq f'(b)$ (the other cases are proven similarly). If $f'(a) = f'(b)$, then there is nothing to prove, since $c = a = b$ can be chosen to meet the intermediate value property in this case. So suppose $f'(a) < f'(b)$, and we choose some k between $f'(a)$ and $f'(b)$.

Define $F : I \to \mathbb{R}$ as $F(x) = f(x) - kx$. F is differentiable on I, and $F'(x) = f'(x) - k$. So we search for a $c \in (a, b)$ with $F'(c) = 0$ as that would imply $f'(c) = k$. Since F is differentiable on I, it is continuous on $[a, b] \subseteq I$ (Theorem 6.2). By the Extreme Value Theorem (Theorem 5.44), there is a global maximum $F(M)$ and minimum $F(m)$, with $m, M \in [a, b]$. If either m or M is an interior point of $[a, b]$, then we are done by Theorem 6.9. We shall eliminate the possibility that both M and m are endpoints in what follows, completing the proof.

Suppose first that $m = a$ and $M = b$ so that $F(a)$ is a global minimum. We compute $F'(a)$ as $x \to a^+$ (which is equal to the full limit as $x \to a$ by Theorem 5.12):

$$\begin{aligned} f'(a) - k \quad = \quad F'(a) \quad &= \quad \lim_{x \to a} \tfrac{F(x) - F(a)}{x - a} \\ &= \quad \lim_{x \to a^+} \tfrac{F(x) - F(a)}{x - a}. \end{aligned}$$

Therefore if $x > a$, then $x - a > 0$, while $F(x) - F(a) \geq 0$ because $F(a)$ is assumed to be a global minimum. So in taking the limit as $x \to a^+$ we have $f'(a) - k = F'(a) \geq 0$ by Theorem 5.22. But by assumption $f'(a) < k$ so $F'(a) = f'(a) - k < 0$, a contradiction.

On the other hand, if we have $m = b$ and $M = a$ so that $F(b)$ is now a global minimum. We establish a contradiction in a similar manner by computing

$F'(b)$ as $x \to b^-$:

$$\begin{aligned} f'(b) - k \;=\; F'(b) &= \lim_{x \to b} \tfrac{F(x)-F(b)}{x-b} \\ &= \lim_{x \to b^-} \tfrac{F(x)-F(b)}{x-b}. \end{aligned}$$

Therefore, if $x < b$ then $x - b < 0$, while $F(x) - F(b) \geq 0$ because $F(b)$ is now assumed to be a global minimum. So in taking the limit as $x \to b^-$ we have $f'(b) - k = F'(b) \leq 0$. But by assumption $k < f'(b)$ so $f'(b) - k > 0$, a contradiction. □

6.1.4 Review of important concepts from Section 6.1

1. The derivative of a function is defined in Definition 6.1.
2. If f is differentiable at c, then it is continuous at c (Theorem 6.2).
3. The basic properties of computing derivatives from basic calculus are presented, including the Product, Quotient, and Chain Rules (see Theorems 6.4, 6.5, and 6.6).
4. Extrema of functions can be found by studying critical points and endpoints of any interval on which they are defined (see Theorem 6.9).
5. Rolle's Theorem (Theorem 6.11) is a fundamental and intuitive result in the theory of differentiation, and will be useful later.
6. For any differentiable function f, its derivative f' has the intermediate value property (Theorem 6.12).

Exercises for Section 6.1

1. Prove the Quotient Rule (Assertion (2) of Theorem 6.5) by using only the Product Rule (Assertion (1) of Theorem 6.5) and the Chain Rule (Theorem 6.6).
2. Suppose $f : I \to \mathbb{R}$, and $f''(x)$ exists at some interior point $x \in I$. Prove that

$$f''(x) = \lim_{h \to 0} \frac{f(x+h) - f(x-h) - 2f(x)}{h^2}.$$

3. Let $f : I \to \mathbb{R}$. If $f(c)$ is a local minimum and f is differentiable at the interior point c, then $f'(c) = 0$. (This completes the proof of Theorem 6.9.)
4. A function f is *odd* if $f(-x) = -f(x)$, and *even* if $f(-x) = f(x)$. Prove that the derivative of an odd function is an even function, and prove that the derivative of an even function is an odd function.

5. Let $y = x^{1/n}$. By solving for x and using the Chain Rule (Theorem 6.6), compute y'. (This method is also sometimes called *implicit differentiation.*)

6. (This exercise appears in [8].) Suppose $f, g : I \to \mathbb{R}$, both f and g are differentiable, and $f(x)g(x) = 1$ for all $x \in I$. Prove that

$$\frac{f'(x)}{f(x)} + \frac{g'(x)}{g(x)} = 0.$$

7. Suppose $f : [a, b] \to \mathbb{R}$ meets the hypotheses of Rolle's Theorem. What is different about simply assuming that f is differentiable on $[a, b]$? Is this a stronger or weaker assumption on f?

8. The purpose of this question is to illustrate the existence of a function that can be differentiated once, but not twice. Let

$$f(x) = \begin{cases} x^2 \sin\left(\frac{1}{x}\right) & \text{for } x \neq 0 \\ 0 & \text{for } x = 0 \end{cases}.$$

   (a) Compute $f'(x)$ for $x \neq 0$.

   (b) Compute $f'(0)$ by using Definition 6.1. Why is it necessary to use the definition here?

   (c) Show that $f'(x)$ is not continuous at 0, and is therefore not differentiable by Theorem 6.2. *[Hint: The sequence $x_n = \frac{1}{\pi n}$ and Theorem 5.36 may be helpful.]*

9. Define $f : \mathbb{R} \to \mathbb{R}$ as

$$f(x) = \begin{cases} x^2 & \text{for } x \in \mathbb{Q} \\ 0 & \text{for } x \notin \mathbb{Q} \end{cases}.$$

   Prove that f is continuous only at $x = 0$ and differentiable only at $x = 0$.

6.2 The Mean Value Theorem

The Mean Value Theorem is probably a result that the average student has a nodding acquaintance with from basic calculus. In fact, this theorem is a fundamental result in the field of analysis relating the behavior of a function to the behavior of its derivative. The usefulness of this result cannot be understated, and this section is devoted to introducing this theorem and providing some useful applications.

6.2.1 Mean Value Theorems

The Mean Value Theorem as described above is meant to provide a link between the behavior of a function and the behavior of its derivative. One way of visualizing this link is to choose two points on a graph and to compute the slope of the secant line between them. The Mean Value Theorem states that there exists a line tangent to some point with that very same slope, provided some conditions on the function are met. See Figure 6.2.1 for a picture of this.

Put symbolically, if $(a, f(a))$ and $(b, f(b))$ are our points, then this conclusion is that there exists some c between a and b with

$$f'(c) = \frac{f(b) - f(a)}{b - a}, \text{ or, } f(b) - f(a) = f'(c)(b - a).$$

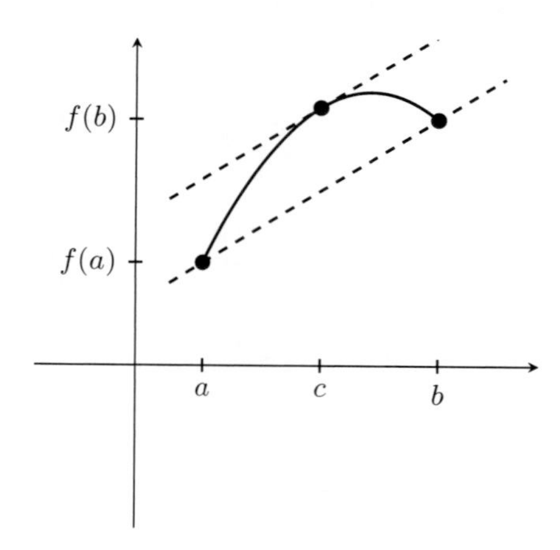

FIGURE 6.2.1
The Mean Value Theorem asserts that there is a tangent line with the same slope as the given secant line.

This will follow from a more general fact that interprets the above in the event that the graph is drawn parametrically as the function $t \mapsto (g(t), f(t))$.

This more general result is sometimes called Cauchy's Mean Value Theorem or the Generalized Mean Value Theorem.

Theorem 6.13 (Generalized Mean Value Theorem). *Let $f, g : [a,b] \to \mathbb{R}$ be continuous on $[a,b]$ and differentiable on (a,b). There exists a $c \in (a,b)$ so that*

$$(f(b) - f(a))g'(c) = (g(b) - g(a))f'(c).$$

Idea of proof. *The idea is to build a function $F : [a,b] \to \mathbb{R}$ from f and g that satisfies the hypotheses of Rolle's Theorem (Theorem 6.11). If this function F is constructed carefully, the c that is generated from Rolle's Theorem will be the c we are looking for here.*

Proof of Theorem 6.13. Define $F : [a,b] \to \mathbb{R}$ as

$$F(x) = (f(b) - f(a))g(x) - (g(b) - g(a))f(x).$$

We check to see that F satisfies the hypotheses of Rolle's Theorem. First, as an allowable combination of the continuous (on $[a,b]$) and differentiable (on (a,b)) functions f and g, we see that F is continuous on $[a,b]$ and differentiable on (a,b). Next,

$$\begin{aligned} F(a) &= (f(b) - f(a))g(a) - (g(b) - g(a))f(a) \\ &= f(b)g(a) - g(b)f(a), \end{aligned}$$

while

$$\begin{aligned} F(b) &= (f(b) - f(a))g(b) - (g(b) - g(a))f(b) \\ &= g(a)f(b) - f(a)g(b). \end{aligned}$$

So, $F(a) = F(b)$, and we apply Rolle's Theorem (Theorem 6.11). There exists a $c \in (a,b)$ with

$$F'(c) = (f(b) - f(a))g'(c) - (g(b) - g(a))f'(c) = 0.$$

Thus, for this c, we have

$$(f(b) - f(a))g'(c) = (g(b) - g(a))f'(c).$$

□

The Mean Value Theorem is now easily proved.

Theorem 6.14 (Mean Value Theorem). *Suppose $f : [a,b] \to \mathbb{R}$ is differentiable on (a,b) and continuous on $[a,b]$. Then there exists a $c \in (a,b)$ so that*

$$f(b) - f(a) = f'(c)(b - a).$$

Idea of proof. *We simply let $g(x) = x$ in Theorem 6.13.*

Proof of Theorem 6.14. Let $g(x) = x$ in Theorem 6.13. □

One can easily verify certain inequalities of functions using the Mean Value Theorem. The next practice problem illustrates this and is used in Examples 3.56 and 3.77 of Chapter 3 to help show that $\left(1+\frac{1}{n}\right)^n \to e$.

Practice 6.15. Establish the following inequality for $x > 0$:

$$\frac{x}{1+x} \leq \ln(1+x) \leq x.$$

Methodology. *Choose any $x > 0$ in advance, and fix it throughout. The idea is to apply the Mean Value Theorem (Theorem 6.14) to the function $f(y) = \ln(1+y)$ on the interval $[0, x]$. We can deduce the left inequality by minimizing f' on $[0, x]$, and the right inequality by maximizing f' on $[0, x]$.*

Solution. Fix $x > 0$, and consider the function $f(y) = \ln(1+y)$ on $[0, x]$. Applying the Mean Value Theorem (Theorem 6.14) to f, there exists a $c \in (0, x)$ with

$$\ln(1+x) = \frac{1}{1+c} \cdot x. \tag{6.3}$$

The function $f'(c) = \frac{1}{1+c}$ is decreasing, and thus

$$f'(x) = \frac{1}{1+x} \leq \frac{1}{1+c} \leq 1.$$

Applying this inequality to Equation (6.3), we have

$$\frac{x}{1+x} \leq \ln(1+x) \leq x$$

as required. ■

6.2.2 Uses for the Mean Value Theorem

The link that the Mean Value Theorem provides between a function and its derivative provides for a number of important results. In this section, we only list a few of the most basic applications of the Mean Value Theorem, it would be impossible to list all of its uses here. Generally, these results follow from the Mean Value Theorem because a relationship between a function and its derivative is required.

The first use of the Mean Value Theorem illustrates how the behavior of a function can be predicted by the sign of its derivative, and follows this familiar definition.

Definition 6.16. Let $f : I \to \mathbb{R}$.

1. If $f(x_1) \leq f(x_2)$ whenever $x_1 < x_2$, then f is *monotone increasing* or *increasing*. If $f(x_1) < f(x_2)$ whenever $x_1 < x_2$, then f is *strictly monotone increasing* or *strictly increasing*.

2. If $f(x_1) \geq f(x_2)$ whenever $x_1 < x_2$, then f is *monotone decreasing* or *decreasing*. If $f(x_1) > f(x_2)$ whenever $x_1 < x_2$, then f is *strictly monotone decreasing* or *strictly decreasing*.

If f satisfies one of the above properties, we may simply describe f as *monotone*.

Theorem 6.17. *Let $f : I \to \mathbb{R}$ be differentiable on I.*

1. If $f'(x) = 0$ for all $x \in I$, then $f(x)$ is a constant function.

2. If $f'(x) \geq 0$ for all $x \in I$, then $f(x)$ is monotone increasing. If $f'(x) > 0$ for all $x \in I$, then f is strictly monotone increasing.

3. If $f'(x) \leq 0$ for all $x \in I$, then $f(x)$ is monotone decreasing. If $f'(x) < 0$ for all $x \in I$, then f is strictly monotone decreasing.

Idea of proof. *All of these assertions will be proven using the Mean Value Theorem (Theorem 6.14). If $x_1 < x_2$ are in I, then there exists a c between them in I with $f(x_2) - f(x_1) = f'(c)(x_2 - x_1)$. Since $x_2 - x_1 > 0$, the sign of $f(x_2) - f(x_1)$ is determined by the sign of $f'(c)$, and we can read off outcome from each hypothesis on $f'(c)$.*

Proof of Theorem 6.17. Let $x_1 < x_2$ be arbitrary elements in I. By the Mean Value Theorem (Theorem 6.14), there exists a $c \in (x_1, x_2)$ with

$$f(x_2) - f(x_1) = f'(c)(x_2 - x_1). \tag{6.4}$$

In Assertion (1), if $f'(c) = 0$ then $f(x_2) = f(x_1)$. In Assertion (2), if $f'(c) \geq 0$ then $f(x_2) - f(x_1) \geq 0$, with strict inequality if $f'(c) > 0$. This implies $f(x_2) \geq f(x_1)$, again with strict inequality if $f'(c) > 0$, so that f is increasing or strictly increasing. The proof of Assertion (3) is similar, and left as Exercise 3. □

As a consequence of the previous result, we can demonstrate that any two antiderivatives[3] of a function must differ only by a constant. This fact will be especially useful when we consider series of functions in Section 7.3.

Theorem 6.18. *Let $f : I \to \mathbb{R}$, and suppose F and G are differentiable functions with $F' = f$ and $G' = f$. There exists a constant C so that $F(x) = G(x) + C$.*

Idea of proof. *Computing the derivative $\frac{d}{dx}[F(x) - G(x)] = 0$ shows that $F - G$ must be constant, according to Assertion (1) of Theorem 6.17.*

Proof of Theorem 6.18. Suppose $F' = G' = f$. Then

$$\frac{d}{dx}[F(x) - G(x)] = F'(x) - G'(x) = f(x) - f(x) = 0.$$

[3]The theory of integration is not meant to be included in this book. One may still separately consider the concept of an antiderivative as a function with a prescribed derivative.

Therefore, by Assertion (1) of Theorem 6.17, there is a constant C with $F(x) - G(x) = C$, so $F(x) = G(x) + C$. □

Recall from Theorem 6.2 that any differentiable function on an interval is also continuous. Using the Mean Value Theorem, we may further show that differentiable functions with a bounded derivative are uniformly continuous.

Theorem 6.19. *Let $f : I \to \mathbb{R}$ be differentiable on I. If $f'(x)$ is bounded on I, then f is uniformly continuous.*

Idea of proof. *We use the boundedness of f' and the Mean Value Theorem to prove f is uniformly continuous by the definition (Definition 5.49).*

Proof of Theorem 6.19. Suppose $|f'(x)| \leq M$ for all $x \in I$. Let $\epsilon > 0$ be given. By the Mean Value Theorem (Theorem 6.14), for each $x, y \in I$, there exists a $c \in I$ between x and y with

$$f(x) - f(y) = f'(c)(x - y).$$

Since $|f'(x)| \leq M$ for every $x \in I$, we have

$$|f(x) - f(y)| = |f'(c)(x - y)| \leq M|x - y|.$$

Now choose $\delta = \frac{\epsilon}{M+1}$, and assume $|x - y| < \delta$. Then

$$|f(x) - f(y)| \leq M|x - y| < M\frac{\epsilon}{M+1} < \epsilon.$$

□

6.2.3 The Inverse Function Theorem

The Inverse Function Theorem is a fundamental result in Analysis. It states that under certain conditions, the derivative of the inverse of a function can be recovered from the derivative of the original function. Graphically speaking, it makes sense that this information should be recoverable: the graph of the inverse of f is the reflection of the graph of f about the origin, thus so must each tangent line of f be reflected into the tangent lines of f^{-1}. This situation is pictured in Figure 6.2.2.

We need a preliminary lemma before we can prove the Inverse Function Theorem. Our approach for this and the proof of the Inverse Function Theorem are taken from [27] as the author could not improve upon them.

Lemma 6.20. *Let $f : I \to \mathbb{R}$ be strictly monotone and continuous. Then f^{-1} is strictly monotone and continuous.*

Idea of proof. *It is a relatively easy proof by contradiction to show that f^{-1} is strictly monotone. We show that f^{-1} is continuous at $y_0 \in f(I) = J$ (which is another interval by the Intermediate Value Theorem (Theorem 5.45)) by considering each two-sided limit $\lim_{y \to y_0^{\pm}} f^{-1}(y) = f^{-1}(y_0)$. This is required since it may be the case that y_0 is an endpoint of the interval J.*

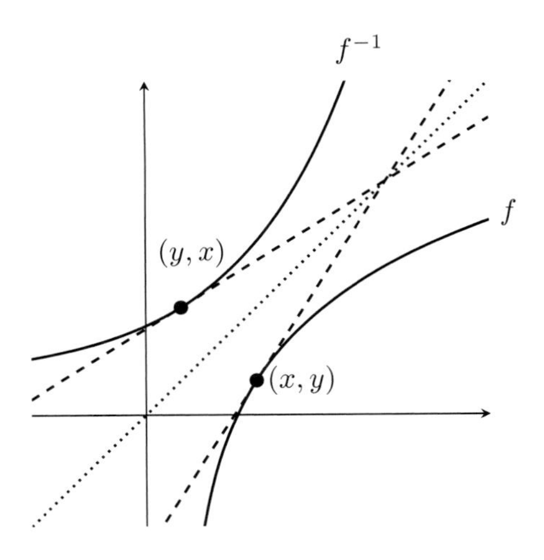

FIGURE 6.2.2
The Inverse Function Theorem formalizes the idea that we can garner information about the derivative of f^{-1} from the derivative of f.

Proof of Lemma 6.20. Assume that f is strictly increasing (the proof if f is strictly decreasing is similar). We first show that the image of f is another interval, f^{-1} is injective, and that f^{-1} is monotone increasing. Let $J = f(I)$. Since f is continuous, the Intermediate Value Theorem (Theorem 5.45) asserts that J is an interval[4]. Since f is strictly monotone, it is a bijection $f : I \to J$, and so f^{-1} exists (Exercise 11 of Section 1.4). Since $(f^{-1})^{-1} = f$, f^{-1} must be a bijection $f^{-1} : J \to I$, so it must be injective. Now suppose $y_1, y_2 \in J$ with $y_1 < y_2$. There exists $x_1, x_2 \in I$ with $f(x_1) = y_1$, and $f(x_2) = y_2$. Since $y_1 \neq y_2$, $x_1 \neq x_2$ and otherwise if $x_1 > x_2$, then $f(x_1) > f(x_2)$ since f is increasing, a violation to our hypotheses. So we must have

$$f^{-1}(y_1) = x_1 < x_2 = f^{-1}(y_2),$$

hence f^{-1} is strictly increasing.

We now prove that f^{-1} is continuous at $y_0 \in J$ by computing

$$\lim_{y \to y_0^-} f^{-1}(y) = f^{-1}(y_0) = \lim_{y \to y_0^+} f^{-1}(y).$$

If y_0 is not an endpoint of J, this would imply $\lim_{y \to y_0} f^{-1}(y) = f^{-1}(y_0)$ by Theorem 5.12. If y_0 is an endpoint of J, then only one of these limits is required to hold.

We first assume that y_0 is not a left endpoint of J and prove $\lim_{y \to y_0^-} f^{-1}(y) = f(y_0)$. Let $x_0 \in I$ satisfy $f(x_0) = y_0$. Since y_0 is not a

[4] J must be an interval, but not necessarily of the same type.

left endpoint of J, the set $(-\infty, y_0) \cap J \neq \emptyset$, and so $(-\infty, x_0) \cap I \neq \emptyset$, making what follows not vacuous.

Let $\epsilon > 0$ be given. If $x_0 - \epsilon \notin I$, then any δ will work. So assume $x_0 - \epsilon \in I$. Set $\delta = f(x_0) - f(x_0 - \epsilon)$. Then $y_0 - \delta = f(x_0 - \epsilon)$. Since f is assumed to be increasing,

$$f((x_0 - \epsilon, x_0]) = (y_0 - \delta, y_0].$$

Now for any $y \in J$ with $y_0 - \delta < y < y_0$,

$$x_0 - \epsilon < f^{-1}(y) < x_0.$$

In other words, when $y_0 - \delta < y < y_0$, we have $|f^{-1}(y) - f^{-1}(y_0)| < \epsilon$, proving that $\lim_{y \to y_0^-} f^{-1}(y) = f^{-1}(y_0)$. The proof that $\lim_{y \to y_0^+} f^{-1}(y) = f^{-1}(y_0)$ in the event that y_0 is not a right endpoint of J is similar and omitted. □

We now use the previous lemma to prove the Inverse Function Theorem.

Theorem 6.21 (The Inverse Function Theorem). *Let $f : I \to \mathbb{R}$ be differentiable, and suppose $f'(x) \neq 0$ for all $x \in I$.*

1. *f is injective on I.*
2. *f^{-1} is differentiable on $J = f(I)$, and for all $y = f(x) \in J$,*

$$(f^{-1})'(y) = \frac{1}{f'(x)}.$$

Idea of proof. *That f is injective follows from Theorems 6.12 and 6.17: Since $f'(x) \neq 0$ and f' satisfies the intermediate value property (Theorem 6.12), we must have $f'(x)$ always greater than or always less than 0. In either event, Theorem 6.17 asserts f is strictly increasing and is therefore injective. The actual computation of the derivative follows from the observation that the difference quotient of f^{-1} is the reciprocal of the difference quotient for f. We use a sequence approach as in Theorem 5.13.*

Proof of Theorem 6.21. We first show that $f'(x)$ is positive for all $x \in I$ or negative for all $x \in I$. If there were $x_1, x_2 \in I$ with $f'(x_1) < 0 < f'(x_2)$, then by Theorem 6.12 there exists a $c \in I$ with $f'(c) = 0$, which the hypotheses forbid. Now since $f'(x)$ has constant sign on I, Theorem 6.17 asserts that f is either strictly increasing on I or strictly decreasing on I. In either event, f is injective.

We now prove the next assertion by computing the difference quotient for f^{-1} and using a sequential approach (Theorem 5.13). Let $d = f(c) \in J = f(I)$, and let $y_n = f(x_n) \to d$, with $y_n \in J$, and $y_n \neq d$. Since f^{-1} is continuous (Lemma 6.20), and $f^{-1}(y_n) = x_n$, it must be that $f^{-1}(y_n) \to f^{-1}(d)$, or rather, $x_n \to c$. Since f is injective and $f(x_n) \neq d$, it must also be that $x_n \neq c$. Then,

$$\begin{aligned} (f^{-1})'(d) &= \lim_{n\to\infty} \frac{f^{-1}(y_n)-f^{-1}(d)}{y_n-d} \\ &= \lim_{n\to\infty} \frac{x_n-c}{f(x_n)-f(c)} \\ &= \lim_{n\to\infty} \frac{1}{\frac{f(x_n)-f(c)}{x_n-c}} \\ &= \frac{1}{f'(c)}. \end{aligned} \tag{6.5}$$

□

With the Inverse Function Theorem, we can compute the derivative of functions that are inverse to those that we already understand, as in the next example.

Example 6.22. This example gives one way of computing the derivative of $g(y) = y^{1/n}$ (for another way, see Exercise 5 in Section 6.1). If $f(x) = x^n$, then $g = f^{-1}$, the inverse to f. Then for any $x \neq 0$ (this is where $f'(x) = 0$ and the Inverse Function Theorem cannot be used),

$$g'(y) = \frac{1}{f'(x)} = \frac{1}{nx^{n-1}}.$$

If $y = f(x) = x^n$, then $y^{1/n} = x$, and so

$$\frac{d}{dy}[y^{1/n}] = \frac{1}{nx^{n-1}} = \frac{1}{n}\frac{1}{y^{\frac{n-1}{n}}} = \frac{1}{n}y^{\frac{1}{n}-1}.$$

One can easily use this result to compute $\frac{d}{dx}[x^{m/n}]$ for any rational $\frac{m}{n}$, see Exercise 10. ■

6.2.4 Review of important concepts from Section 6.2

1. The Mean Value Theorem (Theorem 6.14) provides a link between a function and its derivative, and is a fundamental result in the field of Analysis. It can be easily proven from the Generalized Mean Value Theorem (Theorem 6.13).
2. The sign of the derivative affects the monotonicity of the function, see Theorem 6.17.
3. Any two antiderivatives of a function must differ by a constant (Theorem 6.18).
4. If a function's derivative is bounded, then it is uniformly continuous (Theorem 6.19).
5. The Inverse Function Theorem (Theorem 6.21) provides a way of computing the derivative of f^{-1} from the derivative of f.

Exercises for Section 6.2

1. Either give a proof or a counterexample to the following statement relating to Assertion (1) of Theorem 6.17: If $f : D \to \mathbb{R}$ is a differentiable function and $f'(x) = 0$ for every $x \in D$, then $f(x) = k$ for all $x \in D$, even when D is not an interval.

2. Assume the Mean Value Theorem (Theorem 6.14) holds, and prove Rolle's Theorem (Theorem 6.11). Conclude that the two theorems are logically equivalent.

3. Prove Assertion (3) of Theorem 6.17. That is, if $f : I \to \mathbb{R}$ is differentiable, and $f'(x) \leq 0$ for all $x \in I$, prove that f is decreasing. Also, if $f'(x) < 0$ for all $x \in I$, prove that f is strictly decreasing.

4. Prove that for all $x > 0$, the following inequality holds:

$$\frac{x}{2\sqrt{x+1}} \leq \sqrt{x+1} - 1 \leq \frac{1}{2}x.$$

[Hint: Try using the method of Practice 6.15.]

5. Prove that for all $x \geq 1$, the following inequality holds:

$$\frac{1}{x+1} \leq \ln(x+1) - \ln(x) \leq \frac{1}{x}.$$

[Hint: Apply the Mean Value Theorem (Theorem 6.14) to the function $f(x) = \ln(x)$ on the interval $[x, x+1]$. Then over- and underestimate the derivative.]

6. Prove that for all $x \geq 0$, the following inequality holds:

$$e^x \geq x + 1.$$

[Hint: Try using the method of Practice 6.15.]

7. Suppose $f : (0, 1] \to \mathbb{R}$ is differentiable, and $|f'(x)| \leq 1$ for all $x \in (0, 1]$. Prove that $a_n = f(\frac{1}{n})$ converges. (This exercise appears in [8].)

8. Suppose $f : \mathbb{R} \to \mathbb{R}$ is differentiable.

   (a) Find the error in the following reasoning in attempting to prove the statement "If $\lim_{x\to\infty} f(x) = L \in \mathbb{R}$, then $f'(x) \to 0$ as $x \to \infty$": "Let $x > 0$, and apply the Mean Value Theorem to f on the interval $[x, x+1]$. There exists a $c_x \in [x, x+1]$ so that $f'(c_x) = f(x+1) - f(x)$. Now take the limit on both sides as $x \to \infty$. Since $c_x \in [x, x+1]$, $c_x \to \infty$. On the right-hand side, we have $f(x+1)$ and $f(x)$ both have L as their limit by assumption. Therefore $f'(x) \to 0$ as $x \to \infty$."

   (b) Is the statement above actually true? Prove it, or describe a counterexample.

9. In reference to Theorem 6.19, if f is differentiable on some interval I where f' is unbounded, must it be the case that f is *not* uniformly continuous?

10. Use the result of Practice 6.7 and Example 6.22 to compute the derivative of $g(x) = x^{m/n}$ for nonzero integers m, n.

11. Use the method of Example 6.22 and the identity

$$\cos(\arcsin(x)) = -\sin(\arccos(x)) = \frac{1}{\sqrt{1-x^2}}$$

to compute the derivative of $g_1(y) = \arcsin(y)$ and $g_2(y) = \arccos(y)$.

12. Prove that $\arcsin(y) + \arccos(y) = \frac{\pi}{2}$ for any $y \in [-1, 1]$. *[Hint: Exercise 11 and Theorem 6.18 will be helpful.]*

6.3 Taylor Polynomials

Computing Taylor Polynomials is probably a familiar task from basic calculus: one is given a function and a center to base these polynomials at, and then proceeds to compute derivatives and put it together. These Taylor Polynomials are meant to provide a means to approximate certain functions on a given range.

But why are these polynomials defined in this way? And what specifically about them provides a method for approximation? We begin by giving a motivation for defining Taylor (or Maclaurin) Polynomials in terms of repeated differentiation, and then go on to state and prove Taylor's Theorem (Theorem 6.25), which is the driving force behind the approximations we shall construct.

6.3.1 Defining Taylor Polynomials

For a variety of reasons, working with polynomials seems to be easier than working with other sorts of functions[5], which is why we attempt to find a collection of polynomials P_n ($n = 0, 1, 2, \ldots$) that will approximate a function near a given point $x_0 \in I$, called the *center* of the approximation. These polynomials are called *Taylor Polynomials*, and in the event the center happens to be $x_0 = 0$, they are more specifically referred to as *Maclaurin Polynomials*.

We proceed by degree of the polynomial. Let $f : I \to \mathbb{R}$ be differentiable as many times as we like (for now), and let us center our approximations at a point $x_0 \in I$. A degree zero polynomial is a constant, and since our approximation is to be centered at x_0, we should define

$$P_0(x) = f(x_0).$$

Moving on, a degree one polynomial can be written as

$$P_1(x) = a_0 + a_1(x - x_0). \tag{6.6}$$

As above, at our center x_0 we would want $P_1(x_0) = f(x_0)$. Plugging this into Equation (6.6), we find that $a_0 = f(x_0)$. We want to find a way to determine what a_1 "should be" by only using the center x_0 (as opposed to using other inputs in I) to do so. One way to do this is to proceed by the idea that if P_1 is meant to be a good approximation of f near x_0, then not only should $P_1(x_0) = f(x_0)$, but their derivatives at x_0 should be equal as well: $P_1'(x_0) = f'(x_0)$. Differentiating Equation (6.6) and setting it equal to $f'(x_0)$, we get

$$P_1'(x) = a_1, \text{ so } P_1'(x_0) = a_1 = f'(x_0).$$

[5]Polynomials are not the only sorts of functions analysts might use to approximate a function. For example, Fourier Series use sin and cos.

Using the values for a_0 and a_1 we just found, we conclude that

$$P_1(x) = f(x_0) + f'(x_0)(x - x_0).$$

It should be noted that P_1 is probably familiar as the equation of the tangent line of f at $(x_0, f(x_0))$. The Mean Value Theorem (Theorem 6.14) can give us an intuition about just how close of an approximation P_1 is to f. It states that there is a number c between x_0 and x with

$$f(x) - f(x_0) = f'(c)(x - x_0), \text{ or } f(x) = f(x_0) + f'(c)(x - x_0).$$

Thus the approximation P_1 gives can be measured by how close $f'(c)$ is to $f'(x_0)$. As we shall see in Example 6.27, one usually overestimates this difference and uses that to limit error.

Moving on, we can express an arbitrary polynomial of degree 2 as

$$P_2(x) = a_0 + a_1(x - x_0) + a_2(x - x_0)^2. \tag{6.7}$$

We find a_0, a_1, and a_2 by demanding that $P_2(x_0) = f(x_0)$, $P_2'(x_0) = f'(x_0)$, and $P_2''(x_0) = f''(x_0)$. That $P_2(x_0) = f(x_0)$ implies that $a_0 = f(x_0)$ as before. Differentiating Equation (6.7), we have

$$P_2'(x) = a_1 + 2a_2(x - x_0), \text{ so } P_2'(x_0) = a_1 = f'(x_0) \text{ again.}$$

Differentiating again, we have

$$P_2''(x) = 2a_2, \text{ so } P_2''(x_0) = 2a_2 = f''(x_0).$$

This means that $a_2 = \frac{f''(x_0)}{2}$. To establish a pattern, we note that $2! = 2$, and can express

$$P_2(x) = f(x_0) + f'(x_0)(x - x_0) + \frac{f''(x_0)}{2!}(x - x_0)^2.$$

Determining what the other Taylor Polynomials should be proceeds similarly, motivating the following definition.

Definition 6.23. Let $f : I \to \mathbb{R}$ be n times differentiable at $x_0 \in I$. The nth *Taylor Polynomial* P_n *of* f *centered at* x_0 is defined as

$$P_n(x) = f(x_0) + f'(x_0)(x - x_0) + \frac{f''(x_0)}{2!}(x - x_0)^2 + \cdots + \frac{f^{(n)}(x_0)}{n!}(x - x_0)^n.$$

We illustrate how to find a Taylor Polynomial in the following practice problem.

Practice 6.24. Find the Maclaurin Polynomial $P_3(x)$ for the function $f(x) = \ln(1 + x)$.

Methodology. *The question asks for a Maclaurin Polynomial, so our center is* $x_0 = 0$*. We simply compute the coefficients* $\frac{f^{(i)}(0)}{i!}$ *for* $i = 0, 1, 2, 3$ *and put them together as in Definition 6.23.*

Solution. We first compute the required derivatives:

$$f'(x) = \frac{1}{1+x}, f''(x) = \frac{-1}{(1+x)^2}, \text{ and } f'''(x) = \frac{2}{(1+x)^3}.$$

Therefore, $f(0) = 0, f'(0) = 1, f''(0) = -1,$ and $f'''(0) = 2$. So,

$$P_3(x) = x - \frac{1}{2}x^2 + \frac{1}{3}x^3.$$

■

6.3.2 Taylor's Theorem and applications

Any good theory of approximations includes some measurement of the error involved. Taylor's Theorem provides a first step in this process by specifically relating a function f to any one of its Taylor Polynomials.

Theorem 6.25 (Taylor's Theorem)**.** *Let* $f : I \to \mathbb{R}$*, and let* P_n *be the* n^{th} *Taylor Polynomial of* f *centered at* $x_0 \in I$*. If* f *is* $n+1$ *times differentiable on* I*, then for any* $x \in I$ *there exists a* c *between* x *and* x_0 *with*

$$f(x) = P_n(x) + \frac{f^{(n+1)}(c)}{(n+1)!}(x - x_0)^{n+1}.$$

Remark 6.26. The quantity $R_n(x) = \frac{f^{(n+1)}(c)}{(n+1)!}(x - x_0)^{n+1}$ is sometimes called the *remainder* or more specifically, the *Lagrange form of the remainder*, and so the conclusion of Taylor's Theorem is sometimes written as $f(x) = P_n(x) + R_n(x)$.

Idea of proof. *The idea is to create a function* $g(t)$ *that will satisfy the hypotheses of Rolle's Theorem (Theorem 6.11) repeatedly. The* c *we are looking for will be found as a result of this repeated application.*

Proof of Theorem 6.25. Choose $x, x_0 \in I$. For values of t between x_0 and x, define

$$g(t) = f(t) - P_n(t) - \frac{f(x) - P_n(x)}{(x - x_0)^{n+1}}(t - x_0)^{n+1}.$$

Since $f^{(k)}(x_0) = P_n^{(k)}(x_0)$ for all $k = 0, 1, \ldots, n$, $g^{(k)}(x_0) = 0$. We also check that $g(x) = 0$. Thus, g satisfies the hypotheses of Rolle's Theorem (Theorem 6.11), and so there exists a c_0 between x_0 and x for which $g'(c_0) = 0$. But now g' satisfies the hypotheses of Rolle's Theorem since $g'(x_0) = g'(c_0) = 0$, so there exists a c_1 between x_0 and c_0 with $g''(c_1) = 0$. Continuing in this fashion, we find $c_2, \ldots, c_{n-1}$ with $g^{(k)}(c_{k-1}) = 0$ for $k = 3, \ldots, n+1$. We claim that $c = c_n$ proves our result, which we show presently.

We compute $n+1$ derivatives of g (with respect to the variable t) and evaluate the result at c:

$$0 = g^{(n+1)}(c) = f^{(n+1)}(c) - P_n^{(n+1)}(c) - (n+1)!\frac{f(x) - P_n(x)}{(x-x_0)^{n+1}}.$$

Since P_n is a polynomial of degree n, $P_n^{(n+1)} = 0$. Thus by rearranging the above, we arrive at our result. □

Our original goal was to find a way to approximate functions near a given center, and Taylor's Theorem provides a way to do this. The general idea is to get a sense for how much the remainder term (the remainder measures the difference between f and P_n) can vary as c varies, and to keep in mind that the closer x is to x_0 that generally the approximation would be better. With this information, one could determine how large n must be so that the Taylor Polynomial P_n approximates f to within a given error. The next example illustrates how one goes about this.

Example 6.27. This example illustrates how one can use Taylor's Theorem (6.25) to use Taylor Polynomials to approximate the function $f(x) = \ln(1+x)$ to within 10^{-3} (three decimal places), provided $|x| \leq \frac{1}{10}$. We go on to estimate $\ln(1.1)$ with this approximation.

We begin by differentiating f repeatedly to compute the Maclaurin[6] Polynomials $P_n(x)$. Looking at Practice 6.24, we can see that

$$f^{(n+1)}(x) = (-1)^n \frac{n!}{(x+1)^{n+1}}.$$

Taylor's Theorem ensures that there exists a c between -0.1 and 0.1 so that

$$\begin{aligned} f(x) &= P_n(x) + \tfrac{f^{(n+1)}(c)}{(n+1)!}(x-0)^{n+1}, \text{ so} \\ |f(x) - P_n(x)| = |R_n(x)| &= \left| \frac{(-1)^n \frac{n!}{(c+1)^{n+1}}}{(n+1)!} x^{n+1} \right| \\ &= \tfrac{1}{(n+1)(c+1)^{n+1}} |x|^{n+1}. \end{aligned}$$

We now overestimate $|R_n(x)| = \frac{1}{(n+1)(c+1)^{n+1}}|x|^{n+1}$ by considering two conditions given to us. First, $|x| \leq \frac{1}{10}$. Thus the interval we are considering is $[-\frac{1}{10}, \frac{1}{10}]$, and c must be in this interval: c must be between $x_0 = 0$ and x, while $x \in [-\frac{1}{10}, \frac{1}{10}]$. Minimizing the denominator and maximizing $|x|$ gives us

$$|R_n(x)| = \frac{1}{(n+1)(c+1)^{n+1}}|x|^{n+1}$$

$$\leq \frac{1}{(n+1)\left(\frac{9}{10}\right)^{n+1}} \left(\frac{1}{10}\right)^{n+1} = \frac{1}{n+1}\left(\frac{1}{9}\right)^{n+1}.$$

[6] In this example, our center is $x_0 = 0$.

So for $|R_n(x)| \leq \frac{1}{n+1}\left(\frac{1}{9}\right)^{n+1} \leq 10^{-3}$, we need $10^3 \leq n+1$, so choosing $n = 2$ will do. As a result, we have concluded that the polynomial $P_2(x)$ will be within 3 decimal places of $f(x)$ for values of $x \in [-\frac{1}{10}, \frac{1}{10}]$.

We now estimate $\ln(1.1) = f(\frac{1}{10})$. We know that $P_2(x)$ is within 10^{-3} of $f(x)$ for every $x \in [-\frac{1}{10}, \frac{1}{10}]$, so we plug in $x = \frac{1}{10}$. We find that

$$P_2(x) = x - \frac{1}{2}x^2, \text{ so } P_2\left(\frac{1}{10}\right) = .095.$$

The actual value of $\ln(1.1) = .0953101\ldots$, which gives confirmation that we have succeeded in estimating this value to the needed accuracy. ■

The following practice problem illustrates how to garner useful inequalities from Taylor's Theorem.

Practice 6.28. Show that for $|x| \leq 1$, we have

$$1 + x + \frac{1}{2}x^2 + \frac{1}{6}x^3 \leq e^x \leq 1 + x + \frac{1}{2}x^2 + \frac{1}{6}x^3 + \frac{1}{8}.$$

Methodology. *It appears that the polynomial involved is the Maclaurin Polynomial $P_3(x)$ for $f(x) = e^x$. Thus, the idea is to use Taylor's Theorem (Theorem 6.25) when $n = 3$, and find some way of bounding the remainder $R_3(x)$ by 0 below and $\frac{1}{8}$ above on the interval $[-1, 1]$.*

Solution. Let $f(x) = e^x$. Since $f^{(n)}(x) = e^x$ for all n, the 3rd Maclaurin polynomial for f is given as

$$P_3(x) = 1 + x + \frac{1}{2}x^2 + \frac{1}{6}x^3,$$

with remainder

$$R_3(x) = \frac{e^c}{4!}x^4 \text{ for some } c \in [-1, 1].$$

Since $2 < e < 3$, we have $e^c < 3^c \leq 3$ for $c \in [-1, 1]$. Thus, an overestimate for the remainder occurs when $x = 1$ as $R_3(x) \leq \frac{1}{4!} \cdot 3 = \frac{1}{8}$. To underestimate, we note that $R_3(x) \geq 0$ for every $x \in [-1, 1]$.

By Taylor's Theorem (Theorem 6.25), there is a $c \in [-1, 1]$ with

$$f(x) = e^x = P_3(x) + R_3(x),$$

so since $0 \leq R_3(x) \leq \frac{1}{8}$ and $P_3(x) = 1 + x + \frac{1}{2}x^2 + \frac{1}{6}x^3$, we have

$$1 + x + \frac{1}{2}x^2 + \frac{1}{6}x^3 \leq e^x \leq 1 + x + \frac{1}{2}x^2 + \frac{1}{6}x^3 + \frac{1}{8}$$

for $|x| \leq 1$. ■

6.3.3 Review of important concepts from Section 6.3

1. Taylor Polynomials (Definition 6.23) are polynomials designed to approximate a suitably differentiable function. They are constructed by stipulating that their derivatives at the center (up to a certain point) must agree with the derivatives of the function they are approximating.
2. Taylor's Theorem (Theorem 6.25) provides a way of approximating functions. More specifically, it says that there is some value (inserted into the remainder) in a given interval for which $f(x) = P_n(x) + R_n(x)$. Placing bounds on the remainder is the mechanism that provides these approximations.

Exercises for Section 6.3

1. Compute the Maclaurin Polynomial $P_3(x)$ for the function $f(x) = 3x^3 - 7x^2 + x + 1$.
2. Find constants a_0, a_1, a_2, and a_3 so that
$$1 + 3x^3 = a_0 + a_1(x-1) + a_2(x-1)^2 + a_3(x-1)^3.$$
3. Estimate e to within 6 decimal places using the method of Example 6.27.
4. Show that for $|x| \leq 1$ we have
$$2 + \frac{1}{4}x - \frac{3}{256}x^2 - \frac{3}{16} \leq \sqrt{4+x} \leq 2 + \frac{1}{4}x - \frac{3}{256}x^2 + \frac{3}{16}.$$
Use it to estimate $\sqrt{5}$. How accurate is your estimate?
5. Let $f(x) = \frac{1}{1-x}$, and let $P_n(x)$ be its nth Maclaurin Polynomial. Show that if $x \in (-1, 1)$, then the remainder $R_n(x) \to 0$ as $n \to \infty$.

6.4 L'Hôpital's Rule

Computing the limit of a function can be quite difficult in the event that none of our limit theorems apply. For example, if we were to consider the limit of $\frac{f(x)}{g(x)}$ and both $f(x) \to 0$ and $g(x) \to 0$, we do not yet have any general suggestion as to how to proceed. L'Hôpital's Rule[7] (Theorem 6.29) is a powerful tool to help evaluate limits of this and other types. This section proves the basic form of L'Hôpital's Rule as it applies to the indeterminate forms $\frac{0}{0}$ and $\frac{\infty}{\infty}$. We go on to describe how various other indeterminate forms can be transformed to still apply L'Hôpital's Rule.

6.4.1 Formulating L'Hôpital's Rule

As mentioned above, L'Hôpital's Rule will apply to a number of what are called "indeterminate forms." We will first prove a basic formulation of L'Hôpital's Rule below applying to the $\frac{0}{0}$ and $\frac{\infty}{\infty}$ indeterminate forms. We shall then interpret this in a variety of ways in the context of some of our previous results to extend it to the more familiar L'Hôpital's Rule the student may already be familiar with. We describe in the next subsection how to apply this result to other types of indeterminate forms.

Theorem 6.29 (L'Hôpital's Rule). *Suppose $f, g : (a, b) \to \mathbb{R}$, where $(a, b) \subseteq \mathbb{R}$. If*

1. *f and g are differentiable on (a, b) with $g'(x) \neq 0$ on (a, b),*
2. *$\frac{f'(x)}{g'(x)} \to L \in \mathbb{R}^\sharp$ as $x \to a^+$, and*
3. *$f(x) \to 0$ and $g(x) \to 0$ as $x \to a^+$, or $g(x) \to \infty$ as $x \to a^+$,*

then $\frac{f(x)}{g(x)} \to L$ as $x \to a^+$ well.

Idea of proof. *We follow [24] and prove that for arbitrary p and q with $p < L < q$, that $p < \frac{f(x)}{g(x)} < q$ for x sufficiently close to a. Since p and q were arbitrary, this shows that $\frac{f(x)}{g(x)} \to L$ as $x \to a^+$. This estimation can be achieved through an application of the Generalized Mean Value Theorem (Theorem 6.13).*

Proof of Theorem 6.29. Suppose first that $-\infty \leq L < \infty$. Select arbitrary q

[7] L'Hôpital's Rule was originally discovered by Johann Bernoulli, but named after Marquis de L'Hôpital after it appeared in L'Hôpital's textbook on Infinitesimal Calculus in 1696. This book largely contained results discovered by Bernoulli, and L'Hôpital acknowledged Bernoulli's contributions in the preface of this book. It is for this reason this result should probably have been called "Bernoulli's Rule."

and r with $L < r < q$. Since $\frac{f'(x)}{g'(x)} \to L$ as $x \to a^+$, there is a δ so that if $a < x < a + \delta$, then

$$\frac{f'(x)}{g'(x)} < r.$$

Choose $x, y \in (a, a + \delta)$. Applying the Generalized Mean Value Theorem (Theorem 6.13) on the interval of real numbers between x and y, there exists a c between x and y with

$$\frac{f'(c)}{g'(c)} = \frac{f(y) - f(x)}{g(y) - g(x)} < r. \tag{6.8}$$

If $f(x) \to 0$ and $g(x) \to 0$ as $x \to a^+$, then taking the limit as $x \to a^+$ in Equation (6.8) tells us that for $y \in (a, c)$, we have

$$\frac{f(y)}{g(y)} \leq r < q.$$

If instead it is the case that $g(x) \to \infty$, there exists a $\delta_1 < \delta$ so that if $a < x < a + \delta_1$ then both $g(x) > g(y)$ and $g(x) > 0$. Then by multiplying by $\frac{g(x) - g(y)}{g(x)} > 0$ in Equation (6.8) we have

$$\frac{f(y) - f(x)}{g(y) - g(x)} \cdot \frac{g(x) - g(y)}{g(x)} < \frac{g(x) - g(y)}{g(x)} r. \tag{6.9}$$

But

$$\frac{f(y) - f(x)}{g(y) - g(x)} \cdot \frac{g(x) - g(y)}{g(x)} = \frac{f(x) - f(y)}{g(x)} = \frac{f(x)}{g(x)} - \frac{f(y)}{g(x)},$$

and

$$\frac{g(x) - g(y)}{g(x)} r = \left(1 - \frac{g(y)}{g(x)}\right) r = r - \frac{g(y)}{g(x)} r.$$

So by adding $\frac{f(y)}{g(x)}$ in Equation (6.9), we have

$$\frac{f(x)}{g(x)} < r - \frac{g(y)}{g(x)} r + \frac{f(y)}{g(x)}.$$

Now as $x \to a^+$, we have $\frac{g(y)}{g(x)} r + \frac{f(y)}{g(x)} \to 0$. So there exists a $\delta_2 < \delta_1$ so that if $a < x < a + \delta_2$, then $|\frac{g(y)}{g(x)} r + \frac{f(y)}{g(x)}| < \frac{r+q}{2}$, so that for these x, we still have

$$\frac{f(x)}{g(x)} < q.$$

We have proven that in either situation, if x is sufficiently close to a, then $\frac{f(x)}{g(x)} < q$. A similar argument for $L \in (-\infty, \infty]$ shows that if p is arbitrarily chosen with $p < L$, then for x sufficiently close to a we have $p < \frac{f(x)}{g(x)}$. Since p and q were arbitrary, this shows that $\frac{f(x)}{g(x)} \to L$ as $x \to a^+$. □

Here are some important remarks that will extend Theorem 6.29 to more familiar territory.

1. **The proof of Theorem 6.29 can easily be adapted to consider limits as $x \to b^-$.**

2. **If we replace the one-sided limit hypotheses in Theorem 6.29 with two-sided ones, the result will still hold.** By Theorem 5.12, a limit exists if and only if both one-sided limits exist and are equal. Therefore, if x_0 is an interior point of the common domain of f and g, then we could use Theorem 6.29 to compute the limit as $x \to x_0^+$ and as $x \to x_0^-$.

3. **One can adapt Theorem 6.29 as follows to hold in the event a or b is $-\infty$ or ∞.** Suppose that $f(x) \to 0$, $g(x) \to 0$, and $\frac{f'(x)}{g'(x)} \to L$ as $x \to \infty$. Define $F(x) = f(\frac{1}{x})$ and $G(x) = g(\frac{1}{x})$. By Part 2 of Proposition 5.30, we must have $F(x) \to 0$ and $G(x) \to 0$ as $x \to 0^+$. As the composition of differentiable functions, F and G are differentiable on $(0, b)$ for some b, and we have

$$\frac{F'(x)}{G'(x)} = \frac{f'(\frac{1}{x})\left(\frac{-1}{x^2}\right)}{g'(\frac{1}{x})\left(\frac{-1}{x^2}\right)} = \frac{f'(\frac{1}{x})}{g'(\frac{1}{x})}.$$

   By Proposition 5.30 again, as $x \to 0^+$, we have

$$\frac{F'(x)}{G'(x)} = \frac{f'(\frac{1}{x})}{g'(\frac{1}{x})} \to L.$$

   So Theorem 6.29 applies, and $\frac{F(x)}{G(x)} \to L$ as $x \to 0^+$. Again by Proposition 5.30, we must have $\frac{f(x)}{g(x)} \to L$ as $x \to 0^+$. A similar method could be used in the situation $x \to -\infty$ as well. Thus, Theorem 6.29 holds for limits as $x \to \infty$ or $x \to -\infty$. The above could be phrased equally well to apply to in the case that $g(x) \to \infty$ as $x \to a^+$.

4. **If $g(x) \to \infty$ as $x \to a^+$, we only need to consider situations where $\lim_{x \to a^+} f(x)$ is infinite, which justifies referring to this case as the $\frac{\infty}{\infty}$ indeterminate form.** If f is bounded but $g(x) \to \infty$ as $x \to a^+$, then $\frac{f(x)}{g(x)} \to 0$.

6.4.2 Indeterminate forms

The hypothesis that $f(x) \to 0$ and $g(x) \to 0$ in $\lim_{x \to a^+} \frac{f(x)}{g(x)}$ is a situation that is sometimes referred to as having "$\frac{0}{0}$ indeterminate form." If both $f(x)$ and $g(x)$ approach $\pm\infty$, then similarly $\lim_{x \to a^+} \frac{f(x)}{g(x)}$ is referred to as having "$\frac{\infty}{\infty}$ indeterminate form." These "indeterminate forms" are characterized by some sort of arithmetic in $\mathbb{R}^\sharp$ that is not defined. If clever enough, one can apply L'Hôpital's Rule (Theorem 6.29) to a number of other forms after some clever rearranging. The following example shows how to transform a limit having $0 \cdot \infty$ indeterminate form into a $\frac{\infty}{\infty}$ indeterminate form for which L'Hôpital's Rule applies.

Example 6.30. In this example, we compute $\lim_{x\to 0^+}[-x\ln(x)]$. Since $x \to 0$ and $-\ln(x) \to \infty$ as $x \to 0^+$, this limit takes the indeterminate form $0 \cdot \infty$. We rewrite

$$-x\ln(x) = \frac{-\ln(x)}{\frac{1}{x}},$$

and now both $-\ln(x)$ and $\frac{1}{x}$ have limits of ∞ as $x \to 0^+$. We may now apply L'Hôpital's Rule to this limit, and find that

$$\frac{\frac{d}{dx}[-\ln(x)]}{\frac{d}{dx}\left[\frac{1}{x}\right]} = \frac{-1/x}{-1/x^2} = x \to 0 \text{ as } x \to 0^+.$$

Therefore, $\lim_{x\to 0^+}[-x\ln(x)] = 0$. ■

The next example illustrates how to handle the indeterminate form 0^0.

Example 6.31. In this example, we compute $\lim_{x\to 0^+} x^x$, which has the indeterminate form 0^0. The idea is to instead consider the function $f(x) = \ln(x^x) = x\ln x$, and to use basic limit properties to relate $\lim_{x\to 0^+} f(x)$ to $\lim_{x\to 0^+} x^x$.

We found in Example 6.30 that $x \to 0^+$ we have $-x\ln(x) \to 0$, so $x\ln(x) \to 0$ as well. Since the function $x \mapsto e^x$ is continuous and $x^x = e^{\ln(x^x)} = e^{x\ln(x)}$, we can compute the limit as follows:

$$\lim_{x\to 0^+} x^x = \lim_{x\to 0^+} e^{x\ln(x)} = e^{\lim_{x\to 0^+}[x\ln(x)]} = e^0 = 1.$$

■

In the above example, the key to rearranging the indeterminate form 0^0 to take a different form was to utilize the inverse relationship between the exponential and logarithm functions. Similar transformations will allow us to consider additional indeterminate forms of the type $\infty - \infty, 1^\infty$, and ∞^0. We summarize in the table below an informal suggested way to proceed in each of these cases to transform a given indeterminate form into either $\frac{0}{0}$ or $\frac{\infty}{\infty}$. Keep in mind that some of these transformations may require some more work in actually computing the limit, as in Example 6.31, and that the chart below only offers a suggestion and more work may be required in some cases.

Indeterminate form	Suggestion to $\frac{0}{0}$	Suggestion to $\frac{\infty}{\infty}$
$fg \to 0 \cdot \infty$	$fg = \frac{f}{1/g}$	$fg = \frac{g}{1/f}$
$f - g \to \infty - \infty$	$f - g = \frac{\frac{1}{g} - \frac{1}{f}}{\frac{1}{fg}}$	$f - g = \ln\left(\frac{e^f}{e^g}\right)$
$f^g \to 0^0$	$\ln(f^g) = \frac{g}{\frac{1}{\ln(f)}}$	$\ln(f^g) = \frac{\ln(f)}{1/g}$
$f^g \to 1^\infty$	$\ln(f^g) = \frac{\ln(f)}{1/g}$	$\ln(f^g) = \frac{g}{\frac{1}{\ln(f)}}$
$f^g \to \infty^0$	$\ln(f^g) = \frac{g}{\frac{1}{\ln(f)}}$	$\ln(f^g) = \frac{\ln(f)}{1/g}$

Practice 6.32. Compute the following limits:

(a) $\lim_{x\to\infty} \frac{2^x}{3^x+1}$

(b) $\lim_{x\to\infty} x^2e^{-x}$

(c) $\lim_{x\to\infty} \left(1 + \frac{1}{x}\right)^x$

(d) $\lim_{x\to 0^+} x^{\sin x}$

Methodology. *Each of these is an application of L'Hôpital's Rule (Theorem 6.29), possibly with some manipulation to put it into the correct form. In the order given, the indefinite form we see is* $\frac{\infty}{\infty}$, $0 \cdot \infty$, 1^∞, *and* 0^0. *Before proceeding to the solution below, it may be helpful to check for yourself that these are indeed the forms. Part (a) is a direct use of L'Hôpital's Rule, while the other parts require some manipulation before L'Hôpital's rule can be used. In Part (b), L'Hôpital's Rule must be applied twice. The limit in Part (c) should appear similar to the limit in Exercise 3.56 and should have limit e, although this reasoning alone does not prove that the limit in Part (c) must be e. So, we demonstrate it below with L'Hôpital's Rule*[8]. *Part (d) proceeds similarly to Example 6.31.*

Solution. For each of these parts, we make some use of L'Hôpital's Rule (Theorem 6.29).

(a) This has indeterminate form $\frac{\infty}{\infty}$, and so we directly apply L'Hôpital's Rule:

$$\frac{\frac{d}{dx}[2^x]}{\frac{d}{dx}[3^x+1]} = \frac{\ln 2 \cdot 2^x}{\ln 2 \cdot 3^x} = \frac{\ln 2}{\ln 3}\left(\frac{2}{3}\right)^x \to 0,$$

where $(\frac{2}{3})^x \to 0$ by Exercise 2 in Section 5.3.

(b) This has indeterminate form $0 \cdot \infty$, and therefore requires some manipulation before L'Hôpital's Rule can be applied. We rewrite

$$x^2e^{-x} = \frac{x^2}{e^x} \to \frac{\infty}{\infty}.$$

[8]It does go the other way, though: if we knew the limit in Part (c) were e, then we could deduce that $\lim_{n\to\infty}(1 + \frac{1}{n})^n = e$ by Theorem 5.29, since $x_n = n$ is a sequence with $x_n \to \infty$.

Applying L'Hôpital's Rule,

$$\frac{\frac{d}{dx}[x^2]}{\frac{d}{dx}[e^x]} = \frac{2x}{e^x} \to \frac{\infty}{\infty}.$$

Applying L'Hôpital's Rule again, we find that

$$\frac{\frac{d}{dx}[2x]}{\frac{d}{dx}[e^x]} = \frac{2}{e^x} \to 0 \text{ as } x \to \infty.$$

Going backward through our work, we conclude that

$$\frac{2x}{e^x} \to 0, \text{and so } \frac{x^2}{e^x} \to 0.$$

(c) This has indeterminate form 1^∞, and so we take the suggestion in the chart above to compute

$$\ln\left(\lim_{x\to\infty}\left[\left(1+\frac{1}{x}\right)^x\right]\right) = \lim_{x\to\infty}\left[\ln\left(1+\frac{1}{x}\right)^x\right],$$

and

$$\ln\left(1+\frac{1}{x}\right)^x = x\ln\left(1+\frac{1}{x}\right) = \frac{\ln\left(1+\frac{1}{x}\right)}{1/x} \to \frac{0}{0}.$$

Applying L'Hôpital's Rule to this, we have

$$\frac{\frac{d}{dx}\left[\ln\left(1+\frac{1}{x}\right)\right]}{\frac{d}{dx}[1/x]} = \frac{\frac{1}{1+\frac{1}{x}}\cdot\frac{-1}{x^2}}{-1/x^2} = \frac{1}{1+\frac{1}{x}} \to 1.$$

We have shown that

$$\ln\left(\lim_{x\to\infty}\left[\left(1+\frac{1}{x}\right)^x\right]\right) = 1, \text{so } \lim_{x\to\infty}\left(1+\frac{1}{x}\right)^x = e^1 = e.$$

(d) This has indeterminate form 0^0, and again taking the suggestion in the chart above, we express

$$\ln\left(\lim_{x\to 0+} x^{\sin x}\right) = \lim_{x\to 0+}\left(\ln x^{\sin x}\right),$$

and

$$\ln x^{\sin x} = \sin x \ln x = \frac{\ln x}{\csc x} \to \frac{-\infty}{\infty}.$$

We apply L'Hôpital's Rule to this:

$$\frac{\frac{d}{dx}[\ln x]}{\frac{d}{dx}[\csc x]} = \frac{1/x}{-\csc x\cdot\cot x} = \frac{\sin x}{x}\cdot\tan x. \tag{6.10}$$

Applying L'Hôpital's Rule to $\frac{\sin x}{x} \to \frac{0}{0}$, we have

$$\frac{\frac{d}{dx}[\sin x]}{\frac{d}{dx}[x]} = \frac{\cos x}{1} \to 1 \text{ as } x \to 0^+.$$

Applying this knowledge to Equation (6.10), we have

$$\frac{\sin x}{x} \cdot \tan x \to 1 \cdot 0 \text{ as } x \to 0^+.$$

We have shown that

$$\ln\left(\lim_{x\to 0^+} x^{\sin x}\right) = 0, \text{ so } \lim_{x\to 0^+} x^{\sin x} = e^0 = 1.$$

■

We close this section with an example taken from [3] that illustrates that the hypothesis that $\lim_{x\to a^+} \frac{f'(x)}{g'(x)}$ exist is necessary: we will find that $\lim_{x\to\infty} \frac{f'(x)}{g'(x)}$ does not exist, while $\lim_{x\to\infty} \frac{f(x)}{g(x)}$ does exist.

Example 6.33. We attempt to use L'Hôpital's Rule to compute $\lim_{x\to\infty} \frac{x-\sin(x)}{x+\sin(x)}$, which takes indeterminate form $\frac{\infty}{\infty}$. We have

$$\frac{\frac{d}{dx}[x-\sin(x)]}{\frac{d}{dx}[x+\sin(x)]} = \frac{1-\cos(x)}{1+\cos(x)}.$$

We allow $x \to \infty$ along the sequence $x_n = 2\pi n$, then $\cos(x_n) = 1$, so the above is always 0. However, if we instead proceed along $x_n = \frac{\pi}{2} + 2\pi n$, then $\cos(x_n) = 0$, and the above is always 1. Hence, this limit as $x \to \infty$ does not exist by Theorem 5.29.

Our original limit does exist, however. We can use the Squeeze Theorem to deduce that $\frac{\sin(x)}{x} \to 0$ as $x \to \infty$. Therefore by factoring out an x, we have

$$\frac{x-\sin(x)}{x+\sin(x)} = \frac{1-\frac{\sin(x)}{x}}{1+\frac{\sin(x)}{x}} \to 1 \text{ as } x \to \infty.$$

■

6.4.3 Review of important concepts from Section 6.4

1. L'Hôpital's Rule (Theorem 6.29) is a powerful tool for evaluating limits.
2. The indeterminate forms $\frac{0}{0}$ and $\frac{\infty}{\infty}$ are addressed in Theorem 6.29, while the indeterminate forms $0 \cdot \infty, \infty - \infty, 0^0, 1^\infty$, and ∞^0 can be rearranged to still apply L'Hôpital's Rule (see the table on Page 344).

Exercises for Section 6.4

1. Compute the following limits.

   (a) $\lim_{x\to 0} \frac{\sin x}{x}$

   (b) $\lim_{x\to 0} \frac{\arctan(x)}{x}$

   (c) $\lim_{x\to 0} \frac{e^x - 1}{x}$

   (d) $\lim_{x\to\infty} \ln(x+1) - \ln(x)$

   (e) $\lim_{x\to\infty} x^2 3^{-x}$

   (f) $\lim_{x\to\infty} \frac{\ln(e^x+1)}{x}$

   (g) $\lim_{x\to\infty} \left(1 + \frac{2}{x}\right)^x$

   (h) $\lim_{x\to 1^+} \left(\frac{x}{x-1} - \frac{1}{\ln(x)}\right)$

2. Prove that if $p(x)$ is a polynomial and $a > 0$, then $\lim_{x\to\infty} \frac{p(x)}{a^x} = 0$.
3. Suppose f, g are twice differentiable on a neighborhood of x_0. Suppose also that f'' and g'' are continuous at x_0. Use Taylor's Theorem (Theorem 6.25) to give a different proof of L'Hôpital's Rule (Theorem 6.29).
4. Compute $\lim_{x\to\infty}(\sqrt{x^2+x+1} - \sqrt{x^2+1})$.
5. Compute $\lim_{x\to\infty} \frac{x \sin x}{e^x}$.

6.5 Troubleshooting Guide for Differentiation

Type of difficulty:	*Suggestion for help:*
Section 6.1: The Derivative	
"Why does it seem as if there is a preference to have $(f(x) - f(c))/(x - c)$ as in Definition 6.1. I really preferred $(f(c + h) - f(c))/h$ from calculus that is mentioned later."	The two limits are the same, although the one in Definition 6.1 seems to be more convenient to work with. See, for examples, the results in the next subsection after Definition 6.1.
"I thought that all of the theorems here about computing derivatives were just shortcuts. Why are we proving them?"	These facts can only be taken as true because we have proven them here. Notice in their proofs that we are using Definition 6.1.
"I don't understand why we can't just prove the Chain Rule like it is done on Page 315."	The issue is that both expressions in Equation (6.2) have to be defined for x in a neighborhood of c to be able to use our limit theorems from Section 5.2. This may not be the case, even in very simple instances, as the footnote there points out.
Section 6.2: The Mean Value Theorem	
"What is the deal with the $M + 1$ in the proof of Theorem 6.19?"	This assures that our choice of δ will be defined even in the unlikely event that $M = 0$. By hypothesis, M cannot be negative. This tactic is also used in the proof of Theorem 3.31.
"I don't know why we need to use Lemma 6.20 in the proof of the Inverse Function Theorem (Theorem 6.21)."	This lemma allows us to conclude $x_n \to c$ in that proof. Now in the second to last line in Equation (6.5), we have the quotient of two limits: the numerator is constant, while the denominator converges to $f'(c)$ by Theorem 5.13. If we did not know that $x_n \to c$, we could not compute this limit in that way.

Type of difficulty:	*Suggestion for help:*
"I'm not seeing how they got Equation (6.3) in Practice 6.15."	This comes directly from the Mean Value Theorem (Theorem 6.14), where the function $f(y) = \ln(1+y)$, $a = 0$ and $b = x$. Notice that $f(0) = 0$, so that $f(x) - f(0) = \ln(1+x)$, and $f'(c)(x-0) = \frac{1}{1+c} \cdot x$. The variable y is used to reserve the use of x for the interval $[0, x]$.
Section 6.3: Taylor Polynomials	
"I don't understand the discussion concerning Taylor Polynomials at the beginning of this section."	This discussion is meant to motivate why Taylor Polynomials are defined in the way they are. It probably seems suspiciously convenient that we plug c into everything and that the polynomials P_i are written in terms of $(x-c)$ and not just x. Without this convenience the discussion would veer far off-course.
"Why are there factorials in the denominators of the coefficients in the Taylor Polynomials P_n?"	They are required to offset the repeated differentiation of $(x-c)^n$. Since $2! = 2$, this can be seen in the discussion following Equation (6.7).
Section 6.4: L'Hôpital's Rule	
"I don't understand how to extend Theorem 6.29 to the situation where $x \to \infty$ or $x \to -\infty$."	This was described on Page 342. The main idea is to use Part 2 of Proposition 5.30, which relates the limit of a function $f(x)$ as $x \to \infty$ to the limit of $f(\frac{1}{x})$ as $x \to 0^+$.
"I thought L'Hôpital's Rule applied to other indeterminate forms, not just 0/0."	See the discussion on Page 342, and the subsequent chart of suggestions on Page 344.
"The chart on Page 344 didn't help me in computing my limit. Any suggestions on how to proceed?"	There could be several issues depending on your limit. Sometimes using the chart to change the limit into the $\frac{0}{0}$ form is not as useful as if one changes it instead to $\frac{\infty}{\infty}$, and vice versa. Another possibility is that you may have made a mistake in which indeterminate form you have, or, you may not have any indeterminate form at all, so that using L'Hôpital's Rule may not be appropriate.

7

Sequences and Series of Functions

Chapter Summary:

This chapter studies convergence of sequences and series of functions. These may converge pointwise or uniformly, and we shall find that uniform convergence is a crucial property that provides for several "preservation theorems." We end our study with a discussion of a special kind of series of functions that we may apply this theory to: power series.

Section 7.1: Sequences of Functions. Pointwise and uniform convergence of sequences of functions are introduced, along with the Cauchy Criterion for uniform convergence (Theorem 7.7).

Section 7.2: Consequences of Uniform Convergence. A discussion of what properties are preserved through uniform convergence. Continuity and definite integration are preserved (Theorems 7.9 and 7.10), although differentiation is not unless we consider different hypotheses (Theorem 7.13).

Section 7.3: Infinite Series of Functions. Pointwise and uniform convergence of series of functions is defined in Definition 7.14, and characterized with a Cauchy criterion for uniform convergence in Theorem 7.16. The preservation theorems from Section 7.2 are interpreted here for series concerning continuity (Theorem 7.17), differentiation (Theorem 7.20), and antidifferentiation (Corollary 7.21). The Weierstrass $M-$test (Theorem 7.22) is presented as a tool to help show that a series is uniformly convergent.

Section 7.4: Power Series. We introduce the radius and interval of convergence (Theorem 7.24), and prove that a power series uniformly converges on a suitable subset of its interval of convergence (Theorem 7.26). As a result, we can prove results concerning continuity (including Abel's Theorem, Theorem 7.28), and that term-by-term antidifferentiation and differentiation are allowed. The section ends with a brief introduction to Taylor series.

Section 7.5: Troubleshooting Guide for Series of Real Numbers. A list of frequently asked questions and their answers, arranged by section. Come here if you are struggling with any material from this chapter.

7.1 Sequences of Functions

Sequences of functions seem like a straightforward generalization of sequences of real numbers. Although that may be the case, there are some intricacies in considering what convergence of such a sequence should be. In this section, we present the most basic form of convergence of a sequence of functions, called *pointwise convergence* (Definition 7.1). We go on to define a stronger form of convergence in Definition 7.5 called *uniform convergence.* We close the section with a useful characterization of uniform convergence called the *Cauchy Criterion* (Theorem 7.7).

7.1.1 Sequences of functions and pointwise convergence

Recall from Exercise 14 from Section 3.2 that if $|x| < 1$, then the geometric sequence $x^n \to 0$. In addition, if $x = 1$, then $x^n = 1^n = 1 \to 1$, and for any other value of x, the sequence (x^n) diverges. If we consider the collection of functions $f_n(x) = x^n$, then although the natural domain for each of these is $\mathbb{R}$, only if $x \in (-1, 1]$ does $\lim_{n\to\infty} f(x)$ exist. For this reason, we would only want to consider $f_n : (-1, 1] \to \mathbb{R}$.

This is an example of what we call a *sequence of functions*, and *pointwise convergence*: $f_n(x) = x^n$ is a sequence of functions (one function for each $n \in \mathbb{N}$), and if $x \in (-1, 1]$ is chosen in advance, then $(f_n(x))$ converges. The term "pointwise convergence" is appropriate, since for every point x that is chosen, the subsequent sequence of outputs $(f_n(x))$ converges. The best way to describe the end result of this process is to define a function $f(x) = \lim_{n\to\infty} f_n(x)$, as we do in the definition below.

Definition 7.1. Let $f_n : D \to \mathbb{R}$ be a sequence of functions. We say that $(f_n(x))$ *converges pointwise* on D to $f : D \to \mathbb{R}$ if for every $x \in D$, we have

$$f(x) = \lim_{n\to\infty} f_n(x).$$

This may be denoted as $f_n \to f$ on D.

The following example illustrates that the sequence of functions we considered above converges pointwise on $(-1, 1]$.

Example 7.2. This example illustrates that the sequence of functions $f_n(x) = x^n$ converges pointwise on $D = (-1, 1]$. The function f that it converges to will be determined throughout our discussion.

As we noted above, if $x \in (-1, 1)$, then $x^n \to 0$. And, if $x = 1$, then $x^n = 1^n \to 1$. So we define the following function:

$$f(x) = \begin{cases} 0 & \text{for } x \in (-1, 1) \\ 1 & \text{for } x = 1 \end{cases}.$$

Clearly, for every $x \in (-1, 1]$ that is chosen, $f_n(x) \to f(x)$. ■

Here are several practice problems that ask you to determine the pointwise limit function f of a sequence of functions.

Practice 7.3. For the given sequence of functions, prove that it converges pointwise on the given domain, and find the limit $f(x) = \lim_{n\to\infty} f_n(x)$.

(a) $f_n(x) = \frac{x}{n}$ on $\mathbb{R}$.

(b) $f_n(x) = \frac{nx}{1+n^2x^2}$ on $\mathbb{R}$.

(c) $f_n(x) = x + \frac{x}{n}\sin(nx)$ on $[-a, a]$.

Methodology. *Each of these is an exercise in convergence of sequences. The sequence of functions in Parts (a) and (b) will both converge pointwise to the 0 function, while the sequence in Part (c) will converge to $f(x) = x$. The idea is to choose an x in advance in the given domain and treat it as a constant thereafter.*

Solution. (a) Let $x \in \mathbb{R}$ be given. The sequence $\frac{x}{n} \to 0$, so this sequence converges pointwise to $f(x) = 0$.

(b) Let $x \in \mathbb{R}$ be given. Suppose first that $x \neq 0$. After factoring an n out of the numerator and an n^2 out of the denominator, we have

$$\frac{nx}{1+n^2x^2} = \frac{n}{n^2}\frac{x}{\frac{1}{n^2}+x^2} \to 0 \cdot \frac{x}{x^2} = 0.$$

If $x = 0$, then $f_n(x) = 0$, so $f_n(0) \to 0$ as well. We have shown that $(f_n(x))$ converges pointwise to $f(x) = 0$ for every $x \in \mathbb{R}$.

(c) Let $x \in [-a, a]$ be given. Notice that $\sin(nx)$ is bounded, and that $\frac{x}{n} \to 0$. Thus, as the product of a bounded sequence and one that converges to 0, we have $\frac{x}{n}\sin(nx) \to 0$ (see Exercise 23 of Section 3.2). Thus,

$$f_n(x) = x + \frac{x}{n}\sin(nx) \to x + 0 = x.$$

■

7.1.2 Uniform convergence

There is a stronger and more useful form of convergence of functions called *uniform convergence.* We illustrate how this differs from pointwise convergence through a comparative numerical example.

Example 7.4. In Example 7.2, we showed that $f_n(x) = x^n$ converges pointwise on $(-1, 1]$. Let us look deeper into this by considering the same function on different domains, as shown in Figure 7.1.1.

(a) Suppose we consider the domain $[0, 1]$. We still have pointwise convergence,

$$f_n(x) = x^n \to f(x) = \begin{cases} 0 & \text{for } x \in [0,1) \\ 1 & \text{for } x = 1 \end{cases}.$$

Select some $x \in [0,1]$, say, $x = \frac{1}{2}$. Since $f_n(x) \to f(x)$ for every x, $f_n(\frac{1}{2}) \to f(\frac{1}{2}) = 0$. Let us study the convergence of this sequence more carefully. Let $\epsilon = .001$. One requires at least $N = 10$ to guarantee that if $n \geq N$, then $|(\frac{1}{2})^n| < .001$.

Now select some other x, say, $x = \frac{9}{10}$. We still have $f_n(\frac{9}{10}) \to f(\frac{9}{10}) = 0$, but for $\epsilon = .001$ again, we now require at least $N = 66$ for $|(\frac{9}{10})^n| < .001$ for any $n \geq N$. This illustrates that the N one might find in choosing some x in advance could vary depending on x chosen. This is not unexpected, but in this example, for a chosen ϵ, the closer x gets to 1, the larger N must be.

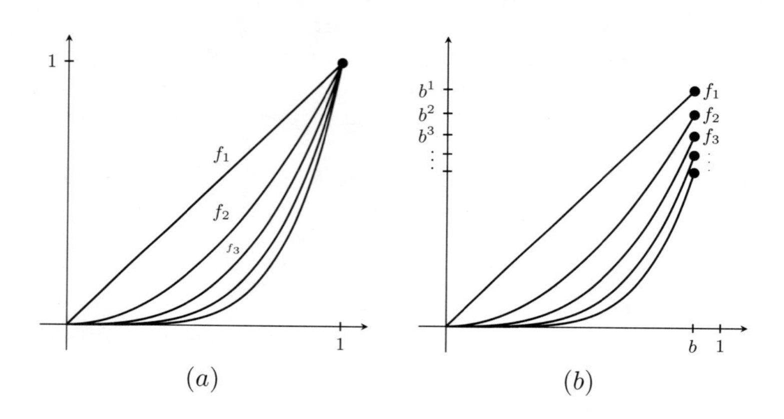

FIGURE 7.1.1
(a) Pointwise but not uniform convergence of $f_n(x)$ on $[0,1]$. (b) Uniform convergence of f_n on $[0,b]$.

(b) Let us now consider the domain $[0,b]$ for some $0 < b < 1$. Now since $b < 1$, we have $f_n(x) \to f(x) = 0$ for all $x \in [0,b]$. We study the relationship between ϵ and N as before on this different domain.

Let x be any number in $[0,b]$, and let $\epsilon > 0$ be given. Since each $f_n(x)$ is increasing[1], $0 \leq f_n(x) \leq f_n(b)$. So,

$$|f_n(x) - f(x)| = x^n \leq b^n.$$

Choose $N > \frac{\ln \epsilon}{\ln b}$. This makes $b^N < \epsilon$ (remember that $\ln b < 0$ since $0 < b < 1$, so the inequality reverses upon multiplication by $\ln b$). Since $0 < b < 1$, $b^n \leq b^N$ for $n \geq N$, this shows that for every $n \geq N$, we have $|f_n(x) - f(x)| < \epsilon$.

What is to be noticed in this example is that the N we found is one that "worked" for every $x \in [0,b]$, as opposed to previously where this N was forced to vary as our choice of x varied. ■

[1]One could find that $f_n'(x) \geq 0$ for $x \in [0,b]$ and use Theorem 6.17.

The previous example illustrates that the convergence of functions might not be "uniform", in the sense that a given ϵ may produce an N that might not make $|f_n(x)-f(x)| < \epsilon$ for every x, but perhaps only an initially chosen x. This stronger form of convergence is understandably called *uniform convergence.* An image of this stronger form of continuity is shown in Figure 7.1.2

Definition 7.5. Let $f_n : D \to \mathbb{R}$. We say that f_n *uniformly converges* to f on D if for every $\epsilon > 0$, there exists an $N \in \mathbb{N}$ so that if $n \geq N$, then

$$|f_n(x) - f(x)| < \epsilon$$

for every $x \in D$. This may be denoted as $f_n \rightrightarrows f$ on D.

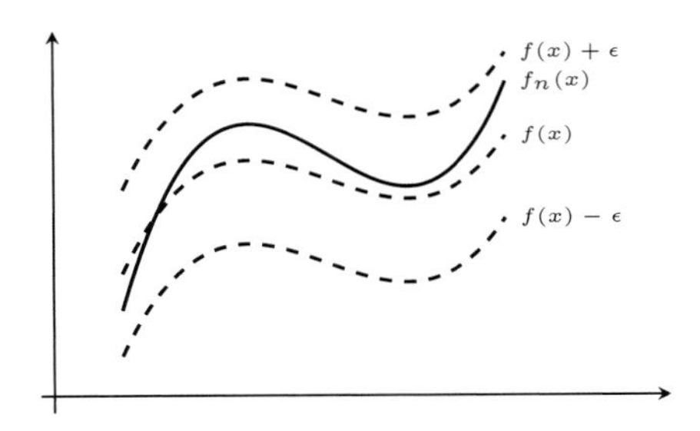

FIGURE 7.1.2
A uniformly convergent sequence of functions (f_n) must be between $f(x) - \epsilon$ and $f(x) + \epsilon$ for every $x \in D$.

Clearly, if a sequence of functions uniformly converges on D, then it converges pointwise on D. We determine in the following practice problem if any of the function sequences in Practice 7.3 converge uniformly.

Practice 7.6. Determine if the following sequences of functions converge uniformly on the given domain.

(a) $f_n(x) = \frac{x}{n}$ on $\mathbb{R}$ (see Figure 7.1.3).
(b) $f_n(x) = \frac{nx}{1+n^2x^2}$ on $\mathbb{R}$ (see Figure 7.1.4).
(c) $f_n(x) = x + \frac{x}{n}\sin(nx)$ on $[-a, a]$ (see Figure 7.1.5).

Methodology. *We show that the sequences in Parts (a) and (b) do not converge uniformly, and that the sequence in Part (c) does converge uniformly. In Parts (a) and (b), choose an ϵ and show that no such N exists according to Definition 7.5 by finding an x in the domain that violates the requirement that $|f_n(x) - f(x)| < \epsilon$. In Part (c) we estimate $|f_n(x) - f(x)|$ to determine an N that will suffice.*

Solution. (a) Let $\epsilon = 1$. If $(f_n(x))$ uniformly converges to $f(x) = 0$ on $\mathbb{R}$, then there exists an $N \in \mathbb{N}$ so that for every $n \geq N$ and $x \in \mathbb{R}$, we must have

$|\frac{x}{n}| < \epsilon$. In particular, this must be the case for $n = N$. Choose $x = 2N \in \mathbb{R}$, and notice that

$$|f_N(2N) - f(2N)| = 2 \geq \epsilon = 1.$$

Thus, the assumption that $(f_n(x))$ uniformly converges on $\mathbb{R}$ is false.

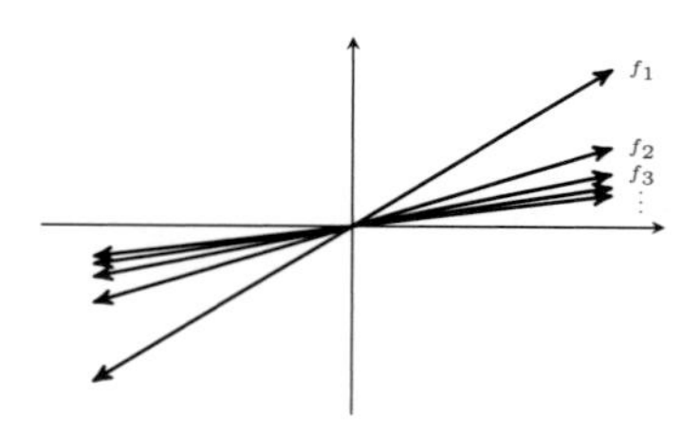

FIGURE 7.1.3
Some graphs of $f_n(x) = \frac{x}{n}$.

(b) Choose $\epsilon = \frac{1}{4}$. If $(f_n(x))$ uniformly converges to $f(x) = 0$ on $\mathbb{R}$, then there exists an $N \in \mathbb{N}$ so that for every $n \geq N$ and $x \in \mathbb{R}$, we must have $|\frac{x}{n}| < \epsilon$. In particular, this must be the case for $n = N$. Choose $x = \frac{1}{N} \in \mathbb{R}$, and notice that

$$\left| f_N\left(\frac{1}{N}\right) - f\left(\frac{1}{N}\right) \right| = \frac{1}{2} \geq \epsilon = \frac{1}{4}.$$

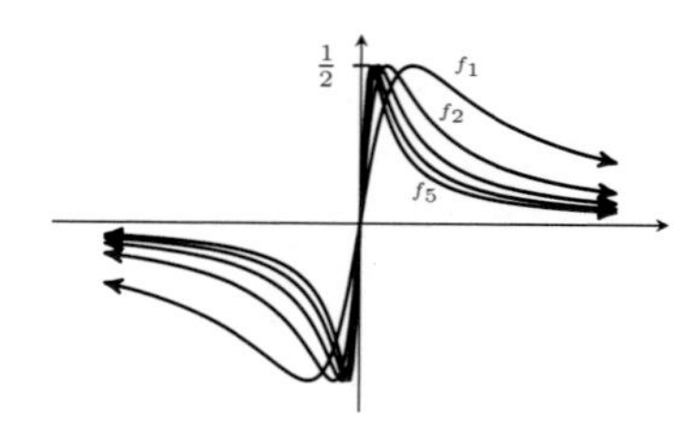

FIGURE 7.1.4
Some graphs of $f_n(x) = \frac{nx}{1+n^2x^2}$.

(c) Let $\epsilon > 0$ be given. Since $|\sin(nx)| \leq 1$, and $|x| \leq a$, we have

$$\begin{aligned} |f_n(x) - f(x)| &= |x + \tfrac{x}{n}\sin(nx) - x| \\ &= |\tfrac{x}{n}\sin(nx)| \\ &\leq \tfrac{|x|}{n} \\ &\leq \tfrac{a}{n}. \end{aligned}$$

Choose N so that $\frac{a}{N} < \epsilon$. Then for $n \geq N$, we have $\frac{a}{n} \leq \frac{a}{N}$, and so for every $x \in [-a, a]$, we have $|f_n(x) - f(x)| < \epsilon$.

■

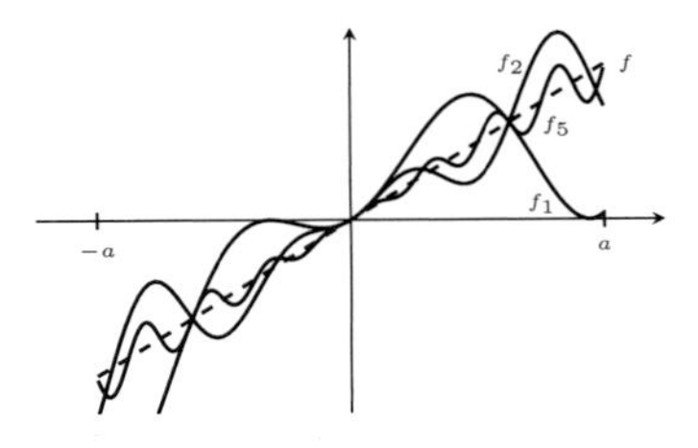

FIGURE 7.1.5
The graphs of f_1, f_2, and f_5 for the functions $f_n(x) = x + \frac{x}{n}\sin(nx)$ on $[-a, a]$.

We close this section with a useful characterization of uniform convergence, called the *Cauchy Criterion.*

Theorem 7.7 (Cauchy Criterion). *The sequence of functions $f_n : D \to \mathbb{R}$ uniformly converges on D if and only if for every $\epsilon > 0$ there exists an $N \in \mathbb{N}$ so that if $n, m \geq N$, then*

$$|f_n(x) - f_m(x)| < \epsilon.$$

Idea of proof. *If (f_n) uniformly converges to f on D, then we can find an N so that $|f_n(x) - f(x)| < \frac{\epsilon}{2}$, and use the triangle inequality to find $|f_n(x) - f_m(x)| < \epsilon$. Conversely, if this Cauchy Criterion holds, we can easily show that $(f_n(x))$ converges pointwise to some function since for a fixed $x \in D$, the sequence $(f_n(x))$ is Cauchy. We can then take the limit as $m \to \infty$ in $|f_n(x) - f_m(x)|$ to produce $|f_n(x) - f(x)|$ for some fixed m. Provided $|f_n(x) - f_m(x)| < \frac{\epsilon}{2}$, this will guarantee $|f_n(x) - f(x)| < \epsilon$ for all $x \in D$.*

Proof of Theorem 7.7. First suppose that (f_n) converges uniformly to f on D. Let $\epsilon > 0$ be given. Choose N so that if $n \geq N$, then $|f_n(x) - f(x)| < \frac{\epsilon}{2}$. Now let $n, m \geq N$. Then

$$|f_n(x) - f_m(x)| \leq |f_n(x) - f(x)| + |f_m(x) - f(x)| < \frac{\epsilon}{2} + \frac{\epsilon}{2} = \epsilon.$$

Conversely, suppose that the Cauchy Criterion holds: for every $\epsilon > 0$, there exists an N so that for every $n, m \geq N$ we have $|f_n(x) - f_m(x)| < \epsilon$ for every $x \in D$. We first show that (f_n) converges pointwise to some function f, and then go on to prove that this convergence is uniform.

Let $x \in D$ be given. The Cauchy Criterion illustrates that $(f_n(x))$ is a Cauchy sequence, and therefore converges to some number $f(x)$. Thus, (f_n) converges pointwise to f.

To show that this convergence is uniform, let $\epsilon > 0$ be given. Choose N so that if $n, m \geq N$ then $|f_n(x) - f_m(x)| < \frac{\epsilon}{2}$ for every $x \in D$. Choose any $n \geq N$. For any $x \in D$, we compute

$$\lim_{m\to\infty} |f_n(x) - f_m(x)| = |f_n(x) - f(x)| \leq \frac{\epsilon}{2}.$$

We have shown that if $n \geq N$, then $|f_n(x) - f(x)| \leq \frac{\epsilon}{2} < \epsilon$ for all $x \in D$, and so (f_n) converges uniformly to f on D. □

7.1.3 Review of important concepts from Section 7.1

1. Pointwise convergence of sequences of functions is defined in Definition 7.1, and serves as the most basic form of convergence of a sequence of functions.
2. Uniform convergence, defined in Definition 7.5, is a stronger form of convergence. Example 7.4 illustrates the main difference between uniform and pointwise convergence.
3. The Cauchy Criterion (Theorem 7.7) is a useful characterization of uniform convergence.

Exercises for Section 7.1

1. Show that $f_n(x) = x^n$ does not converge uniformly to the function

$$f(x) = \begin{cases} 0 & \text{for } 0 \leq x < 1 \\ 1 & \text{for } x = 1 \end{cases}$$

on $[0, 1]$.

2. Show that $f_n(x) = \frac{x}{n}$ converges uniformly on every bounded subset of $\mathbb{R}$. (Compare this with Practices 7.3 and 7.6.)
3. Show that $f_n(x) = \frac{nx}{1+n^2x^2}$ converges uniformly to the 0 function on $[1, \infty)$. (Compare this with Practices 7.3 and 7.6.)

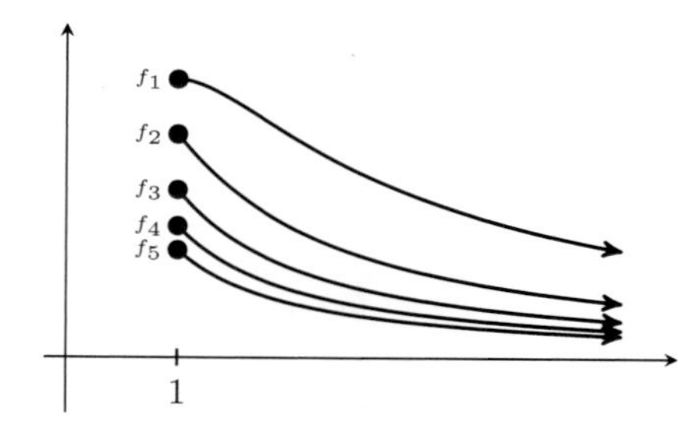

4. Show that $f_n(x) = x + \frac{x}{n}\sin(nx)$ does not converge uniformly to $f(x) = x$ on $\mathbb{R}$. (Compare this with Practices 7.3 and 7.6.)
5. Suppose $f_n, f : D \to \mathbb{R}$. Let $M_n = \sup_{x \in D}\{|f_n(x) - f(x)|\}$. Prove that $(f_n(x))$ converges uniformly to f on D if and only if $M_n \to 0$.
6. Prove that $f_n(x) = \frac{x}{1+nx^2}$ converges uniformly to the 0 function on $[0, \infty)$. *[Hint: Use calculus to find a global maximum M_n of f_n and use Exercise 5.]*

7. Prove that $f_n(x) = \frac{x}{1+n^2x^2}$ converges uniformly to the 0 function on $[0, \infty)$. *[Hint: Use calculus to find a global maximum M_n of f_n and use Exercise 5.]*

8. Suppose (f_n) converges uniformly on D, and that each f_n is bounded. Prove that (f_n) is *uniformly bounded*. That is, prove that there exists an M so that $|f_n(x)| \leq M$ for every $x \in D$ and for every n. Deduce that the limit function $f(x) = \lim_{n\to\infty} f_n(x)$ must be bounded by M on D as well.

9. Determine if $f_n(x) = \frac{n}{1+x^n}$ converges uniformly on $(0, 1)$.

10. Prove that $f_n(x) = nx(1 - x^2)^n$ does not converge uniformly on $[0, 1]$.

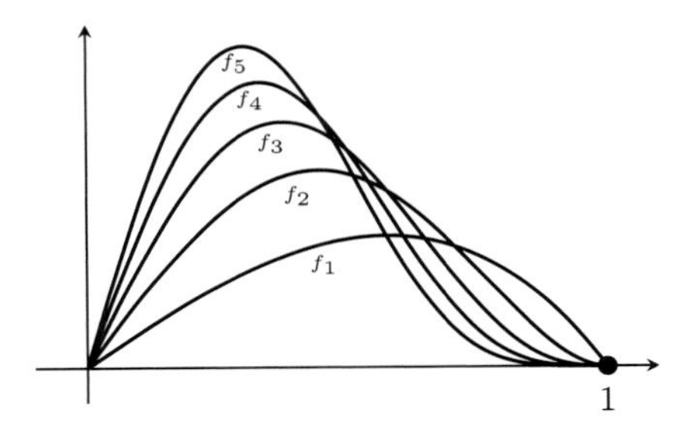

11. Prove that $f_n(x) = nxe^{-nx^2}$ does not converge uniformly on $[0, \infty)$, but does converge uniformly on $[a, \infty)$ for any $a > 0$.

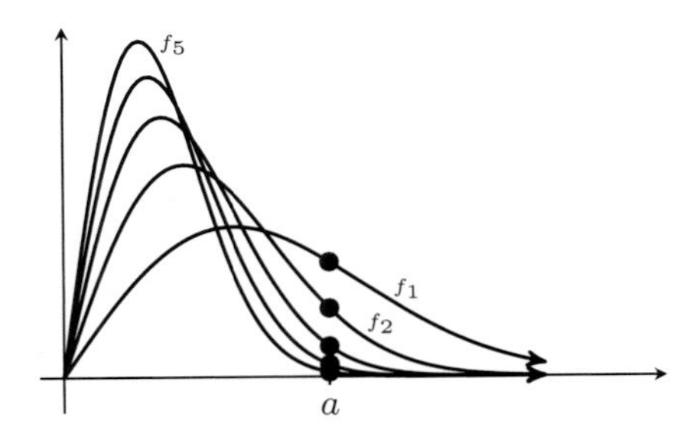

12. Suppose $f_n, g_n : D \to \mathbb{R}$. If (f_n) and (g_n) are bounded (see Exercise 8) and uniformly converge on D, show that (f_ng_n) uniformly converges on D. Then, find an example that shows that the hypothesis of uniform boundedness is required for this result to hold. *[Hint: Use Exercise 8. For the second part, consider the sequence $f_n(x) = x + \frac{1}{n}$.]*

7.2 Consequences of Uniform Convergence

The goal of this section is to investigate what properties a convergent sequence of functions passes on to its limit. We shall see that almost nothing gets passed along when the convergence is pointwise. By contrast, much gets passed along when the convergence is uniform. This "passing on" of data through convergence is sometimes referred to as a *preservation theorem*, and this question is sometimes posed as "what properties are preserved through convergence?".

We start by showing that continuity is preserved through uniform convergence in Theorem 7.9. Dini's Theorem (Theorem 7.11) is a partial converse to this. Definite integration is preserved through uniform convergence (Theorem 7.10), although since we do not discuss integration in this book, we do not provide a proof of this. Differentiation is not necessarily preserved through uniform convergence (Example 7.12), although it is if we consider a different set of hypotheses (see Theorem 7.13).

7.2.1 Continuity is preserved

Our first goal is to determine when continuity is passed on through convergence. That is, if $f_n : D \to \mathbb{R}$ are all continuous, and (f_n) converges to f (either pointwise or uniformly), then is f continuous on D as well?

The following example illustrates that the answer is no when the convergence is pointwise.

Example 7.8. This example illustrates that if $f_n : D \to \mathbb{R}$ converges to f pointwise, and each f_n is continuous on D, then f need not be continuous on D.

Let $f_n(x) = x^n$ on $D = [0, 1]$. We saw in the last section that (f_n) converges pointwise but not uniformly to the function

$$f_n(x) = x^n \to f(x) = \begin{cases} 0 & \text{for } x \in [0,1) \\ 1 & \text{for } x = 1 \end{cases} .$$

Note that each of the functions f_n is continuous, and that f is not. ■

The next theorem shows that if the convergence of (f_n) to f is uniform, then continuity is a property that is preserved. Thus, we can conclude that the convergence in Example 7.8 is not uniform: if it were, then the limit function f would have to be continuous, and it is not.

Theorem 7.9. *Let $f_n : D \to \mathbb{R}$ uniformly converge to f on D. If each f_n is continuous on D, then f is continuous on D.*

Idea of proof. *For any $c \in D$, we need to estimate*[2] $|f(x) - f(c)|$. *We can*

[2]We are really proving that if each f_n is continuous at any $c \in D$, and uniformly converges to f on D, then f is continuous at c.

do so by considering $|f(x) - f_n(x)|$, $|f_n(x) - f_n(c)|$, and $|f_n(c) - f(c)|$. The first of these can be made small since the convergence is uniform. The second can be made small since each f_n is continuous. The last of these can be made small since the convergence of $f_n \to f$ is at least pointwise, but will be taken care of in our first application of the uniform convergence hypothesis.

Proof of Theorem 7.9. Let $\epsilon > 0$ be given, and choose any $c \in D$. Select N so that if $n \geq N$, then $|f_n(x) - f(x)| < \frac{\epsilon}{3}$ for every $x \in D$. Choose any $n \geq N$. Since f_n is continuous at c, there exists a δ so that if $|x - c| < \delta$ and $x \in D$, then $|f_n(x) - f_n(c)| < \frac{\epsilon}{3}$.

Now let $x \in D$ with $|x - c| < \delta$. We write

$$\begin{aligned} |f(x) - f(c)| &\leq |f(x) - f_n(x)| + |f_n(x) - f_n(c)| + |f_n(c) - f(c)| \\ &< \tfrac{\epsilon}{3} + \tfrac{\epsilon}{3} + \tfrac{\epsilon}{3} \\ &= \epsilon. \end{aligned}$$

□

It is important to point out a subtle observation about the conclusion of Theorem 7.9. When the convergence is uniform, then the continuity of each f_n implies the continuity of f. If c is a limit point of D and we express $f(x) = \lim_{n\to\infty} f_n(x)$, the we can express $f(c)$ in two ways:

$$\begin{aligned} f(c) &= \lim_{x\to c}[f(x)] \\ &= \lim_{x\to c}\left[\lim_{n\to\infty} f_n(x)\right], \text{ and} \\ f(c) &= \lim_{n\to\infty}[f_n(c)] \\ &= \lim_{n\to\infty}\left[\lim_{x\to c} f_n(x)\right]. \end{aligned}$$

In other words, when each f_n is continuous, and the convergence of (f_n) to f is uniform, then

$$\lim_{x\to c}\lim_{n\to\infty} f_n(x) = \lim_{n\to\infty}\lim_{x\to c} f_n(x).$$

This exchange of limits is indicative of another sentiment regarding convergence of functions: when can the $\lim_{n\to\infty}$ "pass through" another operation? In the case of Theorem 7.9, this limit can pass through another limit. One can show that the hypothesis of uniform convergence is enough to have this limit pass through a definite integral. This book does not cover integration, and so the proof of this theorem is outside our scope.

Theorem 7.10. *If each $f_n : [a,b] \to \mathbb{R}$ is integrable and $f_n \to f$ uniformly, then f is integrable and*

$$\lim_{n\to\infty}\int_a^b f_n(x)dx = \int_a^b \lim_{n\to\infty} f_n(x)dx = \int_a^b f(x)dx.$$

The next most common "operation" is that of differentiation. We will show in the next subsection that uniform convergence is *not* enough to interchange the limit operation with that of differentiation, although under reasonable hypotheses, it does (see Theorem 7.13).

Before moving on to this result, we present Dini's Theorem, which is a partial converse to Theorem 7.9.

Theorem 7.11 (Dini's Theorem). *Suppose $f_n : [a,b] \to \mathbb{R}$ is a sequence of continuous functions converging pointwise to a continuous function f. If $f_n(x) \geq f_{n+1}(x)$ for all $x \in [a,b]$ and $n \in \mathbb{N}$, then (f_n) converges to f uniformly.*

Idea of proof. *We prove the theorem by contradiction: if the convergence is not uniform, then there is some $\epsilon > 0$ so that for every $N \in \mathbb{N}$, there is some $n \geq N$ and $x \in [a,b]$ so that $|f_n(x) - f(x)| \geq \epsilon$. We repeatedly use this condition to produce a sequence $x_n \in [a,b]$ with $|f_n(x_n) - f(x)| \geq \epsilon$. By the Bolzano-Weierstrass Theorem (Theorem 3.84), there is a convergent subsequence $x_{n_k} \to c \in [a,b]$. We find our contradiction by noting that since the convergence is not uniform*[3]*, $|f_{n_m}(c) - f(c)| \geq \epsilon$. Although, $f_{n_m}(c) - f(c) \to 0$ since (f_n) converges to f pointwise.*

Proof of Theorem 7.11. Notice that since $f_n(x) \geq f_{n+1}(x)$, and (f_n) converges pointwise to f, we must have $f_n(x) \geq f(x)$ for all $x \in [a,b]$. Thus, $|f_n(x) - f(x)| = f_n(x) - f(x)$ for all x.

Suppose this convergence is not uniform. Then there exists an $\epsilon > 0$ so that for every $N \in \mathbb{N}$, there is some $n \geq N$ and $x \in [a,b]$ with $f_n(x) - f(x) \geq \epsilon$. Let $N = 1$, and find $n_1 \geq N$ and $x_1 \in [a,b]$ that satisfies this condition: $f_{n_1}(x_1) - f(x_1) \geq \epsilon$. Next choose $N_2 = n_1 + 1$ and find $n_2 \geq N_2 > n_1$ and $x_2 \in [a,b]$ that satisfies the same condition. Continuing in this manner, we have a strictly increasing (n_k), and $x_k \in [a,b]$ with $f_{n_k}(x_k) - f(x_k) \geq \epsilon$.

By the Bolzano-Weierstrass Theorem (Theorem 3.84), there exists a convergent subsequence $x_{k_i} \to c \in [a,b]$. Fix any $i \in \mathbb{N}$, and consider $j \geq i$. Since $n_j \geq n_i$, we have $f_{n_{k_i}}(x_{k_j}) \geq f_{n_{k_j}}(x_{k_j})$, and so

$$f_{n_{k_i}}(x_{k_j}) - f(x_{k_j}) \geq f_{n_{k_j}}(x_{k_j}) - f(x_{k_j}) \geq \epsilon. \tag{7.1}$$

Now consider taking the limit as $j \to \infty$ in Equation (7.1). Since each $f_{n_{k_i}}$ and f is continuous, the inequality in Equation (7.1) tells us that

$$f_{n_{k_i}}(x_{k_j}) - f(x_{k_j}) \to f_{n_{k_i}}(c) - f(c) \geq \epsilon.$$

But (f_n) converges to f pointwise, and so $f_n(c) \to f(c)$. As a subsequence (of a subsequence), we must have $f_{n_{k_i}}(c) \to f(c)$ as $i \to \infty$, and so eventually, we must have $f_{n_{k_i}}(c) - f(c) < \epsilon$. This contradicts the assumption that this convergence is not uniform and completes the proof. □

[3]Here is where we will use the hypothesis that $f_n(x) \geq f_{n+1}(x)$ for all n and x.

7.2.2 Differentiation is sometimes preserved

Our final topic of this section concerns the role of differentiation in the convergence of functions. It would be nice if differentiation was "preserved" through uniform convergence (as the property of continuity is). That is, if (f_n) converges uniformly to f, it would be nice if

$$\lim_{n\to\infty} f_n'(x) = \lim_{n\to\infty}\left(\frac{d}{dx}f_n(x)\right) = \frac{d}{dx}\left(\lim_{n\to\infty} f_n(x)\right) = \frac{d}{dx}f(x) = f'(x).$$

Regrettably, this need not always be the case. The next example presents a sequence of functions (f_n) that uniformly converges to some function f, but that (f_n') does not even converge pointwise to f'.

Example 7.12. Let $f_n(x) = \frac{x}{1+nx^2}$ on $\mathbb{R}$. It was shown in Exercise 6 of Section 7.1 that (f_n) converges to $f(x) = 0$ uniformly on $\mathbb{R}$. We show here that (f_n') does not even converge pointwise on $\mathbb{R}$ to $f'(x) = 0$.

We differentiate f_n to find

$$f_n'(x) = \frac{1-nx^2}{(1+nx^2)^2}.$$

If $x \neq 0$, then $f_n'(x) \to 0$, while if $x = 0$, $f_n'(0) = 1 \to 1$. So

$$\lim_{n\to\infty} f_n'(x) = \begin{cases} 0 & \text{if } x \neq 0 \\ 1 & \text{if } x = 0 \end{cases},$$

Figure 7.2.1, notice how the value of $f_n'(0)$ seems to be tending to 1, which is evidenced by our computation of $\lim_{n\to\infty} f_n'(0) = 1$ above. But this is not equal to $f'(x) = 0$ for all x. Thus, uniform convergence of (f_n) to f (and differentiability) is not enough to deduce that (f_n') converge to f', even pointwise! ■

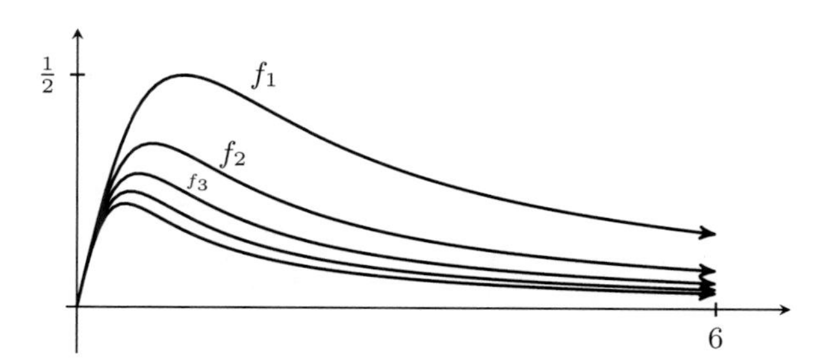

FIGURE 7.2.1
The functions $f_n(x) = \frac{x}{1+nx^2}$ in Example 7.12 converge uniformly to the 0 function on $\mathbb{R}$, but $(f_n'(x))$ does not even converge pointwise to $f'(x)$.

There are different hypotheses one needs before $(f_n'(x))$ converges to $f'(x)$, as outlined in Theorem 7.13.

Theorem 7.13. *Suppose (f_n) is a sequence of differentiable functions on a bounded interval I. Suppose also that (f_n') converges uniformly on I, and that there exists a $c \in I$ for which $(f_n(c))$ converges. Then:*

1. *(f_n) converges uniformly on I to some function $f : I \to \mathbb{R}$,*
2. *f is differentiable, and*
3. *$f'(x) = \lim_{n\to\infty} f_n'(x)$ for all $x \in I$.*

Idea of proof. *To prove the first statement, we use the Cauchy Criterion for uniform convergence (Theorem 7.7). To meet the Cauchy criterion, we use the Mean Value Theorem on the function $f_n(x) - f_m(x)$, and the boundedness of I. Both (2) and (3) are proved at the same time as we compute the limit of the difference quotient for f. We use our estimate from Assertion (1) and a new sequence of functions (the difference quotients of f_n and their limit) to compute $f'(p)$ for some $p \in I$.*

Proof of Theorem 7.13. Suppose that I is a bounded interval, with $\sup(I) = b$ and $\inf(I) = a$. Let $\epsilon > 0$ be given. Since the sequence $(f_n(c))$ converges, it is Cauchy, so there exists an N_1 so that for all $n, m \geq N_1$ we have

$$|f_n(c) - f_m(c)| < \frac{\epsilon}{2}.$$

Since (f_n') uniformly converges, there exists an N_2 so that for every $n, m \geq N_2$ and any $t \in I$, we have

$$|f_n'(t) - f_m'(t)| < \frac{\epsilon}{2(b-a)}.$$

Let $N = \max\{N_1, N_2\}$, and let $n, m \geq N$. For any $x \neq y$ in I, the function $f_n - f_m$ meets the hypotheses of the Mean Value Theorem (Theorem 6.14), so there exists a t between x and y with

$$(f_n(x) - f_m(x)) - (f_n(y) - f_m(y)) = (f_n'(t) - f_m'(t))(x - y).$$

But $|f_n'(t) - f_m'(t)| < \frac{\epsilon}{2(b-a)}$, and since $x, y \in I$ we must have $|x - y| \leq |a - b|$. So,

$$\begin{aligned} |(f_n(x) - f_m(x)) - (f_n(y) - f_m(y))| &= |f_n'(t) - f_m'(t)||x - y| \\ &< \tfrac{\epsilon}{2(b-a)}(b - a) = \tfrac{\epsilon}{2}. \end{aligned} \qquad (7.2)$$

Now using $y = c$ in Equation 7.2, we show that (f_n) meets the Cauchy criterion for uniform convergence (Theorem 7.7) and thus uniformly converges. For any $n, m \geq N$, and $x \in I$, we have

$$\begin{aligned} |f_n(x) - f_m(x)| &\leq |(f_n(x) - f_m(x)) - (f_n(c) - f_m(c))| + |f_n(c) - f_m(c)| \\ &< \tfrac{\epsilon}{2} + \tfrac{\epsilon}{2} \\ &= \epsilon. \end{aligned}$$

We now prove Assertions (2) and (3). Since (f_n) converges uniformly, for every $x \in I$ we can define $f(x) = \lim_{n\to\infty} f_n(x)$ as its limit. Choose any $p \in I$. Define the following functions:

$$g_n(x) = \begin{cases} \frac{f_n(x)-f_n(p)}{x-p} & \text{for } x \neq p \\ f_n'(p) & \text{for } x = p \end{cases} \quad \textit{and} \quad g(x) = \begin{cases} \frac{f(x)-f(p)}{x-p} & \text{for } x \neq p \\ L & \text{for } x = p, \end{cases}$$

where $L = \lim_{n\to\infty} f_n'(p)$. Note that since f_n is differentiable, g_n is continuous at p, so that $g_n(x) \to f_n'(p)$ as $x \to p$. If we now let $y = p$ in Equation 7.2, we have for any $p \neq x \in I$:

$$\frac{|(f_n(x) - f_m(x)) - (f_n(p) - f_m(p))|}{|x - p|} = |f_n'(t) - f_m'(t)| < \frac{\epsilon}{2(b-a)}.$$

But for $x \neq p$,

$$\begin{aligned} \frac{|(f_n(x)-f_m(x))-(f_n(p)-f_m(p))|}{|x-p|} &= \left|\frac{(f_n(x)-f_m(x))-(f_n(p)-f_m(p))}{x-p}\right| \\ &= \left|\frac{f_n(x)-f_n(p)}{x-p} - \frac{f_m(x)-f_m(p)}{x-p}\right| \\ &= |g_n(x) - g_m(x)| \end{aligned}$$

Let $\tilde{\epsilon} > 0$ be given. Select $\epsilon = \tilde{\epsilon} \cdot 2(b-a)$, so that $\frac{\epsilon}{2(b-a)} = \tilde{\epsilon}$. Choose N as we did at the beginning of the proof (with this ϵ we have just found). Now for $n, m \geq N$, we have

$$|g_n(x) - g_m(x)| < \frac{\epsilon}{2(b-a)} = \tilde{\epsilon}$$

for all $x \in I$. Thus, (g_n) meets the Cauchy criterion for convergence (Theorem 7.7) and thus uniformly converges to some function g on I. Since (g_n) is continuous at p and continuity is preserved by uniform continuity (Theorem 7.9), g is continuous at p (see the footnote on Page 360). Recall that $g(p)$ was defined to be $\lim_{n\to\infty} f_n'(p)$. Then

$$\begin{aligned} \lim_{x\to p} \frac{f(x)-f(p)}{x-p} &= \lim_{x\to p} g(x) \\ &= g(p) \\ &= \lim_{n\to\infty} f_n'(p) \end{aligned}$$

Thus, f is differentiable at any point $p \in I$, and $\lim_{n\to\infty} f_n'(p) = f'(p)$. □

7.2.3 Review of important concepts from Section 7.2

1. This section discusses properties that both are and are not preserved by either pointwise or uniform convergence of functions. Generally, pointwise convergence does not preserve very much. For instance, see Example 7.8 that illustrates continuity is not preserved by pointwise convergence.

2. Uniform convergence preserves continuity (see Theorem 7.9).
3. Dini's Theorem (Theorem 7.11) is a partial converse to Theorem 7.9, and concludes that (f_n) converges uniformly, provided each f_n is continuous, their limit is continuous, and that $f_n(x) \geq f_{n+1}(x)$ for all x and n.
4. Definite integration preserved by uniform convergence (Theorem 7.10), although the proof of this is outside the scope of this book and is not included.
5. Differentiation is *not* necessarily preserved by uniform convergence (see Example 7.12). Although, it is under different hypotheses (see Theorem 7.13).

Exercises for Section 7.2

1. Give an example of a sequence of continuous functions (f_n) that converges pointwise but not uniformly to a continuous function f.
2. Let $0 < b < 1$. Use Dini's Theorem (Theorem 7.11) to prove that $f_n(x) = x^n$ converges uniformly on $[0, b]$. Why does Dini's Theorem not apply to this sequence of functions on $[0, 1]$?
3. Find an example of a sequence (f_n) of discontinuous functions that uniformly converges to a continuous function f on $[0, 1]$. By Theorem 7.9, continuity of the (f_n) is preserved, but this exercise shows that discontinuity is *not* preserved[4].
4. Suppose $f_n : D \to \mathbb{R}$ uniformly converges to f. Let $x_n \to c \in D$, where $x_n \in D$. Prove that if f_n is continuous on D, then $f_n(x_n) \to f(c)$. *[Hint: Consider $|f_n(x_n) - f(x_n)|$ and $|f(x_n) - f(c)|$.]*
5. Show that $f_n(x) = \frac{1}{nx+1}$ satisfies $f_n(x) \geq f_{n+1}(x)$ for all n and $x \in (0, 1)$, and that (f_n) converges to the 0 function on $(0, 1)$, but that the convergence is not uniform. (This demonstrates that the hypothesis in Dini's Theorem (Theorem 7.11) must be of the form $[a, b]$.)
6. Suppose f is uniformly continuous on $\mathbb{R}$, and set $f_n(x) = f(x + \frac{1}{n})$. Prove that (f_n) converges uniformly to f on $\mathbb{R}$. Give an example that shows that there are counterexamples if f is only assumed to be continuous on $\mathbb{R}$.
7. Show $f_n(x) = \frac{nx}{1+n^2x^2}$ converges on $\mathbb{R}$ to the 0 function pointwise, but not uniformly, by using Exercise 4 and $x_n = \frac{1}{n}$.
8. Show that $f_n(x) = \frac{x^n}{n}$ converges uniformly to the 0 function on $[0, 1]$, but that (f_n') converges pointwise but not uniformly on $[0, 1]$.

---

[4]Thus, as it appears on page 76 of [5], "Uniform convergence preserves good behavior, not bad behavior."

7.3 Infinite Series of Functions

In the same way that convergence of a series of real numbers is stated in terms of convergence of its sequence of partial sums, we can define convergence of a series of functions in terms of the convergence of its sequence of partial sums. Since a sequence of functions can either converge pointwise or uniformly (or not at all), we can define pointwise convergence and uniform convergence of the series of functions $\sum_{n=1}^{\infty} f_n(x)$.

We cannot interpret each of the "preservation theorems" in the last section in the context of series of functions. Often, the fact that the partial sums are finite sums, these preservation theorems have nicely expected outcomes. For example, a uniformly convergent series of continuous functions converges to a continuous function (Theorem 7.17).

In addition to this result, we present a Cauchy Criterion for uniform convergence of series of functions in Theorem 7.16, and discuss differentiation and antidifferentiation of uniformly convergent series in Theorem 7.20 and Corollary 7.21. We close the section with a useful result known as the Weierstrass $M-$test, which gives straightforward conditions that can determine if a series of functions is uniformly convergent.

7.3.1 Defining infinite series of functions

Given a sequence $f_n : D \to \mathbb{R}$ of functions on D, we can create the infinite series of functions $\sum_{n=1}^{\infty} f_n(x)$ by considering this as a function: $x \mapsto \sum_{n=1}^{\infty} f_n(x)$. Given any $x \in D$, the only real concern in this construction would be whether or not the infinite series of real numbers $\sum_{n=1}^{\infty} f_n(x)$ converges, thus providing a well-defined output. The notion of pointwise and uniform convergence of a series of functions is approached in the same way we considered convergence of infinite series of real numbers: equate convergence of the series with convergence of the associated sequence of partial sums.

Definition 7.14. Given $f_n : D \to \mathbb{R}$, define the *partial sum sequence*

$$s_n(x) = f_1(x) + f_2(x) + \cdots + f_n(x) = \sum_{k=1}^{n} f_k(x).$$

The series of functions *converges pointwise* to f on D if (s_n) converges pointwise to f on D. This series *converges uniformly* on D if (s_n) converges uniformly on D. In either type of convergence, we write

$$f(x) = \sum_{n=1}^{\infty} f_n(x).$$

We illustrate pointwise convergence with a familiar series of functions, even if at this point we have not considered them as such.

Example 7.15. Let $f_n(x) = x^n$ on the interval $I = (-1, 1)$. In Example 4.5 we showed that if $x \in (-1, 1)$, then $\sum_{n=0}^{\infty} x^n$ converges to $f(x) = \frac{1}{1-x}$. Figure 7.3.1 shows the partial sums $s_n(x)$ converging to $f(x) = \frac{1}{1-x}$. This convergence is not uniform on $(-1, 1)$, since if it were, the limit function $\frac{1}{1-x}$ would be bounded on $(-1, 1)$ since each s_n is bounded by Exercise 8 of Section 7.1. It is clearly not bounded, as $\lim_{x \to 1^-} \frac{1}{1-x} = \infty$. We will show in Practice 7.23 that the convergence is in fact uniform on any interval of the form $[-r, r] \subseteq (-1, 1)$ for $0 < r < 1$. ■

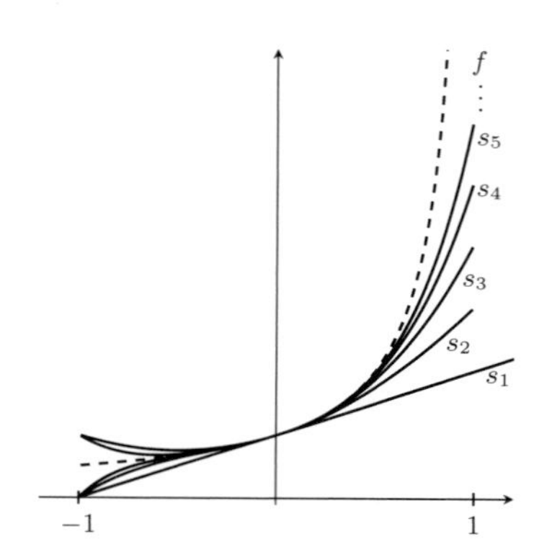

FIGURE 7.3.1
The partial sums s_n of $\sum_{n=0}^{\infty} x^n$ converge pointwise to $f(x) = \frac{1}{1-x}$ on $(-1, 1)$.

As it is with series of real numbers, convergence of $\sum_{n=1}^{\infty} f_n(x)$ is linked with convergence of its sequence $(s_n(x))$ of partial sums, and so we can automatically read off several important results found in the previous section and applied to $(s_n(x))$ that will be useful. First, we list the Cauchy criterion for uniform convergence of infinite series of functions. As in Chapter 4, when the summation limits are unimportant, we write $\sum f_n(x)$ when there is no chance for confusion.

Theorem 7.16 (Cauchy Criterion for Series of Functions)**.** *Suppose $f_n : D \to \mathbb{R}$. The series $\sum f_n(x)$ uniformly converges on D if for every $\epsilon > 0$, there exists an $N \in \mathbb{N}$, so that if $n > m \geq N$, then*

$$\sum_{k=m+1}^{n} f_n(x) < \epsilon$$

for every $x \in D$.

Idea of proof. *We apply the Cauchy Criterion for uniform convergence of sequence of functions (Theorem 7.7) to the partial sum sequence $s_n(x)$, and the result is immediate.*

Proof of Theorem 7.16. Let $s_n(x) = \sum_{k=1}^{n} f_k(x)$ be the partial sum sequence of $\sum_{n=1}^{\infty} f_n(x)$. By definition, the series converges uniformly on D if and only if $(s_n(x))$ converges uniformly on D. By Theorem 7.7, this occurs if and only if for every $\epsilon > 0$, there exists an $N \in \mathbb{N}$ so that for every $n, m \geq N$, we have $|s_n(x) - s_m(x)| < \epsilon$ for all $x \in D$. Without loss of generality, assume that $n > m$. But then

$$\left| \sum_{k=m+1}^{n} f_n(x) \right| = |s_n(x) - s_m(x)| < \epsilon.$$

□

7.3.2 Preservation theorems adapted to series of functions

The "preservation theorems" from the last section can be applied to the partial sum sequence $(s_n(x))$ when we place suitable conditions on the $(f_n(x))$ that define the series. The first of these results concerns continuity.

Theorem 7.17. *Suppose $f_n : D \to \mathbb{R}$ is a sequence of continuous functions, and $\sum f_n$ converges uniformly to f on D. Then f is continuous.*

Idea of proof. *If each f_n is continous, then as a finite sum of continuous functions, $s_n(x) = \sum_{k=1}^{n} f_k(x)$ is also continuous for each n. The assumption that $\sum f_n(x)$ converges uniformly to f on D is, by definition, the same assumption that $(s_n(x))$ converges uniformly to f on D. Theorem 7.9 now guarantees that f is continuous.*

Proof of Theorem 7.17. Let $s_n(x) = \sum_{k=1}^{n} f_k(x)$ be the partial sum sequence of $\sum_{n=1}^{\infty} f_n(x)$. Since every f_n is continuous, as a finite sum of continuous functions s_n is continuous as well. Since $\sum f_n(x)$ converges uniformly to f on D, by definition, (s_n) converges uniformly to f on D. So, by Theorem 7.9, f is continuous on D. □

The following practice concerns power series, which we consider in the next section.

Practice 7.18. Let (a_n) be a sequence of real numbers, and define $f_n(x) = a_n x^n$. Let D be any set so that $\sum a_n x^n = f(x)$ uniformly converges on D. Prove that $f(x)$ is continuous on D.

Methodology. *The idea is to notice that $f_n(x) = a_n x^n$ is just a polynomial, and is therefore continuous on any domain. By Theorem 7.17, $\sum a_n x^n$ is continuous.*

Solution. As a polynomial, $f_n(x) = a_n x^n$ is continuous. Therefore, since we have assumed that $\sum f_n(x) = f(x)$ uniformly converges on D, it follows from Theorem 7.17 that $f(x)$ is continuous. ■

After presenting the fact that continuity is preserved through uniform convergence in the last section (Theorem 7.9), we paused to point out how this result equated to being able to exchange certain operations. We do the same here and point out that if a series of functions uniformly converges, Theorem 7.17 allows for an interchange of a limit and the summation. If f_n is continuous, $\sum f_n(x)$ converges uniformly to f on D, and $c \in D$ is a limit point, then

$$\begin{aligned} f(c) &= \lim_{x\to c} f(x) \\ &= \lim_{x\to c}\left(\sum f_n(x)\right), \text{ and} \\ f(c) &= \sum f_n(c) \\ &= \sum \left(\lim_{x\to c} f_n(x)\right). \end{aligned}$$

Thus, under the hypotheses of continuity and uniform convergence of the series, the limit and the summation can be interchanged.

The next interchanging of operations that we considered last section concerned definite integration. In the event $\sum f_n(x)$ converges uniformly, the sum and the integral may be interchanged and one may "integrate term-by-term." Since integration is not covered in this book, we do not prove this result.

Theorem 7.19. *If each $f_n : [a,b] \to \mathbb{R}$ is integrable, and $\sum_{n=1}^{\infty} f_n(x)$ converges uniformly to $f(x)$ on $[a,b]$, then f is integrable, and*

$$\int_a^b f(x)dx = \int_a^b \sum_{n=1}^{\infty} f_n(x)dx = \sum_{n=1}^{\infty} \int_a^b f_n(x)dx.$$

Our next hope is that one may "differentiate $\sum f_n(x)$ term-by-term," but as Theorem 7.13 points out, we will need to assume something different than just the differentiability of each f_n. Theorem 7.20 provides these conditions which are essentially taken directly from the corresponding differentiation result in the previous section (Theorem 7.13).

Theorem 7.20. *Suppose (f_n) is a sequence of differentiable functions on a bounded interval I. Suppose also that $\sum f_n'(x)$ converges uniformly on I, and that there exists a $c \in I$ for which $\sum f_n(c)$ converges. Then:*

1. *$\sum f_n(x)$ converges uniformly on I to some function $f : I \to \mathbb{R}$,*
2. *f is differentiable, and*
3. *$f'(x) = \sum f_n'(x)$.*

Idea of proof. *This result will follow immediately by applying Theorem 7.13 to the sequence $s_n(x) = \sum_{k=1}^{n} f_k(x)$ of partial sums.*

Proof of Theorem 7.20. Let $s_n(x) = \sum_{k=1}^{n} f_k(x)$ be the sequence of partial sums of $\sum f_n(x)$. We verify that (s_n) meets the hypotheses of Theorem 7.13 on the bounded interval I. As the finite sum of differentiable functions, each s_n is differentiable. By definition, our assumption that $\sum f_n'(x)$ converges uniformly on I is equivalent to the uniform convergence of $\sum_{k=1}^{n} f_k'(x) =$

$\frac{d}{dx}[\sum_{k=1}^{n} f_k(x)] = s_n'(x)$. Finally, the existence of a $c \in I$ for which $\sum f_n(c)$ converges is equivalent to the convergence of $(s_n(c))$. Now that we have verified the hypotheses of Theorem 7.13, the following conclusions can be drawn:

1. $(s_n(x))$ converges uniformly to some function f. That is, $\sum f_n(x)$ converges uniformly on I to some function f.
2. $f(x) = \sum f_n(x)$ is differentiable, and
3. $f'(x) = \lim_{n\to\infty} s_n'(x) = \lim_{n\to\infty} \sum_{k=1}^{n} f_k'(x) = \sum_{n=1}^{\infty} f_n'(x)$.

□

Given a series that is known to uniformly converge to a function, we can integrate each term to find an antiderivative for this function.

Corollary 7.21. *Suppose $\sum f_n(x)$ uniformly converges on a bounded interval I to some function $f(x)$, and that F_n is an antiderivative of f_n for each n. If there exists some $c \in I$ for which $\sum F_n(c)$ converges, then $\sum F_n(x)$ uniformly converges to $F(x)$ on I, and F is an antiderivative for f on I.*

Idea of proof. *We apply Theorem 7.20 to the functions $F_n(x)$, which are differentiable on I by assumption.*

Proof of Corollary 7.21. The functions $F_n(x)$ are differentiable on the bounded interval I, and $\sum F_n' = \sum f_n$ is assumed to converge uniformly on I. Since there is some $c \in I$ so that $\sum F_n(c)$ converges, we use Theorem 7.20 to conclude that $\sum F_n(x)$ uniformly converges on I to some differentiable function F, and $F'(x) = \sum F_n'(x) = \sum f_n(x) = f(x)$. Thus, F is an antiderivative for f. □

7.3.3 The Weierstrass M−test

The Weierstrass M−test is a useful tool in determining uniform convergence of a given series.

Theorem 7.22 (Weierstrass M−test). *If $f_n : D \to \mathbb{R}$, and there are M_n with $|f_n(x)| \le M_n$ for every $n \in \mathbb{N}$, then $\sum f_n(x)$ converges uniformly if $\sum M_n$ converges.*

Idea of proof. *We show the Cauchy Criterion (Theorem 7.16) holds if $\sum M_n$ converges. The Cauchy Criterion for convergence of infinite series of real numbers (Theorem 4.11) will bound $\sum_{k=m+1}^{n} M_k < \epsilon$, and $|\sum_{k=m+1}^{n} f_k(x)| \le \sum_{k=m+1}^{n} |f_k(x)| \le \sum_{k=m+1}^{n} M_k$.*

Proof of Theorem 7.22. Let $\epsilon > 0$ be given. Note that the hypotheses are vacuous unless each of the numbers $M_n \ge 0$. By the Cauchy Criterion for convergence of series of real numbers (Theorem 4.11), there exists an $N \in \mathbb{N}$ so that for every $n > m \geq N$, we have

$$\sum_{k=m+1}^{n} = \left| \sum_{m+1}^{n} M_k \right| < \epsilon.$$

Since each $|f_n(x)| \leq M_n$ for every $n \in \mathbb{N}$ and for all $x \in D$, using the triangle inequality we have for every $n > m \geq N$

$$\left| \sum_{k=m+1}^{n} f_k(x) \right| \leq \sum_{k=m+1}^{n} |f_k(x)| \leq \sum_{k=m+1}^{n} M_k < \epsilon.$$

□

Exercise 3 shows that the converse of the Weierstrass M−test is false: there are uniformly convergent series $\sum f_n(x)$ where f_n do not meet the hypotheses of the Weierstrass M−test. It should also be remarked that the proof of the Weierstrass M−test actually shows that $\sum |f_n(x)|$ uniformly converges on D.

We can use the Weierstrass M−test to show that the geometric series considered in Example 7.15 uniformly converges on $[-r, r] \subseteq (-1, 1)$ for $0 < r < 1$. In Practice 7.23, we go on to study the antiderivative of this series on $[-r, r]$ according to Corollary 7.21. A generalization of this result concerning antiderivatives of any power series can be found in Theorem 7.29.

Practice 7.23. Let $0 < R < 1$ be given, and let $f_n(x) = x^n$ on $[-r, r]$.

(a) Show $\sum_{n=0}^{\infty} x^n$ converges uniformly on $[-r, r]$ to $\frac{1}{1-x}$.

(b) Show that $\sum_{n=0}^{\infty} \frac{1}{n+1} x^{n+1}$ converges uniformly to $-\ln(1-x)$ on $[-r, r]$.

Methodology. *We use the Weierstrass M−test to establish the first assertion. To prove (b), we verify each hypothesis of Corollary 7.21, and conclude that $F(x) = \sum_{n=0}^{\infty} \frac{1}{n+1} x^{n+1}$ is an antiderivative of $\frac{1}{1-x} = \sum_{n=0}^{\infty} x^n$. But $-\ln(1-x)$ is also an antiderivative of $\frac{1}{1-x}$, and so by Theorem 6.18, there is some constant where $F(x) = -\ln(1-x) + C$. Letting $x \in 0 \in [-r, r]$ will show that $C = 0$ so that these functions are equal.*

Solution. (a) We notice that $|f_n(x)| = |x^n| \leq r^n$ for every $x \in [-r, r]$. Since $0 < r < 1$, we set $M_n = r^n$ and find that $\sum M_n$ converges. By the Weierstrass M−test, $\sum_{n=0}^{\infty} x^n$ converges uniformly on $[-r, r]$. Since the series converges pointwise to $\frac{1}{1-x}$, this must also be its uniform limit.

(b) We verify each hypothesis of Corollary 7.21. We showed that $\sum_{n=0}^{\infty} x^n$ converges uniformly on $[-r, r]$ in part (a), and the interval $[-r, r]$ is bounded. For each n, $F_n(x) = \frac{1}{n+1} x^{n+1}$ is an antiderivative of $f_n(x) = x^n$. The number $0 \in [-r, r]$, and $\sum_{n=0}^{\infty} F_n(0) = 0$ converges. According to Corollary 7.21, $\sum_{n=0}^{\infty} F_n(x)$ uniformly converges to some function $F(x)$ on $[-r, r]$, and this F is an antiderivative of $\frac{1}{1-x} = \sum_{n=0}^{\infty} x^n$. The function $-\ln(1-x)$ is also an antiderivative of $\frac{1}{1-x}$, and so by Theorem 6.18 there exists a constant C so that $F(x) = -\ln(1-x) + C$ for every $x \in [-r, r]$. When $x = 0$, we find that $F(0) = 0$, while $-\ln(1-0) + C = C$, so that $C = 0$, and $F(x) = -\ln(1-x)$ for all $x \in [-r, r]$. ■

7.3.4 Review of important concepts from Section 7.3

1. Convergence of a series of functions is described in Definition 7.14. Similar to our definition of convergence of series of real numbers, pointwise and uniform convergence of $\sum f_n(x)$ is determined by the corresponding convergence of the sequence of partial sums.
2. The Cauchy Criterion for uniform convergence of series (Theorem 7.16) is a useful characterization of uniform convergence for series.
3. The preservation theorems from the last section can be adapted to apply to series of functions. Continuity of f_n and uniform convergence imply $\sum f_n(x)$ is continuous as well (Theorem 7.17).
4. If a series of integrable functions converges uniformly, then the sum and definite integral sign may be interchanged (Theorem 7.19). The proof of this is beyond our intended scope and is not included.
5. We can adapt Theorem 7.13 concerning differentiation and uniform convergence from the last section to series of functions. Term-by-term differentiation in $\sum f_n(x)$ is allowed provided the hypotheses of Theorem 7.20 are met.
6. One can interpret Theorem 7.20 to justify term-by-term integration of a uniformly convergent series provided reasonable hypotheses are met (Corollary 7.21).
7. The Weierstrass $M-$test (Theorem 7.22) is a useful tool to demonstrate that a series of bounded functions uniformly converges.

Exercises for Section 7.3

1. Suppose $\sum f_n(x)$ converges uniformly to $f(x)$ on D, and suppose $\varphi : \tilde{D} \to D$. Show that $\sum f_n(\varphi(x))$ converges uniformly to $f(\varphi(x))$ on $\tilde{D}$.
2. Use Exercise 1 to show that $\sum_{n=0}^{\infty}(-1)^n x^{2n}$ converges uniformly to $\frac{1}{1+x^2}$ on $[-r, r]$ for any $r \in (0, 1)$. Prove also that its term-by-term antiderivative $\sum_{n=0}^{\infty} \frac{(-1)^n}{2n+1} x^{2n+1}$ converges uniformly to $\arctan(x)$ on $[-r, r]$.
3. Find functions f_n on some domain $D \subseteq \mathbb{R}$, where $\sum f_n$ converges uniformly, but that there do not exist any bounds M_n so that $|f_n(x)| \leq M_n$ for all n and $x \in D$ with $\sum M_n$ convergent. *[Hint: The*

easiest examples are when f_n are constant, and form a conditionally convergent series.]

4. In Practice 7.23, we showed that $\sum_{n=0}^{\infty} \frac{1}{n+1} x^{n+1}$ converges uniformly to $-\ln(1-x)$ on $[-r, r]$ for every $0 < r < 1$. The series $\sum_{n=0}^{\infty} \frac{1}{n+1} x^{n+1}$ converges when $x = -1$. Is this enough to conclude that $\ln(\frac{1}{2}) = -\ln(2) = \sum_{n=0}^{\infty} \frac{1}{n+1}(-1)^{n+1}$? If not, what additional hypothesis would guarantee this? *[Hint: See Abel's Limit Theorem (Theorem 7.28) in the next section.]*

5. Suppose $f_n, g_n : D \to \mathbb{R}$, $|f_n(x)| \leq g_n(x)$ for all $x \in D$, and that $\sum g_n(x)$ converges uniformly on D. Prove that $\sum f_n(x)$ converges uniformly on D as well. Set $g_n(x) = M_n \in \mathbb{R}$ and notice that the Weierstrass $M-$test (Theorem 7.22) follows immediately. (This is a generalization of the Weierstrass $M-$test called the *Weierstrass Comparison Test for uniform convergence.*)

6. Show that $\sum \sin(n^2 x) x^n$ converges uniformly on $[0, r]$ for $0 < r < 1$. *[Hint: Use Exercise 5.]*

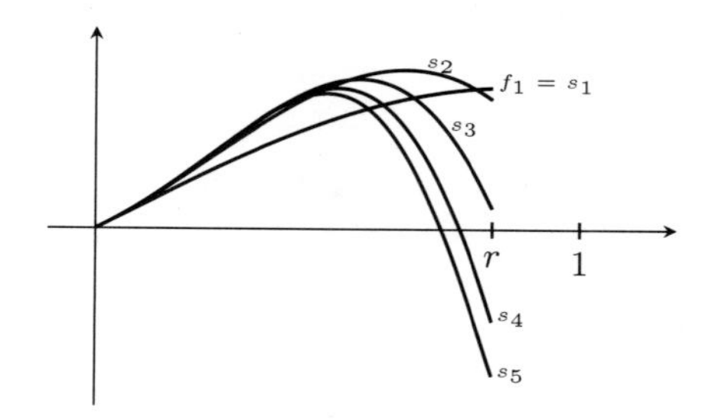

7. Prove Dirichlet's test for uniform convergence: Suppose (f_n) and (g_n) are sequences of functions on D. If

   (a) $\sum f_n$ has uniformly bounded partial sums,

   (b) $g_n \to 0$ uniformly on D,

   (c) $g_{n+1}(x) \leq g_n(x)$ for all $x \in D$ and $n \in \mathbb{N}$,

   then $\sum f_n g_n$ converges uniformly on D. *[Hint: Prove this using the Cauchy Criterion for uniform convergence (Theorem 7.16), and Abel's Lemma (Lemma 4.22).]*

8. Use Exercise 7 to show that $\sum \frac{\sin(nx)}{n}$ converges uniformly on $\left[\frac{\pi}{2}, \frac{3\pi}{2}\right]$. *[Hint: The estimate of the partial sum $\sum_{k=1}^{n} \sin(kx)$ in Example 4.26 will be useful.]*

7.4 Power Series

We have seen that with regard to the series of functions $\sum f_n(x)$, some nice things happen when the $f_n(x)$ are well-behaved. There are few functions more well behaved than polynomials. A special name is given to the situation where $f_n(x)$ is a certain polynomial: a *power series.*

We discuss some remarkable facts about power series, beginning with its radius and interval of convergence (see Theorem 7.24). We will find that a power series uniformly converges on an appropriate subset of its domain (Theorem 7.26). As such, we may use the results from last section to conclude that power series are continuous on the interior of their interval of convergence (Theorem 7.27), and even at the endpoints should the power series converge there (Abel's Theorem, Theorem 7.28). We will also see that term-by-term antidifferentiation and differentiation (Theorems 7.29 and 7.30) are allowed. We close the section with a discussion of Taylor series.

7.4.1 Power series and the radius and interval of convergence

As discussed above, we focus our attention to series of functions where $f_n(x)$ are certain polynomials. It is often the case that we shall use these series of functions to represent other functions, and it will be important to identify a "center" at which to base our study. If (a_n) is a sequence of real numbers, $c \in \mathbb{R}$ is this center, then the choice of $f_n(x) = a_n(x-c)^n$ makes $\sum f_n(x) = \sum a_n(x-c)^n$ a *power series centered at c.*

A remarkable fact about power series is the domain on which they converge.

Theorem 7.24. *Given the power series $\sum a_n(x-c)^n$, there is an extended real number $R \in [0, \infty]$ so that $\sum a_n(x-c)^n$ converges if $|x-c| < R$, and $\sum a_n(x-c)^n$ diverges if $|x-c| > R$. In fact, if $L = \limsup(\sqrt[n]{|a_n|})$, then $R = \frac{1}{L}$ if $L \in (0, \infty)$, $R = \infty$ if $L = 0$, and $R = 0$ if $L = \infty$.*

Idea of proof. *We apply the Root Test (Theorem 4.33) to the power series $\sum a_n(x-c)^n$ and simply read off each result. We first consider $L \in (0, \infty)$, then $L = 0$, and finally $L = \infty$.*

Proof of Theorem 7.24. Let $L = \limsup(\sqrt[n]{|a_n|})$. Since $\sqrt[n]{|a_n|} \geq 0$, we must have $L \geq 0$. We apply the Root Test (Theorem 4.33) to the power series $\sum a_n(x-c)^n$. Suppose first that $L \in (0, \infty)$, and set $R = \frac{1}{L}$. Then by using Exercise 7 of Section 3.9,

$$\limsup(\sqrt[n]{|a_n(x-c)^n|}) = |x-c| \cdot L = \frac{|x-c|}{R}.$$

This quantity is less than 1 if $|x-c| < R$, and so the power series converges (absolutely). It is greater than 1 if $|x-c| > R$, and so the power series diverges.

Now suppose $L = 0$. Using the Root Test and Exercise 7 of Section 3.9 again, we have

$$\limsup(\sqrt[n]{|a_n(x-c)^n|}) = |x-c| \cdot L = 0 < 1.$$

In this case, the power series converges (absolutely) for every x, so indeed $R = \infty$ since if $|x - c| < \infty$ (as it always is for real x), the series converges.

Now suppose that $L = \infty$. If $x = c$, then the power series converges, but we claim that the power series diverges for every other x, so that $R = 0$. If $x \neq c$, then $|x - c| > 0$, and

$$\limsup(\sqrt[n]{|a_n(x-c)^n|}) = |x-c| \cdot L = \infty > 1.$$

The Root Test now guarantees that the power series diverges. □

The number $R \in [0, \infty]$ found in Theorem 7.24 is called the *radius of convergence* of a power series. The term "radius" is appropriate, since those x which make a power series centered at c converge are those that are less than a fixed distance from c. The actual domain of a power series is called its *interval of convergence*. It is important to point out that one needs to manually check the endpoints of the interval of convergence as the Root Test will not determine convergence. This is highlighted in the next practice problem.

Practice 7.25. Compute the radius and interval of convergence for the power series $\sum \frac{x^n}{n}$.

Methodology. *We use Theorem 7.24 to compute the radius of convergence R. The sequence involved converges, so we may use more familiar facts about convergent sequences as opposed to the more general facts concerning the* $\limsup$. *We will find $R = 1$, so the interval of convergence of this power series centered at $c = 0$ is an interval with endpoints -1, and 1. We will find that 1 is not in this interval, but that -1 is, so that the interval of convergence is $[-1, 1)$.*

Solution. Since $\sqrt[n]{n} \to 1$ (see Theorem 3.55), we have

$$\limsup\left(\sqrt[n]{\left|\frac{1}{n}\right|}\right) = \lim_{n\to\infty} \frac{1}{\sqrt[n]{n}} = 1.$$

Thus by Theorem 7.24, our radius of convergence is $\frac{1}{1} = 1$. So if $|x| < 1$, our power series converges, and if $|x| > 1$ our power series diverges. It is not yet clear what happens if $|x| = 1$, and we check $x = \pm 1$ separately.

If $x = 1$, then $\sum \frac{x^n}{n} = \sum \frac{1}{n}$, which diverges. If $x = -1$, then $\sum \frac{x^n}{n} = \sum \frac{(-1)^n}{n}$, which converges by the Alternating Series Test. So, our interval of convergence is $[-1, 1)$. ■

We remind the reader that $\limsup(x_n) = \lim_{n\to\infty} x_n$ in the event that the limit exists. So in the last practice problem, just finding the limit of $\sqrt[n]{|a_n|}$ will suffice if it exists. Of course, this limit may not exist, at which point one is forced to compute $\limsup(\sqrt[n]{|a_n|})$.

7.4.2 Uniform convergence of power series

One benefit of power series is that they converge uniformly on certain subsets of their domain. This will be useful as we discuss continuity, differentiation, and antidifferentiation.

Theorem 7.26. *Let R be the radius of convergence of the power series $\sum a_n(x-c)^n$. If $0 \leq r < R$, then $\sum a_n(x-c)^n$ converges uniformly on $[c-r, c+r]$, and therefore on any interval $[a,b] \subseteq (c-R, c+R)$.*

Idea of proof. *We use the Weierstrass M−test (Theorem 7.22). If $L = \limsup(\sqrt[n]{|a_n|}) \in (0, \infty)$ and $R = \frac{1}{L}$ as in Theorem 7.24, then $M_n = |a_n|r^n$. The case $L = 0$ and $R = \infty$ is handed the same way. If $L = \infty$ then $R = 0$ and there is nothing to show. To prove the last statement, we note that any interval $[a,b] \subseteq [c-r, c+r]$ for some r with $0 \leq r < R$, and the same estimates for M_n will hold there as well.*

Proof of Theorem 7.26. Let $0 \leq r < R$ be given. If $x \in [c-r, c+r]$, then $|a_n(x-c)^n| \leq |a_n|r^n = M_n$. We show that $\sum M_n$ converges, and uniform convergence of the power series will follow by the Weierstrass M−test (Theorem 7.22).

Let $L = \limsup(\sqrt[n]{|a_n|})$ as in Theorem 7.24. If $L = \infty$, then $R = 0$ and there is nothing to prove. Therefore, assume $L \in [0, \infty)$. We apply the Root Test (Theorem 4.33) to $\sum M_n$. By Exercise 7 of Section 3.9,

$$\limsup(\sqrt[n]{|M_n|}) = \limsup(\sqrt[n]{|a_n|r^n}) = Lr.$$

If $L = 0$, then this quantity is less than 1. If $L \in (0, \infty)$, then $R = \frac{1}{L}$, and since $0 \leq r < R$,

$$Lr = \frac{r}{R} < 1.$$

In either case, $\sum M_n$ converges by the Root Test.

Now[5] suppose $[a,b] \subseteq (c-R, c+R)$. If $r = \max\{|a|, |b|\}$, then $[a,b] \subseteq [c-r, c+r]$. Thus for any $x \in [a,b]$ we still have $|a_n(x-c)^n| \leq |a_n|r^n$, and the same argument used above demonstrates that $\sum a_n(x-c)^n$ converges uniformly on $[a,b]$. □

Along with the results of Section 7.3, the conclusion of uniform convergence in Theorem 7.26 will help us to investigate the properties of continuity, and term-by-term differentiation and antidifferentiation of power series.

7.4.3 Continuity of power series

There are two major results concerning continuity of power series. The first (Theorem 7.27) is a simple application of Theorem 7.17 stating that power series are continuous on the interior of their interval of convergence. The second

[5]It is actually less efficient to organize the proof this way, since $[r-c, r+c]$ takes the form $[a,b]$. Organizing the proof this way, however, highlights uniform convergence on $[c-r, c+r]$.

result, called *Abel's Limit Theorem* (Theorem 7.28), discusses the continuity of a power series at the endpoints of its interval of convergence.

Theorem 7.27. *If R is the radius of convergence of the power series $\sum a_n(x-c)^n$, then $\sum a_n(x-c)^n$ is continuous on $(c-R, c+R)$.*

Idea of proof. *If we choose any $t \in (c-R, c+R)$, we show that the power series is continuous at t. Note that t is contained in an interval of the form $[c-r, c+r]$ for some $r \in [0, R)$, as shown in Figure 7.4.1. Since each function $f_n(x) = a_n(x-c)^n$ is continuous on such an interval and the power series uniformly converges there by Theorem 7.26, the continuity of the power series at t now follows by Theorem 7.17.*

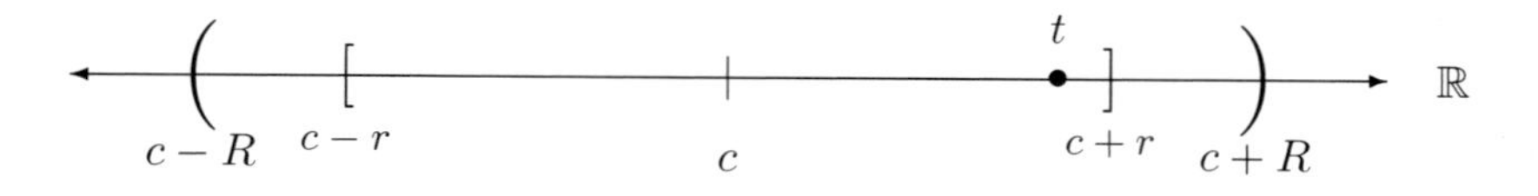

FIGURE 7.4.1
Choosing $r \in [0, R)$ so that $t \in [c-r, c+r] \subseteq (c-R, c+R)$.

Proof of Theorem 7.27. If we choose any $t \in (c-R, c+R)$, then there exists an r with $|t| \leq r < R$ so that

$$t \in [c-r, c+r] \subseteq (c-R, c+R).$$

The functions $f_n(x) = a_n(x-c)^n$ are continuous on $[c-r, c+r]$, and the power series uniformly converges on $[c-r, c+r]$ by Theorem 7.26. By Theorem 7.17, the power series is continuous at t. □

To motivate Abel's Limit Theorem (Theorem 7.28) below, we consider the power series representation for $-\ln(1-x) = \sum_{n=0}^{\infty} \frac{1}{n+1} x^{n+1}$ we encountered in Practice 7.23. By substituting $-x$ for x (see Exercise 1 of Section 7.3), multiplying by -1, and reindexing the sum, we find that for $x \in (-1, 1)$,

$$\ln(1+x) = \sum_{n=1}^{\infty} \frac{(-1)^{n+1} x^n}{n}. \tag{7.3}$$

As an artifact of antidifferentiation, the power series on the right happens to also converge at $x = 1$. It is tempting to just plug in $x = 1$ in Equation (7.3) and to claim that $\ln(2) = \sum_{n=1}^{\infty} \frac{(-1)^{n+1}}{n}$. Although this is indeed the case by Exercise 9 of Section 4.3, the equality of $\ln(1+x)$ with $\sum_{n=1}^{\infty} \frac{(-1)^{n+1} x^n}{n}$ in Equation (7.3) is only valid for $x \in (-1, 1)$ even though the series converges for $x = 1$ as well.

What we need is Abel's Limit Theorem, which states that if a power series converges at an endpoint of its interval of convergence, then it is continuous there. In this case, $\sum_{n=1}^{\infty} \frac{(-1)^{n+1}x^n}{n}$ converges at $x = 1$, and so

$$\lim_{x\to 1^-} \sum_{n=1}^{\infty} \frac{(-1)^{n+1}x^n}{n} = \sum_{n=1}^{\infty} \frac{(-1)^{n+1}}{n}.$$

But for $x < 1$ (and near 1), Equation (7.3) equates $\sum_{n=1}^{\infty} \frac{(-1)^{n+1}x^n}{n}$ with $\ln(1+x)$, so

$$\lim_{x\to 1^-} \sum_{n=1}^{\infty} \frac{(-1)^{n+1}x^n}{n} = \lim_{x\to 1^-} \ln(1+x) = \ln(2).$$

Theorem 7.28 (Abel's Limit Theorem). *Suppose that $\sum_{n=0}^{\infty} a_n(x-c)^n$ has a radius of convergence $R \in (0,\infty)$, and that $f(x) = \sum_{n=0}^{\infty} a_n(x-c)^n$ for $x \in (c-R, c+R)$. If $\sum_{n=0}^{\infty} a_n(x-c)^n$ converges at $x = c+R$, then f is continuous from the left at $x = c+R$, so that*

$$f(c+R) = \lim_{x\to c+R^-} f(x) = \sum_{n=0}^{\infty} a_n R^n.$$

Idea of proof. *The proof will come in three phases. The first phase will be to translate the problem to a power series centered at 0 with radius of convergence 1. This is purely for ease of presentation. The second phase is to express the function $f(x)$ in a different way. The goal here is to "factor out" a $(1-x)$ to aid us in choosing a δ in the limit we must compute. The final phase is to preform an estimation that will lead us to the correct choice of δ.*

Proof of Theorem 7.28. We first translate this problem to an equivalent one involving a power series centered at $c = 0$ with radius of convergence 1. In $\sum_{n=0}^{\infty} a_n(x-c)^n$, set $y = \frac{x-c}{R}$. Then the power series

$$\sum_{n=0}^{\infty} a_n(x-c)^n = \sum_{n=0}^{\infty} a_n R^n y^n.$$

If $L = \limsup(\sqrt[n]{|a_n|})$, then $R = \frac{1}{L}$ since $R \in (0,\infty)$. Computing

$$\limsup(\sqrt[n]{|a_n R^n|}) = LR = 1,$$

so the radius of convergence of $\sum_{n=0}^{\infty} a_n R^n y^n$ is 1 according to Theorem 7.24, and its center is clearly 0. If we could show that the conclusions of the theorem are true for $\sum_{n=0}^{\infty} a_n R^n y^n$ as $y \to 1^-$, then the theorem is true for $\sum_{n=0}^{\infty} a_n(x-c)^n$:

$$\lim_{x\to c+R^-} \sum_{n=0}^{\infty} a_n(x-c)^n = \lim_{y\to 1^-} \sum_{n=0}^{\infty} a_n R^n y^n = \sum_{n=0}^{\infty} a_n R^n.$$

Let $\sum_{n=0}^{\infty} a_n x^n$ be any series with radius of convergence 1, and suppose that when $x = 1$, the series converges to $\sum_{n=0}^{\infty} a_n$. For $x \in (-1, 1)$, define $f(x) = \sum_{n=0}^{\infty} a_n x^n$, let $s_n = \sum_{k=1}^{n} a_k$ be the partial sum sequence of $\sum_{n=0}^{\infty} a_n$, and let $s = \lim_{n\to\infty} s_n = \sum_{k=0}^{\infty} a_k$. We want to show that $\lim_{x\to 1^-} f(x) = s$.

Let $0 < x < 1$ in the estimates that follow. Using $x^k = b_k$ in Abel's Lemma (Lemma 4.22), we express

$$\sum_{k=0}^{n} a_k x^k = \sum_{k=0}^{n} s_k(x^k - x^{k+1}) + s_n x^{n+1}.$$

But

$$s_k(x^k - x^{k+1}) = s_k x^k - s_k x^{k+1} = (1-x)s_k x^k,$$

so

$$\sum_{k=0}^{n} a_k x^k = (1-x)\sum_{k=0}^{n} s_k x^k + s_n x^{n+1}.$$

Since $0 < x < 1$ and $s_n \to s$, as $n \to \infty$, we have $s_n x^{n+1} \to 0$, and so

$$f(x) = \sum_{k=0}^{\infty} a_k x^k = (1-x)\sum_{k=0}^{\infty} s_k x^k.$$

We can now estimate $|f(x) - s|$ and complete the proof. Let $\epsilon > 0$ be given. For any $x \in (0, 1)$, we have $\frac{1}{1-x} = \sum_{k=0}^{\infty} x^k$, so

$$s = \frac{1-x}{1-x}s = (1-x)\sum_{k=0}^{\infty} s x^k.$$

So,

$$\begin{aligned} |f(x) - s| &= |(1-x)\textstyle\sum_{k=0}^{\infty} s_k x^k - (1-x)\sum_{k=0}^{\infty} s x^k| \\ &= (1-x)|\textstyle\sum_{k=0}^{\infty}(s_k - s)x^k|. \end{aligned}$$

Since $s_k \to s$, there exists an N so that if $k \geq N$, then $|s_k - s| < \frac{\epsilon}{2}$. The series $\sum_{k=0}^{\infty}(s_k - s)x^k$ absolutely converges for $x \in (0, 1)$ (Exercise 7), and so

$$(1-x)\left|\sum_{k=0}^{\infty}(s_k - s)x^k\right| \leq (1-x)\sum_{k=0}^{N}|s_k - s|x^k + (1-x)\left|\sum_{k=N+1}^{\infty}(s_k - s)x^k\right|.$$

Since $0 < x < 1$, if we set $M = \sum_{k=0}^{N}|s_k - s|$, we have

$$(1-x)|\sum_{k=0}^{N}|s_k - s| = M(1-x).$$

By adding more terms, we find that since $0 < x < 1$,

$$(1-x)\left|\sum_{k=N+1}^{\infty}(s_k - s)x^k\right| \leq (1-x)\sum_{k=0}^{\infty}\frac{\epsilon}{2}x^k = (1-x)\frac{\epsilon/2}{1-x} = \frac{\epsilon}{2}.$$

We conclude that for any $0 < x < 1$, with M and N as given,

$$|f(x) - s| = \left|\sum_{k=0}^{\infty}(s_k - s)x^k\right| \leq M(1-x) + \frac{\epsilon}{2}.$$

If $M = 0$, we are done with selecting $\delta = 1$. If $\frac{\epsilon}{2M} \geq 1$, then we may again select $\delta = 1$ and we are done[6]. Otherwise, set $\delta = \frac{\epsilon}{2M} < 1$, and we find that if $0 < 1 - x < \delta$, then $|f(x) - s| < \epsilon$. □

7.4.4 Antidifferentiation and differentiation

We can apply the differentiation and antidifferentiation results from the last section quite easily to power series. The first result states that term-by-term antidifferentiation is valid on the interior of the interval of convergence.

Theorem 7.29. *Suppose the power series $\sum a_n(x-c)^n$ has radius of convergence R, and set $f(x) = \sum a_n(x-c)^n$ for every $x \in (c-R, c+R)$. The function $F(x) = \sum a_n \frac{1}{n+1} x^{n+1}$ is an antiderivative of f on $(c-R, c+R)$.*

Idea of proof. *We use $f_n(x) = a_n(x-c)^n$ and $F_n(x) = a_n\frac{1}{n+1}(x-c)^{n+1}$ in Corollary 7.21. If an $x \in (c-R, c+R)$, we may enclose x in a suitable bounded interval to find that $F(x) = \sum F_n(x)$ is an antiderivative for $f(x)$.*

Proof of Theorem 7.29. Let $x \in (c-R, c+R)$, and select $I = [a,b] \subset (c-R, c+R)$ with $x, c \in I$. By Theorem 7.26, $\sum a_n(x-c)^n$ converges uniformly on I. We see that $F_n(x) = a_n\frac{1}{n+1}(x-c)^{n+1}$ is an antiderivative for $a_n(x-c)^n$ on I. Further, if $x = c \in I$, then $\sum F_n(c) = 0$ converges. By Corollary 7.21, $\sum a_n\frac{1}{n+1}(x-c)^{n+1}$ is an antiderivative for $\sum a_n(x-c)^n$. □

We point out that the special case of antidifferentiating $\sum_{n=0}^{\infty} x^n$ considered in Practice 7.23 can now be immediately read off from Theorem 7.29.

The next result concerns differentiation of power series. We will use Theorem 7.20 to conclude that term-by-term differentiation is allowed in the interior of the interval of convergence.

Theorem 7.30. *Suppose the power series $\sum a_n(x-c)^n$ has radius of convergence R, and set $f(x) = \sum a_n(x-c)^n$ for every $x \in (c-R, c+R)$. For all $x \in (c-R, c+R)$, $f'(x) = \sum a_n n(x-c)^{n-1}$.*

Idea of proof. *If $f_n(x) = a_n(x-c)^n$, then we again enclose x in suitable bounded interval, and the result is immediate after we verify each of the hypotheses of Theorem 7.20.*

[6] If δ is any larger, then x could be less than or equal to 0, and our estimates above do not quite work the same way.

Proof of Theorem 7.30. Let $x \in (c-R, c+R)$, and select $I = [a,b] \subset (c-R, c+R)$ with $x, c \in I$. Let $f_n(x) = a_n(x-c)^n$, and note that f_n is differentiable and $f_n'(x) = a_n n(x-c)^{n-1}$. We compute the radius of convergence of $\sum a_n n(x-c)^{n-1}$ using Theorem 7.24. Since $\sqrt[n]{n} \to 1$ (Theorem 3.55), according to Exercise 7 of Section 3.9, we find that

$$\limsup(\sqrt[n]{|a_n n|}) = \limsup(\sqrt[n]{n}) \limsup(\sqrt[n]{|a_n|}) = \limsup(\sqrt[n]{|a_n|}),$$

so the radius of convergence of $\sum a_n n(x-c)^{n-1}$ is equal to R, the radius of convergence of $\sum a_n(x-c)^n$. So $\sum a_n n(x-c)^{n-1}$ uniformly converges on I by Theorem 7.24. Finally, we note that if $x = c$, then $\sum a_n(x-c)^n$ converges. Meeting the hypotheses of Theorem 7.20, we find that $f'(x) = \sum f_n'(x) = \sum a_n n(x-c)^{n-1}$. □

7.4.5 Taylor and Maclaurin series

So far, we have approached power series from the point of view that it exists on its own without any reference to what familiar function it may equal on its interval of convergence. We have already seen that

$$\frac{1}{1-x} = \sum_{n=0}^{\infty} x^n \text{ for } x \in (-1,1).$$

A different perspective would be to start with a familiar function and to attempt to produce a power series that represents it. We assume that a function $f(x)$ is given, and that this function is infinitely differentiable. Then, if we want to center a power series representation for f at c, we need to find each a_n so that

$$f(x) = \sum_{n=0}^{\infty} a_n(x-c)^n = a_0 + a_1(x-c) + a_2(x-c)^2 + a_3(x-c)^3 + \cdots. \quad (7.4)$$

We proceed in a way similar to that of Section 6.3 concerning Taylor Polynomials. We can easily identify a_0 by setting $x = c$ in both sides of Equation (7.4):

$$f(c) = a_0 + 0 + 0 + \cdots = a_0.$$

To reveal a_1, we will need the radius of convergence of our power series to be positive. In this case, we differentiate the function and the power series and use Theorem 7.30 to conclude that

$$f'(x) = a_1 + 2a_2(x-c) + 3a_3(x-c)^2 + \cdots,$$

and so setting $x = c$ again gives

$$f'(c) = a_1 + 0 + 0 + \cdots = a_1.$$

To grasp the pattern, we repeat this process one more time. Differentiating again, we find that

$$f''(x) = a_2 + 2a_2 + 3 \cdot 2a_3(x-c) + \cdots,$$

and so setting $x = c$ again gives

$$f''(c) = 2a_2 + 0 + 0 + \cdots = 2a_2, \text{ so } a_2 = \frac{f''(c)}{2}.$$

Similar to Section 6.3, we will find that

$$a_n = \frac{f^{(n)}(c)}{n!}.$$

Definition 7.31. If f is infinitely differentiable on a neighborhood of $c \in \mathbb{R}$, then the *Taylor series* for f centered at c is

$$\sum_{n=0}^{\infty} \frac{f^{(n)}(c)}{n!}(x-c)^n.$$

If $c = 0$, we call this a *Maclaurin series* for f. If the Taylor series for f has a positive radius of convergence, then we say that f is *equal to* or *represented by its Taylor series* on I when I is some subinterval I of its interval of convergence with c as an interior point, and $f(x)$ is equal to its Taylor series on I.

The following theorem uses Theorem 7.30 to show that a Taylor series for a function is unique.

Theorem 7.32. *If $\sum_{n=0}^{\infty} a_n(x-c)^n$ and $\sum_{n=0}^{\infty} b_n(x-c)^n$ are both Taylor series for f, then $a_n = b_n$ for all n.*

Idea of proof. *If both of these are Taylor series for f, then the previous discussion shows that both a_n and b_n are equal to $\frac{f^{(n)}(c)}{n!}$.*

Proof of Theorem 7.32. As Taylor series for f, both $a_n = \frac{f^{(n)}(c)}{n!} = b_n$. □

We developed the machinery in Taylor's Theorem (Theorem 6.25) to deduce when exactly a function is representable by its Taylor series on some interval.

Theorem 7.33. *Let $f(x)$ be an infinitely differentiable function, and express $f(x) = P_n(x) + R_n(x)$ as in Theorem 6.25. Then if I is some interval containing c as an interior point, f is represented by its Taylor series on I if and only if $\lim_{n\to\infty} R_n(x) = 0$.*

Idea of proof. *Notice that the Taylor polynomials f are just the partial sums of the Taylor series of f. The conclusion that f is represented by its Taylor series on I amounts to just noticing that the Taylor polynomials converge to the Taylor series on I.*

Proof of Theorem 7.33. Let I be given with c as an interior point, and let $P_n(x) = \sum_{k=0}^{n} \frac{f^{(k)}(c)}{k!}(x-c)^k$. Taylor's Theorem (Theorem 6.25) says that $f(x) = P_n(x) + R_n(x)$ for $x \in I$. Notice that f is represented by its Taylor series on I if and only if

$$f(x) = \sum_{n=0}^{\infty} \frac{f^{(n)}(c)}{n!}(x-c)^n = \lim_{n\to\infty} P_n(x),$$

and this occurs if and only if

$$\lim_{n\to\infty} (f(x) - P_n(x)) = \lim_{n\to\infty} R_n(x) = 0.$$

□

We close this section with a demonstration that the function $f(x) = e^x$ is equal to its Maclaurin series on $\mathbb{R}$.

Practice 7.34. Show that $f(x) = e^x$ is equal to its Maclaurin series on $\mathbb{R}$.

Methodology. *We compute the remainder term $R_n(x)$ and find a bound on it. This will show that $\lim_{n\to\infty} R_n(x) = 0$, so that we may apply Theorem 7.33.*

Solution. If $f(x) = e^x$, then $f^{(n+1)}(x) = e^x$, and so according to Taylor's Theorem (Theorem 6.25), there is some[7] $c \in \mathbb{R}$ for which

$$f(x) = P_n(x) + \frac{f^{(n+1)}(c)}{(n+1)!}x^{n+1}.$$

Here, the remainder $R_n(x) = \frac{f^{(n+1)}(c)}{(n+1)!}x^{n+1} = e^c \frac{x^{n+1}}{(n+1)!}$. By Exercise 3 of Section 3.3, $\frac{x^{n+1}}{(n+1)!} \to 0$. As a result,

$$R_n(x) = e^c \frac{x^{n+1}}{(n+1)!} \to 0.$$

By Theorem 7.33, $f(x) = e^x$ is equal to its Maclaurin series on $\mathbb{R}$. ■

7.4.6 Review of important concepts from Section 7.4

1. A power series must only have an interval as its domain, called the *interval of convergence.* This interval extends the same length on either side of the center of the power series by Theorem 7.24, and is called the *radius of convergence.*
2. If R is the radius of convergence of a power series, then it uniformly converges on any interval of the form $[a, b] \subseteq (c - R, c + R)$. See Theorem 7.26.

[7]Careful, c is not the center of this Taylor series, rather, it is the c found in Taylor's Theorem.

3. If R is the radius of convergence of a power series, then it is continuous on $(c-R, c+R)$ (Theorem 7.27), and term-by-term antidifferentiation and differentiation are valid there (Theorems 7.29 and 7.30).
4. Abel's Theorem (Theorem 7.28) asserts that if a power series converges at an endpoint of its interval of convergence, then it is continuous there.
5. A function is represented by its Taylor series when that series has a positive radius of convergence (Definition 7.31). Theorem 7.33 provides a characterization for this in terms of the remainder term found in Taylor's Theorem (Theorem 6.25).

Exercises for Section 7.4

1. Compute the radius and interval of convergence of the following power series.

   (a) $\sum \frac{x^n}{n}$
   (b) $\sum 2^n x^n$
   (c) $\sum \frac{x^{2n}}{9^n}$
   (d) $\sum \left(1-\frac{1}{n}\right)^n x^n$
   (e) $\sum \left(1-\frac{1}{n}\right)^{n^2} x^n$

2. Let $\sum a_n(x-c)^n$ be a power series, and suppose that

$$L = \lim_{n\to\infty}\left|\frac{a_{n+1}}{a_n}\right|.$$

   Prove that the radius of convergence R of the power series is $R=\frac{1}{L}$ if $L \in (0,\infty)$, it is $R=\infty$ if $L=0$, and it is $R=0$ if $L=\infty$. Thus, one can use the Ratio Test in the event the limit L exists.

3. Compute the radius of convergence of

$$\sum_{n=1}^{\infty} a_n x^n = 2x + x^2 + 2x^3 + x^4 + 2x^5 + \cdots.$$

   [Hint: This series was considered in Example 4.36 and illustrates that the Ratio Test in Exercise 2 is insufficient to measure the radius of convergence of all power series.]

4. Prove that the radius of convergence R of the power series[8] $\sum a_n x^n$ can be computed as

$$R = \sup\{|x| \,|\, a_n x^n \to 0\}.$$

[8] The result is still true for the more general power series $\sum a_n(x-c)^n$, but we center it at 0 in this problem for convenience.

5. Let $f(x) = \sum_{n=0}^{\infty} a_n x^n$, and suppose the radius of convergence of this power series is greater than zero.

   (a) Prove that f is even if and only if $a_n = 0$ for all odd n.

   (b) Prove that f is odd if and only if $a_n = 0$ for all even n.

6. In Exercise 2 of Section 7.3 we showed that for every $x \in (-1, 1)$, we have

$$\arctan(x) = \sum_{n=0}^{\infty} \frac{(-1)^n}{2n+1} x^{2n+1}.$$

   Use Abel's Limit Theorem (Theorem 7.28) to prove that

$$\pi = 4 \sum_{n=0}^{\infty} \frac{(-1)^n}{2n+1}.$$

7. Prove that if $\sum_{k=0}^{\infty} a_k x^k$ converges for $x \in (-1, 1]$ and $s_n = \sum_{k=0}^{n} a_k$ is the partial sum sequence of $\sum_{k=0}^{\infty} a_k$, then when $s = \sum_{k=0}^{\infty} a_k$, the series $\sum_{k=0}^{\infty}(s_k - s)x^k$ absolutely converges for $|x| < 1$.

8. Suppose f is an infinitely differentiable function, and that there is a number M so that $|f^{(n)}(x)| \leq M^n$ for every $x \in I$. If c is an interior point of I, show that f is equal to its Taylor series centered at c on I.

9. Prove that

   (a) $\sin(x)$ is equal to its Maclaurin series on $\mathbb{R}$.

   (b) $\cos(x)$ is equal to its Maclaurin series on $\mathbb{R}$.

10. Suppose f, g, and $\frac{f}{g}$ are equal to their Maclaurin series on a neighborhood of 0. If $\frac{f(x)}{g(x)} = a_0 + a_1 x + \cdots$, expand the product

$$\begin{aligned} f(0) + f'(0)x + \cdots &= f(x) = \tfrac{f(x)}{g(x)} \cdot g(x) \\ &= (a_0 + a_1 x + \cdots)(g(0) + g'(0)x + \cdots) \end{aligned}$$

   and solve for $a_1 = \frac{d}{dx}\left(\frac{f}{g}\right)|_{x=0}$ and recover the Quotient Rule (Theorem 6.5).

7.5 Troubleshooting Guide for Series of Real Numbers

Type of difficulty:	***Suggestion for help:***
Section 7.1: Sequences of Functions	
"I still don't really see the difference between pointwise and uniform convergence."	In pointwise convergence, one chooses x in advance, and finds that $f_n(x) \to f(x)$ independent of any other considerations. In the definition of uniform convergence, the basic definition of convergence appears to be there, but with the added "for all $x \in D$", so that the N one finds is independent of any initial choice of x. See Example 7.4 for an illustration of this difference.
"How did they know to choose $x = \frac{1}{N}$ in Part (b) of Practice 7.6? How am I supposed to know what to choose?"	Generally, a careful look at the graphs of a few of the $f_n(x)$ will reveal some problematic behavior if the convergence is not uniform. Specifically, thinking of all of the f_n as lying between $f(x) - \epsilon$ and $f(x) + \epsilon$ can be helpful, as in Figure 7.1.2. In Part (b) of Practice 7.6, we noticed that xn appears in the numerator, and in the denominator as $n^2x^2 = (xn)^2$, so choosing $x = \frac{1}{N}$ seemed like a natural choice when $n = N$.
Section 7.2: Consequences of Uniform Convergence	
"Why does nothing seem to be preserved by pointwise convergence?"	This is difficult to briefly answer in any technical sense, but those properties considered concern the behavior of a function that compare various outputs of that function: for example, continuity and differentiation. Pointwise convergence can be loosely described as "convergence one point at a time" and does not consider these comparisons of different outputs.
Section 7.3: Infinite Series of Functions	

Type of difficulty:	***Suggestion for help:***
"Can't you always differentiate and antidifferentiate term-by-term in a series?"	In general, the answer is no: you cannot. You can under special (but not unusual) hypotheses, as the theorems in this section illustrate (see Theorem 7.20 and Corollary 7.21). Power series often meet these criteria, and it may be the case that you are basing this intuition off of your experience with these sorts of series of functions.
"It seems as though the Weierstrass M−test (Theorem 7.22) is a great tool, but what if I can't use it?"	It is possible that the Weierstrass M−test does not apply in situations of uniform or nonuniform convergence, see Exercise 3. Perhaps an estimate for M_n is out there but you do not yet see it?
Section 7.4: Power Series	
"Why is the $\limsup$ back all of a sudden? I thought you could just use the Ratio Test (Theorem 4.31) to find the radius of convergence of a power series?"	Yes, the Ratio Test will find the radius of convergence, but only when the limit in the Ratio Test exists (see Exercise 2 from this section). Generally, the only sure thing is what is outlined in Theorem 7.24. See the discussion after Practice 7.25 and Exercises 2 and 3.
"Why do we need all of these hypotheses on I when talking about Taylor series?"	We want to apply this theory to a number of different intervals, and we need c to be an interior point of I to be able to apply Taylor's Theorem (Theorem 6.25).

Appendix A

List of Commonly Used Symbols

Here is a list of commonly used notations in analysis. The first column is the symbol as found in this book. The second column might give an alternate symbolism found elsewhere. The third column is a description of what the symbol means, and where it can be found in this book.

Symbol:	Also as:	Description:
$\neg P$	$\sim P$	The negation of the statement P. See page 3.
$P \wedge Q$		The conjunction of statements, "P and Q." See page 3.
$P \vee Q$		The disjunction of statements, "P or Q." See page 3.
$P \Rightarrow Q$		The implication "P implies Q." See page 4.
$P \Leftrightarrow Q$		Logical equivalence of the statements P and Q. See page 4.
$\mathbb{N}$		The set of natural numbers. See Definition 1.5.
$\mathbb{Z}$		The set of integers. See Definition 1.5.
$\mathbb{Q}$		The set of rational numbers, see Definition 1.5, or page 61.
$\mathbb{R}$		The set of real numbers, see Definition 1.5, or page 60.
$\{\cdots \mid \cdots\}$		Set notation, including intervals of real numbers, are described beginning on page 12.
$S \cup T$		The union of sets S and T. See Definition 1.8 on page 15.
$S \cap T$		The intersection of sets S and T. See Definition 1.8 on page 15.
$T - S$	S^c	The complement of S in T. See page 18.

Symbol:	Also as:	Description:
$S \times T$		The Cartesian or cross-product of S and T. See page 19.
$\mathcal{P}(X)$		The power set of S. See Exercise 6 in Section 1.2, and Exercise 6 in Section 1.5.
$f : X \to Y$		f is a function with domain X and range Y. See Definition 1.20 on page 29.
$f(S)$		The image of S under the function f. See Definition 1.23 on page 30.
$f^{-1}(T)$		The inverse image of T under f. See Definition 1.23 on page 30.
$S \sim T$		Sets S and T are equivalent. See Definition 1.37 on page 39.
$\lvert S\rvert$	n(S), #S, card(S)	The cardinality of S. See Definition 1.40 on page 41 for finite sets, and the discussion on page 52 for infinite sets.
$\mathbb{R}^\sharp$	$\mathbb{R}^*$, $\mathbb{R}^\natural$, $[-\infty, \infty]$	The set of extended real numbers. See Definition 2.10 on page 69.
$U_\epsilon, U_\epsilon(x)$		The $\epsilon-$neighborhood about x. See page 71.
$\sup(S), \inf(S)$	$lub(S), glb(S)$	The supremum (infimum) of the set S. See Definitions 2.18 and 2.22.
$(a_n)_{n\in\mathbb{N}}$	$(a_n), (a_n)_n$, etc.	A sequence of numbers, a_n is the nth term of the sequence. See page 95.
$a_n \to L$	$\lim\limits_{n\to\infty} a_n = L, a_n \xrightarrow{n} L$	The sequence a_n converges to L. See Definition 3.13 on page 108.
(a_{n_k})		Notation for a subsequence of (a_n). See Section 3.4.
A'		The derived set of A, A' is the set of all limit points of A. See page 169.
$\bar{A}$		The closure of A, $\bar{A} = A \cup A'$. See page 169.

Symbol:	Also as:	Description:
$a_n \to \pm\infty$	$\lim_{n\to\infty} a_n = \pm\infty$	The sequence (a_n) diverges to $\pm\infty$. See Definition 3.106 on page 184.
$\limsup(x_n)$ and $\liminf(x_n)$	$\overline{\lim}(x_n)$ and $\underline{\lim}(x_n)$	The limit superior of (x_n) ($\limsup(x_n)$), and limit inferior of (x_n) ($\liminf(x_n)$). See Definition 3.119 on page 196.
$\sum a_n$		An infinite series, see Definition 4.1.
$\lim_{x\to c} f(x) = L$	$f(x) \to L$ as $x \to c$	The limit of f as x approaches c, or also, the limit of f at c. See Definition 5.1 on page 255.
U^-		Deleted neighborhood, see Definition 5.8.
$\lim_{x\to c^\pm} f(x)$		A one-sided limit, see Definition 5.11.
$f'(c)$	$\frac{df}{dx}$, Df, etc.	The derivative of f at c, see Definition 6.1.
$P_n(x)$	$T_n(x)$	The Taylor Polynomial of a function of degree n. See Definition 6.23.
$f_n \to f$		(f_n) converges to f pointwise. See Definition 7.1.
$f_n \rightrightarrows f$		(f_n) uniformly converges to f on some domain. See Definition 7.5.
$\sum f_n(x)$		Series of functions, see Definition 7.14.
$s_n(x)$	$\sum_{k=1}^n f_k(x)$	Partial sum of a series of functions, see Definition 7.14.

Bibliography

[1] S. Abbott. *Understanding Analysis.* Undergraduate Texts in Mathematics. Springer, 2nd edition, July 2002.

[2] David Applebaum. *Limits, Limits Everywhere: The Tools of Mathematical Analysis.* Oxford University Press, 2012.

[3] R.G. Bartle and D.R. Sherbert. *Introduction to Real Analysis.* John Wiley & Sons, New York, 4th edition, 2013.

[4] E.T. Bell. *Men of Mathematics.* Simon & Shuster, Inc., 1937.

[5] John Olmsted Bernard Gelbaum. *Counterexamples in Analysis.* Holden-Day, San Fransisco, CA, 3rd printing edition, 1966.

[6] Carl B. Boyer and Uta C. Merzbach. *A History of Mathematics.* John Wiley & Sons, New York, 1991.

[7] G. Chartland, A. Polimeni, and P. Zhang. *Mathematical Proofs.* Pearson, 3rd edition, 2013.

[8] Frank Dangello and Michael Seyfried. *Introductory Real Analysis.* Houghton Mifflin, 2000.

[9] Martin Davis. *Applied Nonstandard Analysis.* Wiley, 1977.

[10] Keith Devlin. *The Man of Numbers: Fibonacci's Arithmetic Revolution.* Walker & Company, New York, 2011.

[11] Emanuel Fischer. *Intermediate Real Analysis.* Springer-Verlag, NY, 1983.

[12] Stewart Galanor. Riemann's rearrangement theorem. *The Mathematics Teacher*, 80(8):675–681, November 1987.

[13] Joseph A. Gallian. *Contemporary Abstract Algebra.* Cengage, 8th edition, 2013.

[14] Yang Hansheng and Bin Lu. Another proof for the p-series test. *The College Mathematics Journal*, 36(3):235–237, May 2005.

[15] H.E. Huntley. *The Divine Proportion: A Study in Mathematical Beauty.* Dover, 1970.

[16] R. Johnsonbaugh and W.E. Pfaffenberger. *Foundations of Mathematical Analysis.* Dover, 2010.

[17] Morris Kline. *Mathematical Thought from Ancient to Modern Times.* Oxford University Press, March 1990.

[18] Ron Larson and Bruce Edwards. *Calculus of a Single Variable.* Brooks/Cole, 10th edition, 2014.

[19] Steven R. Lay. *Analysis With an Introduction to Proof.* Pearson Prentice Hall, 4th edition, 2005.

[20] Jonathan Lewin and Myrtle Lewin. *An Introduction to Mathematical Analysis.* McGraw-Hill, 2nd edition, 1993.

[21] Henry Parker Manning. *Irrational numbers and their representation by sequences and series.* Wiley, 1906.

[22] Thomas Osler. Rearranging terms of harmonic-like series. *Mathematics and Computer Education*, 35:136–139, 2001.

[23] Michal Potter. *Set Theory and its Philosophy: A Critical Introduction.* Oxford University Press, 2004.

[24] Walter Rudin. *Principles of Mathematical Analysis.* McGraw-Hill, 1976.

[25] Michael Spivak. *Calculus.* Publish or Perish, 2nd edition, 1980.

[26] John Stillwell. *The Real Numbers: An Introduction to Set Theory and Analysis.* Springer, 2013.

[27] Manfred Stoll. *Introduction to Real Analysis.* Addison-Wesley Higher Mathematics, 1997.

[28] William F. Trench. *Introduction to Real Analysis.* Pearson, 2013.

Index